DIGITAL PROCESSING
OF
SPEECH SIGNALS

PRENTICE-HALL SIGNAL PROCESSING SERIES

Alan V. Oppenheim, Editor

ANDREWS AND HUNT   *Digital Image Restoration*
BRIGHAM   *The Fast Fourier Transform*
HAMMING   *Digital Filters*
OPPENHEIM AND SCHAFER   *Digital Signal Processing*
OPPENHEIM et al.   *Applications of Digital Signal Processing*
RABINER AND GOLD   *Theory and Application of Digital Signal Processing*
RABINER AND SCHAFER   *Digital Processing of Speech Signals*

Advanced Monographs

McCLELLAN AND RADER   *Number Theory in Digital Signal Processing*
TRIBOLET   *Seismic Applications of Homomorphic Signal Processing*

DIGITAL PROCESSING
OF
SPEECH SIGNALS

Lawrence R. Rabiner

Acoustics Research Laboratory
Bell Telephone Laboratories
Murray Hill, New Jersey

Ronald W. Schafer

School of Electrical Engineering
Georgia Institute of Technology
Atlanta, Georgia

Prentice-Hall, Inc., Englewood Cliffs, New Jersey 07632

Library of Congress Cataloging in Publication Data

Rabiner, Lawrence R (date)
  Digital processing of speech signals.

  (Prentice-Hall signal processing series)
  Includes bibliographies and index.
  1. Speech processing systems.   2. Digital elec-
tronics.   I. Schafer, Ronald W., joint author.
II. Title.
TK7882.S65R3   621.38 '0412   78-8555
ISBN 0–13–213603–1

Printed in the United States of America

10   9   8   7   6

PRENTICE-HALL INTERNATIONAL, INC., *London*
PRENTICE-HALL OF AUSTRALIA PTY., LIMITED, *Sydney*
PRENTICE-HALL OF CANADA, LTD., *Toronto*
PRENTICE-HALL OF INDIA PRIVATE LIMITED, *New Delhi*
PRENTICE-HALL OF JAPAN, INC., *Tokyo*
PRENTICE-HALL OF SOUTHEAST ASIA PTE. LTD., *Singapore*
WHITEHALL BOOKS LIMITED, *Wellington, New Zealand*

Contents

Preface

This book is an outgrowth of an association between the authors which started as fellow graduate students at MIT, was nurtured by a close collaboration at Bell Laboratories for slightly over 6 years, and has continued ever since as colleagues and close friends. The spark which ignited formal work on this book was a tutorial paper on digital representations of speech signals which we prepared for an IEEE Proceedings special issue on Digital Signal Processing, edited by Professor Alan Oppenheim of MIT. At the time we wrote that paper we realized that the field of digital speech processing had matured sufficiently that a book was warranted on the subject.

Once we convinced ourselves that we were both capable of and ready to write such a text, a fundamental question concerning organization had to be resolved. We considered at least 3 distinct ways of organizing such a text and the problem was deciding which, if any, would provide the most cohesive treatment of this field. The 3 organizations considered were

1. According to digital representations
2. According to parameter estimation problems
3. According to individual applications areas.

After much discussion it was felt that the most fundamental notions were those related to digital speech representations and that a sound understanding of such representations would allow the reader both to understand and to advance the methods and techniques for parameter estimation and for designing speech processing systems. Therefore, we have chosen to organize this book around several basic approaches to digital representations of speech signals, with discussions of specific parameter estimation techniques and applications serving as examples of the utility of each representation.

The formal organization of this book is as follows. Chapter 1 provides an introduction to the area of speech processing, and gives a brief discussion of application areas which are directly related to topics discussed throughout the book. Chapter 2 provides a brief review of the fundamentals of digital signal processing. It is expected that the reader has a solid understanding of linear systems and Fourier transforms and has taken, at least, an introductory course in digital signal processing. Chapter 2 is not meant to provide such background, but rather to establish a notation for discussing digital speech processing, and to provide the reader with handy access to the key equations of digital signal processing. In addition, this chapter provides an extensive discussion of sampling, and decimation and interpolation, key processes that are fundamental to most speech processing systems. Chapter 3 deals with digital models for the speech signal. This chapter discusses the physical basis for sound production in the vocal tract, and this leads to various types of digital models to approximate this process. In addition this chapter gives a brief introduction to acoustic phonetics; that is, a discussion of the sounds of speech and some of their physical properties.

Chapter 4 deals with time domain methods in speech processing. Included in this chapter are discussions of some fundamental ideas of digital speech processing— e.g., short-time energy, average magnitude, short-time average zero-crossing rate, and short-time autocorrelation. The chapter concludes with a section on a nonlinear smoothing technique which is especially appropriate for smoothing the time-domain measurements discussed in this chapter. Chapter 5 deals with the topic of direct digital representations of the speech waveform—i.e., waveform coders. In this chapter the ideas of instantaneous quantization (both uniform and nonuniform), adaptive quantization, differential quantization, and predictive coding (both fixed and adaptive) are discussed and are shown to form the basis of a variety of coders from simple pulse code modulation (PCM) to adaptive differential PCM (ADPCM) coding.

Chapter 6 is the first of two chapters that deal with spectral representations of speech. This chapter concerns the ideas behind short-time Fourier analysis and synthesis of speech. This area has traditionally been the one which has received most attention by speech researchers since some of the key speech processing systems, such as the sound spectrograph and the channel vocoder, are directly related to the concepts discussed in this chapter. Here it is shown how a fairly general approach to speech spectral analysis and synthesis provides a framework for discussing a wide variety of speech processing systems, including those mentioned above. Chapter 7, the second chapter on spectral representations of speech, deals with the area of homomorphic speech processing. The idea behind homomorphic processing of speech is to transform the speech waveform (which is naturally represented as a convolution) to the frequency domain as a sum of terms which can be separated by ordinary linear filtering techniques. Techniques for carrying out this procedure are discussed in this chapter, as are several examples of applications of homomorphic speech processing.

Chapter 8 deals with the topic of linear predictive coding of speech. This representation is based upon a minimum mean-squared error approximation to the time-varying speech waveform, subject to an assumed linear system model of the speech signal. This method has been found to be a robust, reliable, and accurate method for representing speech signals for a wide variety of conditions.

The final chapter, Chapter 9, provides a discussion of several speech processing systems in the area of man-machine communication by voice. The purpose of this chapter is twofold: first, to give concrete examples of specific speech processing systems which are used in real world applications, and second, to show how the ideas developed throughout the book are applied in representative speech processing systems. The systems discussed in this chapter deal with computer voice response, speaker verification and identification, and speech recognition.

The material in this book is intended as a one-semester course in speech processing. To aid the teaching process, each chapter (from Chapter 2 to Chapter 8) contains a set of representative homework problems which are intended to reinforce the ideas discussed in each chapter. Successful completion of a reasonable percentage of these homework problems is essential for a good understanding of the mathematical and theoretical concepts of speech processing. However, as the reader will see, much of speech processing is, by its very nature, empirical. Thus, some "hands on" experience is essential to learning about digital speech processing. In teaching courses based on this book, we have found that a first order approximation to this experience can be obtained by assigning students a term project in one of the following three broad categories:

1. A literature survey and report
2. A hardware design project
3. A computer project

Some guidelines and lists of suggested topics for the three types of projects are given at the end of Chapter 9. We have found that these projects, although demanding, have been popular with our students. We strongly encourage other instructors to incorporate such projects into courses using this book.

Acknowledgements

Several people have had a significant impact, both directly and indirectly, on the material presented in this book. Our biggest debt of gratitude goes to Dr. James L. Flanagan, head of the Acoustics Research Department at Bell Laboratories. Jim has had the dual roles of supervisor and mentor for both authors of this book. For a number of years he has provided us with a model both for how to conduct research and how to report on the research in a meaningful way. His influence on both this book, and our respective careers, has been profound.

Other people with whom we have had the good fortune to collaborate and learn from include Dr. Ben Gold of MIT Lincoln Laboratory, Professor Alan Oppenheim of MIT, and Professor Kenneth Stevens of MIT. These men have served as our teachers and our colleagues and we are grateful to them for their guidance.

Colleagues who have been involved directly with the preparation of this book include Professor Peter Noll of Bremen University, who provided critical comments on Chapter 5, Dr. Ronald Crochiere of Bell Laboratories, who reviewed the entire first draft of this book, and Professor Tom Barnwell of Georgia Tech, who provided valuable comments and criticisms of the entire text. Mr. Gary Shaw carefully worked all the homework problems. His solutions have provided the basis for a solutions manual that is available to instructors. Messrs. J. M. Tribolet, D. Dlugos, P. Papamichalis, S. Gaglio, M. Richards, and L. Kizer provided valuable comments on the final draft of the book. Finally, we wish to thank the Bell Laboratories book review board for overseeing the production of the book, and Ms. Carmela Patuto who was primarily responsible for a superb job of typing of the text of this book throughout its many revisions. In addition we acknowledge the assistance of Ms. Penny Blaine, Jeanette Reinbold, Janie Evans, Nancy Kennell, and Chris Tillery in preparing early drafts of some of the chapters. The generous support of the John and Mary Franklin Foundation to one of us (RWS) is gratefully acknowledged. The authors also wish to acknowledge use of the phototypesetting services at Bell Laboratories on which the entire text of this book was set.

LAWRENCE R. RABINER and RONALD W. SCHAFER

1

Introduction

1.0 Purpose of This Book

The purpose of this book is to show how digital signal processing techniques can be applied in problems related to speech communication. Therefore, this introductory chapter is devoted to a general discussion of questions such as: what is the nature of the speech signal, how can digital signal processing techniques play a role in learning about the speech signal, and what are some of the important application areas of speech communication in which digital signal processing techniques have been used?

1.1 The Speech Signal

The purpose of speech is communication. There are several ways of characterizing the communications potential of speech. One highly quantitative approach is in terms of information theory ideas as introduced by Shannon [1]. According to information theory, speech can be represented in terms of its *message content*, or *information*. An alternative way of characterizing speech is in terms of the *signal* carrying the message information, i.e., the acoustic waveform. Although information theoretic ideas have played a major role in sophisticated communications systems, we shall see throughout this book that it is the speech representation based on the waveform, or some parametric model, which has been most useful in practical applications.

1

In considering the process of speech communication, it is helpful to begin by thinking of a message represented in some abstract form in the brain of the speaker. Through the complex process of producing speech, the information in that message is ultimately converted to an acoustic signal. The message information can be thought of as being represented in a number of different ways in the process of speech production. For example, the message information is first converted into a set of neural signals which control the articulatory mechanism (that is, the motions of the tongue, lips, vocal cords, etc.). The articulators move in response to these neural signals to perform a sequence of gestures, the end result of which is an acoustic waveform which contains the information in the original message.

The information that is communicated through speech is intrinsically of a discrete nature; i.e., it can be represented by a concatenation of elements from a finite set of symbols. The symbols from which every sound can be classified are called *phonemes*. Each language has its own distinctive set of phonemes, typically numbering between 30 and 50. For example, English can be represented by a set of around 42 phonemes. (See Chapter 3.)

A central concern of information theory is the rate at which information is conveyed. For speech a crude estimate of the information rate can be obtained by noting that physical limitations on the rate of motion of the articulators require that humans produce speech at an average rate of about 10 phonemes per second. If each phoneme is represented by a binary number, then a six-bit numerical code is more than sufficient to represent all of the phonemes of English. Assuming an average rate of 10 phonemes per second and neglecting any correlation between pairs of adjacent phonemes we get an estimate of 60 bits/sec for the average information rate of speech. In other words, the *written* equivalent of speech contains information equivalent to 60 bits/sec at normal speaking rates. Of course a lower bound on the "true" information content of speech is considerably higher than this rate. The above estimate does not take into account factors such as the identity and emotional state of the speaker, the rate of speaking, the loudness of the speech, etc.

In speech communication systems, the speech signal is transmitted, stored, and processed in many ways. Technical concerns lead to a wide variety of representations of the speech signal. In general, there are two major concerns in any system:

1. Preservation of the message content in the speech signal.
2. Representation of the speech signal in a form that is convenient for transmission or storage, or in a form that is flexible so that modifications may be made to the speech signal without seriously degrading the message content.

The representation of the speech signal must be such that the information content can easily be extracted by human listeners, or automatically by machine. Throughout this book we shall see that representations of the speech signal (rather than message content) may require from 500 to upwards of 1 million bits per second. In the design and implementation of these representations, the methods of signal processing play a fundamental role.

1.2 Signal Processing

The general problem of information manipulation and processing is depicted in Figure 1.1. In the case of speech signals the human speaker is the information source. The measurement or observation is generally the acoustic waveform.

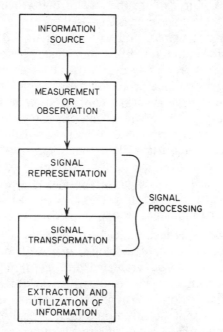

Fig. 1.1 General view of information manipulation and processing.

Signal processing involves first obtaining a representation of the signal based on a given model and then the application of some higher level transformation in order to put the signal into a more convenient form. The last step in the process is the extraction and utilization of the message information. This step may be performed either by human listeners or automatically by machines. By way of example, a system whose function is to automatically identify a speaker from a given set of speakers might use a time-dependent spectral representation of the speech signal. One possible signal transformation would be to average spectra across an entire sentence, compare the average spectrum to a stored averaged spectrum template for each possible speaker, and then based on a spectral similarity measurement choose the identity of the speaker. For this example the "information" in the signal is the identity of the speaker.

Thus, processing of speech signals generally involves two tasks. First, it is a vehicle for obtaining a general representation of a speech signal in either waveform or parametric form. Second, signal processing serves the function of aiding in the process of transforming the signal representation into alternate forms which are less general in nature, but more appropriate to specific applications. Throughout this book we will see numerous specific examples of the importance of signal processing in the area of speech communication.

1.3 Digital Signal Processing

The focus of this book is to explore the role of digital techniques in processing speech signals. Digital signal processing is concerned both with obtaining discrete representations of signals, and with the theory, design, and implementation of numerical procedures for processing the discrete representation. The objectives in digital signal processing are identical to those in analog signal processing. Therefore, it is reasonable to ask why digital signal processing techniques should be singled out for special consideration in the context of speech communication. A number of very good reasons can be cited. First, and probably most important, is the fact that extremely sophisticated signal processing functions can be implemented using digital techniques. The algorithms that we shall describe in this book are intrinsically discrete-time, signal processing systems. For the most part, it is not appropriate to view these systems as approximations to analog systems. Indeed in many cases there is no realizable counterpart available with analog implementation.

Digital signal processing techniques were first applied in speech processing problems, as simulations of complex analog systems. The point of view initially was that analog systems could be simulated on a computer to avoid the necessity of building the system in order to experiment with choices of parameters and other design considerations. When digital simulations of analog systems were first applied, the computations required a great deal of time. For example, as much as an hour might have been required to process only a few seconds of speech. In the mid 1960's a revolution in digital signal processing occurred. The major catalysts were the development of faster computers and rapid advances in the theory of digital signal processing techniques. Thus, it became clear that digital signal processing systems had virtues far beyond their ability to simulate analog systems. Indeed the present attitude toward laboratory computer implementations of speech processing systems is to view them as exact simulations of a digital system that could be implemented either with special purpose digital hardware or with a dedicated computer system.

In addition to theoretical developments, concomitant developments in the area of digital hardware have led to further strengthening of the advantage of digital processing techniques over analog systems. Digital systems are reliable and very compact. Integrated circuit technology has advanced to a state where extremely complex systems can be implemented on a single chip. Logic speeds are fast enough so that the tremendous number of computations required in many signal processing functions can be implemented in real-time at speech sampling rates.

There are many other reasons for using digital techniques in speech communication systems. For example, if suitable coding is used, speech in digital form can be reliably transmitted over very noisy channels. Also, if the speech signal is in digital form it is identical to data of other forms. Thus a communications network can be used to transmit both speech and data with no need to distinguish between them except in the decoding. Also, with regard to transmission of voice signals requiring security, the digital representation has a distinct advantage over analog systems. For secrecy, the information bits can be scrambled in a manner which can ultimately be unscrambled at the receiver. For these and numerous other reasons digital techniques are being increasingly applied in speech communication problems [3].

1.4 Digital Speech Processing

In considering the application of digital signal processing techniques to speech communication problems, it is helpful to focus on three main topics: the representation of speech signals in digital form, the implementation of sophisticated processing techniques, and the classes of applications which rely heavily on digital processing.

The representation of speech signals in digital form is, of course, of fundamental concern. In this regard we are guided by the well-known sampling theorem [4] which states that a bandlimited signal can be represented by samples taken periodically in time — provided that the samples are taken at a high enough rate. Thus, the process of sampling underlies all of the theory and application of digital speech processing. There are many possibilities for discrete representations of speech signals. As shown in Figure 1.2, these representations can be classified into two broad groups, namely waveform representations and parametric representations. Waveform representations, as

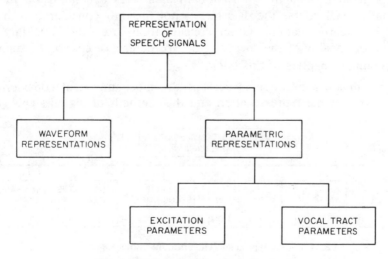

Fig. 1.2 Representations of speech signals.

5

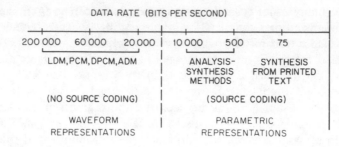

DATA RATE (BITS PER SECOND)

| 200 000 | 60 000 | 20 000 | 10 000 | 500 | 75 |

LDM, PCM, DPCM, ADM | ANALYSIS-   SYNTHESIS
                          SYNTHESIS   FROM PRINTED
                          METHODS   TEXT

(NO SOURCE CODING) | (SOURCE CODING)

WAVEFORM | PARAMETRIC
REPRESENTATIONS | REPRESENTATIONS

Fig. 1.3 Range of bit rates for various types of speech representations. (After Flanagan [3]).

the name implies, are concerned with simply preserving the "wave shape" of the analog speech signal through a sampling and quantization process. Parametric representations, on the other hand, are concerned with representing the speech signal as the output of a model for speech production. The first step in obtaining a parametric representation is often a digital waveform representation; that is, the speech signal is sampled and quantized and then further processed to obtain the parameters of the model for speech production. The parameters of this model are conveniently classified as either excitation parameters (i.e., related to the source of speech sounds) or vocal tract response parameters (i.e., related to the individual speech sounds).[1]

Figure 1.3 shows a comparison of a number of different representations of speech signals according to the data rate required. The dotted line, at a data rate of about 15,000 bits per second, separates the high data rate waveform representations at the left from the lower data rate parametric representations at the right. This figure shows variations in data rate from 75 bits per second (approximately the basic message information of the text) to data rates upward of 200,000 bits per second for simple waveform representations. This represents about a 3000 to 1 variation in data rates depending on the signal representation. Of course the data rate is not the only consideration in choosing a speech representation. Other considerations are cost, flexibility of the representation, quality of the speech, etc. We defer a discussion of such issues to the remaining chapters of this book.

The ultimate application is perhaps the most important consideration in the choice of a signal representation and the methods of digital signal process-

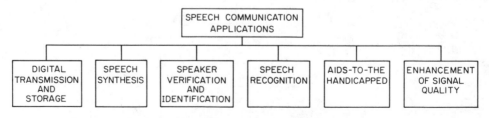

Fig. 1.4 Some typical speech communications applications.

[1]Chapter 3 provides a detailed discussion of parametric speech models.

6

ing subsequently applied. Figure 1.4 shows just a few of the many applications areas in speech communications. Although we have already referred to several of these areas, it is worthwhile giving a brief discussion of each of these areas as a means for motivating the techniques to be discussed in subsequent chapters.

1.4.1 Digital transmission and storage of speech [3]

One of the earliest and most important applications of speech processing was the vocoder or *voice coder,* invented by Homer Dudley in the 1930's [5]. The purpose of the vocoder was to reduce the bandwidth required to transmit the speech signal. The need to conserve bandwidth remains, in many situations, in spite of the increased bandwidth provided by sattelite, microwave, and optical communications systems. Furthermore, a need has arisen for systems which digitize speech at as low a bit rate as possible, consistent with low terminal cost for future applications in the all-digital telephone plant. Also, the possibility of extremely sophisticated encryption of the speech signal is sufficient motivation for the use of digital transmission in many applications.

1.4.2 Speech synthesis systems

Much of the interest in speech synthesis systems is stimulated by the need for economical digital storage of speech for computer voice response systems [6]. A computer voice response system is basically an all-digital, automatic information service which can be queried by a person from a keyboard or terminal, and which responds with the desired information by voice. Since an ordinary Touch-Tone® telephone can be the keyboard for such a system, the capabilities of such automatic information services can be made universally available over the switched telephone facilities without the need for any additional specialized equipment [3]. Speech synthesis systems also play a fundamental role in learning about the process of human speech production [7].

1.4.3 Speaker verification and identification systems [8]

The techniques of speaker verification and identification involve the authentication or identification of a speaker from a large ensemble of possible speakers. A speaker verification system must decide if a speaker is the person he claims to be. Such a system is potentially applicable to situations requiring control of access to information or restricted areas and to various kinds of automated credit transactions. A speaker identification system must decide which speaker among an ensemble of speakers produced a given speech utterance. Such systems have potential forensic applications.

1.4.4 Speech recognition systems [9]

Speech recognition is, in its most general form, a conversion from an acoustic waveform to a written equivalent of the message information. The nature of the speech recognition problem is heavily dependent upon the constraints placed on speaker, speaking situation and message context. The potential applications of speech recognition systems are many and varied; e.g. a voice operated typewriter and voice communication with computers. Also, a speech recognizing system combined with a speech synthesizing system comprises the ultimate low bit rate communication system.

1.4.5 Aids-to-the-handicapped

This application concerns processing of a speech signal to make the information available in a form which is better matched to a handicapped person than is normally available. For example variable rate playback of prerecorded tapes provides an opportunity for a blind "reader" to proceed at any desired pace through given speech material. Also a variety of signal processing techniques have been applied to design sensory aids and visual displays of speech information as aids in teaching deaf persons to speak [10].

1.4.6 Enhancement of signal quality

In many situations, speech signals are degraded in ways that limit their effectiveness for communication. In such cases digital signal processing techniques can be applied to improve the speech quality. Examples include such applications as the removal of reverberation (or echos) from speech, or the removal of noise from speech, or the restoration of speech recorded in a helium-oxygen mixture as used by divers.

1.5 Summary

In this chapter we have introduced the ways in which digital signal processing techniques are applied in speech communication. It is clear that we have selected a very wide range of topics, and to cover them in complete depth would be extremely difficult. There are a number of ways in which a book of this type could be organized. For example, it could be organized with respect to the signal representations of Figure 1.2. Alternatively, a book could be written that would emphasize applications areas. Indeed, a book could be written about each area shown in Figure 1.4. A third possibility, which we have chosen, is to organize the book with respect to signal processing methods. We feel that this approach offers the greatest opportunity to focus on topics that will be of continued importance. As such, the remaining chapters of this book provide a review of digital signal processing methods (Chapter 2), an introduction to the digital speech model (Chapter 3), discussions of time domain representations of speech (Chapter 4), waveform representations (Chapter 5), short-time spectral representations (Chapter 6), homomorphic representations (Chapter 7), and linear predictive representations (Chapter 8). These chapters detail the basic theory of digital speech processing. This theory is widely applicable in many applications areas. To illustrate such applications, the final chapter (Chapter 9) discusses several examples of man-machine communications systems which involve extensive use of the digital signal processing methods discussed in this book.

REFERENCES

1. C. E. Shannon, "A Mathematical Theory of Communication," *Bell System Tech. J.,* Vol. 27, pp. 623-656, October 1968.

2. J. L. Flanagan, *Speech Analysis, Synthesis, and Perception,* 2nd Edition, Springer Verlag, New York, 1972.

3. J. L. Flanagan, "Computers That Talk and Listen: Man-Machine Communication by Voice," *Proc. IEEE,* Vol. 64, No. 4, pp. 416-432, April 1976.

4. H. Nyquist, "Certain Topics in Telegraph Transmission Theory," *Trans. AIEE,* Vol. 47, pp. 617-644, February 1928.

5. H. Dudley, "Remaking Speech," *J. Acoust. Soc. Am.,* Vol. 11, pp. 169-177, 1939.

6. L. R. Rabiner and R. W. Schafer, "Digital Techniques for Computer Voice Response: Implementations and Applications," *Proc. IEEE,* Vol. 64, pp. 416-433, April 1976.

7. C. H. Coker, "A Model of Articulatory Dynamics and Control," *Proc. IEEE,* Vol. 64, No. 4, pp. 452-460, April 1976.

8. B. S. Atal, "Automatic Recognition of Speakers from Their Voices," *Proc. IEEE,* Vol. 64, No. 4, pp. 460-475, April 1976.

9. D. R. Reddy, "Speech Recognition by Machine: A Review," *Proc. IEEE,* Vol. 64, No. 4, pp. 501-531, April 1976.

10. H. Levitt, "Speech Processing Aids for the Deaf: An Overview," *IEEE Trans. on Audio and Electroacoustics,* Vol. AU-21, pp. 269-273, June 1973.

2

Fundamentals
of Digital Signal Processing

2.0 Introduction

Since the speech processing schemes and techniques that we shall discuss in this book are intrinsically discrete-time signal processing systems, it is essential that a reader have a good understanding of the basic techniques of digital signal processing. In this chapter we present a brief review of the important concepts. This review is intended to serve as a convenient reference for later chapters and to establish the notation that will be used throughout the book. Those readers who are completely unfamiliar with techniques for representation and analysis of discrete-time signals and systems may find it worthwhile to consult a textbook on digital signal processing [1-3] when this chapter does not provide sufficient detail.

2.1 Discrete-Time Signals and Systems

In almost every situation involving information processing or communication, it is natural to begin with a representation of the signal as a continuously varying pattern. The acoustic wave produced in human speech is most certainly of this nature. It is mathematically convenient to represent such continuously varying patterns as functions of a continuous variable t, which represents time. In this book we shall use notation of the form $x_a(t)$ to denote continuously varying (or analog) time waveforms. As we shall see, it is also possible to represent

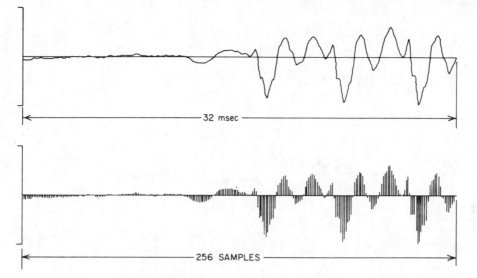

Fig. 2.1 Representations of a speech signal.

the speech signal as a sequence of numbers; indeed, that is what this book is all about. In general we shall use notation of the form, $x(n)$, to denote sequences. If, as is the case for sampled speech signals, a sequence can be thought of as a sequence of samples of an analog signal taken periodically with sampling period, T, then we may find it useful to explicitly indicate this by using the notation, $x_a(nT)$. Figure 2.1 shows an example of a speech signal represented both as an analog signal and as a sequence of samples at a sampling rate of 8 kHz. In subsequent figures, convenience in plotting may dictate the use of the analog representation (i.e., continuous functions) even when the discrete representation is being considered. In such cases, the continuous curve can simply be viewed as the envelope of the sequence of samples.

In our study of digital speech processing systems we will find a number of special sequences repeatedly arising. Several of these sequences are depicted in Fig. 2.2. The unit sample or unit impulse sequence is defined as

$$\delta(n) = 1 \quad n = 0$$
$$= 0 \quad \textit{otherwise} \tag{2.1}$$

The unit step sequence is

$$u(n) = 1 \quad n \geqslant 0$$
$$= 0 \quad n < 0 \tag{2.2}$$

An exponential sequence is of the form

$$x(n) = a^n \tag{2.3}$$

If a is complex, i.e., $a = re^{j\omega_0}$, then

$$x(n) = r^n e^{j\omega_0 n} = r^n(\cos \omega_0 n + j \sin \omega_0 n) \tag{2.4}$$

11

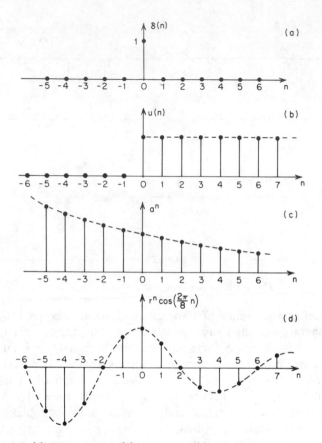

Fig. 2.2 (a) Unit sample; (b) unit step; (c) real exponential; and (d) damped cosine.

If $r = 1$ and $\omega_0 \neq 0$, $x(n)$ is a complex sinusoid; if $\omega_0 = 0$, $x(n)$ is real; and if $r < 1$ and $\omega_0 \neq 0$, then $x(n)$ is an exponentially decaying oscillatory sequence. Sequences of this type arise especially in the representation of linear systems and in modelling the speech waveform.

Signal processing involves the transformation of a signal into a form which is in some sense more desirable. Thus we are concerned with discrete systems, or equivalently, transformations of an input sequence into an output sequence. We shall depict such transformations by block diagrams such as Fig. 2.3a. Many speech analysis systems are designed to estimate several time-varying parameters from samples of the speech wave. Such systems therefore

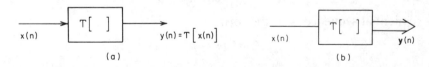

Fig. 2.3 Block diagram representations of: (a) single input/single output system; (b) single input/multiple output system.

12

have a multiplicity of outputs; i.e., a single input sequence representing the speech signal is transformed into a vector of output sequences as depicted in Fig. 2.3b. In this book, we shall discuss both single output and multiple output speech processing systems.

The special class of linear shift-invariant systems is especially useful in speech processing. Such systems are completely characterized by their response to a unit sample input. For such systems, the output can be computed from the input, $x(n)$, and the unit sample response, $h(n)$, using the convolution sum expression

$$y(n) = \sum_{k=-\infty}^{\infty} x(k)h(n-k) = x(n) * h(n) \tag{2.5a}$$

where the symbol $*$ stands for discrete convolution. An equivalent expression is

$$y(n) = \sum_{k=-\infty}^{\infty} h(k)x(n-k) = h(n) * x(n) \tag{2.5b}$$

Linear shift invariant systems are useful for performing filtering operations on speech signals and, perhaps more importantly, they are useful as models for speech production.

2.2 Transform Representation of Signals and Systems

The analysis and design of linear systems are greatly facilitated by frequency-domain representations of both signals and systems. Thus, it is useful to review Fourier and z-transform representations of discrete-time signals and systems.

2.2.1 The z-transform

The *z-transform* representation of a sequence is defined by the pair of equations

$$X(z) = \sum_{n=-\infty}^{\infty} x(n)z^{-n} \tag{2.6a}$$

$$x(n) = \frac{1}{2\pi j} \oint_C X(z)\, z^{n-1} dz \tag{2.6b}$$

The "z-transform" or "direct transform" of $x(n)$ is defined by Eq. (2.6a). It can be seen that in general $X(z)$ is an infinite power series in the variable z^{-1}, where the sequence values, $x(n)$, play the role of coefficients in the power series. In general such a power series will converge (add up) to a finite value only for certain values of z. A sufficient condition for convergence is

$$\sum_{n=-\infty}^{\infty} |x(n)||z^{-n}| < \infty \tag{2.7}$$

13

The set of values for which the series converges defines a region in the complex z-plane known as the *region of convergence*. In general this region is of the form

$$R_1 < |z| < R_2 \qquad (2.8)$$

To see the relationship of the region of convergence to the nature of the sequence, let us consider some examples.

2.2.1a Example 1

Let $x(n) = \delta(n-n_0)$. Then by substitution into Eq. (2.6a)

$$X(z) = z^{-n_0}$$

2.2.1b Example 2

Let $x(n) = u(n) - u(n-N)$. Then

$$X(z) = \sum_{n=0}^{N-1} (1)z^{-n} = \frac{1 - z^{-N}}{1 - z^{-1}}$$

In both of these cases, $x(n)$ is of finite duration. Therefore $X(z)$ is simply a polynomial in the variable z^{-1}, and the region of convergence is everywhere but $z = 0$. All finite length sequences have a region of convergence that is at least the region $0 < |z| < \infty$.

2.2.1c Example 3

Let $x(n) = a^n u(n)$. Then

$$X(z) = \sum_{n=0}^{\infty} a^n z^{-n} = \frac{1}{1 - az^{-1}}, \qquad |a| < |z|$$

In this case the power series is recognized as a geometric series for which a convenient closed form expression exists for the sum. This result is typical of infinite duration sequences which are nonzero for $n > 0$. In this general case, the region of convergence is of the form $|z| > R_1$.

2.2.1d Example 4

Let $x(n) = - b^n u(-n-1)$. Then

$$X(z) = \sum_{n=-\infty}^{-1} b^n z^{-n} = \frac{1}{1 - bz^{-1}}, \qquad |z| < |b|$$

This is typical of infinite duration sequences that are nonzero for $n < 0$, where the region of convergence is, in general, $|z| < R_2$. The most general case in which $x(n)$ is nonzero for $-\infty < n < \infty$ can be viewed as a combination of the cases illustrated by Examples 3 and 4. Thus for this case, the region of convergence is of the form $R_1 < |z| < R_2$.

14

The "inverse transform" is given by the contour integral in Eq. (2.6b), where C is a closed contour that encircles the origin of the z-plane and lies inside the region of convergence of $X(z)$. For the special case of rational transforms, a partial fraction expansion provides a convenient means for finding inverse transforms [1].

There are many theorems and properties of the z-transform representation that are useful in the study of discrete-time systems. A working familiarity with these theorems and properties is essential for complete understanding of the material in subsequent chapters. A list of important theorems is given in Table 2.1. These theorems can be seen to be similar in form to corresponding theorems for Laplace transforms of continuous time functions. However, this similarity should not be construed to mean that the z-transform is in any sense an approximation to the Laplace transform. The Laplace transform is an *exact* representation of a continuous-time function, and the z-transform is an *exact* representation of a sequence of numbers. The appropriate way to relate the continuous and discrete representations of a signal is through the sampling theorem as discussed in Section 2.4.

Table 2.1 Sequences and Their Corresponding z-Transforms

	Sequence	*z-Transform*
1. Linearity	$ax_1(n) + bx_2(n)$	$aX_1(z) + bX_2(z)$
2. Shift	$x(n+n_0)$	$z^{n_0}X(z)$
3. Exponential Weighting	$a^n x(n)$	$X(a^{-1}z)$
4. Linear Weighting	$nx(n)$	$-z\dfrac{dX(z)}{dz}$
5. Time Reversal	$x(-n)$	$X(z^{-1})$
6. Convolution	$x(n)*h(n)$	$X(z)H(z)$
7. Multiplication of Sequences	$x(n)w(n)$	$\dfrac{1}{2\pi j}\oint_C X(v)\,W(z/v)\,v^{-1}dv$

2.2.2 The Fourier transform

The Fourier transform representation of a discrete-time signal is given by the equations

$$X(e^{j\omega}) = \sum_{n=-\infty}^{\infty} x(n)e^{-j\omega n} \qquad (2.9a)$$

$$x(n) = \frac{1}{2\pi}\int_{-\pi}^{\pi} X(e^{j\omega})e^{j\omega n}d\omega \qquad (2.9b)$$

These equations can easily be seen to be a special case of Eqs. (2.6). Specifically the Fourier representation is obtained by restricting the z-transform to the unit circle of the z-plane; i.e., by setting $z = e^{j\omega}$. As depicted in Fig. 2.4, the digital frequency variable, ω, also has the interpretation as angle in the

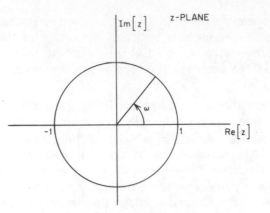

Fig. 2.4 The unit circle of the z-plane.

z-plane. A sufficient condition for the existence of a Fourier transform representation can be obtained by setting $|z| = 1$ in Eq. (2.7), thus obtaining

$$\sum_{n=-\infty}^{\infty} |x(n)| < \infty \qquad (2.10)$$

As examples of typical Fourier transforms, we can return to the examples of Section 2.2.1. The Fourier transform is obtained simply by setting $z = e^{j\omega}$ in the given expression. In the first two examples, the result is clearly the Fourier transform since the region of convergence of $X(z)$ includes the unit circle. However, in Examples 3 and 4, the Fourier transform will exist only if $|a| < 1$ and $|b| > 1$ respectively. These conditions, of course, correspond to decaying sequences for which Eq. (2.10) holds.

An important feature of the Fourier transform of a sequence is that $X(e^{j\omega})$ is a periodic function of ω, with period 2π. This follows easily by substituting $\omega + 2\pi$ into Eq. (2.9a). Alternatively, since $X(e^{j\omega})$ is the evaluation of $X(z)$ on the unit circle, we can see that $X(e^{j\omega})$ must repeat each time we go completely around the unit circle; i.e., ω has gone through 2π radians.

By setting $z = e^{j\omega}$ in each of the theorems in Table 2.1, we obtain a corresponding set of theorems for the Fourier transform. Of course, these results are valid only if the Fourier transforms that are involved do indeed exist.

2.2.3 The discrete Fourier transform

As in the case of analog signals, if a sequence is periodic with period N; i.e.,

$$\tilde{x}(n) = \tilde{x}(n+N) \qquad -\infty < n < \infty \qquad (2.11)$$

then $\tilde{x}(n)$ can be represented by a discrete sum of sinusoids rather than an integral as in Eq. (2.9b). The Fourier series representation for a periodic sequence is

$$\tilde{X}(k) = \sum_{n=0}^{N-1} \tilde{x}(n) e^{-j\frac{2\pi}{N}kn} \qquad (2.12a)$$

$$\tilde{x}(n) = \frac{1}{N} \sum_{k=0}^{N-1} \tilde{X}(k) e^{j\frac{2\pi}{N}kn} \qquad (2.12b)$$

This is an *exact* representation of a periodic sequence. However, the great utility of this representation lies in imposing a different interpretation upon Eqs. (2.12). Let us consider a finite length sequence, $x(n)$, that is zero outside the interval $0 \leqslant n \leqslant N-1$. Then the z-transform is

$$X(z) = \sum_{n=0}^{N-1} x(n) z^{-n} \qquad (2.13)$$

If we evaluate $X(z)$ at N equally spaced points on the unit circle, i.e., $z_k = e^{j2\pi k/N}$, $k = 0, 1, ..., N-1$, then we obtain

$$X(e^{j\frac{2\pi}{N}k}) = \sum_{n=0}^{N-1} x(n) e^{-j\frac{2\pi}{N}kn} \qquad k = 0, 1, ..., N-1 \qquad (2.14)$$

If we construct a periodic sequence as an infinite sequence of replicas of $x(n)$,

$$\tilde{x}(n) = \sum_{r=-\infty}^{\infty} x(n+rN) \qquad (2.15)$$

then, the samples $X(e^{j2\pi k/N})$ are easily seen from Eqs. (2.12a) and (2.14) to be the Fourier coefficients of the periodic sequence $\tilde{x}(n)$ in Eq. (2.15). Thus a sequence of length N can be exactly represented by a discrete Fourier transform (DFT) representation of the form

$$X(k) = \sum_{n=0}^{N-1} x(n) e^{-j\frac{2\pi}{N}kn} \qquad k = 0, 1, \ldots, N-1 \qquad (2.16a)$$

$$x(n) = \frac{1}{N} \sum_{k=0}^{N-1} X(k) e^{j\frac{2\pi}{N}kn} \qquad n = 0, 1, \ldots, N-1 \qquad (2.16b)$$

Clearly the only difference between Eqs. (2.16) and (2.12) is a slight modification of notation (removing the $\sim$ symbols which indicate periodicity) and the explicit restriction to the finite intervals $0 \leqslant k \leqslant N-1$ and $0 \leqslant n \leqslant N-1$. It is extremely important, however, to bear in mind when using the DFT representation that all sequences behave as if they were periodic when represented by a DFT representation. That is, the DFT is really a representation of the periodic sequence given in Eq. (2.15). An alternative point of view is that when DFT representations are used, sequence indices must be interpreted modulo N. This follows from the fact that if $x(n)$ is of length N

$$\tilde{x}(n) = \sum_{k=-\infty}^{\infty} x(n+rN) = x(n \ modulo \ N)$$

$$= x((n))_N. \qquad (2.17)$$

17

The double parenthesis notation provides a convenient expression of the inherent periodicity of the DFT representation. This built-in periodicity has a significant effect on the properties of the DFT representation. Some of the more important theorems are listed in Table 2.2. The most obvious feature is that shifted sequences are shifted modulo N. This leads, for example, to significant differences in the discrete convolution.

The DFT representation, with all its peculiarities, is important for a number of reasons:

1. The DFT, $X(k)$, can be viewed as a sampled version of the z-transform (or Fourier transform) of a finite length sequence
2. The DFT has properties very similar (with modifications due to the inherent periodicity) to many of the useful properties of z-transforms and Fourier transforms.
3. The N values of $X(k)$ can be computed very efficiently (with time proportional to $N \log N$) by a set of computational algorithms known collectively as the fast Fourier transform (FFT) [1-4].

The DFT is widely used for computing spectrum estimates, correlation functions and for implementing digital filters [5-6]. We shall have frequency occasion to apply DFT representations in speech processing.

Table. 2.2 Sequences and Their Corresponding Discrete Fourier Transforms

	Sequence	N-point DFT
1. Linearity	$ax_1(n) + bx_2(n)$	$aX_1(k) + bX_2(k)$
2. Shift	$x((n+n_o))_N$	$e^{j\frac{2\pi}{N}kn_0}X(k)$
3. Time Reversal	$x((-n))_N$	$X^*(k)$
4. Convolution	$\sum_{m=0}^{N-1} x(m)h((n-m))_N$	$X(k)H(k)$
5. Multiplication of Sequences	$x(n)w(n)$	$\frac{1}{N}\sum_{r=0}^{N-1} X(r)W((k-r))_N$

2.3 Fundamentals of Digital Filters

A digital filter is a discrete-time linear shift-invariant system. Recall that for such a system the input and output are related by the convolution sum expression of Eqs. (2.5). The corresponding relation between the z-transform of the sequences involved is as given in Table 2.1,

$$Y(z) = H(z)X(z) \tag{2.18}$$

The z-transform of the unit sample response, $H(z)$, is called the *system function* of the system. The Fourier transform of the unit impulse response, $H(e^{j\omega})$, is called the *frequency response*. $H(e^{j\omega})$ is in general a complex function of ω,

18

which can be expressed in terms of real and imaginary parts as

$$H(e^{j\omega}) = H_r(e^{j\omega}) + jH_i(e^{j\omega}) \qquad (2.19)$$

or in terms of magnitude and phase angle as

$$H(e^{j\omega}) = |H(e^{j\omega})|e^{j\arg[H(e^{j\omega})]} \qquad (2.20)$$

A *causal* linear shift invariant-system is one for which $h(n) = 0$ for $n < 0$. A *stable* system is one for which every bounded input produces a bounded output. A necessary and sufficient condition for a linear shift-invariant system to be stable is

$$\sum_{n=-\infty}^{\infty} |h(n)| < \infty \qquad (2.21)$$

This condition is identical to Eq. (2.10) and thus is sufficient for the existence of $H(e^{j\omega})$.

In addition to the convolution sum expression of Eq. (2.5), all linear shift invariant systems of interest for implementation as filters have the property that the input and output satisfy a linear difference equation of the form

$$y(n) - \sum_{k=1}^{N} a_k y(n-k) = \sum_{r=0}^{M} b_r x(n-r) \qquad (2.22)$$

By evaluating the z-transform of both sides of this equation we can show that

$$H(z) = \frac{Y(z)}{X(z)} = \frac{\sum_{r=0}^{M} b_r z^{-r}}{1 - \sum_{k=1}^{N} a_k z^{-k}} \qquad (2.23)$$

A useful observation results from comparing Eq. (2.22) to Eq. (2.23). That is, given a difference equation in the form of Eq. (2.22) we can obtain $H(z)$ directly by simply identifying the coefficients of the delayed input in Eq. (2.22) with corresponding powers of z^{-1} in the numerator and coefficients of the delayed output with corresponding powers of z^{-1} in the denominator.

The system function, $H(z)$, is in general a rational function of z^{-1}. As such it is characterized by the locations of its poles and zeros in the z-plane. Specifically $H(z)$ can be expressed as

$$H(z) = \frac{A \prod_{r=1}^{M} (1 - c_r z^{-1})}{\prod_{k=1}^{N} (1 - d_k z^{-1})} \qquad (2.24)$$

From our discussion of z-transforms, we recall that a causal system will have a region of convergence of the form $|z| > R_1$. If the system is also stable, then R_1 must be less than unity so that the region of convergence contains the unit

19

circle. Therefore the poles of $H(z)$ must all be inside the unit circle for a stable and causal system.

It is convenient to define two classes of linear shift invariant systems. These are the class of finite duration impulse response (FIR) systems and the class of infinite duration impulse response (IIR) systems. These classes have distinct properties which we shall summarize below.

2.3.1 FIR systems

If all the coefficients, a_k, in Eq. (2.22) are zero, the difference equation becomes

$$y(n) = \sum_{r=0}^{M} b_r x(n-r) \tag{2.25}$$

Comparing Eq. (2.25) to Eq. (2.5b) we observe that

$$h(n) = b_n \quad 0 \leqslant n \leqslant M$$
$$= 0 \quad otherwise \tag{2.26}$$

FIR systems have a number of important properties. First, we note that $H(z)$ is a polynomial in z^{-1}, and thus $H(z)$ has no nonzero poles, only zeros. Also, FIR systems can have exactly linear phase. If $h(n)$ satisfies the relation

$$h(n) = \pm h(M-n) \tag{2.27}$$

then $H(e^{j\omega})$ has the form

$$H(e^{j\omega}) = A(e^{j\omega}) e^{-j\omega(M/2)} \tag{2.28}$$

where $A(e^{j\omega})$ is either purely real or imaginary depending upon whether Eq. (2.27) is satisfied with + or − respectively.

The possibility of *exactly* linear phase is often very useful in speech processing applications where precise time alignment is essential. This property of FIR filters also can greatly simplify the approximation problem since it is only necessary to be concerned with approximating a, desired magnitude response. The penalty that is paid for being able to design filters with an exact linear phase response is that a large impulse response duration is required to adequately approximate sharp cutoff filters.

Based on the properties associated with linear phase FIR filters there have developed three well known design methods for approximating an arbitrary set of specifications with an FIR filter. These three methods are:

1. Window design [1,2,5,7]
2. Frequency sampling design [1,2,8]
3. Optimal (minimax error) design [1,2,9-11]

Only the first of these techniques is an analytical design technique, i.e., a closed

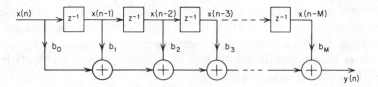

Fig. 2.5 Digital network for FIR system.

form set of equations can be solved to obtain the filter coefficients. The second and third design methods are optimization methods which use iterative (rather than closed form) approaches to obtain the desired filter. Although the window method is simple to apply, the third method is also widely used. This is in part due to a series of intensive investigations into the properties of the optimal FIR filters, and in part due to the general availability of a well-documented design program which enables the user to approximate any desired set of specifications [2,10].

In considering the implementation of digital filters, it is often useful to represent the filter in block diagram form. The difference equation of Eq. (2.25) is depicted in Fig. 2.5. Such a diagram, often called a digital filter structure, graphically depicts the operations required to compute each value of the output sequence from values of the input sequence. The basic elements of the diagram depict means for addition, multiplication of sequence values by constants (constants indicated on branches imply multiplication), and storage of past values of the input sequence. Thus the block diagram gives a clear indication of the complexity of the system. When the system has linear phase, further significant simplifications can be incorporated into the implementation. (See Problem 2.7).

2.3.2 IIR systems

If the system function of Eq. (2.24) has poles as well as zeros, then the difference equation of Eq. (2.22) can be written as

$$y(n) = \sum_{k=1}^{N} a_k y(n-k) + \sum_{r=0}^{M} b_r x(n-r) \tag{2.29}$$

This equation is a recurrence formula that can be used sequentially to compute the values of the output sequence from past values of the output and present and past values of the input sequence. If $M < N$ in Eq. (2.24), $H(z)$ can be expanded in a partial fraction expansion as in

$$H(z) = \sum_{k=1}^{N} \frac{A_k}{1 - d_k z^{-1}} \tag{2.30}$$

For a causal system, it is easily shown (See Problem 2.9) that

$$h(n) = \sum_{k=1}^{N} A_k (d_k)^n u(n) \tag{2.31}$$

21

Thus, we see that $h(n)$ has infinite duration. However, because of the recurrence formula of Eq. (2.29), it is often possible to implement an IIR filter that approximates a given set of specifications more efficiently (i.e., using fewer computations) than is possible with an FIR system. This is particularly true for sharp cutoff frequency selective filters.

A wide variety of design methods are available for IIR filters. Design methods for frequency selective filters (lowpass, bandpass, etc.) are generally based on transformations of classical analog design procedures that are straightforward to implement. Included in this class are

1. Butterworth designs - (maximally flat amplitude)
2. Bessel designs - (maximally flat group delay)
3. Chebyshev designs (equiripple in either passband or stopband)
4. Elliptic designs - (equiripple in both passband and stopband)

All the above methods are analytical in nature and have been widely applied to the design of IIR digital filters [1,2]. In addition a variety of IIR optimization

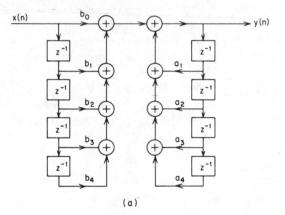

(a)

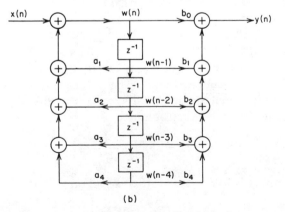

(b)

Fig. 2.6 (a) Direct form IIR structure; (b) direct form structure with minimum storage.

methods have been developed for approximating design specifications which are not easily adapted to one of the above approximation methods [12].

The major difference between FIR and IIR filters is that IIR filters cannot be designed to have exact linear phase, whereas FIR filters can have this property. In exchange, the IIR filter is often orders of magnitude more efficient in realizing sharp cutoff filters than FIR filters [13].

There is considerable flexibility in the implementation of IIR systems. The network implied by Eq. (2.29) is depicted in Fig. 2.6a, for the case $M = N = 4$. This is often called the direct form implementation. The generalization to arbitrary M and N is obvious. The difference equation Eq. (2.29) can be transformed into many equivalent forms. Particularly useful among these is the set of equations

$$w(n) = \sum_{k=1}^{N} a_k w(n-k) + x(n)$$

$$y(n) = \sum_{r=0}^{M} b_r w(n-r) \qquad (2.32)$$

(See Problem 2.10). This set of equations can be implemented as shown in Fig. 2.6b, with a significant saving of memory required to store the delayed sequence values.

Equation (2.24) shows that $H(z)$ can be expressed as a product of poles and zeros. These poles and zeros occur in complex conjugate pairs since the coefficients a_k and b_r are real. By grouping the complex conjugate poles and zeros into complex conjugate pairs it is possible to express $H(z)$ as a product of elementary second-order system functions, of the form

$$H(z) = A \prod_{k=1}^{K} \left[\frac{1 + b_{1k}z^{-1} + b_{2k}z^{-2}}{1 - a_{1k}z^{-1} - a_{2k}z^{-2}} \right] \qquad (2.33)$$

where K is the integer part of $(N+1)/2$. Each second order system can be implemented as in Fig. 2.6 and the systems cascaded to implement $H(z)$. This is depicted in Fig. 2.7a for $N = M = 4$. Again the generalization to higher orders is obvious. The partial fraction expansion of Eq. (2.30) suggests still another approach to implementation. By combining terms involving complex conjugate poles, $H(z)$ can be expressed as

$$H(z) = \sum_{k=1}^{K} \frac{c_{0k} + c_{1k}z^{-1}}{1 - a_{1k}z^{-1} - a_{2k}z^{-2}} \qquad (2.34)$$

This suggests a parallel form implementation as depicted in Fig. 2.7b for $N = 4$.

All of the implementations discussed are used in speech processing. For linear filtering applications, the cascade form generally exhibits superior performance with respect to roundoff noise, coefficient inaccuracies, and stability [1,2]. All of the above forms have been used in speech synthesis applications, with the direct form being particularly important in synthesis from linear prediction parameters (See Chapter 8).

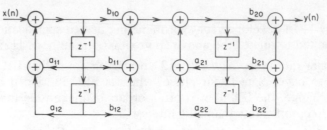

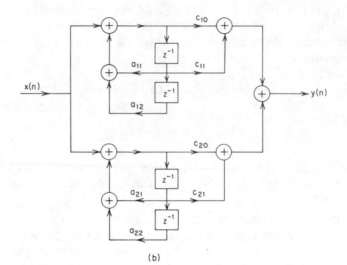

Fig. 2.7 (a) Cascade form; (b) parallel form.

2.4 Sampling

To use digital signal processing methods on an analog signal such as speech, it is necessary to represent the signal as a sequence of numbers. This is commonly done by sampling the analog signal, denoted $x_a(t)$, periodically to produce the sequence

$$x(n) = x_a(nT) \quad -\infty < n < \infty \tag{2.35}$$

where n, of course, takes on only integer values. Figure 2.1 shows a speech waveform and the corresponding set of samples with period $T = 1/8000$ sec.

2.4.1 The sampling theorem

The conditions under which the sequence of samples in Eq. (2.35) is a unique representation of the original analog signal are well known and are often summarized as follows:

The Sampling Theorem: If a signal $x_a(t)$ has a bandlimited Fourier transform $X_a(j\Omega)$, such that $X_a(j\Omega) = 0$ for $\Omega \geqslant 2\pi F_N$, then

24

$x_a(t)$ can be uniquely reconstructed from equally spaced samples $x_a(nT)$, $-\infty < n < \infty$, if $1/T > 2F_N$.

The above theorem follows from the fact that if the Fourier transform of $x_a(t)$ is defined as

$$X_a(j\Omega) = \int_{-\infty}^{\infty} x_a(t)e^{-j\Omega t}dt \qquad (2.36)$$

and the Fourier transform of the sequence $x(n)$ is defined as in Eq. (2.9a), then if $X(e^{j\omega})$ is evaluated for frequencies $\omega = \Omega T$, then $X(e^{j\Omega T})$ is related to $X_a(j\Omega)$ by [1,2]

$$X(e^{j\Omega T}) = \frac{1}{T} \sum_{k=-\infty}^{\infty} X_a(j\Omega + j\frac{2\pi}{T}k) \qquad (2.37)$$

To see the implications of Eq. (2.37), let us assume that $X_a(j\Omega)$ is as shown in Fig. 2.8a; i.e., assume that $X_a(j\Omega) = 0$ for $|\Omega| > \Omega_N = 2\pi F_N$. The frequency F_N is called the *Nyquist frequency*. Now according to Eq. (2.37), $X(e^{j\Omega T})$ is the sum of an infinite number of replicas of $X_a(j\Omega)$, each centered at integer multiples of $2\pi/T$. Fig. 2.8b depicts the case when $1/T > 2F_N$ so that the images of the Fourier transform do not overlap into the base band $|\Omega| < 2\pi F_N$. Figure 2.8c, on the other hand, shows the case $1/T < 2F_N$. In

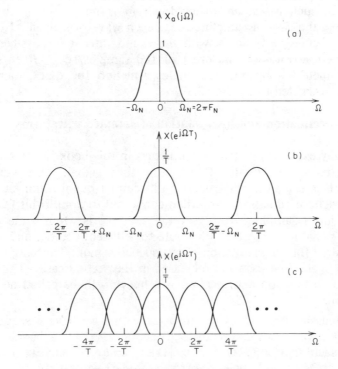

Fig. 2.8 Illustration of sampling.

25

this case, the image centered at $2\pi/T$ overlaps into the base band. This condition, where a high frequency seemingly takes on the identity of a lower frequency, is called *aliasing*. Clearly, aliasing can be avoided only if the Fourier transform is bandlimited and if the sampling frequency $(1/T)$ is equal to at least twice the Nyquist frequency $(1/T > 2F_N)$.

Under the condition $1/T > 2F_N$, it is clear that the Fourier transform of the sequence of samples is proportional to the Fourier transform of the analog signal in the base band; i.e.,

$$X(e^{j\Omega T}) = \frac{1}{T} X_a(j\Omega) \quad |\Omega| < \frac{\pi}{T} \tag{2.38}$$

Using this result, it can be shown that [1,2] the original signal can be related to the sequence of samples by the interpolation formula

$$x_a(t) = \sum_{n=-\infty}^{\infty} x_a(nT) \left[\frac{\sin[\pi(t-nT)/T]}{\pi(t-nT)/T} \right] \tag{2.39}$$

Thus, given samples of a bandlimited analog signal taken at a rate at least twice the Nyquist frequency, it is possible to reconstruct the original analog signal using Eq.(2.39). Practical digital-to-analog converters seek to approximate Eq. (2.39).

Sampling is implicit in many speech processing algorithms that seek to estimate basic parameters of speech production such as pitch and formant frequencies. In such cases, an analog function is not available to be sampled directly, as in the case of sampling the speech waveform itself. However, such parameters change very slowly with time, and thus it is possible to estimate (sample) them at rates on the order of 100 samples/sec. Given samples of a speech parameter, a bandlimited analog function for that parameter can of course be constructed using Eq. (2.39).

2.4.2 Decimation and interpolation of sampled waveforms

In many examples that we shall discuss in this book, there arises the need to change the sampling rate of a discrete time signal. One example occurs when speech is sampled using one-bit differential quantization at a high sampling rate (delta modulation), and then converted to a multi-bit PCM representation at a lower sampling rate. Another example is when some parameter of the speech signal is sampled at a low rate for efficient coding, and then a higher rate is required for reconstruction of the speech signal. The sampling rate must be reduced in the first case and increased in the second case. The processes of sampling rate reduction and increase will henceforth be called decimation and interpolation.

In discussing both cases, let us assume that we have a sequence of samples $x(n) = x_a(nT)$, where the analog function $x_a(t)$ has a bandlimited Fourier transform such that $X_a(j\Omega) = 0$ for $|\Omega| > 2\pi F_N$. Then we have just seen that if $1/T > 2F_N$, the Fourier transform of $x(n)$ will satisfy

26

$$X(e^{j\Omega T}) = \frac{1}{T} X_a(j\Omega) \quad |\Omega| < \frac{\pi}{T} \tag{2.40}$$

2.4.2a Decimation

Let us suppose that we wish to reduce the sampling rate by a factor M; i.e., we wish to compute a new sequence corresponding to samples of $x_a(t)$ taken with period $T' = MT$, i.e.,

$$y(n) = x_a(nT') = x_a(nTM) \tag{2.41}$$

It is easily seen that

$$y(n) = x(Mn) \quad -\infty < n < \infty. \tag{2.42}$$

That is, $y(n)$ is obtained simply by periodically retaining only one out of every M samples. From our previous discussion of the sampling theorem we note that if $1/T' > 2F_N$, then the samples $y(n)$ will also be adequate to uniquely represent the original analog signal. The Fourier transforms of $x(n)$ and $y(n)$ are related by the expression [14]

$$Y(e^{j\Omega T'}) = \frac{1}{M} \sum_{k=0}^{M-1} X(e^{\frac{j(\Omega T' - 2\pi k)}{M}}) \tag{2.43}$$

From Eq. (2.43) it can be seen that in order that there be no overlap between the images of $X(e^{j\Omega T})$, we must have $1/T' > 2F_N$. If this condition holds, then we see that

$$Y(e^{j\Omega T'}) = \frac{1}{M} X(e^{\frac{j\Omega T'}{M}})$$

$$= \frac{1}{M} \frac{1}{T} X_a(j\Omega)$$

$$= \frac{1}{T'} X_a(j\Omega) \quad -\frac{\pi}{T'} < \Omega < \frac{\pi}{T'} \tag{2.44}$$

Figure 2.9 shows an example of sampling rate reduction. Figure 2.9a shows the Fourier transform of the original analog signal. Figure 2.9b shows the Fourier transform of $x(n) = x_a(nT)$ where the sampling rate $(1/T)$ is somewhat greater than the Nyquist rate $(2F_N)$. Figure 2.9c shows the case of sampling rate reduction by a factor of 3; i.e., $T' = 3T$. For this case aliasing occurs because $1/T' < 2F_N$. However, suppose $x(n)$ is filtered with a digital lowpass filter with cutoff frequency $\pi/T' = \pi/(3T)$ producing a sequence $w(n)$. For our example, the Fourier transform of the output of the lowpass filter is shown in Figure 2.9d. Aliasing does not occur when the sampling rate of the filtered signal is reduced by a factor of 3 as depicted in Figure 2.9e; however, the samples $y(n)$ no longer represent $x_a(t)$ but rather a new signal $y_a(t)$ which is a lowpass filtered version of $x_a(t)$. A block diagram of a general decimation system is given in Figure 2.10.

27

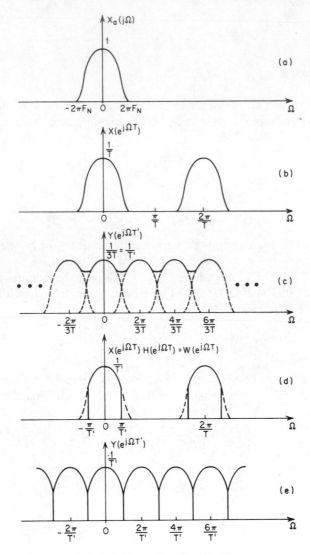

Fig. 2.9 Illustration of decimation.

2.4.2b Interpolation

Now suppose that we have samples of an analog waveform $x(n) = x_a(nT)$. If we wish to increase the sampling rate by an integer factor L, we must compute a new sequence corresponding to samples of $x_a(t)$ taken with period $T' = T/L$; i.e.,

$$y(n) = x_a(nT') = x_a(nT/L) \qquad (2.45)$$

Clearly, $y(n) = x(n/L)$ for $n = 0, \pm L, \pm 2L, \ldots$ but we must fill in the unknown samples for all other values of n by an interpolation process [14]. To see how this can be done using a digital filter, consider the sequence

28

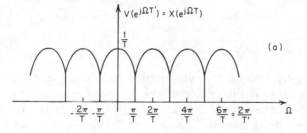

Fig. 2.10 Block diagram representation of decimation.

$$v(n) = x\left[\frac{n}{L}\right] \qquad n = 0, \pm L, \pm 2L, \ldots$$

$$= 0 \qquad otherwise \qquad (2.46)$$

The Fourier transform of $v(n)$ is easily shown to be [14]

$$V(e^{j\Omega T'}) = X(e^{j\Omega T'L})$$

$$= X(e^{j\Omega T}). \qquad (2.47)$$

Thus $V(e^{j\Omega T'})$ is periodic with period $2\pi/T = 2\pi/(LT')$, as well as with period $2\pi/T'$ as is the case in general for sequences associated with a sampling period T'. Figure 2.11a shows $V(e^{j\Omega T'})$ [and $X(e^{j\Omega T})$] for the case $T' = T/3$. In order to obtain the sequence

$$y(n) = x_a(nT')$$

from the sequence $v(n)$, we must ensure that

$$Y(e^{j\Omega T'}) = \frac{1}{T'} X_a(j\Omega) \qquad -\frac{\pi}{T'} \leqslant \Omega \leqslant \frac{\pi}{T'} \qquad (2.48)$$

Assuming that

$$X(e^{j\Omega T}) = \frac{1}{T} X_a(j\Omega) \qquad -\frac{\pi}{T} \leqslant \Omega \leqslant \frac{\pi}{T} \qquad (2.49)$$

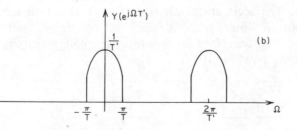

Fig. 2.11 Illustration of interpolation.

29

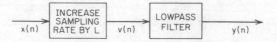

Fig. 2.12 Block diagram representation of interpolation.

then it is clear from Figure 2.11a that what is required is that the images of $X_a(j\Omega)$ in $V(e^{j\Omega T})$, that are centered at $\Omega = 2\pi/T$ and $\Omega = 4\pi/T$, must be removed by a digital lowpass filter that rejects all frequency components in the range $\pi/T \leqslant \Omega \leqslant \pi/T'$. Moreover, to ensure that the amplitude is correct for sampling interval T', the gain of the filter must be $L = T/T'$. That is

$$Y(e^{j\Omega T'}) = H(e^{j\Omega T'})V(e^{j\Omega T'}) = H(e^{j\Omega T'})X(e^{j\Omega T})$$

$$= H(e^{j\Omega T'}) \frac{1}{T} X_a(j\Omega) \tag{2.50}$$

Thus, in order that $Y(e^{j\Omega T'}) = (1/T') X_a(j\Omega)$ for $\Omega \leqslant \pi/T'$ we require that

$$H(e^{j\Omega T'}) = L \qquad |\Omega| \leqslant \frac{\pi}{T}$$

$$= 0 \qquad otherwise \tag{2.51}$$

The general interpolation system is depicted in Fig. 2.12.

2.4.2c Non-Integer Sampling Rate Changes

It is readily seen that samples corresponding to a sampling period $T' = MT/L$ can be obtained by a combination of interpolation by a factor L followed by decimation by a factor M. By suitable choice of the integers M and L, we can approach arbitrarily close to any desired ratio of sampling rates. By combining Figures 2.9 and 2.11, we observe that a single lowpass filter suffices for both the interpolation and decimation filter. This is depicted in Fig. 2.13.

2.4.2d Advantages of FIR Filters

An extremely important consideration in the implementation of decimators and interpolators is the choice of the type of lowpass filter. For these systems, a significant savings in computation over alternative filter types can be obtained by using finite impulse response (FIR) filters in a standard direct form implementation. The savings in computations for FIR filters is due to the observation that for decimators only one of each M output samples needs to be calculated, while for interpolators, $L - 1$ out of every L samples of the input are zero valued, and therefore do not affect the computation. These facts cannot be fully exploited using IIR filters [14].

Fig. 2.13 Block diagram representation of sampling rate increase by a factor of L/M.

Assuming that the required filtering is being performed using FIR filters, then for large changes in the sampling rate (i.e, large M for decimators, or large L for interpolators) it has been shown that it is more efficient to reduce (or increase) the sampling rate with a series of decimation stages than to make the entire rate reduction with one stage. In this way the sampling rate is reduced gradually resulting in much less severe filtering requirements on the lowpass filters at each stage. The details of multistage implementation of decimation, interpolation and narrowband filtering are given in Refs. [15-18].

2.5 Summary

In this chapter we have presented a review of the fundamentals of discrete-time signal processing. The notions of discrete convolution, difference equations, and frequency domain representations of signals and systems will be used extensively in this book. Also the concepts of sampling of analog signals and digital alteration of the sampling rate discussed in Section 2.4 are extremely important in all types of digital speech processing systems.

REFERENCES

1. A. V. Oppenheim and R. W. Schafer, *Digital Signal Processing*, Prentice-Hall, Inc., Englewood Cliffs, N.J., 1975.

2. L. R. Rabiner and B. Gold, *Theory and Application of Digital Signal Processing*, Prentice-Hall, Inc., Englewood Cliffs, N.J., 1975.

3. A. Peled and B. Liu, *Digital Signal Processing, Theory, Design and Implementation*, John Wiley and Sons, New York, 1976.

4. J. W. Cooley and J. W. Tukey, "An Algorithm for the Machine Computation of Complex Fourier Series," *Math Computation*, Vol. 19, pp. 297-381, April 1965.

5. H. D. Helms, "Fast Fourier Transform Method of Computing Difference Equations and Simulating Filters," *IEEE Trans. Audio and Electroacoustics*, Vol. 15, No. 2, pp. 85-90, 1967.

6. T. G. Stockham, "High-Speed Convolution and Correlation," *1966 Spring Joint Computer Conference*, AFIPS Proc., Vol. 28, pp. 229-233, 1966.

7. J. F. Kaiser, "Nonrecursive Digital Filter Design Using the I_o-Sinh Window Function," *Proc. 1974 IEEE Int. Symp. on Circuits and Systems*, San Francisco, pp. 20-23, April 1974.

8. L. R. Rabiner, B. Gold, and C. A. McGonegal, "An Approach to the Approximation Problem for Nonrecursive Digital Filters," *IEEE Trans. Audio and Electroacoustics*, Vol. 19, No. 3, pp. 200-207, September 1971.

9. T. W. Parks and J. H. McClellan, "Chebyshev Approximation for Nonre-

cursive Digital Filter with Linear Phase," *IEEE Trans. Circuit Theory,* Vol. CT-19, pp. 189-194, March 1972.

10. J. H. McClellan, T. W. Parks, and L. R. Rabiner, "A Computer Program for Designing Optimum FIR Linear Phase Digital Filters," *IEEE Trans. Audio and Electroacoustics,* Vol. AU-21, pp. 506-526, December 1973.

11. L. R. Rabiner, J. H. McClellan, and T. W. Parks, "FIR Digital Filter Design Techniques Using Weighted Chebyshev Approximation," *Proc. IEEE,* Vol. 63, No. 4, pp. 595-609, April 1975.

12. A. G. Deczky, "Synthesis of Recursive Digital Filters Using the Minimum p-Error Criterion," *IEEE Trans. Audio and Electroacoustics,* Vol. AU-20, No. 5, pp. 257-263, October 1972.

13. L. R. Rabiner, J. F. Kaiser, O. Herrmann, and M. T. Dolan, "Some Comparisons Between FIR and IIR Digital Filters," *Bell Syst. Tech. J.,* Vol. 53, No. 2, pp. 305-331, February 1974.

14. R. W. Schafer and L. R. Rabiner, "A Digital Signal Processing Approach to Interpolation," *Proc. IEEE,* Vol. 61, No. 6, pp. 692-702, June 1973.

15. L. R. Rabiner and R. E. Crochiere, "A Novel Implementation for FIR Digital Filters," *IEEE Trans. Acoustics, Speech, and Signal Proc.,* Vol. ASSP-23, pp. 457-464, October 1975.

16. R. E. Crochiere and L. R. Rabiner, "Optimum FIR Digital Filter Implementation for Decimation, Interpolation and Narrowband Filters," *IEEE Trans. Acoust. Speech, and Signal Proc.,* Vol. ASSP-23, pp. 444-456, October 1975.

17. R. E. Crochiere and L. R. Rabiner, "Further Considerations in the Design of Decimators and Interpolators," *IEEE Trans. Acoustics, Speech, and Signal Processing,* Vol. ASSP-24, No. 4, pp. 269-311, August 1976.

18. D. J. Goodman, "Digital Filters for Code Format Conversion," *Electronics Letters,* Vol. 11, February 1975.

PROBLEMS

2.1 Consider the sequence

$$x(n) = a^n \quad n \geqslant n_0$$
$$= 0 \quad n < n_0$$

(a) Find the z-transform of $x(n)$.
(b) Find the Fourier transform of $x(n)$. Under what conditions does the Fourier transform exist?

2.2 The input to a linear, time-invariant system is

$$x(n) = 1 \quad 0 \leqslant n \leqslant N - 1$$
$$= 0 \quad otherwise$$

The impulse response of the system is

$$h(n) = a^n \quad n \geqslant 0$$
$$= 0 \quad n < 0.$$

(a) Using discrete convolution, find the output, $y(n)$, of the system for all n.

(b) Find the output using z-transforms.

2.3 Find the z-transform and the Fourier transform of each of the following sequences. (Each of these are commonly used as "windows" in speech processing systems.)

(1) Exponential window

$$w_1(n) = a^n \quad 0 \leqslant n \leqslant N - 1$$
$$= 0 \quad otherwise$$

(2) Rectangular window

$$w_2(n) = 1 \quad 0 \leqslant n \leqslant N - 1$$
$$= 0 \quad otherwise$$

(3) Hamming window

$$w_3(n) = 0.54 - 0.46 \cos[2\pi n/(N-1)] \quad 0 \leqslant n \leqslant N - 1$$
$$= 0 \quad\quad\quad\quad\quad\quad\quad\quad\quad\quad otherwise$$

Sketch the magnitude of the Fourier transforms in each case. Hint: obtain a relationship between $W_3(e^{j\omega})$ and $W_2(e^{j\omega})$.

2.4 The frequency response of an ideal lowpass filter is

$$H(e^{j\omega}) = 1 \quad |\omega| < \omega_c$$
$$= 0 \quad \omega_c < |\omega| \leqslant \pi$$

($H(e^{j\omega})$ is, of course, periodic with period 2π.)
(a) Find the impulse response of the ideal lowpass filter.
(b) Sketch the impulse response for $\omega_c = \pi/4$.
The frequency response of an ideal bandpass filter is

$$H(e^{j\omega}) = 1 \quad \omega_a < |\omega| < \omega_b$$
$$= 0 \quad |\omega| < \omega_a \quad \text{and} \quad \omega_b < |\omega| \leqslant \pi,$$

(c) Find the impulse response of the ideal bandpass filter.
(d) Sketch the impulse response for $\omega_a = \pi/4$ and $\omega_b = 3\pi/4$.

2.5 The frequency response of an ideal differentiator is

$$H(e^{j\omega}) = j\omega e^{-j\omega\tau} \quad -\pi < \omega < \pi$$

(This response is repeated with period 2π.) The quantity τ is the delay of the system in samples.
(a) Sketch the magnitude and phase response of this system.
(b) Find the impulse response, $h(n)$, of this system.
(c) The impulse response of this ideal system can be truncated to a

33

length N samples by a window such as those in Problem 2.3. In so doing the delay is set equal to $\tau = (N-1)/2$ so that the ideal impulse response can be truncated symmetrically [1]. If $\tau = (N-1)/2$ and N is an odd integer, show that the ideal impulse response decreases as $1/n$. Sketch the ideal impulse response for the case $N = 11$.

(d) In the case that N is even, show that $h(n)$ decreases as $1/n^2$. Sketch the ideal impulse response for the case $N = 10$.

2.6 The frequency response of an ideal Hilbert transformer (90° phase shifter) with delay τ is

$$H(e^{j\omega}) = -je^{-j\omega\tau} \quad 0 < \omega < \pi$$
$$= je^{-j\omega\tau} \quad -\pi < \omega < 0 .$$

Find and sketch the impulse response of this system.

2.7 Consider a linear phase FIR digital filter. The impulse response of such a filter has the property

$$h(n) = h(N-1-n) \quad 0 \leqslant n \leqslant N - 1$$
$$= 0 \quad otherwise$$

(a) Show that if N is an even integer the convolution sum expression for the output of such a system can be expressed as

$$y(n) = \sum_{k=0}^{(N-2)/2} h(k)[x(n-k)+x(n-N+1+k)]$$

and if N is odd

$$y(n) = \sum_{k=0}^{(N-3)/2} h(k)[x(n-k)+x(n-N+1+k)]+h((N-1)/2)x(n-(N-1)/2) .$$

Thus, the number of multiplications required to compute each output sample is essentially halved.

(b) Draw the digital filter structures for each of the above equations.

2.8 Consider the first order system

$$y(n) = \alpha y(n-1) + x(n)$$

(a) Find the system function, $H(z)$, for this system.
(b) Find the impulse response of this system.
(c) For what values of α will the system be stable?
(d) Assume that the input is obtained by sampling with period T. Find the value of α such that

$$h(n) < e^{-1} \quad for \quad nT < 2 \text{ msec}$$

i.e., find the value of α that gives a time constant of 2 msec.

2.9 Consider a system function of the form of Eq. (2.24)
(a) Show that if $M < N$, $H(z)$ can be expressed as a partial fraction

expansion as in Eq. (2.30), where the coefficients A_m can be found from

$$A_m = H(z)(1-d_mz^{-1})|_{z=d_m} \quad m = 1,2,...,N$$

(b) Show that the z-transform of the sequence $A_k(d_k)^n u(n)$ is

$$\frac{A_k}{1 - d_kz^{-1}} \quad |z| > |d_k|,$$

and thus $h(n)$ is given by Eq. (2.31).

2.10 Consider two linear shift-invariant systems in cascade as shown in Fig. P2.10 — i.e., the output of the first system is the input to the second.

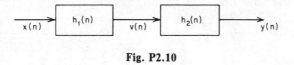

Fig. P2.10

(a) Show that the impulse response of the overall system is

$$h(n) = h_1(n)*h_2(n).$$

(b) Show that

$$h_1(n)*h_2(n) = h_2(n)*h_1(n)$$

and thus that the overall response does not depend on the order in which the systems are cascaded.

(c) Consider the system function of Eq. (2.23) written as

$$H(z) = \left(\sum_{r=0}^{M} b_rz^{-r}\right)\left[\frac{1}{1 - \sum_{k=1}^{N} a_kz^{-k}}\right]$$

$$= H_1(z)\cdot H_2(z)$$

i.e., as a cascade of two systems. Write the difference equations for the overall system from this point of view.

(d) Now consider the two systems of part (c) in the opposite order; i.e.,

$$H(z) = H_2(z)H_1(z)$$

Show that the difference equations of Eq. (2.32) result.

2.11 For the difference equation

$$y(n) = 2\cos(bT)y(n-1) - y(n-2)$$

find the two initial conditions $y(-1)$ and $y(-2)$ such that

(a) $\quad y(n) = \cos(bTn) \quad n \geqslant 0$

(b) $\quad y(n) = \sin(bTn) \quad n \geqslant 0$

35

2.12 Consider the set of difference equations

$$y_1(n) = Ay_1(n-1) + By_2(n-1) + x(n)$$
$$y_2(n) = Cy_1(n-1) + Dy_2(n-1)$$

(a) Draw the network diagram for this system.

(b) Find the transfer functions

$$H_1(z) = \frac{Y_1(z)}{X(z)} \quad \text{and} \quad H_2(z) = \frac{Y_2(z)}{X(z)}$$

(c) For the case $A = D = r \cos \theta$ and $C = -B = r\sin\theta$, find the impulse responses $h_1(n)$ and $h_2(n)$ that result when the system is excited by $x(n) = \delta(n)$.

2.13 A causal linear shift invariant system has the system function

$$H(z) = \frac{(1+2z^{-1}+z^{-2})(1+2z^{-1}+z^{-2})}{(1 + \frac{7}{8} z^{-1} + \frac{5}{16} z^{-2})(1 + \frac{3}{4} z^{-1} + \frac{7}{8} z^{-2})}$$

(a) Draw a digital network diagram of an implementation of this system in
(i) Cascade form
(ii) Direct form.

(b) In this system stable? Explain.

2.14 For the system of Fig. P2.14,
(a) Write the difference equations represented by the network.
(b) Find the system function for the network.

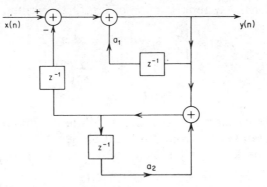

Fig. P2.14

2.15 Find a_1, a_2, and a_3 in terms of b_1, and b_2 so that the two networks of Fig P2.15 have the same transfer function.

2.16 The system function for a simple resonator is of the form

$$H(z) = \frac{1 - 2e^{-aT}\cos(bT) + e^{-2aT}}{1 - 2e^{-aT}\cos(bT)z^{-1} + e^{-2aT}z^{-2}}$$

36

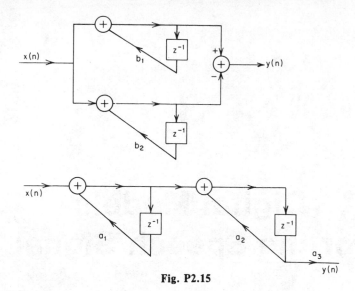

Fig. P2.15

(a) Find the poles and zeros of $H(z)$ and plot them in the z-plane.

(b) Find the impulse response of this system and sketch it for the constants

$$T = 10^{-4}$$
$$b = 1000\pi$$
$$a = 200\pi$$

(c) Sketch the frequency response of this system as a function of analog frequency, Ω.

2.17 Consider the finite length sequence

$$x(n) = \delta(n) + 0.5\delta(n-5)$$

(a) Find the z-transform and Fourier transform of $x(n)$.

(b) Find the N-point DFT of $x(n)$ for $N = 50$, 10 and 5.

(c) How are the DFT values for $N = 5$ related to those of the DFT for $N = 50$?

(d) What is the relationship between the N-point DFT of $x(n)$ and the Fourier transform of $x(n)$?

2.18 A speech signal is sampled at a rate of 20000 samples/sec (20 kHz). A segment of length 1024 samples is selected and the 1024-point DFT is computed.

(a) What is the time duration of the segment of speech?

(b) What is the frequency resolution (spacing in Hz) between the DFT values?

(c) How do your answers to parts (a) and (b) change if we compute the 1024-point DFT of 512 samples of the speech signal. (The 512 samples would be augmented with 512 zero samples before the transform was computed.)

37

3

Digital Models
for the Speech Signal

3.0 Introduction

In order to apply digital signal processing techniques to speech communication problems, it is essential to understand the fundamentals of the speech production process as well as the fundamentals of digital signal processing. This chapter provides a review of the acoustic theory of speech production and shows how this theory leads to a variety of ways of representing the speech signal. Specifically, we shall be concerned with obtaining discrete-time models for representing sampled speech signals. These models will serve as a basis for application of digital processing techniques.

This chapter plays a role similar to that of Chapter 2 in serving as a review of an established area of knowledge. Several excellent references provide much more detail on many of the topics of this chapter [1-5]. Particularly noteworthy are the books by Fant [1] and Flanagan [2]. Fant's book deals primarily with the acoustics of speech production and contains a great deal of useful data on vocal system measurements and models. Flanagan's book, which is much broader in scope, contains a wealth of valuable insights into the physical modelling of the speech production process and the way that such models are used in representing and processing speech signals. These books are indispensable to the serious student of speech communication.

Before discussing the acoustic theory and the resulting mathematical models for speech production, it is necessary to consider the various types of sounds that make up human speech. Thus, this chapter begins with a very brief introduction to acoustic phonetics in the form of a summary of the phonemes of English and a discussion of the place and manner of articulation for each of the major phoneme classes. Then the fundamentals of the acoustic theory of speech production are presented. Topics considered include sound propagation in the vocal tract, transmission line analogies, and the steady state behaviour of the vocal system in the production of a single sustained sound. This theory provides the basis for the classical approach to modelling the speech signal as the output of a time-varying linear system (vocal tract) excited by either random noise or a quasi-periodic sequence of pulses. This approach is applied to obtain discrete time models for the speech signal. These models, which are justified in terms of the acoustic theory and formulated in terms of digital filtering principles, serve as the basis for discussion of speech processing techniques throughout the remainder of this book.

3.1 The Process of Speech Production

Speech signals are composed of a sequence of sounds. These sounds and the transitions between them serve as a symbolic representation of information. The arrangement of these sounds (symbols) is governed by the rules of language. The study of these rules and their implications in human communication is the domain of *linguistics,* and the study and classification of the sounds of speech is called *phonetics.* A detailed discussion of phonetics and linguistics would take us too far afield. However, in processing speech signals to enhance or extract information, it is helpful to have as much knowledge as possible about the structure of the signal; i.e., about the way in which information is encoded in the signal. Thus, it is worthwhile to discuss the main classes of speech sounds before proceeding to a detailed discussion of mathematical models of the production of speech signals. Although this will be all that we shall have to say about linguistics and phonetics, this is not meant to minimize their importance − especially in the areas of speech recognition and speech synthesis.

3.1.1 The mechanism of speech production

Figure 3.1 is an X-ray photograph which places in evidence the important features of the human vocal system [6]. The *vocal tract,* outlined by the dotted lines in Fig. 3.1, begins at the opening between the vocal cords, or *glottis,* and ends at the lips. The vocal tract thus consists of the *pharynx* (the connection from the esophagus to the mouth) and the mouth or *oral cavity.* In the average male, the total length of the vocal tract is about 17 cm. The cross-sectional area of the vocal tract, determined by the positions of the tongue, lips, jaw, and velum varies from zero (complete closure) to about $20cm^2$. The *nasal tract* begins at the velum and ends at the nostrils. When the velum is lowered, the nasal tract is acoustically coupled to the vocal tract to produce the nasal sounds of speech.

39

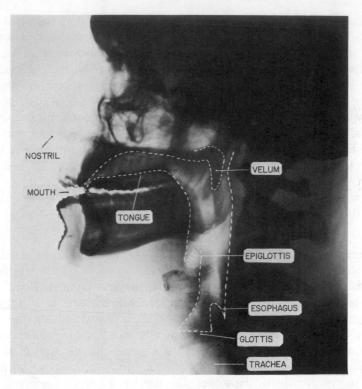

Fig. 3.1 Saggital plane X-ray of the human vocal apparatus. (After Flanagan et al. [6].)

In studying the speech production process, it is helpful to abstract the important features of the physical system in a manner which leads to a realistic yet tractable mathematical model. Figure 3.2 shows such a schematic diagram of the vocal system [6]. For completeness the diagram includes the sub-glottal system composed of the lungs, bronchi and trachea. This sub-glottal system serves as a source of energy for the production of speech. Speech is simply the acoustic wave that is radiated from this system when air is expelled from the lungs and the resulting flow of air is perturbed by a constriction somewhere in the vocal tract. As an example of a speech wave, Figure 3.3a shows the waveform of the utterance, "should we cha(se)," spoken by a male speaker. The general features of this waveform can be readily explained by a more detailed consideration of the mechanism of speech production.

Speech sounds can be classified into 3 distinct classes according to their mode of excitation. *Voiced* sounds are produced by forcing air through the glottis with the tension of the vocal cords adjusted so that they vibrate in a relaxation oscillation, thereby producing quasi-periodic pulses of air which excite the vocal tract. Voiced segments are labelled /U/, /d/, /w/, /i/ and /e/ in Fig. 3.3a. *Fricative or unvoiced sounds* are generated by forming a constriction at some point in the vocal tract (usually toward the mouth end), and forcing air through the constriction at a high enough velocity to produce turbulence. This creates a broad-spectrum noise source to excite the vocal tract. The segment labelled /ʃ/ in Fig. 3.3a is the fricative "sh." *Plosive sounds* result from making a complete closure (again, usually toward the front of the vocal tract), building up pressure behind the closure, and abruptly releasing it. Plosive excitation is involved in creating the sound labelled /tʃ/ at the beginning of the fourth line of Fig. 3.3a. Note the gap (region of very small amplitude) at the end of the third line which precedes the burst of noise-like waveform. This gap corresponds to the time of complete closure of the vocal tract.

The vocal tract and nasal tract are shown in Figure 3.2 as tubes of nonuniform cross-sectional area. As sound, generated as discussed above, propagates down these tubes, the frequency spectrum is shaped by the frequency selectivity of the tube. This effect is very similar to the resonance effects observed with organ pipes or wind instruments. In the context of speech production, the resonance frequencies of the vocal tract tube are called *formant frequencies* or simply *formants*. The formant frequencies depend upon the shape and dimensions of the vocal tract; each shape is characterized by a set of formant frequencies. Different sounds are formed by varying the shape of the vocal tract. Thus, the spectral properties of the speech signal vary with time as the vocal tract shape varies.

The time-varying spectral characteristics of the speech signal can be graphically displayed through the use of the sound spectrograph [2,7]. This device produces a two-dimensional pattern called a *spectrogram* in which the vertical dimension corresponds to frequency and the horizontal dimension to time. The darkness of the pattern is proportional to signal energy. Thus, the resonance frequencies of the vocal tract show up as dark bands in the spectrogram. Voiced regions are characterized by a striated appearance due to the periodicity of the time waveform, while unvoiced intervals are more solidly filled in. A spectrogram of the utterance of Fig. 3.3a is shown in Figure 3.3b. The spectrogram is labelled to correspond to the labelling of Fig. 3.3a so that the time domain and frequency domain features can be correlated.

The sound spectrograph has long been a principal tool in speech research, and although more flexible displays can be generated using digital processing techniques (see Chapter 6), its basic principles are still widely used. An early,

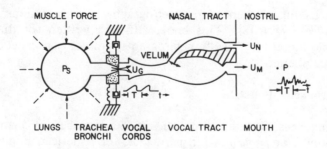

Fig. 3.2 Schematized diagram of the vocal apparatus (After Flanagan et al. [6].)

41

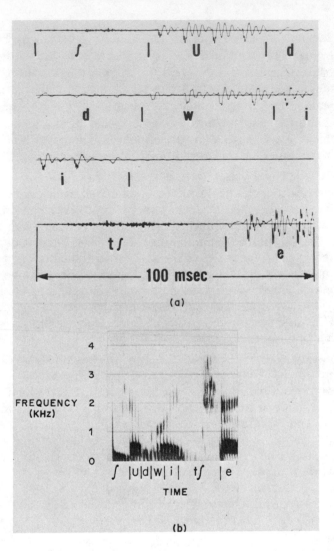

Fig. 3.3 (a) Waveform of the utterance "Should we cha(se)"; (b) corresponding spectrogram.

but still very useful, reference on spectrographic representations of speech is the book *Visible Speech* [8]. This book, although written for the purpose of teaching people literally to "read" spectrograms, provides an excellent introduction to acoustic phonetics.

3.1.2 Acoustic phonetics

Most languages, including English, can be described in terms of a set of distinctive sounds, or *phonemes*. In particular, for American English, there are about 42 phonemes including vowels, diphthongs, semivowels and consonants. There are a variety of ways of studying phonetics; e.g., linguists study the distinctive features or characteristics of the phonemes [9,10]. For our purposes it is sufficient to consider an acoustic characterization of the various sounds including the place and manner of articulation, waveforms, and spectrographic characterizations of these sounds.

Table 3.1 shows how the sounds of American English are broken into phoneme classes.[1] The four broad classes of sounds are vowels, diphthongs, semivowels, and consonants. Each of these classes may be further broken down into sub-classes which are related to the manner, and place of articulation of the sound within the vocal tract.

Each of the phonemes in Table 3.1 can be classified as either a continuant, or a noncontinuant sound. Continuant sounds are produced by a fixed (non-time-varying) vocal tract configuration excited by the appropriate source. The class of continuant sounds includes the vowels, the fricatives (both unvoiced and voiced), and the nasals. The remaining sounds (diphthongs, semivowels, stops and affricates) are produced by a changing vocal tract configuration. These are therefore classed as noncontinuants.

3.1.2a Vowels

Vowels are produced by exciting a fixed vocal tract with quasi-periodic pulses of air caused by vibration of the vocal cords. As we shall see later in this chapter, the way in which the cross-sectional area varies along the vocal

Table 3.1 Phonemes in American English.

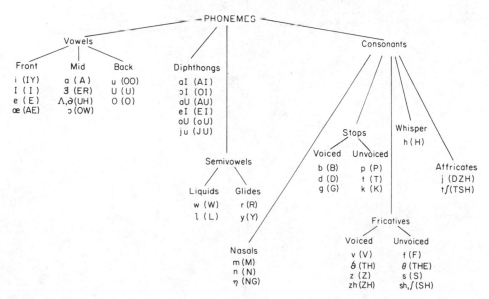

[1]Table 3.1 gives both a phonetic representation and an orthographic representation for each phoneme. The phonetic and orthographic representations are used interchangably throughout this text.

tract determines the resonant frequencies of the tract (formants) and thus the sound that is produced. The dependence of cross-sectional area upon distance along the tract is called the *area function* of the vocal tract. The area function for a particular vowel is determined primarily by the position of the tongue, but the positions of the jaw, lips, and, to a small extent, the velum also influence the resulting sound. For example, in forming the vowel /a/ as in "father," the vocal tract is open at the front and somewhat constricted at the back by the main body of the tongue. In contrast, the vowel /i/ as in "eve" is formed by raising the tongue toward the palate, thus causing a constriction at the front and increasing the opening at the back of the vocal tract. Thus, each vowel sound can be characterized by the vocal tract configuration (area function) that is used in its production. It is obvious that this is a rather imprecise characterization because of the inherent differences between the vocal tracts of speakers. An alternative representation is in terms of the resonance frequencies of the vocal tract. Again a great deal of variability is to be expected among speakers producing the same vowel. Peterson and Barney [11] measured the formant (resonance) frequencies (using a sound spectrograph) of vowels that were perceived to be equivalent. Their results are shown in Fig. 3.4 which is a plot of second

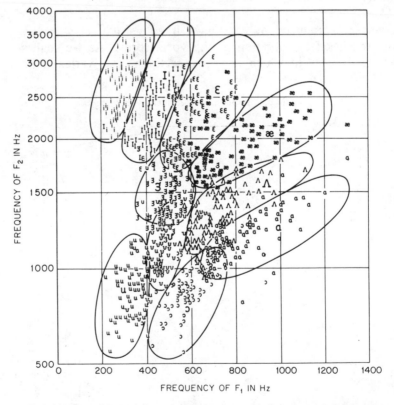

Fig. 3.4 Plot of second formant frequency versus first formant frequency for vowels by a wide range of speakers. (After Peterson and Barney [11].)

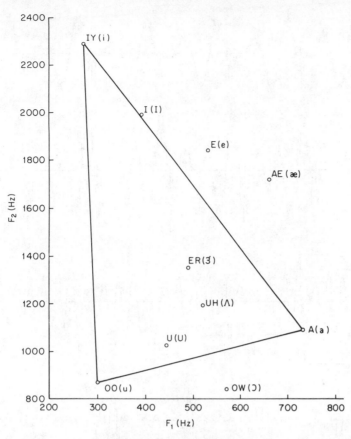

Fig. 3.5 The vowel triangle.

formant frequency as a function of first formant frequency for several vowels spoken by men and children. The broad ellipses in Figure 3.4 show the approximate range of variation in formant frequencies for each of these vowels. Table 3.2 gives average values of the first three formant frequencies of the vowels for male speakers. Although a great deal of variation clearly exists in the vowel formants, the data of Table 3.2 serve as a useful characterization of the vowels.

Table 3.2 Average Formant Frequencies for the Vowels. (After Peterson and Barney [11].)

FORMANT FREQUENCIES FOR THE VOWELS					
Typewritten Symbol for Vowel	IPA Symbol	Typical Word	F_1	F_2	F_3
IY	i	(beet)	270	2290	3010
I	I	(bit)	390	1990	2550
E	ɛ	(bet)	530	1840	2480
AE	œ	(bat)	660	1720	2410
UH	ʌ	(but)	520	1190	2390
A	ɑ	(hot)	730	1090	2440
OW	ɔ	(bought)	570	840	2410
U	U	(foot)	440	1020	2240
OO	u	(boot)	300	870	2240
ER	ʒ	(bird)	490	1350	1690

45

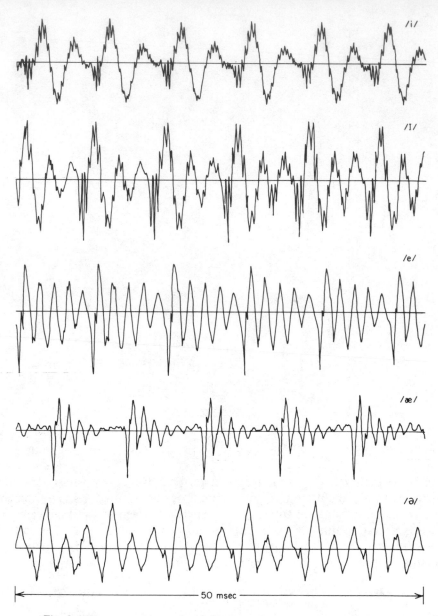

Fig. 3.6 The acoustic waveforms for several American English vowels and corresponding spectrograms.

Figure 3.5 shows a plot of the second formant frequency versus the first formant frequency for the vowels of Table 3.2. The so-called "vowel triangle" is readily seen in this figure. At the upper left hand corner of the triangle is the vowel /i/ with a low first formant, and a high second formant. At the lower left hand corner is the vowel /u/ with low first and second formants. The third vertex of the triangle is the vowel /a/ with a high first formant, and a low second formant. Later in this chapter we shall see how vocal tract shape affects the formant frequencies of vowels.

46

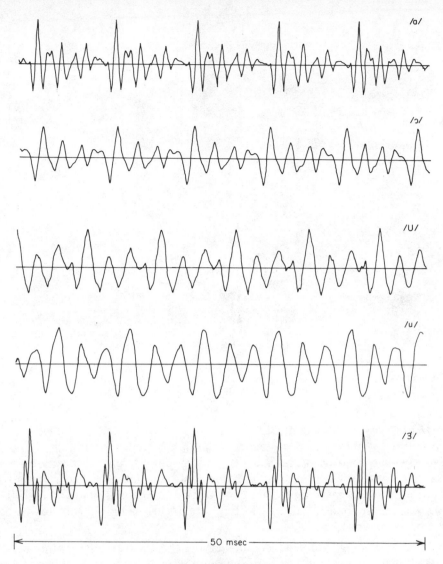

/ɑ/

/ɔ/

/ʊ/

/u/

/ɜ/

|← ——————————— 50 msec ——————————— →|

Fig. 3.6 (Continued)

The acoustic waveforms and spectrograms for each of the vowels of English are shown in Fig. 3.6. The spectrograms clearly show a different pattern of resonances for each vowel. The acoustic waveforms, in addition to showing the periodicity characteristic of voiced sounds, also display the gross spectral properties if a single "period" is considered. For example, the vowel /i/ shows a low frequency damped oscillation upon which is superimposed a relatively strong high frequency oscillation. This is consistent with a low first formant and high second and third formant (see Table 3.2). (Two resonances in proximity tend to boost the spectrum.) In contrast the vowel /u/ shows relatively little high frequency energy as a consequence of the low first and second formant frequencies. Similar correspondences can be observed for all the vowels in Fig. 3.6.

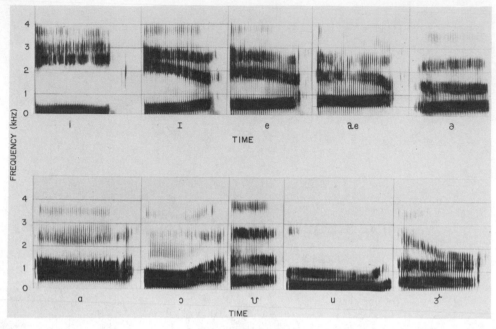

Fig. 3.6 (Continued)

3.1.2b Diphthongs

Although there is some ambiguity and disagreement as to what is and what is not a diphthong, a reasonable definition is that a diphthong is a gliding monosyllabic speech item that starts at or near the articulatory position for one vowel and moves to or toward the position for another. According to this definition, there are six diphthongs in American English including /eI/ (as in b<u>ay</u>), /oU/ as in (b<u>oa</u>t), /aI/ (as in b<u>uy</u>), /aU/ (as in h<u>ow</u>), /oI/ (as in b<u>oy</u>), and /ju/ (as in y<u>ou</u>).

The diphthongs are produced by varying the vocal tract smoothly between vowel configurations appropriate to the diphthong. To illustrate this point, Figure 3.7 shows a plot of measurements of the second formant versus the first formant (as a function of time) for the diphthongs [12]. The arrows in this figure indicate the direction of motion of the formants (in the $(F_1 - F_2)$ plane) as time increases. The dashed circles in this figure indicate average positions of the vowels. Based on these data, and other measurements, the diphthongs can be characterized by a time varying vocal tract area function which varies between two vowel configurations.

3.1.2c Semivowels

The group of sounds consisting of /w/, /l/, /r/, and /y/ is quite difficult to characterize. These sounds are called semivowels because of their vowel-like nature. They are generally characterized by a gliding transition in vocal tract

48

area function between adjacent phonemes. Thus the acoustic characteristics of these sounds are strongly influenced by the context in which they occur. For our purposes they are best described as transitional, vowel-like sounds, and hence are similar in nature to the vowels and diphthongs. An example of the semivowel /w/ is shown in Figure 3.3.

3.1.2d Nasals

The nasal consonants /m/, /n/, and /ŋ/ are produced with glottal excitation and the vocal tract totally constricted at some point along the oral passageway. The velum is lowered so that air flows through the nasal tract, with sound being radiated at the nostrils. The oral cavity, although constricted toward the front, is still acoustically coupled to the pharynx. Thus, the mouth serves as a resonant cavity that traps acoustic energy at certain natural frequencies. As far as the radiated sound is concerned these resonant frequencies of the oral cavity appear as anti-resonances, or zeros of sound transmission [2]. Furthermore, nasal consonants and nasalized vowels (i.e., some vowels preceding or following nasal consonants) are characterized by resonances which are spectrally broader, or more highly damped, than those for vowels. The broadening of the nasal

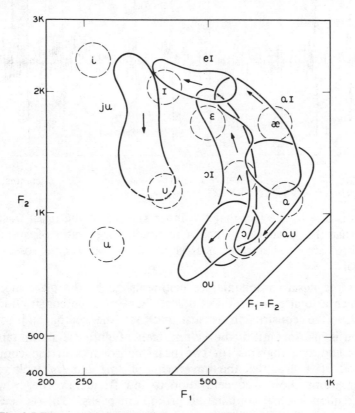

Fig. 3.7 Time variations of the first two formants for diphthongs. (After Holbrook and Fairbanks [27].)

49

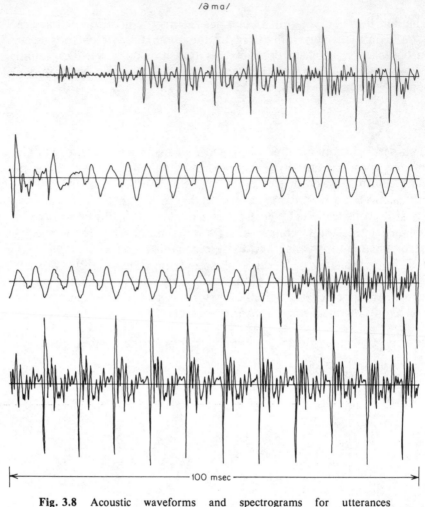

/ə mɑ/

— 100 msec —

Fig. 3.8 Acoustic waveforms and spectrograms for utterances /UH−M−A/ and /UH−N−A/.

resonances is due to the fact that the inner surface of the nasal tract is convoluted, so that the nasal cavity has a relatively large ratio of surface area to cross-sectional area. Therefore, heat conduction and viscous losses are larger than normal.

The three nasal consonants are distinguished by the place along the oral tract at which a total constriction is made. For /m/, the constriction, is at the lips; for /n/ the constriction is just back of the teeth; and for /η/ the constriction is just forward of the velum itself. Figure 3.8 shows typical speech waveforms and spectrograms for two nasal consonants in the context vowel-nasal-vowel. It is clear that the waveforms of /m/ and /n/ look very similar. The spectrograms show a concentration of low frequency energy with a midrange of frequencies that contains no prominent peaks. This is because of the particular combination of resonances and anti-resonances that result from the coupling of the nasal and oral tracts [13].

50

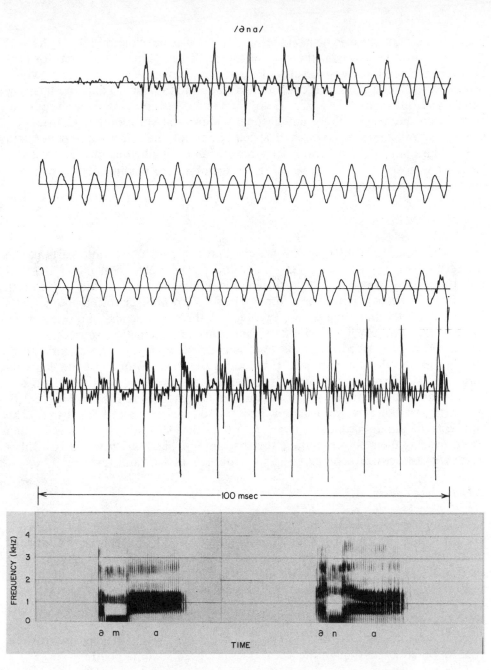

Fig. 3.8 (Continued)

3.1.2e Unvoiced Fricatives

The unvoiced fricatives /f/, /θ/, /s/, and /sh/ are produced by exciting the vocal tract by a steady air flow which becomes turbulent in the region of a constriction in the vocal tract. The location of the constriction serves to determine which fricative sound is produced. For the fricative /f/ the constriction is

51

near the lips; for /θ/ it is near the teeth; for /s/ it is near the middle of the oral tract; and for /sh/ it is near the back of the oral tract. Thus the system for producing unvoiced fricatives consists of a source of noise at a constriction, which separates the vocal tract into two cavities. Sound is radiated from the lips; i.e. from the front cavity. The back cavity serves, as in the case of nasals, to trap energy and thereby introduce anti-resonances into the vocal output [2,14]. Figure 3.9 shows the waveforms and spectrograms of the fricatives /f/, /s/ and /sh/. The nonperiodic nature of fricative excitation is obvious in the waveform plots. The spectral differences among the fricatives are readily seen by comparing the three spectrograms.

3.1.2f Voiced Fricatives

The voiced fricatives /v/, /th/, /z/ and /zh/ are the counterparts of the unvoiced fricatives /f/, /θ/, /s/, and /sh/, respectively, in that the place of constriction for each of the corresponding phonemes is essentially identical. However, the voiced fricatives differ markedly from their unvoiced counterparts in that two excitation sources are involved in their production. For voiced fricatives the vocal cords are vibrating, and thus one excitation source is at the glottis. However, since the vocal tract is constricted at some point forward of the glottis, the air flow becomes turbulent in the neighborhood of the constriction. Thus the spectra of voiced fricatives can be expected to display two distinct components. These excitation features are readily observed in Figure 3.10 which shows typical waveforms and spectra for several voiced fricatives. The similarity of the unvoiced fricative /f/ to the voiced fricative /v/ is easily seen by comparing their corresponding spectrograms in Figures 3.9 and 3.10. Likewise it is instructive to compare the spectrograms of /sh/ and /zh/.

3.1.2g Voiced Stops

The voiced stop consonants /b/, /d/, and /g/, are transient, noncontinuant sounds which are produced by building up pressure behind a total constriction somewhere in the oral tract, and suddenly releasing the pressure. For /b/ the constriction is at the lips; for /d/ the constriction is back of the teeth; and for /g/ it is near the velum. During the period when there is a total constriction in the tract there is no sound radiated from the lips. However, there is often a small amount of low frequency energy radiated through the walls of the throat (sometimes called a voice bar). This occurs when the vocal cords are able to vibrate even though the vocal tract is closed at some point.

Since the stop sounds are dynamical in nature, their properties are highly influenced by the vowel which follows the stop consonant [15]. As such, the waveforms for stop consonants give little information about the particular stop consonant. Figure 3.11 shows the waveform and spectrogram of the syllable /UH-B-A/. The waveform of /b/ shows few distinguishing features except for the voiced excitation and lack of high frequency energy.

52

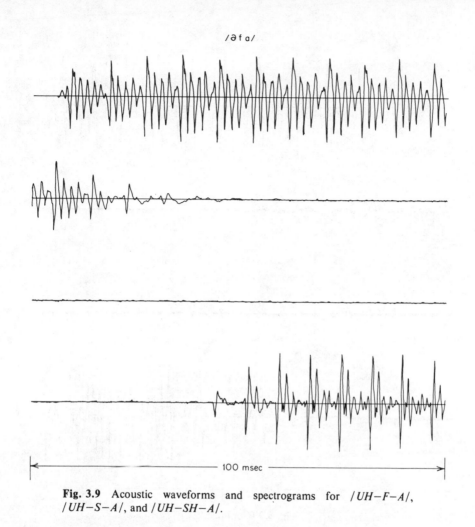

/ ə f ɑ /

100 msec

Fig. 3.9 Acoustic waveforms and spectrograms for /UH−F−A/, /UH−S−A/, and /UH−SH−A/.

3.1.2h Unvoiced Stops

The unvoiced stop consonants /p/, /t/, and /k/ are similar to their voiced counterparts /b/, /d/, and /g/ with one major exception. During the period of total closure of the tract, as the pressure builds up, the vocal cords do not vibrate. Thus, following the period of closure, as the air pressure is released, there is a brief interval of friction (due to sudden turbulence of the escaping air) followed by a period of aspiration (steady air flow from the glottis exciting the resonances of the vocal tract) before voiced excitation begins.

Figure 3.12 shows waveforms and spectrograms of the voiceless stop consonants /p/ and /t/. The "stop gap," or time interval during which the pressure is built up is clearly in evidence. Also, it is readily seen that the duration and frequency content of the frication noise and aspiration varies greatly with the stop consonant.

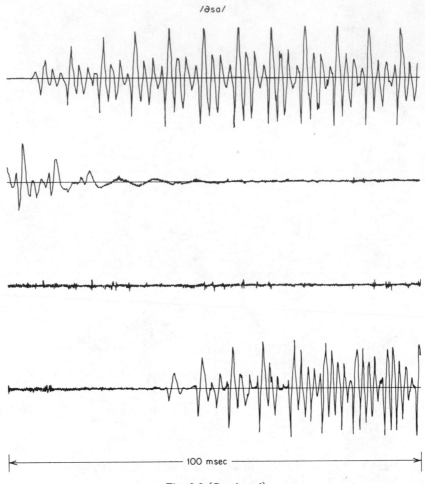

/əsɑ/

100 msec

Fig. 3.9 (Continued)

3.1.2i Affricates and /h/

The remaining consonants of American English are the affricates /t∫/ and /j/, and the phoneme /h/. The unvoiced affricate /t∫/ is a dynamical sound which can be modelled as the concatenation of the stop /t/ and the fricative /∫/. (See Fig. 3.3a for an example.) The voiced affricate /j/ can be modelled as the concatenation of the stop /d/ and the fricative /zh/. Finally, the phoneme /h/ is produced by exciting the vocal tract by a steady air flow — i.e., without the vocal cords vibrating, but with turbulent flow being produced at the glottis.[2] The characteristics of /h/ are invariably those of the vowel which follows /h/ since the vocal tract assumes the position for the following vowel during the production of /h/.

[2]Note that this is also the mode of excitation for whispered speech.

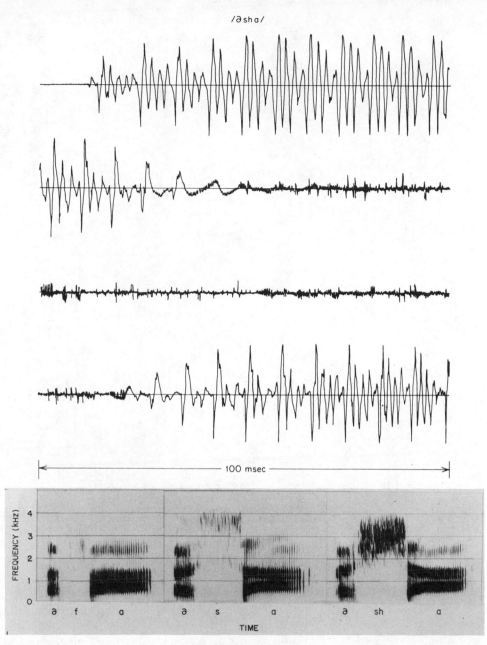

Fig. 3.9 (Continued)

3.2 The Acoustic Theory of Speech Production

The previous section was a review of the qualitative description of the sounds of speech and the way that they are produced. In this section we shall consider mathematical representations of the process of speech production. Such mathematical representations serve as the basis for the analysis and synthesis of speech.

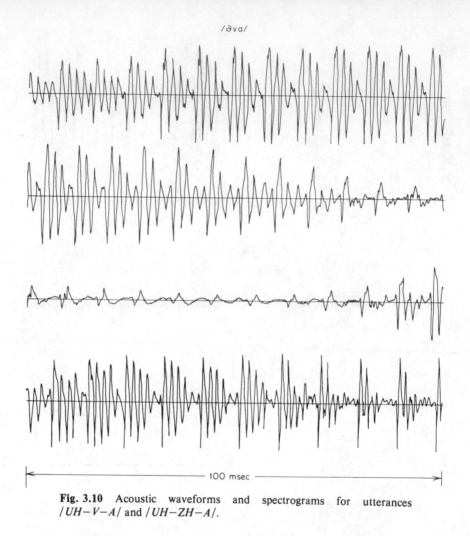

Fig. 3.10 Acoustic waveforms and spectrograms for utterances /UH−V−A/ and /UH−ZH−A/.

3.2.1 Sound propagation

Sound is almost synonymous with vibration. Sound waves are created by vibration and are propagated in air or other media by vibrations of the particles of the media. Thus, the laws of physics are the basis for describing the generation and propagation of sound in the vocal system. In particular, the fundamental laws of conservation of mass, conservation of momentum, and conservation of energy along with the laws of thermodynamics and fluid mechanics, all apply to the compressible, low viscosity fluid (air) that is the medium for sound propagation in speech. Using these physical principles, a set of partial differential equations can be obtained that describe the motion of air in the vocal system [16-20]. The formulation and solution of these equations is extremely difficult except under very simple assumptions about vocal tract shape and energy losses in the vocal system. A detailed acoustic theory must consider the effects of the following:

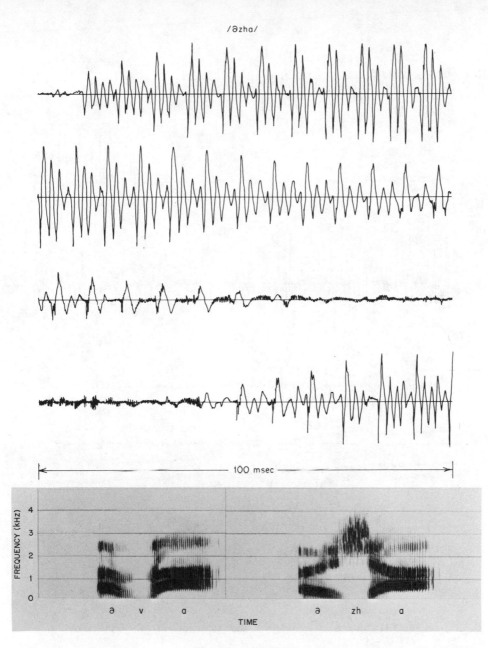

/əʒɑ/

100 msec

Fig. 3.10 (Continued)

1. Time variation of the vocal tract shape.
2. Losses due to heat conduction and viscous friction at the vocal tract walls.
3. Softness of the vocal tract walls.
4. Radiation of sound at the lips.
5. Nasal coupling.
6. Excitation of sound in the vocal tract.

57

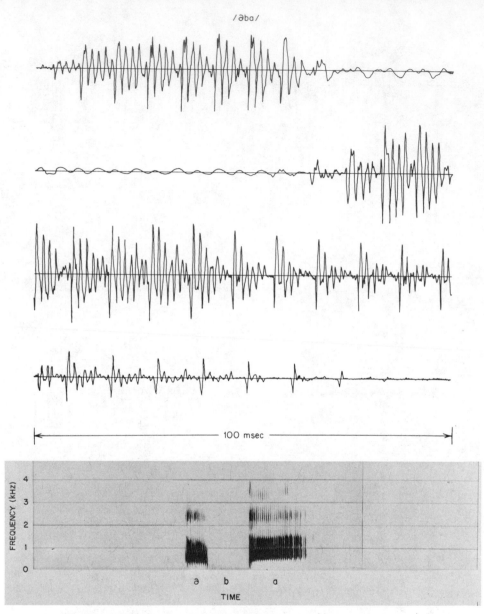

Fig. 3.11 Acoustic waveform and spectrogram for utterance /*UH−B−A*/.

A completely detailed acoustic theory incorporating all the above effects is beyond the scope of this chapter, and indeed, such a theory is not yet available. We must be content to survey these factors, providing references to details when available, and qualitative discussions when suitable references are unavailable.

The simplest physical configuration that has a useful interpretation in terms of the speech production process is depicted in Figure 3.13a. The vocal

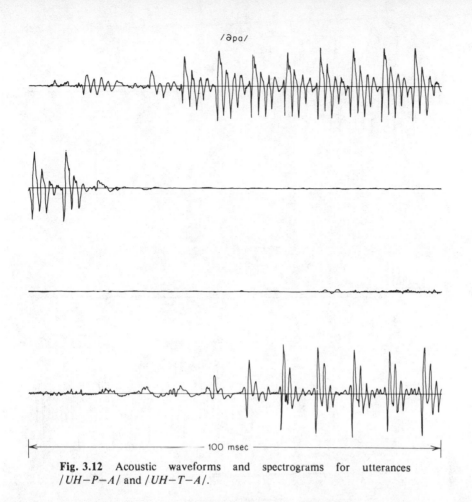

/əpɑ/

100 msec

Fig. 3.12 Acoustic waveforms and spectrograms for utterances /UH−P−A/ and /UH−T−A/.

tract is modeled as a tube of nonuniform, time-varying, cross-section. For frequencies corresponding to wavelengths that are long compared to the dimensions of the vocal tract (less than about 4000 Hz), it is reasonable to assume plane wave propagation along the axis of the tube. A further simplifying assumption is that there are no losses due to viscosity or thermal conduction, either in the bulk of the fluid or at the walls of the tube. With these assumptions, and the laws of conservation of mass, momentum and energy, Portnoff [18] has shown that sound waves in the tube satisfy the following pair of equations:

$$-\frac{\partial p}{\partial x} = \rho \frac{\partial (u/A)}{\partial t} \tag{3.1a}$$

$$-\frac{\partial u}{\partial x} = \frac{1}{\rho c^2} \frac{\partial (pA)}{\partial t} + \frac{\partial A}{\partial t} \tag{3.1b}$$

where

$p = p(x,t)$ is the variation in sound pressure in the tube at position x and time t.

59

/ ətɑ /

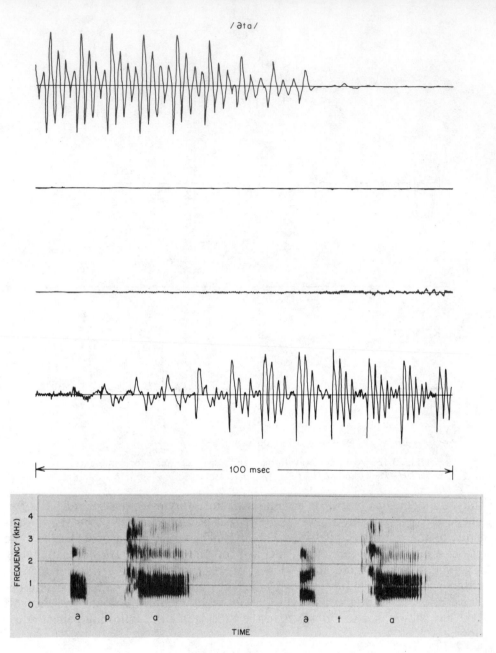

Fig. 3.12 (Continued)

$u = u(x,t)$ is the variation in volume velocity flow at position x and time t.

ρ is the density of air in the tube.

c is the velocity of sound

$A = A(x,t)$ is the "area function" of the tube; i.e., the value of cross-sectional area

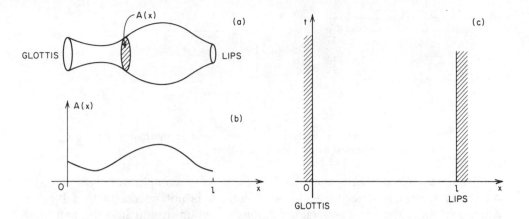

Fig. 3.13 (a) Schematic vocal tract; (b) corresponding area function; (c) $x-t$ plane for solution of wave equation.

> normal to the axis of the tube as a
> function of a distance along the
> tube and as a function of time.

A similar set of equations has been derived by Sondhi [20].

Closed form solutions to Eqs. (3.1) are not possible except for the simplest configurations. Numerical solutions can be obtained, however. Complete solution of the differential equations requires that pressure and volume velocity be found for values of x and t in the region bounded by the glottis and the lips. To obtain the solution, boundary conditions must be given at each end of the tube. At the lip end, the boundary condition must account for the effects of sound radiation. At the glottis (or possibly some internal point), the boundary condition is imposed by the nature of the excitation.

In addition to the boundary conditions, the vocal tract area function, $A(x,t)$, must be known. Figure 3.13b shows the area function for the tube in Fig. 3.13a, at a particular time. For continuant sounds, it is reasonable to assume that $A(x,t)$ does not change with time; however this is not the case for noncontinuants. Detailed measurements of $A(x,t)$ are extremely difficult to obtain even for continuant sounds. One approach to such measurements is through the use of X-ray motion pictures. Fant [1] and Perkell [21] provide some data of this form; however, such measurements can only be obtained on a limited scale. Another approach is to infer the vocal tract shape from acoustic measurements. Sondhi and Gopinath [22] have described an approach which involves the excitation of the vocal tract by an external source. Both of these approaches are useful for obtaining knowledge of the dynamics of speech production, but they are not directly applicable to the representation of speech signals (e.g. for purposes of transmission). Atal [23] has described investigations directed toward obtaining $A(x,t)$ directly from the speech signal produced under normal speaking conditions.

61

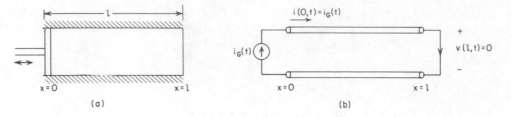

Fig. 3.14 (a) Uniform lossless tube with ideal terminations; (b) corresponding electrical transmission line analogy.

The complete solution of Eqs. (3.1) is very complicated [18] even if $A(x,t)$ is accurately determined. Fortunately, it is not necessary to solve the equations under the most general conditions to obtain insight into the nature of the speech signal. A variety of reasonable approximations and simplifications can be invoked to make the solution possible.

3.2.2 Example: uniform lossless tube

Useful insight into the nature of the speech signal can be obtained by considering a very simple model in which the vocal tract area function is assumed constant in both x and t (time invariant with uniform cross-section). This configuration is approximately correct for the neutral vowel /UH/. We shall examine this model first, returning later to examine more realistic models. Figure 3.14a depicts a tube of uniform cross-section being excited by an ideal source of volume velocity flow. This ideal source is represented by a piston that can be caused to move in any desired fashion, independent of pressure variations in the tube. A further assumption is that at the open end of the tube, there are no variations in air pressure — only in volume velocity. These are obviously gross simplifications which in fact are impossible to achieve in reality; however, we are justified in considering this example since the basic approach of the analysis and the essential features of the resulting solution have much in common with more realistic models. Furthermore we shall show that more general models can be constructed by concatenation of uniform tubes.

If $A(x,t) = A$ is a constant, then the partial differential equations Eqs. (3.1) reduce to the form

$$-\frac{\partial p}{\partial x} = \frac{\rho}{A} \frac{\partial u}{\partial t} \tag{3.2a}$$

$$-\frac{\partial u}{\partial x} = \frac{A}{\rho c^2} \frac{\partial p}{\partial t} \tag{3.2b}$$

It can be shown (see Problem 3.3) that the solution to Eqs. (3.2) has the form

$$u(x,t) = [u^+(t-x/c) - u^-(t+x/c)] \tag{3.3a}$$

$$p(x,t) = \frac{\rho c}{A} [u^+(t-x/c) + u^-(t+x/c)] \tag{3.3b}$$

In Eqs. (3.3) the functions $u^+(t-x/c)$ and $u^-(t+x/c)$ can be interpreted as traveling waves in the positive and negative directions respectively. The relationship between these traveling waves is determined by the boundary conditions.

Anyone familiar with the theory of electrical transmission lines will recall that for a lossless uniform line the voltage $v(x,t)$ and current $i(x,t)$ on the line satisfy the equations

$$-\frac{\partial v}{\partial x} = L\,\frac{\partial i}{\partial t} \tag{3.4a}$$

$$-\frac{\partial i}{\partial x} = C\,\frac{\partial v}{\partial t} \tag{3.4b}$$

where L and C are the inductance and capacitance per unit length respectively. Thus the theory of lossless uniform electric transmission lines [24,25] applies directly to the uniform acoustic tube if we make the analogies shown in Table 3.3.

Table 3.3 Analogies Between Acoustic and Electric Quantities

Acoustic Quantity	Analogous Electric Quantity
p - pressure	v - voltage
u - volume velocity	i - current
ρ/A - acoustic inductance	L - inductance
$A/(\rho c^2)$ - acoustic capacitance	C - capacitance

Using these analogies, the uniform acoustic tube behaves identically to a lossless uniform transmission line terminated in a short circuit $(v(l,t)=0)$ at one end and excited by a current source $(i(0,t)=i_G(t))$ at the other end. This is depicted in Fig. 3.14b.

Frequency domain representations of linear systems such as transmission lines and circuits are exceedingly useful. By analogy we can obtain similar representations of the lossless uniform tube. The frequency-domain representation of this model is obtained by assuming a boundary condition at $x = 0$ of

$$u(0,t) = u_G(t) = U_G(\Omega)e^{j\Omega t} \tag{3.5}$$

That is, the tube is excited by a complex exponential variation of volume velocity of radian frequency Ω and complex amplitude, $U_G(\Omega)$. Since Equations (3.2) are linear, the solution $u^+(t-x/c)$ and $u^-(t+x/c)$ must be of the form

$$u^+(t-x/c) = K^+ e^{j\Omega(t-x/c)} \tag{3.6a}$$

$$u^-(t+x/c) = K^- e^{j\Omega(t+x/c)} \tag{3.6b}$$

Substituting these equations into Eqs. (3.3) and applying the boundary condition

$$p(l,t) = 0 \tag{3.7}$$

63

at the lip end of the tube and Eq. (3.5) at the glottis end we can solve for the unknown constants K^+ and K^-. The resulting sinusoidal steady state solutions for $p(x,t)$ and $u(x,t)$ are

$$p(x,t) = jZ_0 \frac{\sin[\Omega(l-x)/c]}{\cos[\Omega l/c]} U_G(\Omega)e^{j\Omega t} \tag{3.8a}$$

$$u(x,t) = \frac{\cos[\Omega(l-x)/c]}{\cos[\Omega l/c]} U_G(\Omega)e^{j\Omega t} \tag{3.8b}$$

where

$$Z_0 = \frac{\rho c}{A} \tag{3.9}$$

is by analogy called the *characteristic acoustic impedance* of the tube.

An alternative approach which we will use subsequently avoids solution for the forward and backward traveling waves by expressing $p(x,t)$ and $u(x,t)$ for a complex exponential excitation directly as[3]

$$p(x,t) = P(x,\Omega)e^{j\Omega t} \tag{3.10a}$$

$$u(x,t) = U(x,\Omega)e^{j\Omega t} \tag{3.10b}$$

Substituting these solutions into Eqs. (3.1) gives the ordinary differential equations relating the complex amplitudes

$$-\frac{dP}{dx} = ZU \tag{3.11a}$$

$$-\frac{dU}{dx} = YP \tag{3.11b}$$

where

$$Z = j\Omega \frac{\rho}{A} \tag{3.12}$$

can be called the *acoustic impedance* per unit length and

$$Y = j\Omega \frac{A}{\rho c^2} \tag{3.13}$$

is the *acoustic admittance* per unit length. The differential equations of Eqs. (3.11) have solutions of the form

$$P(x,\Omega) = Ae^{\gamma x} + Be^{-\gamma x} \tag{3.14a}$$

$$U(x,\Omega) = Ce^{\gamma x} + De^{-\gamma x} \tag{3.14b}$$

where

$$\gamma = \sqrt{ZY} = j\Omega/c \tag{3.14c}$$

---

[3]Henceforth our convention will be to denote time domain variables with lower case letters (e.g. $u(x,t)$) and their corresponding frequency domain representations with capital letters (i.e. $U(x,\Omega)$).

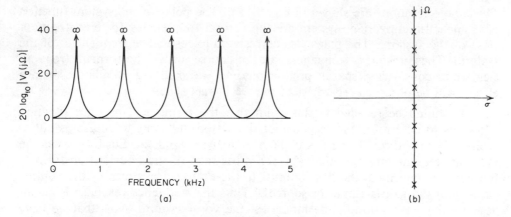

Fig. 3.15 (a) Frequency response; and (b) pole locations for a uniform lossless tube.

The unknown coefficients can be found by applying the boundary conditions

$$P(l, \Omega) = 0 \tag{3.15a}$$

$$U(0, \Omega) = U_G(\Omega) \tag{3.15b}$$

The result is, of course, the same as Eqs. (3.8). Equations (3.8) express the relationship between the sinusoidal volume velocity source and the pressure and volume velocity at any point in the tube. In particular, if we consider the relationship between the volume velocity at the lips and the volume velocity source, we obtain from Eq. (3.8b),

$$u(l,t) = U(l, \Omega) e^{j\Omega t}$$

$$= \frac{1}{\cos(\Omega l/c)} U_G(\Omega) e^{j\Omega t} \tag{3.16}$$

The ratio

$$\frac{U(l, \Omega)}{U_G(\Omega)} = V_a(j\Omega) = \frac{1}{\cos(\Omega l/c)} \tag{3.17}$$

is the frequency response relating the input and output volume velocities. This function is plotted in Figure 3.15a for values $l = 17.5$ cm and $c = 35000$ cm/sec. Replacing Ω by s/j, we obtain the Laplace transform or system function

$$V_a(s) = \frac{2e^{-sl/c}}{1 + e^{-s2l/c}} \tag{3.18}$$

Note that $V_a(s)$ has an infinite number of poles equally spaced on the $j\Omega$ axis at

$$s_n = \pm j \left| \frac{(2n+1)\pi c}{2l} \right| \qquad n = 0, \pm 1, \pm 2, \ldots \tag{3.19}$$

These pole locations are shown in Fig. 3.15b. The poles of the system function of a linear time-invariant system are the natural frequencies (or eigenfrequencies) of the system. The poles also correspond to resonance frequencies of the system. These resonant frequencies are, of course, called the formant frequencies when considering speech production. As we shall see, similar resonance effects will be observed regardless of the vocal tract shape.

It should be recalled at this point that the frequency response function allows us to determine the response of the system not only to sinusoids but to arbitrary inputs through the use of Fourier analysis. Indeed, Eq. (3.17) has the more general intepretation that $V_a(j\Omega)$ is the ratio of the Fourier transform of the volume velocity at the lips (output) to the Fourier transform of the volume velocity at the glottis (input or source). Thus the frequency response is a convenient characterization of the model for the vocal system. Now that we have demonstrated a method for determining the frequency response of acoustic models for speech production by considering the simplest possible model, we can begin to consider more realistic models.

3.2.3 Effects of losses in the vocal tract

The equations of motion for sound propagation in the vocal tract that we have given were derived under the assumption of no energy loss in the tube. In reality, energy will be lost as a result of viscous friction between the air and the walls of the tube, heat conduction through the walls of the tube, and vibration of the tube walls. To include these effects, we might attempt to return to the basic laws of physics and derive a new set of equations of motion. This is made extremely difficult by the frequency dependence of these losses. As a result, a common approach is to modify the frequency domain representation of the equations of motion [2,18]. We shall survey the results of this approach in this section.

Let us first consider the effects of the vibration of the vocal tract wall. The variations of air pressure inside the tract will cause the walls to experience a varying force. Thus, if the walls are elastic, the cross-sectional area of the tube will change depending upon the pressure in the tube. Assuming that the walls are "locally reacting" [17,18], then the area $A(x,t)$ will be a function of $p(x,t)$. Since the pressure variations are very small, the resulting variation in cross-sectional area can be treated as a small perturbation of the "nominal" area; i.e., we can assume that

$$A(x,t) = A_0(x,t) + \delta A(x,t) \tag{3.20}$$

where $A_0(x,t)$ is the nominal area and $\delta A(x,t)$ is a small perturbation. This is depicted in Fig. 3.16. Because of the mass and elasticity of the vocal tract wall, the relationship between the area perturbation $\delta A(x,t)$, and the pressure variations, $p(x,t)$, can be modeled by a differential equation of the form

$$m_w \frac{d^2(\delta A)}{dt^2} + b_w \frac{d(\delta A)}{dt} + k_w(\delta A) = p(x,t) \tag{3.21}$$

66

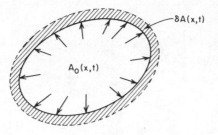

Fig. 3.16 Illustration of the effects of wall vibration.

where

$m_w(x)$ is the mass/unit length of the vocal tract wall
$b_w(x)$ is the damping/unit length of the vocal tract wall
$k_w(x)$ is the stiffness/unit length of the vocal tract wall.

Neglecting second order terms in the quantities u/A and pA, we can write Eqs. (3.1) as

$$-\frac{\partial p}{\partial x} = \rho \, \frac{\partial(u/A_0)}{\partial t} \tag{3.22a}$$

$$-\frac{\partial u}{\partial x} = \frac{1}{\rho c^2} \frac{\partial(pA_0)}{\partial t} + \frac{\partial A_0}{\partial t} + \frac{\partial(\delta A)}{\partial t} \tag{3.22b}$$

Thus, sound propagation in a soft walled tube such as the vocal tract is described by the set of equations, Eqs. (3.20), (3.21) and (3.22).

To examine this effect in more detail let us obtain a frequency domain representation, as before, by considering a time invariant tube, excited by a complex volume velocity source; i.e., the boundary condition at the glottis is

$$u(0,t) = U_G(\Omega)e^{j\Omega t} \tag{3.23}$$

Then because the differential equations Eqs. (3.21) and (3.22) are linear and time invariant for this case, the volume velocity and pressure are also of the form

$$p(x,t) = P(x, \Omega)e^{j\Omega t} \tag{3.24a}$$

$$u(x,t) = U(x, \Omega)e^{j\Omega t} \tag{3.24b}$$

Substituting Eqs. (3.24) into Eqs. (3.21) and (3.22) yields the equations

$$-\frac{\partial P}{\partial x} = ZU \tag{3.25a}$$

$$-\frac{\partial U}{\partial x} = YP + Y_w P \tag{3.25b}$$

where

$$Z(x, \Omega) = j\Omega \, \frac{\rho}{A_0(x)} \tag{3.26a}$$

67

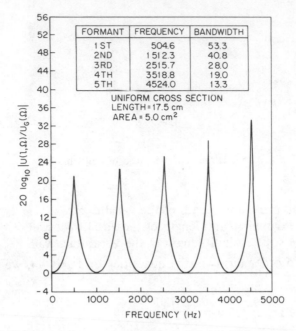

FORMANT	FREQUENCY	BANDWIDTH
1 ST	504.6	53.3
2ND	1512.3	40.8
3RD	2515.7	28.0
4TH	3518.8	19.0
5TH	4524.0	13.3

UNIFORM CROSS SECTION
LENGTH=17.5 cm
AREA = 5.0 cm^2

Fig. 3.17 Frequency response of uniform tube with yielding walls and no other losses. Terminated in a short circuit ($p(l,t)=0$). (After Portnoff [18].)

$$Y(x, \Omega) = j\Omega \frac{A_0(x)}{\rho c^2} \qquad (3.26b)$$

and

$$Y_w(x, \Omega) = \frac{1}{j\Omega m_w(x) + b_w(x) + \dfrac{k_w(x)}{j\Omega}} \qquad (3.26c)$$

Note that Eqs. (3.25) are identical to Eqs. (3.11) except for the addition of the wall admittance term Y_w and for the fact that the acoustic impedance and admittances are in this case functions of x. If we consider a uniform tube, then $A_0(x)$ is constant, and Eqs. (3.12) and (3.13) are identical to Eqs. (3.26a) and (3.26b).

Using estimates obtained from measurements on body tissues [2], the parameters in Eq. (3.26c) were estimated and the differential equations, Eqs. (3.25), were solved with boundary condition $p(l,t) = 0$ at the lip end [18,19]. The ratio

$$V_a(j\Omega) = \frac{U(l, \Omega)}{U_G(\Omega)} \qquad (3.27)$$

is plotted as a function of Ω in Fig. 3.17 for the case of a uniform tube of length 17.5 cm [18]. The results are similar to Fig. 3.15 but different in an important way. It is clear that the resonances are no longer exactly on the $j\Omega$ axis of the s-plane. This is evident since the frequency response no longer is infinite at frequencies 500 Hz, 1500 Hz, 2500 Hz, etc., although the response is peaked in the vicinity of these frequencies. The center frequencies and bandwidths[4] of the resonances in Figure 3.17 are given in the associated table. Several important effects are evident in this example. First we note that the center frequencies are slightly higher than for the lossless case. Second, the bandwidths of the resonances are no longer zero as in the lossless case, since the peak value is no longer infinite. It can be seen that the effect of yielding walls is most pronounced at low frequencies. This is to be expected since we would expect very little motion of the massive walls at high frequencies. The results of this example are typical of the general effects of vocal tract wall vibration; i.e., the center frequencies are slightly increased and the low frequency resonances are broadened as compared to the rigid wall case.

The effects of viscous friction and thermal conduction at the walls are much less pronounced than the effects of wall vibration. Flanagan [2] has considered these losses in detail and has shown that the effect of viscous friction can be accounted for in the frequency domain representation (Eq. (3.25)) by including a real, frequency dependent term in the expression for the acoustic impedance, Z; i.e.,

$$Z(x, \Omega) = \frac{S(x)}{[A_0(x)]^2} \sqrt{\Omega \rho \mu / 2} + j\Omega \frac{\rho}{A_0(x)} \qquad (3.28a)$$

where $S(x)$ is the circumference of the tube, μ is the coefficient of friction, and ρ is the density of air in the tube. The effects of heat conduction through the vocal tract wall can likewise be accounted for by adding a real frequency dependent term to the acoustic admittance, $Y(x, \Omega)$; i.e.,

$$Y(x, \Omega) = \frac{S(x)(\eta - 1)}{\rho c^2} \sqrt{\frac{\lambda \Omega}{2 c_p \rho}} + j\Omega \frac{A_0(x)}{\rho c^2} \qquad (3.28b)$$

where c_p is the specific heat at constant pressure, η is the ratio of specific heat at constant pressure to that at constant volume, and λ is the coefficient of heat conduction [2]. Typical values for the constants in Eqs. (3.28) are given by Flanagan [2]. For our purposes, it is sufficient to note that the loss due to friction is proportional to the real part of $Z(x, \Omega)$, and thus to $\Omega^{1/2}$. Likewise the thermal loss is proportional to the real part of $Y(x, \Omega)$, which in turn is proportional to $\Omega^{1/2}$. Using the values given by Eqs. (3.28) for $Z(x, \Omega)$ and $Y(x, \Omega)$ and the values of $Y_w(x, \Omega)$ given by Eq. (3.26c), Eqs. (3.25) were again solved numerically [18]. The resulting frequency response for the boundary condition of $p(l,t) = 0$ is shown in Fig. 3.18. Again the center frequencies and bandwidths were determined and are shown in the associated table. Comparing Fig. 3.18 with Fig. 3.17, we observe that the center frequencies are decreased by the addition of friction and thermal loss, while the bandwidths are increased. Since friction and thermal losses increase with $\Omega^{1/2}$, the higher frequency resonances experience a greater broadening than do the lower resonances.

---

[4]The bandwidth of a resonance is defined as the frequency interval around a resonance in which the frequency response is greater than 0.707 times the peak value at the center frequency [26].

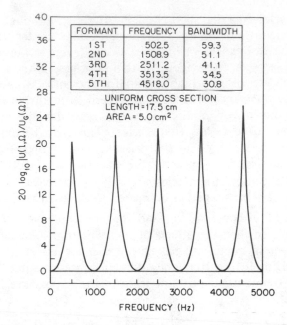

FORMANT	FREQUENCY	BANDWIDTH
1 ST	502.5	59.3
2 ND	1508.9	51.1
3 RD	2511.2	41.1
4 TH	3513.5	34.5
5 TH	4518.0	30.8

UNIFORM CROSS SECTION
LENGTH = 17.5 cm
AREA = 5.0 cm²

Fig. 3.18 Frequency response of uniform tube with yielding walls, friction and thermal losses, and terminated in a short circuit $(p(l,t)=0)$. (After Portnoff [18].)

The examples depicted in Figs. 3.17 and 3.18 are typical of the general effects of losses in the vocal tract. To summarize, viscous and thermal losses increase with frequency and have their greatest effect in the high frequency resonances, while wall loss is most pronounced at low frequencies. The yielding walls tend to raise the resonant frequencies while the viscous and thermal losses tend to lower them. The net effect for the lower resonances is a slight upward shift as compared to the lossless, rigid walled model. The effect of friction and thermal loss is small compared to the effects of wall vibration for frequencies below 3-4 kHz. Thus, Eqs. (3.21) and (3.22), which neglect these losses, are nevertheless a good representation of sound transmission in the vocal tract. As we shall see in the next section, the radiation termination at the lips is a much greater source of high frequency loss. This provides further justification for neglecting friction and thermal loss in models or simulations of speech production.

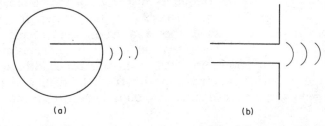

Fig. 3.19 (a) Radiation from a spherical baffle; (b) radiation from an infinite plane baffle.

3.2.4 Effects of radiation at the lips

So far we have discussed the way that internal losses affect the sound transmission properties of the vocal tract. In our examples we have assumed the boundary condition $p(l,t) = 0$ at the lips. In the electric transmission line analogy this corresponds to a short circuit. The acoustic counterpart of a short circuit is as difficult to achieve as an electrical short circuit since it requires a configuration in which volume velocity changes can occur at the end of the vocal tract tube without corresponding pressure changes. In reality, the vocal tract tube terminates with the opening between the lips (or the nostrils in the case of nasals). Thus a reasonable model is as depicted in Fig. 3.19a, which shows the lip opening as an orifice in a sphere. In this model, at low frequencies, the opening can be considered a radiating surface, with the radiated sound waves being diffracted by the spherical baffle that represents the head.

The resulting diffraction effects are complicated and difficult to represent; however, for determining the boundary condition at the lips, all that is needed is a relationship between pressure and volume velocity at the radiating surface. Even this is very complicated for the configuration of Fig. 3.19a. However, if the radiating surface (lip opening) is small compared to the size of the sphere, a reasonable approximation assumes that the radiating surface is set in a plane baffle of infinite extent as depicted in Fig. 3.19b. In this case, it can be shown [2,17,18] that the sinusoidal steady state relation between the complex amplitudes of pressure and volume velocity at the lips is

$$P(l, \Omega) = Z_L(\Omega) \cdot U(l, \Omega) \tag{3.29a}$$

where the "radiation impedance" or "radiation load" at the lips is approximately of the form

$$Z_L(\Omega) = \frac{j\Omega L_r R_r}{R_r + j\Omega L_r} \tag{3.29b}$$

The electrical analog to this radiation load is a parallel connection of a radiation resistance, R_r, and radiation inductance, L_r. Values of R_r and L_r that provide a good approximation to the infinite plane baffle are [2]

$$R_r = \frac{128}{9\pi^2} \tag{3.30a}$$

$$L_r = \frac{8a}{3\pi c} \tag{3.30b}$$

where a is the radius of the opening and c is the velocity of sound.

The behavior of the radiation load influences the nature of wave propagation in the vocal tract through the boundary condition of Eqs. (3.29). Note that it is easily seen from Eq. (3.29b) that at very low frequencies $Z_L(\Omega) \approx 0$; i.e., at very low frequencies the radiation impedance approximates the ideal short circuit termination that has been assumed up to this point. Likewise, it is clear from Eq. (3.29b) that for a mid range of frequencies, (when

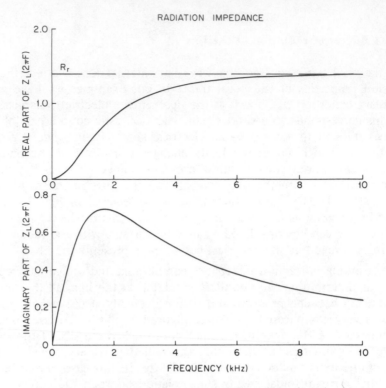

Fig. 3.20 Real and imaginary parts of the radiation impedance.

$\Omega L_r << R_r$), $Z_L(\Omega) \approx j\Omega L_r$. At higher frequencies $(\Omega L_r >> R_r)$, $Z_L(\Omega) \approx R_r$. This is readily seen in Fig. 3.20 which shows the real and imaginary parts of $Z_L(\Omega)$ as a function of Ω for typical values of the parameters. The energy dissipated due to radiation is proportional to the real part of the radiation impedance. Thus we can see that for the complete speech production system (vocal tract and radiation), the radiation losses will be most significant at higher frequencies. To assess the magnitude of this effect, Eqs. (3.25), (3.26c) and (3.29) were solved simultaneously for the case of a uniform time invariant tube with yielding walls, friction and thermal losses, and radiation loss corresponding to an infinite plane baffle. Figure 3.21 shows the resulting frequency response.

$$V_a(j\Omega) = \frac{U(l, \Omega)}{U_G(\Omega)} \tag{3.31}$$

for an input $U(0,t) = U_G(\Omega)e^{j\Omega t}$. Comparing Figure 3.21 to Figures 3.17 and 3.18 shows that the major effect is to broaden the resonances (increase loss) and to lower the resonance frequencies (formant frequencies). As expected the major effect on the resonance bandwidths occurs at higher frequencies. The first resonance (formant) bandwidth is primarily determined by the wall loss, while the higher formant bandwidths are primarily determined by radiation loss. The second and third formant bandwidths can be said to be determined by a combination of these two loss mechanisms.

72

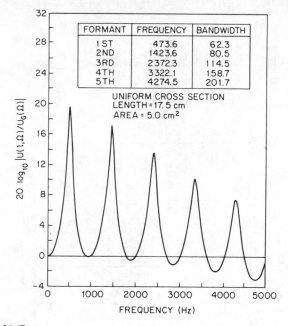

FORMANT	FREQUENCY	BANDWIDTH
1 ST	473.6	62.3
2ND	1423.6	80.5
3RD	2372.3	114.5
4TH	3322.1	158.7
5TH	4274.5	201.7

UNIFORM CROSS SECTION
LENGTH = 17.5 cm
AREA = 5.0 cm²

Fig. 3.21 Frequency response of uniform tube with yielding walls, friction and thermal loss. (After Portnoff [18].)

The frequency response shown in Figure 3.21 relates the volume velocity at the lips to the input volume velocity at the lips. The relationship between pressure at lips and volume velocity at the glottis may be of interest, especially if a pressure sensitive microphone is used in converting the acoustic wave to an electrical wave. Since $P(l, \Omega)$ and $U(l, \Omega)$ are related by Eq. (3.29a), the pressure transfer function is simply

$$H_a(\Omega) = \frac{P(l, \Omega)}{U_G(\Omega)} = \frac{P(l, \Omega)}{U(l, \Omega)} \cdot \frac{U(l, \Omega)}{U_G(\Omega)}$$

$$= Z_L(\Omega) \cdot V_a(\Omega) \tag{3.32}$$

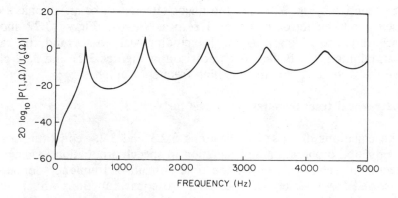

Fig. 3.22 Frequency response relating pressure at lips to volume velocity at glottis for uniform tube.

73

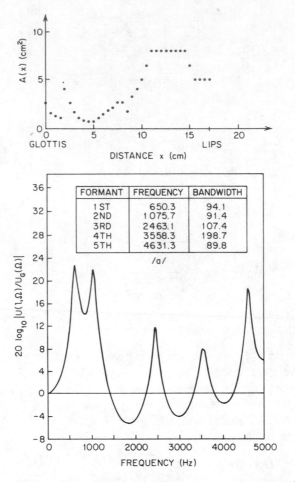

Fig. 3.23 Area function (after Fant [1]) and frequency response (after Portnoff [18]) for the Russian vowel /a/.

It can be seen from Fig. 3.21 that the major effects will be an emphasis of high frequencies and the introduction of a zero at. $\Omega = 0$. Figure 3.22 shows the frequency response $20 \log_{10} |H_a(\Omega)|$ including wall losses and the radiation loss of an infinite plane baffle. A comparison of Figures 3.21 and 3.22 places in evidence the zero at $\Omega = 0$ and the high frequency emphasis.

3.2.5 Vocal tract transfer functions for vowels

The equations discussed in Sections 3.2.3 and 3.2.4 constitute a detailed model for sound propagation and radiation in speech production. Using numerical integration techniques, either the time domain or frequency domain forms can be solved for a variety of vocal tract response functions. Such solutions provide considerable insight into the nature of the speech production process and the speech signal.

74

As an example [18], the frequency domain equations, Eqs. (3.25), (3.26c), (3.28), and (3.29), were used to compute frequency response functions for a set of area functions measured by Fant [1]. Figures 3.23-3.26 show the vocal tract area functions and corresponding frequency responses $(U(l, \Omega)/U_G(\Omega))$ for the Russian vowels /a/, /e/, /i/, and /u/. These figures illustrate the effects of all the loss mechanisms discussed in Sections 3.2.3 and 3.2.4. The formant frequencies and bandwidths compare favorably with measurements on natural vowels of formant frequencies obtained by Peterson and Barney [11] and formant bandwidths by Dunn [27].

In summary, we may conclude from these examples and those of the previous sections that:

1. The vocal system is characterized by a set of resonances (formants) that depend primarily upon the vocal tract area function, although there is some shift due to losses, as compared to the lossless case.

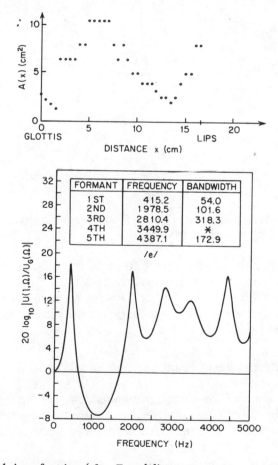

FORMANT	FREQUENCY	BANDWIDTH
1ST	415.2	54.0
2ND	1978.5	101.6
3RD	2810.4	318.3
4TH	3449.9	*
5TH	4387.1	172.9

/e/

Fig. 3.24 Area function (after Fant [1]) and frequency response (after Portnoff [18]) for the Russian vowel /e/.

75

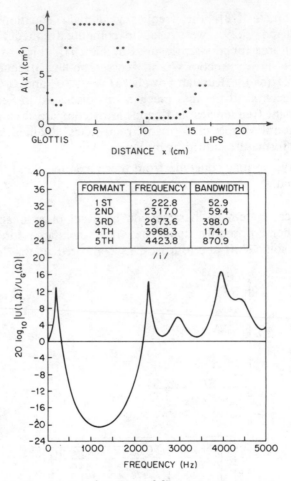

Fig. 3.25 Area function (after Fant [1]) and frequency response (after Portnoff [18]) for the Russian vowel /i/.

2. The bandwidths of the lowest formant frequencies (first and second) depend primarily upon the vocal tract wall loss.[5]
3. The bandwidths of the higher formant frequencies depend primarily upon the viscous friction and thermal losses in the vocal tract and the radiation loss.

3.2.6 The effect of nasal coupling

In the production of the nasal consonants /m/, /n/, and /η/ the velum is lowered like a trap-door to couple the nasal tract to the pharynx. Simultaneously a complete closure is formed in the oral tract (e.g., at the lips for /m/).

---

[5]We shall see in Section 3.2.7 that loss associated with the excitation source also effects the lower formants.

This configuration can be represented as in Fig. 3.27a, which shows two branches, one of which is completely closed. At the point of branching the sound pressure is the same at the input to each tube, while the volume velocity must be continuous at the branching point; i.e., the volume velocity at the output of the pharynx tube must be the sum of the volume velocities at the inputs to the nasal and oral cavities. The corresponding electrical transmission line analog is shown in Fig. 3.27b. Note that continuity of volume velocity at the junction of the 3 tubes corresponds to Kirchoff's current law at the junction of the transmission lines.

For nasal consonants the radiation of sound occurs primarily at the nostrils. Thus the nasal tube is terminated with a radiation impedance appropriate for the size of the nostril openings. The oral tract, which is completely closed, is terminated by the equivalent of an open electrical circuit; i.e., no flow occurs. Nasalized vowels are produced by the same system with the oral tract ter-

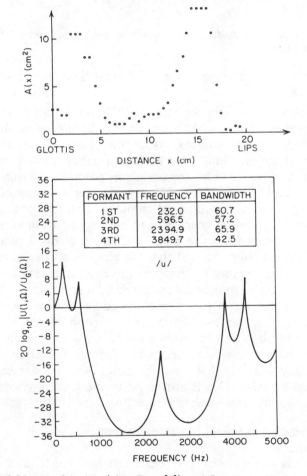

Fig. 3.26 Area function (after Fant [1]) and frequency response (after Portnoff [18]) for the Russian vowel /u/.

77

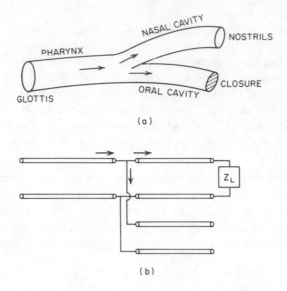

Fig. 3.27 (a) Tube model for production of nasals; (b) corresponding electrical analog.

minated as for vowels. The speech signal would then be the superposition of the nasal and oral outputs.

The mathematical model for this configuration consists of three sets of partial differential equations with boundary conditions being imposed by the form of glottal excitation, terminations of the nasal and oral tracts, and continuity relations at the junction. This leads to a rather complicated set of equations which could in principle be solved, given adequate measurements of area functions for all three tubes. However, the transfer function of the complete system would have many features in common with the previous examples. That is, the system would be characterized by a set of resonances or formants that would be dependent upon the shape and length of the 3 tubes. An important difference results from the fact that the closed oral cavity can trap energy at certain frequencies, preventing those frequencies from appearing in the nasal output. In the electrical transmission line analogy, these are frequencies at which the input impedance of the open circuited line is zero. At these frequencies the junction is short circuited by the transmission line corresponding to the oral cavity. The result is that for nasal sounds, the vocal system transfer function will be characterized by anti-resonances (zeros) as well as resonances. It has also been observed [13] that nasal formants have broader bandwidths than non-nasal voiced sounds. This is attributed to the greater viscous friction and thermal loss due to the large surface area of the nasal cavity.

3.2.7 Excitation of sound in the vocal tract

The previous sub-sections have described how the laws of physics can be applied to describe the propagation and radiation of sound in speech production. To complete our discussion of acoustic principles we must now consider the

78

mechanisms whereby sound waves are generated in the vocal system. Recall that in our general overview of speech production in Section 3.1.1, we identified 3 major mechanisms of excitation. These are:

1. Air flow from the lungs is modulated by the vocal cord vibration, resulting in a quasi-periodic pulse-like excitation.
2. Air flow from the lungs becomes turbulent as the air passes through a constriction in the vocal tract, resulting in a noise-like excitation.
3. Air flow builds up pressure behind a point of total closure in the vocal tract. The rapid release of this pressure, by removing the constriction, causes a transient excitation.

A detailed model of excitation of sound in the vocal system involves the sub-glottal system (lungs, bronchi, and trachea), the glottis, and the vocal tract. Indeed, a model which is complete in all necessary details is also fully capable of simulating breathing as well as speech production! [2]. The first comprehensive effort toward a detailed physical model of sound generation in the vocal system was by Flanagan [2,28]. Subsequent research has produced a much refined model that provides a very detailed representation of the process of generation of both voiced and unvoiced speech [28-31]. This model, which is based upon classical mechanics and fluid mechanics, is beyond the scope of our discussion here. However, a brief qualitative discussion of the basic principles of sound generation will be helpful in pointing the way toward the simple models that are widely used as the basis for speech processing.

The vibration of the vocal cords in voiced speech production can be explained by considering the schematic representation of the vocal system shown in Fig. 3.28. The vocal cords constrict the path from the lungs to the vocal tract. As lung pressure is increased, air flows out of the lungs and through the opening between the vocal cords (glottis). Bernoulli's law states that when a fluid flows through an orifice, the pressure is lower in the constriction than on either side. If the tension in the vocal cords is properly adjusted, the reduced pressure allows the cords to come together, thereby completely constricting air flow. (This is indicated by the dotted lines in Figure 3.28.) As a result, pressure increases behind the vocal cords. Eventually it builds up to a level sufficient to force the vocal cords to open and thus allow air to flow through the glottis again. Again the air pressure in the glottis falls, and the

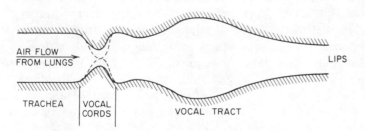

Fig. 3.28 Schematic representation of the vocal system.

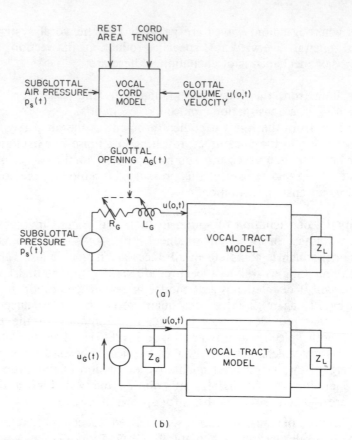

Fig. 3.29 (a) Diagram of vocal cord model; (b) approximate model for vocal cords.

cycle is repeated. Thus, the vocal cords enter a condition of sustained oscillation. The rate at which the glottis opens and closes is controlled by the air pressure in the lungs, the tension and stiffness of the vocal cords, and the area of the glottal opening under rest conditions. These are the control parameters of a detailed model of vocal cord behavior. Such models must also include the effects of the vocal tract, since pressure variations in the vocal tract influence the pressure variations in the glottis. In terms of the electrical analog, the vocal tract acts as a load on the vocal cord oscillator. A schematic diagram of the vocal cord model (adapted from [30]) is shown in Figure 3.29a. The vocal cord model consists of a set of complicated nonlinear differential equations. The coupling of these differential equations to the partial differential equations describing vocal tract transmission can be represented by a time varying acoustic resistance and inductance as shown [30]. These impedance elements are functions of $1/A_G(t)$. For example, when $A_G(t) = 0$ (glottis closed) the impedance is infinite and the volume velocity is zero. Thus, the glottal flow is automatically chopped up into pulses. An example of the signals generated by such a model is shown in Fig. 3.30 [30]. The upper waveform is the volume velocity and the lower waveform is the pressure at the lips for a vocal tract

80

configuration appropriate for the vowel /a/. The pulse-like nature of the glottal flow is certainly consistent with our previous discussion and with direct observation through the use of high-speed motion pictures [2]. The damped oscillations of the output are, of course, consistent with our previous discussion of the nature of sound propagation in the vocal tract.

Since glottal area is a function of the flow into the vocal tract, the overall system of Fig. 3.29a is nonlinear, even though the vocal tract transmission and radiation systems are linear. The coupling between the vocal tract and the glottis is weak, however, and it is common to neglect this interaction. This leads to a separation and linearization of the excitation and transmission system as depicted in Figure 3.29b. In this case $u_G(t)$ is a volume velocity source whose wave shape is of the form of the upper waveform in Fig. 3.30. The glottal acoustic impedance, Z_G, is obtained by linearization of the relations between pressure and volume velocity in the glottis [2]. This impedance is of the form

$$Z_G(\Omega) = R_G + j\Omega L_G \tag{3.33}$$

where R_G and L_G are constants. With this configuration the ideal frequency domain boundary condition of $U(0, \Omega) = U_G(\Omega)$ is replaced by

$$U(0, \Omega) = U_G(\Omega) - P(0, \Omega)/Z_G(\Omega) \tag{3.34}$$

The glottal source impedance has significant effects upon resonance bandwidths for the speech production system. The major effect is a broadening of the lowest resonance. This is because $Z_G(\Omega)$ increases with frequency so that at high frequencies Z_G appears as an open circuit and all of the glottal source flows into the vocal tract system. Thus, yielding walls and glottal loss control the bandwidths of the lower formants while radiation, friction, and thermal losses control the bandwidths of the higher formants.

The mechanism of production of voiceless sounds involves the turbulent flow of air. This can occur at a constriction whenever the volume velocity exceeds a certain critical value [2,29]. Such excitation can be modeled by

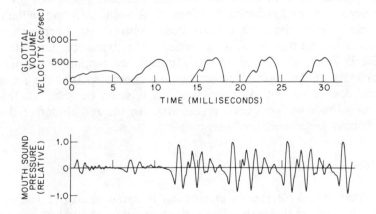

Fig. 3.30 Glottal volume velocity and sound pressure at the mouth for vowel /a/. (After Ishizaka and Flanagan [30].)

81

inserting a randomly time varying source at the point of constriction. The strength of the source is made dependent (nonlinearly) upon the volume velocity in the tube. In this way, frication is automatically inserted when needed [2,29,31]. For fricative sounds, the vocal cord parameters are adjusted so that the cords do not vibrate. For voiced fricatives, the vocal cords vibrate and turbulent flow occurs at a constriction whenever the volume velocity exceeds the critical value. This usually occurs at the peaks of the volume velocity pulses. For plosives, the vocal tract is closed for a period of time while pressure is built up behind the closure with the vocal cords not vibrating. When the constriction is released, the air rushes out at a high velocity thus causing turbulent flow.

3.2.8 Models based upon the acoustic theory

Section 3.2 has discussed in some detail the important features of the acoustic theory of speech production. The detailed models for sound generation, propagation, and radiation can in principle be solved with suitable values of the excitation and vocal tract parameters to compute an output speech waveform. Indeed, it can be argued effectively that this may be the best

Fig. 3.31 Source-system model of speech production.

approach to the synthesis of natural sounding synthetic speech [31]. However, for many purposes such detail is impractical or unnecessary. In such cases the acoustic theory points the way to a simplified approach to modeling speech signals. Figure 3.31 shows a general block diagram that is representative of numerous models that have been used as the basis for speech processing. These models all have in common that the excitation features are separated from the vocal tract and radiation features. The vocal tract and radiation effects are accounted for by the time–varying linear system. Its purpose is to model the resonance effects that we have discussed. The excitation generator creates a signal that is either a train of (glottal) pulses, or randomly varying (noise). The parameters of the source and system are chosen so that the resulting output has the desired speech-like properties. If this can be done, the model may serve as a useful basis for speech processing. In the remainder of this chapter we shall discuss some models of this type.

3.3 Lossless Tube Models

A widely used model for speech production is based upon the assumption that the vocal tract can be represented as a concatenation of lossless acoustic tubes, as depicted in Fig. 3.32. The constant cross-sectional areas $\{A_k\}$, of the tubes

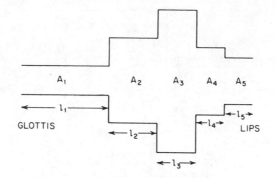

Fig. 3.32 Concatenation of 5 lossless acoustic tubes.

are chosen so as to approximate the area function, $A(x)$, of the vocal tract. If a large number of tubes of short length is used, we can reasonably expect the resonant frequencies of the concatenated tubes to be close to those of a tube with continuously varying area function. However, since this approximation neglects the losses due to friction, heat conduction, and wall vibration, we may also reasonably expect the bandwidths of the resonances to differ from those of a detailed model which includes these losses. However, losses can be accounted for at the glottis and lips, and as we shall see here and in Chapter 8, this can be done so as to accurately represent the resonance properties of the speech signal.

More important for our present discussion is the fact that lossless tube models provide a convenient transition between continuous-time models and discrete-time models. Thus we shall consider models of the form of Figure 3.32 in considerable detail.

3.3.1 Wave propagation in concatenated lossless tubes

Since each tube in Figure 3.32 is assumed lossless, sound propagation in each tube is described by Equations (3.2) with appropriate values of the cross-sectional area. Thus if we consider the kth tube with cross-sectional area, A_k, the pressure and volume velocity in that tube have the form

$$p_k(x,t) = \frac{\rho c}{A_k}\left[u_k^+(t-x/c) + u_k^-(t+x/c) \right] \tag{3.35a}$$

$$u_k(x,t) = u_k^+(t-x/c) - u_k^-(t+x/c) \tag{3.35b}$$

where x is distance measured from the left-hand end of the kth tube ($0 \leqslant x \leqslant l_k$) and $u_k^+(\ )$ and $u_k^-(\ )$ are positive–going and negative–going traveling waves in the kth tube. The relationship between the traveling waves in adjacent tubes can be obtained by applying the physical principle that pressure and volume velocity must be continuous in both time and space everywhere in the system. This provides boundary conditions that can be applied at both ends of each tube.

Consider in particular the junction between the kth and $(k+1)^{st}$ tubes as

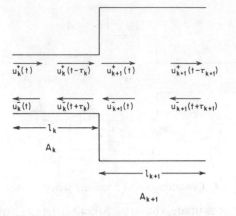

Fig. 3.33 Illustration of the junction between two lossless tubes.

depicted in Figure 3.33. Applying the continuity conditions at the junction gives

$$p_k(l_k,t) = p_{k+1}(0,t) \tag{3.36a}$$

$$u_k(l_k,t) = u_{k+1}(0,t) \tag{3.36b}$$

Substituting Eqs. (3.35) into Eqs. (3.36) gives

$$\frac{A_{k+1}}{A_k}\left[u_k^+(t-\tau_k) + u_k^-(t+\tau_k)\right] = u_{k+1}^+(t) + u_{k+1}^-(t) \tag{3.37a}$$

$$u_k^+(t-\tau_k) - u_k^-(t+\tau_k) = u_{k+1}^+ - u_{k+1}^-(t) \tag{3.37b}$$

where $\tau_k = l_k/c$ is the time for a wave to travel the length of the kth tube. From Figure 3.33 we observe that part of the positive going wave that reaches the junction is propagated on to the right while part is reflected back to the left. Likewise part of the backward traveling wave is propagated on to the left while part is reflected back to the right. Thus, if we solve for $u_{k+1}^+(t)$ and $u_k^-(t+\tau_k)$ in terms of $u_{k+1}^-(t)$ and $u_k^+(t-\tau_k)$ we will be able to see how the forward and reverse traveling waves propagate in the overall system. Solving Eq. (3.37b) for $u_k^-(t+\tau_k)$ and substituting the result into Eq. (3.37a) yields

$$u_{k+1}^+(t) = \left[\frac{2A_{k+1}}{A_{k+1}+A_k}\right]u_k^+(t-\tau_k) + \left[\frac{A_{k+1}-A_k}{A_{k+1}+A_k}\right]u_{k+1}^-(t) \tag{3.38a}$$

Subtracting Eq. (3.37b) from Eq. (3.37a) gives

$$u_k^-(t+\tau_k) = -\left[\frac{A_{k+1}-A_k}{A_{k+1}+A_k}\right]u_k^+(t-\tau_k) + \left[\frac{2A_k}{A_{k+1}+A_k}\right]u_{k+1}^-(t) \tag{3.38b}$$

It can be seen from Eq. (3.38a) that the quantity

$$r_k = \frac{A_{k+1} - A_k}{A_{k+1} + A_k} \tag{3.39}$$

is the amount of $u_{k+1}^-(t)$ that is reflected at the junction. Thus, the quantity r_k is called the reflection coefficient for the kth junction. It is easily shown that since the areas are all positive (see Problem 3.4),

$$-1 \leqslant r_k \leqslant 1 \qquad (3.40)$$

Using this definition of r_k, Eqs. (3.38) can be expressed as

$$u_{k+1}^+(t) = (1+r_k)u_k^+(t-\tau_k) + r_k u_{k+1}^-(t) \qquad (3.41a)$$

$$u_k^-(t+\tau_k) = -r_k u_k^+(t-\tau_k) + (1-r_k)u_{k+1}^-(t) . \qquad (3.41b)$$

Equations of this form were first used for speech synthesis by Kelly and Lochbaum [32]. It is useful to depict these equations graphically as in Figure 3.34. In this figure, signal flow-graph conventions[6] are used to represent the multiplications and additions of Eqs. (3.41). Clearly, each junction of a system such as that depicted in Fig. 3.32 can be represented by a system such as Fig. 3.34, as long as our interest is only in values of pressure and volume velocity at the input and output of the tubes. This is not restrictive since we are primarily interested only in the relationship between the output of the last tube and the input of the first tube. Thus, a 5 tube model such as Fig. 3.32, would have 5 sets of forward and backward delays and 4 junctions, each characterized by a reflection coefficient. To complete the representation of wave propagation in such a system we must consider boundary conditions at the "lips" and "glottis" of the system.

3.3.2 Boundary conditions

Let us assume that there are N sections indexed from 1 to N starting at the glottis. Then the boundary condition at the lips will relate pressure, $p_N(l_N,t)$, and volume velocity, $u_N(l_N,t)$, at the output of the Nth tube to the

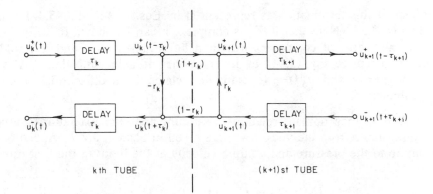

Fig. 3.34 Signal-flow representation of the junction between two lossless tubes.

[6]See Ref. [33] for an introduction to the use of signal flow graphs in signal processing.

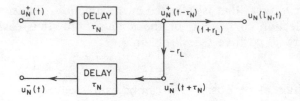

Fig. 3.35 Termination at lip end of a concatenation of lossless tubes.

radiated pressure and volume velocity. If we use the frequency-domain relations of Section 3.2.4 we obtain a relation of the form

$$P_N(l_N, \Omega) = Z_L \cdot U_N(l_N, \Omega) \tag{3.42}$$

If we assume for the moment that Z_L is real, then we obtain the time domain relation

$$\frac{\rho c}{A_N}\left[u_N^+(t-\tau_N) + u_N^-(t+\tau_N)\right] = Z_L\left[u_N^+(t-\tau_N) - u_N^-(t+\tau_N)\right] \tag{3.43}$$

(If Z_L is complex Eq. (3.43) would be replaced by a differential equation relating $p_N(l_N,t)$ and $u_N(l_N,t)$.) Solving for $u_N^-(t+\tau_N)$ we obtain

$$u_N^-(t+\tau_N) = -r_L u_N^+(t-\tau_N) \tag{3.44}$$

where the reflection coefficient at the lips is

$$r_L = \left|\frac{\rho c/A_N - Z_L}{\rho c/A_N + Z_L}\right| \tag{3.45}$$

The output volume velocity at the lips is

$$u_N(l_N,t) = u_N^+(t-\tau_N) - u_N^-(t+\tau_N)$$
$$= (1+r_L)u_N^+(t-\tau_N) \tag{3.46}$$

The effect of this termination as represented by Eqs. (3.44) and (3.46) is depicted in Fig. 3.35. Note that if Z_L is complex, it can be shown that Eq. (3.45) remains valid, but, of course, r_L will then be complex also, and it would be necessary to replace Eq. (3.44) by its frequency-domain equivalent. Alternatively, $u_N^-(t+\tau_N)$ and $u_N^+(t-\tau_N)$ could be related by a differential equation. (See Problem 3.5.)

The frequency domain relations, assuming that the excitation source is linearly separable from the vocal tract, are given in Section 3.2.7. Applying this assumption to the pressure and volume velocity at the input to the first tube we get

$$U_1(0, \Omega) = U_G(\Omega) - P_1(0, \Omega)/Z_G \tag{3.47}$$

Assuming again that Z_G is real,

$$u_1^+(t) - u_1^-(t) = u_G(t) - \frac{\rho c}{A_1}\left[\frac{u_1^+(t) + u_1^-(t)}{Z_G}\right] \tag{3.48}$$

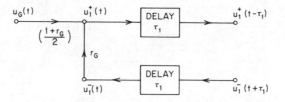

Fig. 3.36 Termination at glottal end of a concatenation of lossless tubes.

Solving for $u_1^+(t)$ we obtain (see Problem 3.6)

$$u_1^+(t) = \frac{(1+r_G)}{2} u_G(t) + r_G u_1^-(t) \tag{3.49}$$

where the glottal reflection coefficient is

$$r_G = \left| \frac{Z_G - \dfrac{\rho c}{A_1}}{Z_G + \dfrac{\rho c}{A_1}} \right| \tag{3.50}$$

Equation 3.49 can be depicted as in Fig. 3.36. As in the case of the radiation termination, if Z_G is complex, then Eq. (3.50) still holds. However, r_G would then be complex and Eq. (3.49) would be replaced by its frequency domain equivalent or $u_1^+(t)$ would be related to $u_G(t)$ and $u_1^-(t)$ by a differential equation. Normally the impedances Z_G and Z_L are taken to be real for simplicity.

As an example, the complete diagram representing wave propagation in a two tube model is shown in Fig. 3.37. The volume velocity at the lips is defined as $u_L(t) = u_2(l_2,t)$. Writing the equations for this system in the frequency domain, the frequency response of the system can be shown to be

$$V_a(\Omega) = \frac{U_L(\Omega)}{U_G(\Omega)}$$

$$= \frac{0.5(1+r_G)(1+r_L)(1+r_1)e^{-j\Omega(\tau_1+\tau_2)}}{1 + r_1 r_G e^{-j\Omega 2\tau_1} + r_1 r_L e^{-j\Omega 2\tau_2} + r_L r_G e^{-j\Omega 2(\tau_1+\tau_2)}} \tag{3.51}$$

(See Problem 3.7.) Several features of $V_a(\Omega)$ are worth pointing out. First, note the factor $e^{-j\Omega(\tau_1+\tau_2)}$ in the numerator. This represents simply the total

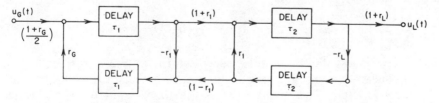

Fig. 3.37 Complete flow diagram of a two-tube model.

87

propagation delay in the system from glottis to lips. The system function of the system is found by replacing $j\Omega$ by s in Eq. (3.51), with the result

$$V_a(s) = \frac{0.5(1+r_G)(1+r_L)(1+r_1)e^{-s(\tau_1+\tau_2)}}{1 + r_1 r_G e^{-s2\tau_1} + r_1 r_L e^{-s2\tau_2} + r_L r_G e^{-s2(\tau_1+\tau_2)}} \quad (3.52)$$

The poles of $V_a(s)$ are the complex resonance frequencies of the system. We see that there will be an infinite number of poles because of the exponential dependence upon s. Fant [1] and Flanagan [2] show that through proper choice of section lengths and cross-sectional areas, realistic formant frequency distributions can be obtained for vowels. (Also see Problem 3.8.)

3.3.3 Relationship to digital filters

The form of $V_a(s)$ for the two tube model suggests that lossless tube models have many properties in common with digital filters. To see this, let us consider a system composed of N lossless tubes each of length $\Delta x = l/N$,

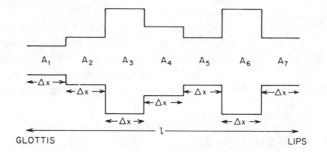

Fig. 3.38 Concatenation of ($N=7$) lossless tubes of equal length.

where l is the overall length of the vocal tract. Such a system is depicted in Figure 3.38 for $N = 7$. Wave propagation in this system can be represented as in Fig. 3.34 with all the delays being equal to $\tau = \Delta x/c$, the time to propagate the length of one tube. It is instructive to begin by considering the response of the system to a unit impulse source, $u_G(t) = \delta(t)$. The impulse propagates down the series of tubes, being partially reflected and partially propagated at the junctions. A detailed consideration of this process will confirm that the impulse response (i.e., the volume velocity at the lips due to an impulse at the glottis) will be of the form

$$v_a(t) = \alpha_0 \delta(t-N\tau) + \sum_{k=1}^{\infty} \alpha_k \delta(t-N\tau-2k\tau) \quad (3.53)$$

Clearly, the soonest that an impulse can reach the output is $N\tau$ sec. Then successive impulses due to reflections at the junctions reach the output at multiples of 2τ seconds later. The quantity 2τ is the time required to propagate both ways in one section. The system function of such a system will be of the form

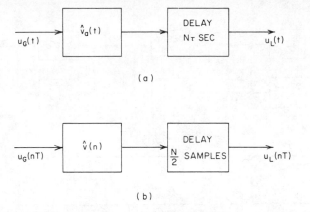

(a)

(b)

Fig. 3.39 (a) Block diagram representation of lossless acoustic tube model; (b) equivalent discrete-time system.

$$V_a(s) = \sum_{k=0}^{\infty} \alpha_k e^{-s(N+2k)\tau}$$

$$= e^{-sN\tau} \sum_{k=0}^{\infty} \alpha_k e^{-s2\tau k} \tag{3.54}$$

The factor $e^{-sN\tau}$ corresponds to the delay time required to propagate through all N sections. The quantity

$$\hat{V}_a(s) = \sum_{k=0}^{\infty} \alpha_k e^{-sk2\tau} \tag{3.55}$$

is the system function of a linear system whose impulse response is simply $\hat{v}_a(t) = v_a(t+N\tau)$. This part represents the resonance properties of the system. Figure 3.39a is a block diagram representation of the lossless tube model showing the separation of the system $\hat{v}_a(t)$ from the delay. The frequency response $\hat{V}_a(\Omega)$ is

$$\hat{V}_a(\Omega) = \sum_{k=0}^{\infty} \alpha_k e^{-j\Omega k2\tau} \tag{3.56}$$

It is easily shown that

$$\hat{V}_a(\Omega + \frac{2\pi}{2\tau}) = \hat{V}_a(\Omega) \tag{3.57}$$

This is, of course, very reminiscent of the frequency response of a discrete-time system. In fact, if the input to the system (i.e., the excitation) is bandlimited to frequencies below $\pi/(2\tau)$, then we can sample the input with period $T = 2\tau$ and filter the sampled signal with a digital filter whose impulse response is

$$\hat{v}(n) = \alpha_n \quad n \geqslant 0$$

$$= 0 \quad n < 0 \tag{3.58}$$

For a sampling period of $T = 2\tau$, the delay of $N\tau$ sec corresponds to a shift of

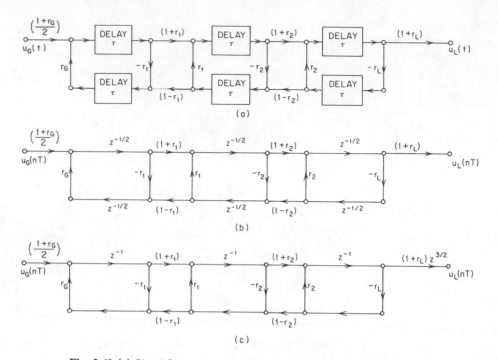

Fig. 3.40 (a) Signal flow graph for lossless tube model of the vocal tract; (b) equivalent discrete-time system; (c) equivalent discrete-time system using only whole delays in ladder part.

$N/2$ samples. Thus, the equivalent discrete time system for bandlimited inputs is shown in Fig. 3.39b. Note that if N is even, $N/2$ is an integer and the delay can be implemented by simply shifting the output sequence of the first system. If N is odd, however, an interpolation would be required to obtain samples of the output of Fig. 3.39a. This delay would most likely be ignored or avoided in some way (see below) since it is of no consequence in most applications of speech models.

The z-transform of $\hat{v}(n)$ is simply $\hat{V}_a(s)$ with e^{sT} replaced by z. Thus,

$$\hat{V}(z) = \sum_{k=0}^{\infty} \alpha_k z^{-k} \qquad (3.59)$$

A signal flow graph for the equivalent discrete-time system can be obtained from the flow graph of the analog system in an analogous way. Specifically, each node variable in the analog system is replaced by the corresponding sequence of samples. Also each τ sec delay is replaced by a 1/2 sample delay, since $\tau = T/2$. An example is depicted in Figure 3.40. Note in particular that the propagation delay is represented in Fig. 3.40b by a transmittance of $z^{-1/2}$.

The 1/2 sample delays in Fig. 3.40b imply an interpolation half-way between sample values. Such interpolation is impossible to implement exactly. A more desirable configuration can be obtained by observing that the structure of Fig. 3.40b has the form of a ladder, with the delay elements only in the upper and lower paths. Signals propagate to the right in the upper path and to the left in the lower path. We can see that the delay around any closed path in Fig. 3.40b will be preserved if the delays in the lower branches are literally moved up to the corresponding branches directly above. The overall delay from input to output will then be wrong but this is of minor significance in practice and theoretically can be compensated by the insertion of the correct amount of advance (in general $z^{N/2}$).[7] Figure 3.40c shows how this is done for the three tube example. The advantage of this form is that difference equations can be written for this system and these difference equations can be used iteratively to compute samples of the output from samples of the input.

Digital networks [33] such as Fig. 3.40c can be used to compute samples of a synthetic speech signal from samples of an appropriate excitation signal [32]. In such applications, the structure of the network representation determines the complexity of the operations required to compute each output sample. Each branch whose transmittance is not unity requires a multiplication. We see that each junction requires 4 multiplications and 2 additions. Generalizing from Fig. 3.40c, we see that $4N$ multiplications and $2N$ additions are required to implement an N-tube model. Since multiplications often are the most time consuming operation, it is of interest to consider other structures (literally, other organizations of the computations) which may require fewer multiplications. These can easily be derived by considering a typical junction as depicted in Fig. 3.41a. The difference equations represented by this diagram are

$$u^+(n) = (1+r)w^+(n) + ru^-(n) \qquad (3.60a)$$

$$w^-(n) = -rw^+(n) + (1-r)u^-(n) \qquad (3.60b)$$

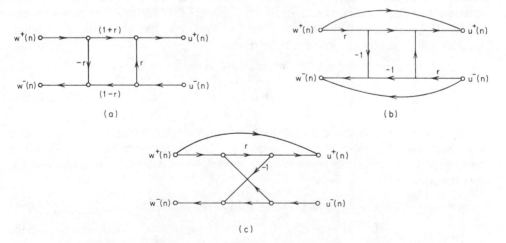

Fig. 3.41 (a) 4 multiplier representation of lossless tube junction; (b) 2 multiplier configuration; (c) 1 multiplier configuration.

[7]Note that we could also move all the delay to the lower branches. In this case, the delay through the system could be corrected by inserting a *delay* of $N/2$ samples.

91

These equations can be written as

$$u^+(n) = w^+(n) + rw^+(n) + ru^-(n) \qquad (3.61a)$$

$$w^-(n) = - rw^+(n) - ru^-(n) + u^-(n) \qquad (3.61b)$$

Noting that the terms $rw^+(n)$ and $ru^-(n)$ occur in both equations, 2 out of the 4 multiplications in Eqs. (3.60) can be eliminated as shown in Fig. 3.41b. Note that this configuration requires 2 multiplications and 4 additions. Still another implementation follows from grouping terms involving r as in

$$u^+(n) = w^+(n) + r[w^+(n)+u^-(n)] \qquad (3.62a)$$

$$w^-(n) = u^-(n) - r[w^+(n)+u^-(n)] \qquad (3.62b)$$

Now, since the term $r[w^+(n)+u^-(n)]$ occurs in both equations, this configuration requires only 1 multiplication and 3 additions as shown in Fig. 3.41c. This form of the lossless tube model was first obtained by Itakura and Saito [34]. When using the lossless tube model for speech synthesis, the choice of computational structure depends on the speed with which multiplications and additions can be done, and the ease of controlling the computation.

3.3.4 Transfer function of the lossless tube model

To complete our discussion of lossless tube discrete-time models for speech production it is instructive to derive a general expression for the transfer function in terms of the reflection coefficients. Equations of the type that we shall derive have been obtained before by Atal and Hanauer [35], Markel and Gray [36], and Wakita [37] in the context of linear predictive analysis of speech. We shall return to a consideration of lossless tube models and their relation to linear predictive analysis in Chapter 8. Our main concern at this point is the general form of the transfer function and the variety of other models suggested by the lossless tube model.

Let us begin by noting that we seek the transfer function

$$V(z) = \frac{U_L(z)}{U_G(z)} \qquad (3.63)$$

To find $V(z)$, it is most convenient to express $U_G(z)$ in terms of $U_L(z)$ and then solve for the ratio above. To do this, let us consider Figure 3.42 which depicts a junction in the lossless tube model. The z-transform equations for this junction are

$$U_{k+1}^+(z) = (1+r_k)z^{-1/2}U_k^+(z) + r_k U_{k+1}^-(z) \qquad (3.64a)$$

$$U_k^-(z) = - r_k z^{-1}U_k^+(z) + (1-r_k)z^{-1/2}U_{k+1}^-(z) \qquad (3.64b)$$

Solving for $U_k^+(z)$ and $U_k^-(z)$ we obtain

$$U_k^+(z) = \frac{z^{1/2}}{1 + r_k} U_{k+1}^+(z) - \frac{r_k z^{1/2}}{1 + r_k} U_{k+1}^-(z) \qquad (3.65a)$$

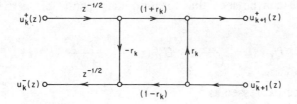

Fig. 3.42 Flow graph representing relationship among z-transforms at a junction.

$$U_k^-(z) = \frac{-r_k z^{-1/2}}{1 + r_k} U_{k+1}^+(z) + \frac{z^{-1/2}}{1 + r_k} U_{k+1}^-(z) \qquad (3.65b)$$

Equations (3.65) permit us to work backwards from the output of the lossless tube model to obtain $U_G(z)$ in terms of $U_L(z)$.

To make the result more compact it is helpful to represent the boundary condition at the lips in the same manner as all the junctions in the system. Toward this end, we define $U_{N+1}(z)$ to be the z-transform of the input to a fictitious $(N+1)^{st}$ tube that is infinitely long so that there is no negative-going wave in the $(N+1)^{st}$ tube. An equivalent point of view is that the $(N+1)^{st}$ tube is terminated in its characteristic impedance. In any case, $U_{N+1}^+(z) = U_L(z)$ and $U_{N+1}^-(z) = 0$. Then from Eqs. (3.39) and (3.45) we see that if $A_{N+1} = \rho c / Z_L$, we can define $r_N = r_L$.

Now, Eqs. (3.65) can be expressed in matrix form as

$$\mathbf{U}_k = \mathbf{Q}_k \mathbf{U}_{k+1} \qquad (3.66)$$

where

$$\mathbf{U}_k = \begin{bmatrix} U_k^+(z) \\ U_k^-(z) \end{bmatrix} \qquad (3.67)$$

and

$$\mathbf{Q}_k = \begin{bmatrix} \dfrac{z^{1/2}}{1 + r_k} & \dfrac{-r_k z^{1/2}}{1 + r_k} \\ \dfrac{-r_k z^{-1/2}}{1 + r_k} & \dfrac{z^{-1/2}}{1 + r_k} \end{bmatrix} \qquad (3.68)$$

By repeatedly applying Eq. (3.66), it can be easily shown that the variables at the input to the first tube can be expressed in terms of the variables at the output by the matrix product

$$\begin{aligned} \mathbf{U}_1 &= \mathbf{Q}_1 \cdot \mathbf{Q}_2 \cdots \mathbf{Q}_N \mathbf{U}_{N+1} \\ &= \prod_{k=1}^{N} \mathbf{Q}_k \cdot \mathbf{U}_{N+1} \end{aligned} \qquad (3.69)$$

From Fig. 3.36 it can be seen that the boundary condition at the glottis can be expressed as

$$U_G(z) = \frac{2}{(1+r_G)} U_1^+(z) - \frac{2r_G}{1+r_G} U_1^-(z) \tag{3.70}$$

which can also be expressed as

$$U_G(z) = \left[\frac{2}{1+r_G} \ , \ -\frac{2r_G}{1+r_G} \right] U_1 \tag{3.71}$$

Thus, since

$$\mathbf{U}_{N+1} = \begin{bmatrix} U_L(z) \\ 0 \end{bmatrix} = \begin{bmatrix} 1 \\ 0 \end{bmatrix} U_L(z) \tag{3.72}$$

we can at last write

$$\frac{U_G(z)}{U_L(z)} = \left[\frac{2}{1+r_G} \ , \ -\frac{2r_G}{1+r_G} \right] \prod_{k=1}^{N} \mathbf{Q}_k \begin{bmatrix} 1 \\ 0 \end{bmatrix} \tag{3.73}$$

which is equal to $1/V(z)$.

To examine the properties of $V(z)$, it is helpful to first express $\mathbf{Q}_k$ as

$$\mathbf{Q}_k = z^{1/2} \begin{bmatrix} \dfrac{1}{1+r_k} & \dfrac{-r_k}{1+r_k} \\ \dfrac{-r_k z^{-1}}{1+r_k} & \dfrac{z^{-1}}{1+r_k} \end{bmatrix}$$

$$= z^{1/2} \hat{\mathbf{Q}}_k \tag{3.74}$$

Thus, Eq. (3.73) can be expressed as

$$\frac{1}{V(z)} = z^{N/2} \left[\frac{2}{1+r_G} \ , \ -\frac{2r_G}{1+r_G} \right] \prod_{k=1}^{N} \hat{\mathbf{Q}}_k \begin{bmatrix} 1 \\ 0 \end{bmatrix} \tag{3.75}$$

First, we note that since the elements of the matrices $\hat{\mathbf{Q}}_k$ are either constant or proportional to z^{-1}, the complete matrix product will reduce to a polynomial in the variable z^{-1} of order N. For example it can be shown (see Problem 3.9) that for $N = 2$,

$$\frac{1}{V(z)} = \frac{2(1 + r_1 r_2 z^{-1} + r_1 r_G z^{-1} + r_2 r_G z^{-2})z}{(1+r_G)(1+r_1)(1+r_2)} \tag{3.76}$$

or

$$V(z) = \frac{0.5(1+r_G)(1+r_1)(1+r_2)z^{-1}}{1 + (r_1 r_2 + r_1 r_G)z^{-1} + r_2 r_G z^{-2}} \tag{3.77}$$

94

In general, it can be seen from Eqs. (3.74) and (3.75) that for a lossless tube model, the transfer function can always be expressed as

$$V(z) = \frac{0.5(1+r_G) \prod\limits_{k=1}^{N} (1+r_k)z^{-N/2}}{D(z)} \qquad (3.78a)$$

where $D(z)$ is a polynomial in z^{-1} given by the matrix

$$D(z) = [1, -r_G] \begin{bmatrix} 1 & -r_1 \\ -r_1 z^{-1} & z^{-1} \end{bmatrix} \cdots \begin{bmatrix} 1 & -r_N \\ -r_N z^{-1} & z^{-1} \end{bmatrix} \begin{bmatrix} 1 \\ 0 \end{bmatrix} \qquad (3.78b)$$

It can be seen from Eq. (3.78b) that $D(z)$ will have the form

$$D(z) = 1 - \sum_{k=1}^{N} \alpha_k z^{-k} \qquad (3.79)$$

In other words, the transfer function of a lossless tube model has a delay corresponding to the number of sections of the model and it has no zeros — only poles. These poles, of course, define the resonances or formants of the lossless tube model.

In the special case $r_G = 1$ ($Z_G = \infty$), the polynomial $D(z)$ can be found using a recursion formula that can be derived from Eq. (3.78b). If we begin by evaluating the matrix product from the left, we will always be multiplying a 1×2 row matrix by a 2×2 matrix until finally we multiply by the 2×1 column vector on the right in Eq. (3.78b). The desired recursion formula becomes evident after evaluating the first few matrix products. Let us define

$$\mathbf{P}_1 = [1, -1] \begin{bmatrix} 1 & -r_1 \\ -r_1 z^{-1} & z^{-1} \end{bmatrix} = \left[(1+r_1 z^{-1}), - (r_1 + z^{-1}) \right] \qquad (3.80)$$

If we define

$$D_1(z) = 1 + r_1 z^{-1} \qquad (3.81)$$

then it is easily shown that

$$\mathbf{P}_1 = [D_1(z), - z^{-1} D_1(z^{-1})] \qquad (3.82)$$

Similarly, the row matrix $\mathbf{P}_2$ is defined as

$$\mathbf{P}_2 = \mathbf{P}_1 \begin{bmatrix} 1 & -r_2 \\ -r_2 z^{-1} & z^{-1} \end{bmatrix} \qquad (3.83)$$

If the indicated multiplication is carried out it is easily shown that

$$\mathbf{P}_2 = [D_2(z), - z^{-2} D_2(z^{-1})] \qquad (3.84)$$

where

$$D_2(z) = D_1(z) + r_2 z^{-2} D_1(z^{-1}) \qquad (3.85)$$

By induction it can be shown that

$$\mathbf{P}_k = \mathbf{P}_{k-1} \begin{bmatrix} 1 & -r_k \\ -r_k z^{-1} & z^{-1} \end{bmatrix}$$

$$= [D_k(z), -z^{-k}D_k(z^{-1})] \tag{3.86}$$

where

$$D_k(z) = D_{k-1}(z) + r_k z^{-k} D_{k-1}(z^{-1}) \tag{3.87}$$

Finally, the desired polynomial $D(z)$ is

$$D(z) = \mathbf{P}_N \begin{bmatrix} 1 \\ 0 \end{bmatrix} = D_N(z) \tag{3.88}$$

Thus, we can see that it is not necessary to carry out all the matrix multiplies but we can simply evaluate the recursion

$$D_0(z) = 1 \tag{3.89a}$$

$$D_k(z) = D_{k-1}(z) + r_k z^{-k} D_{k-1}(z^{-1}) \quad k = 1, 2, \ldots, N \tag{3.89b}$$

$$D(z) = D_N(z) \tag{3.89c}$$

The effectiveness of the lossless tube model can be demonstrated by computing the transfer function for the area function data used to compute Figures 3.23-3.26. To do this we must decide upon the termination at the lips and the number of sections to use. In our derivations, we have represented the radiation load as a tube of area A_{N+1} which has no reflected wave. The value of A_{N+1} is chosen to give the desired reflection coefficient at the output. This is the only source of loss in the system (if $r_G = 1$), and thus it is to be expected that the choice of A_{N+1} will control the bandwidths of the resonances of $V(z)$. For example, $A_{N+1} = \infty$ gives $r_N = r_L = 1$, the reflection coefficient for an acoustic short circuit. This, of course, is the completely lossless case. Usually A_{N+1} would be chosen to give a reflection coefficient at the lips which produces reasonable bandwidths for the resonances. An example is presented below.

The choice of number of sections depends upon the sampling rate chosen to represent the speech signal. Recall that the frequency response of the lossless tube model is periodic; and thus, the model can only approximate the vocal tract behavior in a band of frequencies $|F| < 1/(2T)$, where T is the sampling period. We have seen that this requires $T = 2\tau$, where τ is the one-way propagation time in a single section. If there are N sections, for a total length, l, then $\tau = l/(cN)$. Since the order of the denominator polynomial is N, there can be at most $N/2$ complex conjugate poles to provide resonances in the band $|F| < 1/(2T)$. Using the above value for τ with $l = 17.5$ cm and $c = 35000$ cm/sec, we see that

$$\frac{1}{2T} = \frac{1}{4\tau} = \frac{Nc}{4l} = \frac{N}{2} (1000) \text{ Hz} \tag{3.90}$$

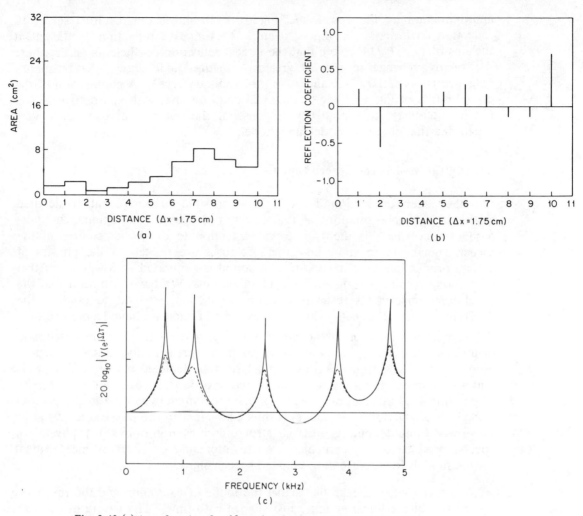

Fig. 3.43 (a) Area function for 10 section lossless tube terminated with reflectionless section of area 30 cm^2; (b) reflection coefficients for 10 section tube; (c) frequency response of 10 section tube; dotted curve corresponds to conditions of (b); solid curve corresponds to short-circuit termination. (Note area data of (a) estimated from data given by Fant [1] for the Russian vowel /a/.)

This implies that there will be about $N/2$ resonances (formants) per 1000 Hz of frequency for a vocal tract of total length 17.5 cm. For example, if $1/T = 10000$ Hz, then the baseband is 5000 Hz. This implies that N should be 10. A glance at Figures 3.21 through 3.26 confirm that vocal tract resonances seem to occur with a density of about one formant per 1000 Hz. Shorter overall vocal tract lengths will have fewer resonances per kiloHertz and vice versa.

Figure 3.43 shows an example for $N = 10$ and $1/T = 10$ kHz. Figure 3.43a shows the area function data of Fig. 3.23 sampled to give a 10 tube approximation for the vowel /a/. Figure 3.43b shows the resulting set of 10 reflection coefficients for $A_{11} = 30$ cm^2. This gives a reflection coefficient at the lips of $r_N = 0.714$. Note that the largest reflection coefficients occur where the relative change in area is greatest. Figure 3.43c shows the frequency response curves for $r_N = 1$ and $r_N = .714$ (dotted curve). A comparison of the dotted curve of Figure 3.43c to Fig. 3.23 confirms that with appropriate loss at the lip boundary, the frequency response of the lossless tube model is very much like that of the more detailed model.

3.4 Digital Models for Speech Signals

We have seen in Section 3.2 that it is possible to derive rather detailed mathematical representations of the acoustics of speech production. Our purpose in surveying this theory is to call attention to the basic features of the speech signal and to show how these features are related to the physics of speech production. We have seen that sound is generated in 3 ways, and that each mode results in a distinctive type of output. We have also seen that the vocal tract imposes its resonances upon the excitation so as to produce the different sounds of speech. This is the essence of what we have learned so far.

An important idea should now be emerging from this lengthy discussion of models. It is simply that a valid approach to representation of speech signals is in terms of a "terminal analog" model such as depicted before in Fig. 3.31; that is, a linear system whose output has the desired speech-like properties when controlled by a set of parameters that are somehow related to the process of speech production. The model is thus equivalent to the physical model at its terminals (output) but its internal structure does not mimic the physics of speech production. In particular, we are interested in discrete-time terminal analog models for representing sampled speech signals.

To produce a speech-like signal the mode of excitation and the resonance properties of the linear system must change with time. The nature of this time variation can be seen in Section 3.1. In particular, waveform plots such as Fig. 3.3a show that the properties of the speech signal change relatively slowly with time. For many speech sounds it is reasonable to assume that the general properties of the excitation and vocal tract remain fixed for periods of 10-20 msec. Thus, a terminal analog model involves a slowly time-varying linear system excited by an excitation signal whose basic nature changes from quasi-periodic pulses for voiced speech to random noise for unvoiced speech.

The lossless tube discrete-time model of the previous section serves as an example of what we mean. The essential features of that model are depicted in Fig. 3.44a. Recall that the vocal tract system was characterized by a set of areas or, equivalently, reflection coefficients. Systems of the form of Fig. 3.40c can thus be used to compute the speech output given an appropriate input. We showed that the relationship between the input and output could be represented by a transfer function, $V(z)$, of the form

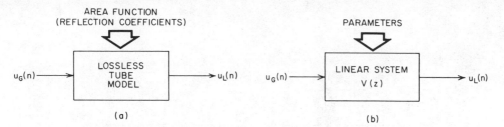

Fig. 3.44 (a) Block diagram representation of the lossless tube model; (b) terminal analog model.

$$V(z) = \frac{G}{1 - \sum_{k=1}^{N} \alpha_k z^{-k}} \qquad (3.91)$$

where G and $\{\alpha_k\}$ depend upon the area function. (Note that the fixed delay in Eq. (3.78a) has been dropped.) Insofar as the output is concerned, any system having this transfer function will produce the same output in response to a given input. (This is not strictly true for time-varying systems, but differences can be minimized by careful implementation.) Thus, discrete-time terminal analog models take the general form of Fig. 3.44b. This leads to a consideration of alternative implementations of the vocal tract filter.

In addition to the vocal tract response a complete terminal analog model includes a representation of the changing excitation function and the effects of sound radiation at the lips. In the remainder of this section we shall examine each of the model components separately, and then combine them into a complete model.

3.4.1 Vocal tract

The resonances (formants) of speech correspond to the poles of the transfer function $V(z)$. An all-pole model is a very good representation of vocal tract effects for a majority of speech sounds; however, the acoustic theory tells us that nasals and fricatives require both resonances and anti-resonances (poles and zeros). In these cases, we may include zeros in the transfer function or we may reason with Atal [35] that effect of a zero of the transfer function can be achieved by including more poles. (See Problem 3.10.) In most cases this approach is to be preferred.

Since the coefficients of the denominator of $V(z)$ in Eq. (3.91) are real, the roots of the denominator polynomial will be either real or occur in complex conjugate pairs. A typical complex resonant frequency of the vocal tract is

$$s_k, s_k^* = -\sigma_k \pm j2\pi F_k \qquad (3.92)$$

The corresponding complex conjugate poles in the discrete-time representation would be

$$z_k, z_k^* = e^{-\sigma_k T} e^{\pm j2\pi F_k T}$$

$$= e^{-\sigma_k T} \cos(2\pi F_k T) \pm j e^{-\sigma_k T} \sin(2\pi F_k T) \qquad (3.93)$$

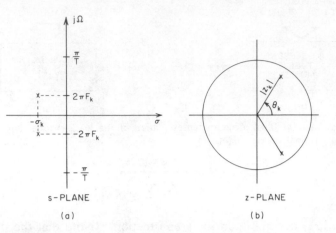

Fig. 3.45 (a) s-plane; and (b) z-plane representations of a vocal tract resonance.

The bandwidth of the vocal tract resonance is approximately $2\sigma_k$ and the center frequency is $2\pi F_k$ [26]. In the z-plane, the radius from the origin to the pole determines the bandwidth, i.e.,

$$|z_k| = e^{-\sigma_k T} \tag{3.94a}$$

and the z-plane angle is

$$\theta_k = 2\pi F_k T \tag{3.94b}$$

Thus if the denominator of $V(z)$ is factored, the corresponding analog formant frequencies and bandwidths can be found using Eqs. (3.94). As shown in Figure 3.45 the complex natural frequencies of the human vocal tract are all in the left half of the s-plane since it is a stable system. Thus, $\sigma_k > 0$, and therefore $|z_k| < 1$; i.e., all of the corresponding poles of the discrete-time model must be inside the unit circle as required for stability. Figure 3.45 depicts typical complex resonant frequencies in both the s-plane and the z-plane.

In Section 3.3 we showed how a lossless tube model leads to a transfer function of the form of Eq. (3.91). It can be shown [35,36] that as long as the areas of the tube model are positive, all the poles of the corresponding $V(z)$ will be inside the unit circle. Conversely, it can be shown that given a transfer function, $V(z)$, as in Eq. (3.91), a lossless tube model can be found [35,36]. Thus, one way to implement a given transfer function is to use a ladder structure as in Fig. 3.40c, possibly incorporating one of the junction forms of Fig. 3.41. Another approach is to use one of the standard digital filter implementation structures given in Chapter 2. For example we could use a direct form implementation of $V(z)$ as depicted in Fig. 3.46a. Alternatively, we can represent $V(z)$ as a cascade of second order systems (resonators); i.e.,

$$V(z) = \prod_{k=1}^{M} V_k(z) \tag{3.95}$$

100

where M is the largest integer in $((N+1)/2)$, and

$$V_k(z) = \frac{(1-2|z_k|\cos(2\pi F_k T) + |z_k|^2)}{(1-2|z_k|\cos(2\pi F_k T)z^{-1} + |z_k|^2 z^{-2})} \qquad (3.96)$$

The numerator of $V_k(z)$ is chosen so that the product will have the same gain as the lossless tube model. Note that at zero frequency ($z = 1$), $V_k(1) = 1$. A cascade model is depicted in Fig. 3.46b. Problem 3.11 shows a novel way of eliminating multiplications in cascade models. Still another approach to implementing the system $V(z)$ is to make a partial fraction expansion of $V(z)$ and thus obtain a parallel form model. This approach is explored in Problem 3.12.

It is interesting to note that cascade and parallel models were first considered as analog models. In this context there is a serious limitation, since analog second order systems (resonators) have frequency responses that die away with frequency. This led Fant [1] to derive "higher pole correction" factors that were cascaded with the analog formant resonators to achieve proper high frequency spectral balance. When digital simulations began to be used, Gold and Rabiner [38] observed that digital resonators had, by virtue of their inherent periodicity, the correct high frequency behavior. We have, of course, already seen this in the context of the lossless tube model. Thus no "higher pole correction" network is required in digital simulations.

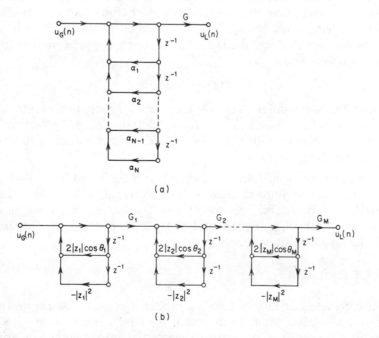

Fig. 3.46 (a) Direct form implementation of all-pole transfer function; (b) cascade implementation of all-pole transfer function ($G_k = 1 - 2|z_k|\cos\theta_k + |z_k|^2$).

101

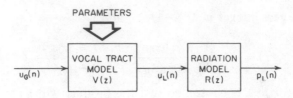

Fig. 3.47 Terminal analog model including radiation effects.

3.4.2 Radiation

So far we have considered the transfer function $V(z)$ which relates volume velocity at the source to volume velocity at the lips. If we wish to obtain a model for pressure at the lips (as is usually the case), then the effects of radiation must be included. We saw in Section 3.2.4 that in the analog model, the pressure and volume velocity are related by Eqs. (3.29). We desire a similar z-transform relation of the form

$$P_L(z) = R(z) U_L(z) \tag{3.97}$$

It can be seen from the discussion of Section 3.2.4 and from Fig. 3.20 that pressure is related to volume velocity by a highpass filtering operation. In fact, at low frequencies it can be argued that the pressure is approximately the derivative of the volume velocity. Thus, to obtain a discrete-time representation of this relationship we must use a digitization technique that avoids aliasing. For example, by using the bilinear transform method of digital filter design [33] it can be shown (see Problem 3.13) that a reasonable approximation to the radiation effects is obtained with

$$R(z) = R_0(1 - z^{-1}) \tag{3.98}$$

i.e., a first backward difference. (A more accurate approximation is also considered in Problem 3.13.) The crude "differentiation" effect of the first difference is consistent with the approximate differentiation at low frequencies that is commonly assumed.

This radiation "load" can be cascaded with the vocal tract model as in Fig. 3.47. $V(z)$ can be implemented in any convenient way and the required parameters will, of course, be appropriate for the chosen configuration; e.g., area function for the lossless tube model or formant frequencies and bandwidths for the cascade model.

3.4.3 Excitation

To complete our terminal analog model, we must discuss means for generating an appropriate input to the vocal tract radiation system. Recalling that the majority of speech sounds can be classed as either voiced or voiceless, we see that in general terms what is required is a source that can produce either a quasi-periodic pulse waveform or a random noise waveform.

In the case of voiced speech, the excitation waveform must appear some-what like the upper waveform in Fig. 3.30. A convenient way to represent the generation of the glottal wave is shown in Fig. 3.48. The impulse train genera-tor produces a sequence of unit impulses which are spaced by the desired fun-damental period. This signal in turn excites a linear system whose impulse response $g(n)$ has the desired glottal wave shape. A gain control, A_v, controls the intensity of the voiced excitation.

The choice of the form of $g(n)$ is probably not critical as long as its Fourier transform has the right properties. Rosenberg [39], in a study of the effect of glottal pulse shape on speech quality, found that the natural glottal pulse waveform could be replaced by a synthetic pulse waveform of the form

$$g(n) = \frac{1}{2}[1 - \cos(\pi n/N_1)] \quad 0 \leqslant n \leqslant N_1 .$$

$$= \cos(\pi(n-N_1)/2N_2) \quad N_1 \leqslant n \leqslant N_1 + N_2$$

$$= 0 \qquad \qquad otherwise \qquad (3.99)$$

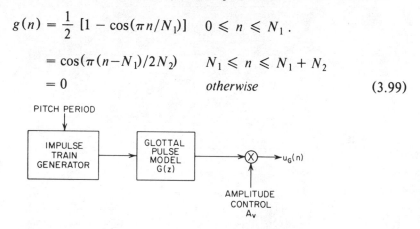

Fig. 3.48 Generation of the excitation signal for voiced speech.

This wave shape is very similar in appearance to the pulses in Fig. 3.30. Figure 3.49 shows the pulse waveform and its Fourier transform magnitude for typical values of N_1 and N_2. It can be seen that, as would be expected, the effect of the glottal pulse in the frequency domain is to introduce a lowpass filtering effect.

Since $g(n)$ in Eq. (3.99) has finite length, its z-transform, $G(z)$, has only zeros. An all-pole model is often more desirable. Good success has also been achieved using a two-pole model for $G(z)$ [36].

For voiceless sounds the excitation model is much simpler. All that is required is a source of random noise and a gain parameter to control the inten-sity of the unvoiced excitation. For discrete-time models, a random number generator provides a source of flat-spectrum noise. The probability distribution of the noise samples does not appear to be critical.

3.4.4 The complete model

Putting all the ingredients together we obtain the model of Figure 3.50. By switching between the voiced and unvoiced excitation generators we can

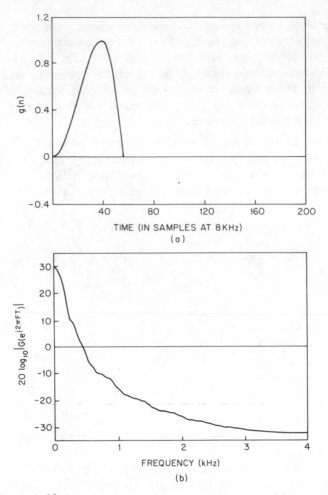

Fig. 3.49 (a) Rosenberg approximation to glottal pulse; (b) corresponding Fourier transform.

model the changing mode of excitation. The vocal tract can be modeled in a wide variety of ways as we have discussed. In some cases it is convenient to combine the glottal pulse and radiation models into a single system. In fact, we shall see that in the case of linear predictive analysis it is convenient to combine the glottal pulse, radiation and vocal tract components all together and represent them as a single transfer function

$$H(z) = G(z) V(z) R(z) \qquad (3.100)$$

of the all-pole type. In other words Figure 3.50 is only a general representation. There is much latitude for modification.

A natural question at this point concerns the limitations of such a model. Certainly the model is far from the partial differential equations with which we began. Fortunately none of the deficiencies of this model severely limits its applicability. First, there is the question of time variation of the parameters.

In continuant sounds such as vowels, the parameters change very slowly and the model works very well. With transient sounds such as stops, the model is not as good but still adequate. It should be emphasized that our use of transfer functions and frequency response functions implicitly assumes that we can represent the speech signal on a "short-time" basis. That is, the parameters of the model are assumed to be constant over time intervals typically 10-20 msec long. The transfer function $V(z)$, then, really serves to define the structure of a model whose parameters vary slowly with time. We shall repeatedly invoke this principle of quasi-stationarity in subsequent chapters. A second limitation is the lack of provision for zeros as required theoretically for nasals and fricatives. This is definitely a limitation for nasals, but not too severe for fricatives. Zeros can be included in the model if desired. Third, the simple dichotomy of voiced-unvoiced excitation is inadequate for voiced fricatives. Simply adding the voiced and unvoiced excitations is inadequate since frication is correlated with the peaks of the glottal flow. A more sophisticated model for voiced fricatives has been developed [40] and can be employed when needed. Finally, a relatively minor concern is that the model of Fig. 3.50 requires that the glottal pulses be spaced by an integer multiple of the sampling period, T. Winham and Steiglitz [41] have considered ways of eliminating this limitation in situations requiring precise pitch control.

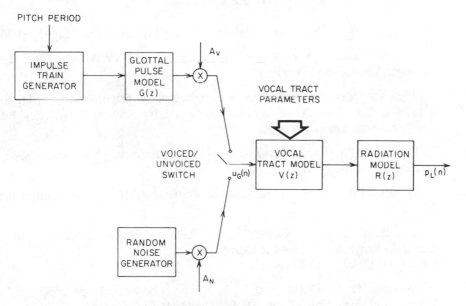

Fig. 3.50 General discrete-time model for speech production.

3.5 Summary

This chapter has focused upon three main areas: the sounds of speech, the physics of speech production, and discrete-time models for speech production. Our review of acoustic phonetics and the acoustic theory of speech production has been lengthy but far from complete. Our purpose has been to provide adequate knowledge about the general properties of speech signals so as to motivate and suggest models that are useful for speech processing.

The models discussed in Sections 3.3 and 3.4 will be the basis for our discussion in the remainder of this book. We shall think of these models in two ways. One point of view is called speech analysis; the other is called speech synthesis. In speech analysis we are interested in techniques for estimating the parameters of the model from a natural speech signal that is assumed to be the output of the model. In speech synthesis, we wish to use the model to create a synthetic speech signal by controlling the model with suitable parameters. These two points of view will become intermingled in many cases and will arise in many problem areas. Underlying all our subsequent discussion will be models of the type discussed in this chapter. Having reviewed the subject of digital signal processing in Chapter 2 and the acoustic theory of speech production here, we are now ready to begin to see how digital signal processing techniques can be applied in processing speech signals.

REFERENCES

1. G. Fant, *Acoustic Theory of Speech Production,* Mouton, The Hague, 1970.

2. J. L. Flanagan, *Speech Analysis, Synthesis and Perception,* 2nd Ed., Springer-Verlag, New York, 1972.

3. H. Fletcher, *Speech and Hearing in Communcation,* original edition, D. Van Nostrand Co., New York, 1953. Reprinted by Robert E. Krieger Pub. Co. Inc., New York, 1972.

4. T. Chiba and M. Kajiyama, *The Vowel, Its Nature and Structure,* Phonetic Society of Japan, 1958.

5. I. Lehiste, Ed., *Readings in Acoustic Phonetics,* MIT Press, Cambridge, Mass., 1967.

6. J. L. Flanagan, C. H. Coker, L. R. Rabiner, R. W. Schafer, and N. Umeda, "Synthetic Voices for Computers," *IEEE Spectrum,* Vol. 7, No. 10, pp. 22-45, October 1970.

7. W. Koenig, H. K. Dunn, and L. Y. Lacy, "The Sound Spectrograph," *J. Acoust. Soc. Am.,* Vol. 17, pp. 19-49, July 1946.

8. R. K. Potter, G. A. Kopp, and H. C. Green, *Visible Speech,* D. Van Nostrand Co., New York, 1947. Republished by Dover Publications, Inc., 1966.

9. R. Jakobson, C. G. M. Fant, and M. Halle, *Preliminaries to Speech Analysis: The Distinctive Features and Their Correlates,* M.I.T. Press, Cambridge, Mass., 1963.

10. N. Chomsky and M. Halle, *The Sound Pattern of English,* Harper & Row, Publishers, New York, 1968.

11. G. E. Peterson and H. L. Barney, "Control Methods Used in a Study of the Vowels," *J. Acoust. Soc. Am.,* Vol. 24, No. 2, pp. 175-184, March 1952.

12. A. Holbrook and G. Fairbanks, "Diphthong Formants and Their Movements," *J. of Speech and Hearing Research,* Vol. 5, No. 1, pp. 38-58, March 1962.

13. O. Fujimura, "Analysis of Nasal Consonants," *J. Acoust. Soc. Am.,* Vol. 34, No. 12, pp. 1865-1875, December 1962.

14. J. M. Heinz and K. N. Stevens, "On the Properties of Voiceless Fricative Consonants," *J. Acoust. Soc. Am.,* Vol. 33, No. 5, pp. 589-596, May 1961.

15. P. C. Delattre, A. M. Liberman, and F. S. Cooper, "Acoustic Loci and Transitional Cues for Consonants," *J. Acoust. Soc. Am.,* Vol. 27, No. 4, pp. 769-773, July 1955.

16. L. L. Beranek, *Acoustics,* McGraw-Hill Book Co., New York, 1954.

17. P. M. Morse and K. U. Ingard, *Theoretical Acoustics,* McGraw-Hill Book Co., New York, 1968.

18. M. R. Portnoff, "A Quasi-One-Dimensional Digital Simulation for the Time-Varying Vocal Tract," M. S. Thesis, Dept. of Elect. Engr., MIT, Cambridge, Mass., June 1973.

19. M. R. Portnoff and R. W. Schafer, "Mathematical Considerations in Digital Simulations of the Vocal Tract," *J. Acoust. Soc. Am.,* Vol. 53, No. 1 (Abstract), p. 294, January 1973.

20. M. M. Sondhi, "Model for Wave Propagation in a Lossy Vocal Tract," *J. Acoust. Soc. Am.,* Vol. 55, No. 5, pp. 1070-1075, May 1974.

21. J. S. Perkell, *Physiology of Speech Production: Results and Implications of a Quantitative Cineradiographic Study,* MIT Press, Cambridge, Mass., 1969.

22. M. M. Sondhi and B. Gopinath, "Determination of Vocal-Tract Shape from Impulse Response at the Lips," *J. Acoust. Soc. Am.,* Vol. 49, No. 6 (Part 2), pp. 1847-1873, June 1971.

23. B. S. Atal, "Towards Determining Articulator Positions from the Speech Signal," *Proc. Speech Comm. Seminar,* Stockholm, Sweden, pp. 1-9, 1974.

24. R. B. Adler, L. J. Chu, and R. M. Fano, *Electromagnetic Energy Transmission and Radiation,* John Wiley and Sons, Inc., New York, 1963.

25. D. T. Paris and F. K. Hurd, *Basic Electromagnetic Theory,* McGraw-Hill Book Co., New York, 1969.

26. A. M. Bose and K. N. Stevens, *Introductory Network Theory,* Harper and Row, New York, 1965.

27. H. K. Dunn, "Methods of Measuring Vowel Formant Bandwidths," *J. Acoust. Soc. Am.,* Vol. 33, pp. 1737-1746, 1961.

28. J. L. Flanagan and L. L. Landgraf, "Self Oscillating Source for Vocal-Tract Synthesizers," *IEEE Trans. Audio and Electroacoustics,* Vol. AU-16, pp. 57-64, March 1968.

29. J. L. Flanagan and L. Cherry, "Excitation of Vocal-Tract Synthesizer," *J. Acoust. Soc. Am.,* Vol. 45, No. 3, pp. 764-769, March 1969.

30. K. Ishizaka and J. L. Flanagan, "Synthesis of Voiced Sounds from a Two-Mass Model of the Vocal Cords," *Bell Syst. Tech. J.,* Vol. 50, No. 6, pp. 1233-1268, July-August 1972.

31. J. L. Flanagan, K. Ishizaka, and K. L. Shipley, "Synthesis of Speech from a Dynamic Model of the Vocal Cords and Vocal Tract," *Bell Sys. Tech J.,* Vol. 54, No. 3, pp. 485-506, March 1975.

32. J. L. Kelly, Jr. and C. Lochbaum, "Speech Synthesis," *Proc. Stockholm Speech Communications Seminar,* R.I.T., Stockholm, Sweden, September 1962.

33. A. V. Oppenheim and R. W. Schafer, *Digital Signal Processing,* Prentice-Hall, Inc., Englewood Cliffs, N.J., 1975.

34. F. Itakura and S. Saito, "Digital Filtering Techniques for Speech Analysis and Synthesis," *7th Int. Cong. on Acoustics,* Budapest, Paper 25 C1, 1971.

35. B. S. Atal and S. L. Hanauer, "Speech Analysis and Synthesis by Linear Prediction of the Speech Wave," *J. Acoust. Soc. Am.,* Vol. 50, No. 2 (Part 2), pp. 637-655, August 1971.

36. J. D. Markel and A. H. Gray, Jr., *Linear Prediction of Speech,* Springer-Verlag, New York, 1976.

37. H. Wakita, "Direct Estimation of the Vocal Tract Shape by Inverse Filtering of Acoustic Speech Waveforms," *IEEE Trans. Audio and Electroacoustics,* Vol. AU-21, No. 5, pp. 417-427, October 1973.

38. B. Gold and L. R. Rabiner, "Analysis of Digital and Analog Formant Synthesizers," *IEEE Trans. Audio and Electroacoustics,* Vol. AU-16, pp. 81-94, March 1968.

39. A. E. Rosenberg, "Effect of Glottal Pulse Shape on the Quality of Natural Vowels," *J. Acoust. Soc. Am.,* Vol. 49, No. 2, pp. 583-590, February 1971.

40. L. R. Rabiner, "Digital Formant Synthesizer for Speech Synthesis Studies," *J. Acoust. Soc. Am.,* Vol. 43, No. 4, pp. 822-828, April 1968.

41. G. Winham and K. Steiglitz, "Input Generators for Digital Sound Synthesis," *J. Acoust. Soc. Am.,* Vol. 47, No. 2, pp. 665-666, February 1970.

PROBLEMS

3.1 The waveform plot of Fig. P3.1 shows a 500 msec section (100 msec/line) of a speech waveform.

    (a)   Indicate the regions of voiced speech, unvoiced speech, and silence (background noise).

    (b)   For the voiced regions estimate the pitch period on a period-by-period basis and plot the pitch period versus time for this section of speech. (Let the period be indicated as zero during unvoiced and silence intervals.)

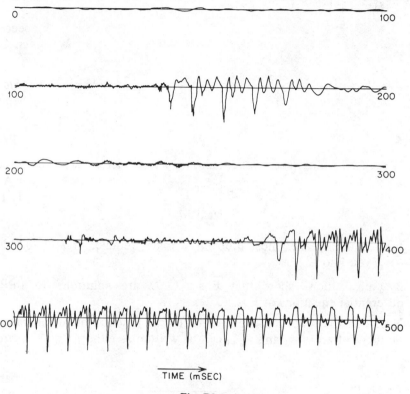

Fig. P3.1

3.2 The waveform plot of Fig. P3.2 is for the word "cattle." Note that each line of the plot corresponds to 100 msec of the signal.

    (a)   Indicate the boundaries between the phonemes; i.e. give the times corresponding to the boundaries /c/a/tt/le/.

    (b)   Indicate the point where the voice pitch frequency is (i) the highest; and (ii) the lowest. What are the approximate pitch frequencies at these points?

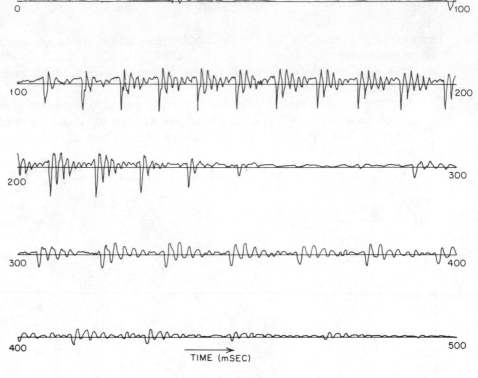

Fig. P3.2

(c) Is the speaker most probably a male, female, or a child? How do you know?

3.3 By substitution, show that Eqs. (3.3) are solutions to the partial differential equations of Eqs. (3.2).

3.4 Note that the reflection coefficients for the junction of two lossless acoustic tubes of areas A_k and A_{k+1} can be written as either

$$r_k = \frac{\dfrac{A_{k+1}}{A_k} - 1}{\dfrac{A_{k+1}}{A_k} + 1}$$

or

$$r_k = \frac{1 - \dfrac{A_k}{A_{k+1}}}{1 + \dfrac{A_k}{A_{k+1}}}$$

Show that since both A_k and A_{k+1} are positive,

$$-1 \leqslant r_k \leqslant 1$$

110

3.5 In determining the effect of the radiation load termination on a lossless tube model, it was assumed that Z_L was real and constant. A more realistic model is given by Eq. (3.29b).

(a) Beginning with the boundary condition

$$P_N(l_N, \Omega) = Z_L \cdot U_N(l_N, \Omega)$$

find a relation between the Fourier transforms of $u_N^-(t+\tau_N)$ and $u_N^+(t-\tau_N)$.

(b) From the frequency domain relation found in (a) and Eq. (3.29b), show that $u_N^-(t+\tau_N)$ and $u_N^+(t-\tau_N)$ satisfy the ordinary differential equation

$$L_r\left[R_r + \frac{\rho c}{A_N}\right] \frac{du_N^-(t+\tau_N)}{dt} + \frac{\rho c}{A_N} R_r u_N^-(t+\tau_N)$$

$$= L_r\left[R_r - \frac{\rho c}{A_N}\right] \frac{du_N^+(t-\tau_N)}{dt} - \frac{\rho c}{A_N} R_r u_N^+(t-\tau_N)$$

3.6 By substitution of Eq. (3.50) into Eq. (3.49), show that Eq. (3.48) and Eq. (3.49) are equivalent.

3.7 Consider the two-tube model of Fig. 3.37. Write the frequency domain equations for this model and show that the transfer function between the input and output volume velocities is given by Eq. (3.51).

3.8 Consider an ideal lossless tube model for the production of vowels consisting of 2 sections as shown in Fig. P3.8. Assume that the terminations at the glottis and lips are completely lossless. For the above conditions the system function of the lossless tube model will be obtained from Eq. (3.52) by substituting $r_G = r_L = 1$ and

$$r_1 = \frac{A_2 - A_1}{A_2 + A_1}.$$

(a) Show that the poles of the system are on the $j\Omega$ axis and are located at values of Ω satisfying the equations

$$\cos \Omega (\tau_1 + \tau_2) + r_1 \cos \Omega (\tau_2 - \tau_1) = 0$$

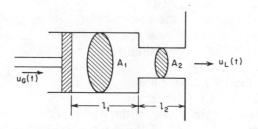

Fig. P3.8

111

or equivalently

$$\frac{A_1}{A_2} \tan (\Omega \tau_2) = \cot (\Omega \tau_1)$$

where $\tau_1 = l_1/c$, $\tau_2 = l_2/c$, and c is the velocity of sound.

(b) The values of Ω that satisfy the equations derived in (a) are the formant frequencies of the lossless tube model. By judicious choice of the parameters l_1, l_2, A_1, and A_2 we can approximate the vocal tract configurations of vowels, and by solving the above equations obtain the formant frequencies for the model. The following table gives parameters for several vowel configurations [2]. Solve for the formant frequencies for each case. (Note that the nonlinear equations must be solved graphically or iteratively.) Use $c = 35000$ cm/sec.

Vowel	l_1	A_1	l_2	A_2
/i/	9 cm	8 cm²	6 cm	1 cm²
/ae/	4 cm	1 cm²	13 cm	8 cm²
/a/	9 cm	1 cm²	8 cm	7 cm²
/ʌ/	17 cm	6 cm²	0	6 cm²

3.9 By substituting the appropriate matrices $\hat{Q}_1$ and $\hat{Q}_2$ into Eq. (3.75) show that the transfer function of a two-tube discrete-time vocal tract model is given by Eq. (3.77).

3.10 Show that if $|a| < 1$,

$$1 - az^{-1} = \frac{1}{\displaystyle\sum_{n=0}^{\infty} a^n z^{-n}}$$

and thus, that a zero can be approximated as closely as desired by multiple poles.

3.11 The transfer function of a digital formant resonator is of the form

$$V_k(z) = \frac{1 - 2|z_k|\cos \theta_k + |z_k|^2}{1 - 2|z_k|\cos \theta_k z^{-1} + |z_k|^2 z^{-2}}$$

where $|z_k| = e^{-\sigma_k T}$ and $\theta_k = 2\pi F_k T$.

(a) Plot the locations of the poles of $V_k(z)$ in the z-plane. Also plot the corresponding analog poles in the s-plane.

(b) Write the difference equation relating the output, $y_k(n)$, of $V_k(z)$ to its input, $x_k(n)$.

(c) Draw a digital network implementation of the digital formant network with three multipliers.

(d) By rearranging the terms in the difference equation obtained in (b), draw a digital network implementation of the digital formant network that only requires two multiplications.

3.12 Consider the system function for a discrete-time vocal tract model

$$V(z) = \frac{G}{\displaystyle\prod_{k=1}^{N} (1 - z_k z^{-1})}$$

(a) Show that $V(z)$ can be expressed as the partial-fraction expansion

$$V(z) = \sum_{k=1}^{M} \left[\frac{G_k}{1 - z_k z^{-1}} + \frac{G_k^*}{1 - z_k^* z^{-1}} \right]$$

where M is the largest integer contained in $(N+1)/2$, and it is assumed that all the poles of $V(z)$ are complex. Give an expression for the G_k's in the above expression.

(b) Combine terms in the above partial fraction expansion to show that

$$V(z) = \sum_{k=1}^{M} \frac{B_k - C_k z^{-1}}{1 - 2|z_k| \cos \theta_k z^{-1} + |z_k|^2 z^{-2}}$$

where $z_k = |z_k| e^{j\theta_k}$. Give expressions for B_k and C_k in terms of G_k and z_k. This expression is the *parallel form* representation of $V(z)$.

(c) Draw the digital network diagram for the parallel form implementation of $V(z)$ for $M = 3$.

(d) For a given all-pole system function $V(z)$ which implementation would require the most multiplications — the parallel form or the cascade form as suggested in Problem 3.11?

3.13 The relationship between pressure and volume velocity at the lips is given by

$$P(l,s) = Z_L(s) U(l,s)$$

where $P(l,s)$ and $U(l,s)$ are the Laplace transforms of $p(l,t)$ and $u(l,t)$ respectively, and

$$Z_L(s) = \frac{s R_r L_r}{R_r + s L_r}$$

where

$$R_r = \frac{128}{9\pi^2} \quad \text{and} \quad L_r = \frac{8a}{3\pi c}$$

and c is the velocity of sound and a is the radius of the lip opening. In a discrete-time model, we desire a corresponding relationship of the form (Eq. (3.97))

$$P_L(z) = R(z) U_L(z)$$

where $P_L(z)$ and $U_L(z)$ are z-transforms of $p_L(n)$ and $u_L(n)$, the sampled versions of the bandlimited pressure and volume velocity.

113

One approach to obtaining $R(z)$ is to use the bilinear transformation, [33] i.e.,

$$R(z) = Z_L(s)\Big|_{s = \frac{2}{T}\left[\frac{1-z^{-1}}{1+z^{-1}}\right]}$$

(a) For $Z_L(s)$ as given above determine $R(z)$.

(b) Write the corresponding difference equation that relates $p_L(n)$ and $u_L(n)$.

(c) Give the locations of the pole and zero of $R(z)$.

(d) If $c = 35000$ cm/sec, $T = 10^{-4}$ sec^{-1}, and 0.5 cm $< a < 1.3$ cm, what is the range of pole values?

(e) A simple approximation to $R(z)$ obtained above is obtained by neglecting the pole; i.e.,

$$\hat{R}(z) = R_0(1-z^{-1})$$

For $a = 1$ cm and $T = 10^{-4}$ find R_0 such that $\hat{R}(-1) = Z_L(\infty) = R(-1)$.

(f) Sketch the frequency responses $Z_L(\Omega)$, $R(e^{j\Omega T})$, and $\hat{R}(e^{j\Omega T})$ as a function of Ω for $a = 1$ cm and $T = 10^{-4}$ for $0 \leqslant \Omega \leqslant \pi/T$.

3.14 A simple approximate model for a glottal pulse is given in Fig. P3.14a.

(a) Find the z-transform, $G_1(z)$, of the above sequence. (Hint: Note that $g_1(n)$ can be expressed as the convolution of the sequence

$$p(n) = 1 \quad 0 \leqslant n \leqslant N - 1$$
$$= 0 \quad otherwise$$

with itself).

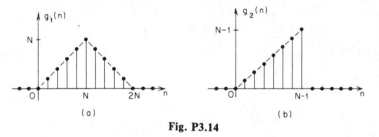

(a)   (b)

Fig. P3.14

(b) Plot the poles and zeros of $G_1(z)$ in the z-plane for $N = 10$.

(c) Sketch the magnitude of the Fourier transform of $g_1(n)$ as a function of ω.

Now consider the glottal pulse model $g_2(n)$ as given in Fig. P3.14b.

(d) Show that the z-transform, $G_2(z)$, is given by

$$G_2(z) = z^{-1} \sum_{n=0}^{N-2} (n+1)z^{-n}$$

$$= z^{-1} \left[\frac{1 - Nz^{-(N-1)} + (N-1)z^{-N}}{(1-z^{-1})^2}\right]$$

(Hint: Use the fact that the z-transform of $nx(n)$ is $-z \, \dfrac{dX(z)}{dz}$.)

(e) Show that in general $G_2(z)$ must have at least one zero outside the unit circle. Find the zeros of $G_2(z)$ for $N = 4$.

3.15 A commonly used approximation to the glottal pulse is

$$
\begin{aligned}
g(n) &= na^n & n \geqslant 0 \\
&= 0 & n < 0
\end{aligned}
$$

(a) Find the z-transform of $g(n)$.
(b) Sketch the Fourier transform, $G(e^{j\omega})$, as a function of ω.
(c) Show how a should be chosen so that

$$
20 \log_{10}|G(e^{j0})| - 20 \log_{10}|G(e^{j\pi})| = 60 \text{ dB}.
$$

4

Time-Domain Methods
for Speech Processing

4.0 Introduction

We have reviewed the most important digital signal processing techniques in Chapter 2, and we discussed the properties of the speech signal in some detail in Chapter 3. We are now ready to begin to see how digital signal processing methods can be applied to speech signals.

Our goal in processing the speech signal is to obtain a more convenient or more useful representation of the information carried by the speech signal. The required precision of this representation is dictated by the particular information in the speech signal that is to be preserved or, in some cases, made more prominent. For example, the purpose of the digital processing may be to facilitate the determination of whether a particular waveform corresponds to speech or not. In a similar but somewhat more complicated vein, we may wish to make a 3-way classification as to whether a section of the signal is voiced speech, unvoiced speech, or silence (background noise). In such cases, a representation which discards "irrelevant" information and places the desired features clearly in evidence is to be preferred over a more detailed representation that retains all the inherent information. Other situations (e.g., digital transmission) may require the most accurate representation of the speech signal that can be obtained with a given set of constraints.

In this chapter we shall be interested in a set of processing techniques that are reasonably termed *time-domain* methods. By this we mean simply that the processing methods involve the waveform of the speech signal directly. This is in contrast, for example, to the techniques described in Chapters 6-8 which we classify as frequency-domain methods since they involve (either explicitly or implicitly) some form of spectrum representation.[1]

Some examples of representations of the speech signal in terms of time-domain measurements include average zero-crossing rate, energy, and the auto-correlation function. Such representations are attractive because the required digital processing is very simple to implement, and, in spite of this simplicity, the resulting representations provide a useful basis for estimating important features of the speech signal.

We shall begin this chapter by presenting a general framework for discussing time-domain processing techniques. Several important methods will be discussed as examples of the general time-domain representation and processing approach. Finally we shall discuss a number of schemes for estimating features of the speech waveform such as voiced/unvoiced classification, pitch, and intensity from the time-domain representations. Many more examples exist than can be covered here. Our purpose, however, is to show how time-domain representations can be used effectively in speech processing — not to provide an exhaustive survey of such applications.

4.1 Time-Dependent Processing of Speech

A sequence of samples (8000 samples/sec) representing a typical speech signal is shown in Figure 4.1. It is evident from this figure that the properties of the speech signal change with time. For example, the excitation changes between voiced and unvoiced speech, there is significant variation in the peak amplitude of the signal, and there is considerable variation of fundamental frequency within voiced regions. The fact that these variations are so evident in a waveform plot suggests that simple time-domain processing techniques should be capable of providing useful representations of such signal features as intensity, excitation mode, pitch, and possibly even vocal tract parameters such as formant frequencies.

The underlying assumption in most speech processing schemes is that the properties of the speech signal change relatively slowly with time. This assumption leads to a variety of "short-time" processing methods in which short segments of the speech signal are isolated and processed as if they were short segments from a sustained sound with fixed properties. This is repeated (usually periodically) as often as desired. Often these short segments, which are sometimes called *analysis frames,* overlap one another. The result of the processing on each frame may be either a single number, or a set of numbers. Therefore,

---

[1]It should be pointed out that in all cases, we shall assume that the speech signal has been band-limited and sampled at least at the Nyquist rate. Further, we shall assume that the resulting samples are finely quantized so that quantization error is negligible. (See Chapter 5 for a discussion of the effects of quantization.)

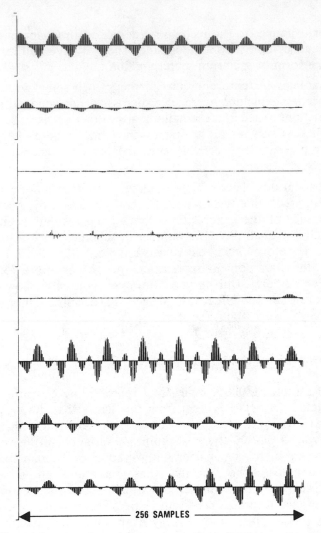

Fig. 4.1 Samples of a typical speech waveform (8 kHz sampling rate).

such processing produces a new time-dependent sequence which can serve as a representation of the speech signal.

Most of the short-time processing techniques that we shall discuss in this chapter, as well as the short-time Fourier representation of Chapter 6, can be represented mathematically in the form

$$Q_n = \sum_{m=-\infty}^{\infty} T[x(m)]\, w(n-m) \tag{4.1}$$

The speech signal (possibly after linear filtering to isolate a desired frequency band) is subjected to a transformation, $T[\ \ ]$, which may be either linear or nonlinear, and which may depend upon some adjustable parameter or set of parameters. The resulting sequence is then multiplied by a window sequence positioned at a time corresponding to sample index n. The product is then

118

summed over all nonzero values. Usually the window sequence will be of finite duration, although this is not always the case. The values Q_n are therefore a sequence of local weighted average values of the sequence $T[x(m)]$.

The short-time energy of a signal is a simple example which illustrates the ideas discussed above. The <u>energy</u> of a discrete-time signal is defined as

$$E = \sum_{m=-\infty}^{\infty} x^2(m) \tag{4.2}$$

Such a quantity has little meaning or utility for speech since it gives little information about the time-dependent properties of the speech signal. A simple definition of the <u>short-time</u> energy is

$$E_n = \sum_{m=n-N+1}^{n} x^2(m) \tag{4.3}$$

That is, the short-time energy at sample n is simply the sum of squares of the N samples $n - N + 1$ through n. In terms of our general expression, Eq. (4.1), the operation $T[\quad]$ is simply the square, and

$$w(n) = 1 \quad 0 \leqslant n \leqslant N - 1$$
$$= 0 \quad otherwise \tag{4.4}$$

Figure 4.2 depicts the computation of the short-time energy sequence. Note that the window literally slides along the sequence of squared values (in general, $T[x(m)]$) selecting the interval to be involved in the computation.

We shall discuss the short-time energy of speech in more detail in the next section, but first we shall point out an important feature of the general expression, Eq. (4.1). It can be seen that Eq. (4.1) is <u>exactly in the form</u> of a <u>discrete convolution</u> of the window, $w(n)$, with the sequence, $T[x(n)]$. Thus, Q_n can be interpreted as the output of a linear time-invariant system with impulse response, $\underline{h(n) = w(n)}$.[2] This is depicted in the block diagram of Fig.

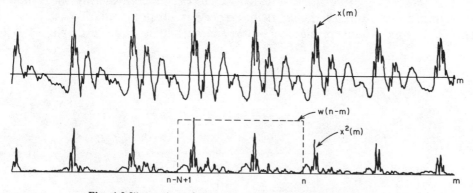

Fig. 4.2 Illustration of the computation of short-time energy.

[2]Note that we have used a subscript notation for the values of the short-time representation, i.e. the output of the filter. Although this may at first be confusing, it should not cause any great difficulty and in later definitions it will lead to clearer notation.

Fig. 4.3 General representation of the short-time analysis principle.

4.3. The importance of this viewpoint will become clear in the remainder of this chapter and in Chapter 6 where we discuss several different representations that are of the form of Eq. (4.1).

4.2 Short-Time Energy and Average Magnitude

We have observed that the amplitude of the speech signal varies appreciably with time. In particular, the amplitude of unvoiced segments is generally much lower than the amplitude of voiced segments. The short-time energy of the speech signal provides a convenient representation that reflects these amplitude variations. In general, we can define the short-time energy as

$$E_n = \sum_{m=-\infty}^{\infty} [x(m)\,w(n-m)]^2 \tag{4.5}$$

This expression can be written as

$$E_n = \sum_{m=-\infty}^{\infty} x^2(m) \cdot h(n-m) \tag{4.6}$$

where

$$h(n) = w^2(n) \tag{4.7}$$

Equation (4.6) can thus be interpreted as depicted in Fig. 4.4a. That is, the signal $x^2(n)$ is filtered by a linear filter with impulse response $h(n)$ as given by Eq. (4.7)

The choice of the impulse response, $h(n)$, or equivalently the window, determines the nature of the short-time energy representation. To see how the choice of window affects the short-time energy, let us observe that if $h(n)$ in

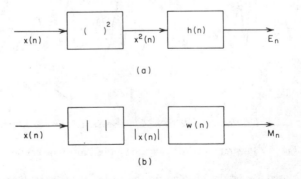

Fig. 4.4 Block diagram representation of (a) the short-time energy; and (b) the short-time average magnitude.

120

ie. N big

Eq. (4.6) were very long, and of constant amplitude, E_n would change very little with time. Such a window would be the equivalent of a very narrowband lowpass filter. Clearly what is desired is some lowpass filtering but not so much that the output is constant; i.e., we want the short-time energy to reflect the amplitude variations of the speech signal. Thus, we encounter for the first time a conflict that will arise repeatedly in the study of short-time representations of speech signals. That is, we wish to have a short duration window (impulse response) to be responsive to rapid amplitude changes, but a window that is too short will not provide sufficient averaging to produce a smooth energy function.

The effect of the window on the time-dependent energy representation can be illustrated by discussing the properties of two representative windows, i.e., the rectangular window

$$h(n) = 1 \quad 0 \leqslant n \leqslant N - 1$$
$$= 0 \quad otherwise \tag{4.8}$$

and the Hamming window

$$h(n) = 0.54 - 0.46 \cos(2\pi n/(N-1)), \quad 0 \leqslant n \leqslant N - 1$$
$$= 0 \quad otherwise \tag{4.9}$$

The rectangular window, as we have seen in Eq. (4.3), corresponds to applying equal weight to all the samples in the interval $(n-N+1)$ to n. The frequency response of a rectangular window (impulse response given by Eq. (4.8)) is easily shown to be (see Problem 4.1)

$$H(e^{j\Omega T}) = \frac{\sin(\Omega NT/2)}{\sin(\Omega T/2)} e^{-j\Omega T(N-1)/2} \tag{4.10}$$

The log magnitude of this response is shown in Fig. 4.5a for a 51 sample window $(N = 51)$. Note that the first zero of Eq. (4.10) occurs at analog frequency

$$F = F_s/N \tag{4.11}$$

where $F_s = 1/T$ is the sampling frequency. This is nominally the cutoff frequency of the lowpass filter corresponding to the rectangular window. The frequency response of a 51 point Hamming window is shown in Fig. 4.5b. It can be seen that the bandwidth of the Hamming window is about twice the bandwidth of a rectangular window of the same length. It is also clear that the Hamming window gives much greater attenuation outside the passband than the comparable rectangular window. The attenuation of both these windows is essentially independent of the window duration. Thus, increasing the length, N, simply decreases the bandwidth.[3] If N is too small, i.e., on the order of a pitch period or less, E_n will fluctuate very rapidly depending on exact details of the waveform. If N is too large, i.e., on the order of several pitch periods, E_n

---

[3]A detailed discussion of the properties of windows is not required for the short-time representations of this chapter. Further discussion is given in Chapter 6.

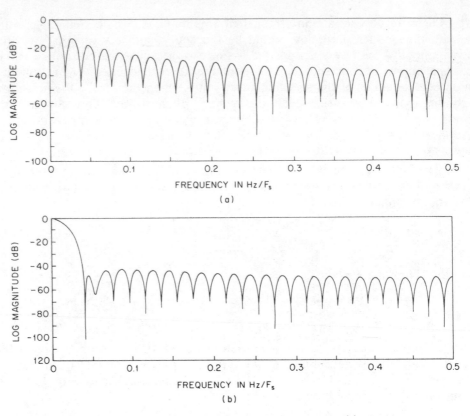

Fig. 4.5 Fourier transform of (a) rectangular window; (b) Hamming window.

will change very slowly and thus will not adequately reflect the changing properties of the speech signal. Unfortunately this implies that no single value of N is entirely satisfactory because the duration of a pitch period varies from about 20 samples (at a 10 kHz sampling rate) for a high pitch female or a child, up to 250 samples for a very low pitch male. With these shortcomings in mind, a suitable practical choice for N is on the order of 100-200 for a 10 kHz sampling rate (i.e., 10-20 msec duration).

Figures 4.6 and 4.7 show the effects of varying the duration of the window (for the rectangular and Hamming windows, respectively) on the energy computation for the utterance /What, she said/ spoken by a male speaker. It is readily seen that as N increases, the energy becomes smoother for both windows.

The major significance of E_n is that it provides a basis for distinguishing voiced speech segments from unvoiced speech segments. As can be seen in Figs. 4.6 and 4.7, the values of E_n for the unvoiced segments are significantly smaller than for voiced segments. The energy function can also be used to locate approximately the time at which voiced speech becomes unvoiced, and vice versa, and, for very high quality speech (high signal-to-noise ratio), the energy can be used to distinguish speech from silence.

Sampling freq.
10 kHz ≥ 100 k̸s
∴ Pitch Period
is ∼ 2 to 2.5 ms.
ie 20 to 250 samples

122

One difficulty with the short-time energy function as defined by Eq. (4.6) is that it is very sensitive to large signal levels (since they enter the computation in Eq. (4.6) as a square), thereby emphasizing large sample-to-sample variations in $x(n)$. A simple way of alleviating this problem is to define an "average magnitude" function

$$M_n = \sum_{m=-\infty}^{\infty} |x(m)| w(n-m) \qquad (4.12)$$

where the weighted sum of absolute values of the signal is computed instead of the sum of squares. Figure 4.4b shows how Eq. (4.12) can be implemented as a linear filtering operation on $|x(n)|$. Note that a simplification in arithmetic is achieved by eliminating the squaring operation.

Figures 4.8 and 4.9 show average magnitude plots corresponding to Figs. 4.6 and 4.7. The differences are particularly noticeable in the unvoiced regions. For the average magnitude computation of Eq. (4.12), the dynamic range (ratio of maximum to minimum) is approximately the square root of the dynamic range for the standard energy computation. Thus the differences in level between voiced and unvoiced regions are not as pronounced as for the short-time energy.

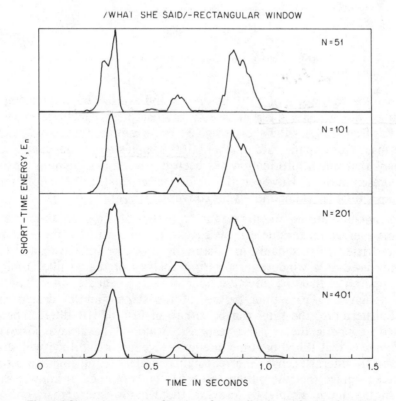

/WHAI SHE SAID/-RECTANGULAR WINDOW

N = 51

N = 101

N = 201

N = 401

SHORT-TIME ENERGY, E_n

TIME IN SECONDS

Fig. 4.6 Short-time energy functions for rectangular windows of various lengths.

123

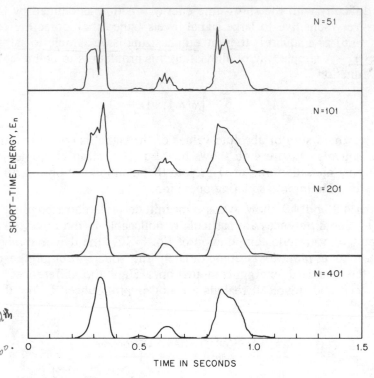

SHORT–TIME ENERGY, E_n

N = 51

N = 101

N = 201

N = 401

TIME IN SECONDS

Fig. 4.7 Short-time energy functions for Hamming windows of various lengths. $\propto F_s/N$

[handwritten margin notes: 20 ms ⇒ N=200; 10 kHz ÷ 200 = 50 Hz bandwidth; ∴ sample at twice that, 100.]

Since the <u>bandwidth</u> of both the energy and average magnitude function is <u>just that of the lowpass filter</u>, it is evident that these functions need not be sampled as frequently as the speech signal. For example, for a window of duration 20 msec, a sampling rate of about <u>100 samples/sec</u> is adequate. Clearly, this means that much information has been discarded in obtaining these short-time representations. However, it is also clear that information regarding speech amplitude is retained in a very convenient form.

To conclude our comments on the properties of the short-time energy and short-time average magnitude it is instructive to point out that the window need not be restricted to rectangular or Hamming form, or indeed to any function commonly used as a window in spectrum analysis or digital filter design. All that is required is that the effective filter provide adequate smoothing. Thus, we can design a lowpass filter by any of the standard filter design methods [1,2]. Furthermore, the filter can be either an FIR or IIR filter. There is an advantage in having the impulse response (window) be always positive since this guarantees that the short-time energy or average magnitude will always be positive. FIR filters (such as the rectangular or Hamming impulse responses) have the advantage that the output can easily be computed at a lower sampling rate than the input simply by moving the window more than one sample between computations. For example, if the speech signal is sampled at 10000

124

samples/sec, and a window of duration 20 msec (200 samples) is used, the short-time energy can be computed at a sampling rate of about 100 samples/sec, or once every 100 samples at the input sampling rate.

It is not necessary to use a finite length window. Although this may seem contradictory, it is possible to implement the filtering implied by an infinite length window if its z-transform is a rational function. A simple example is a window of the form

$$h(n) = a^n \quad n \geqslant 0$$
$$= 0 \quad n < 0 \tag{4.13}$$

A value of $0 < a < 1$ gives a window whose effective duration can be adjusted as desired. The corresponding z-transform of the window is

$$H(z) = \frac{1}{1 - az^{-1}} \quad |z| > |a| \tag{4.14}$$

from which it is easily seen that the frequency response, $H(e^{j\Omega T})$, has the desired lowpass property. Such a filter can be implemented by a simple

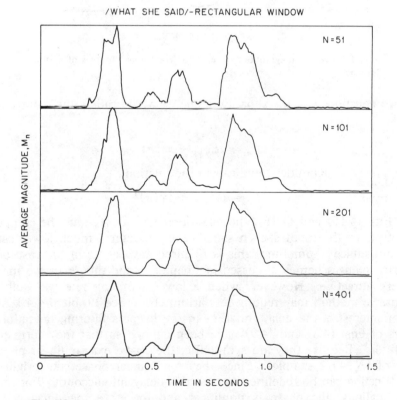

/WHAT SHE SAID/-RECTANGULAR WINDOW

N = 51

N = 101

N = 201

N = 401

AVERAGE MAGNITUDE, M_n

TIME IN SECONDS

Fig. 4.8 Average magnitude functions for rectangular windows of various lengths.

125

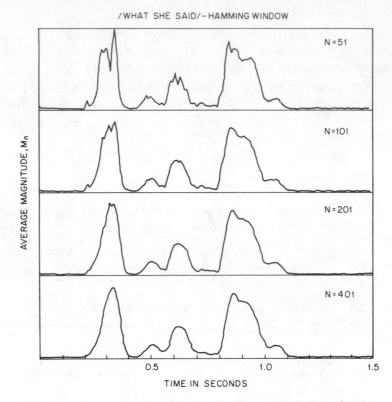

Fig. 4.9 Average magnitude functions for Hamming windows of various lengths.

difference equation; i.e., the short-time energy would satisfy the recurrence formula

$$E_n = aE_{n-1} + x^2(n) \tag{4.15}$$

and the average magnitude would satisfy the relation

$$M_n = aM_{n-1} + |x(n)| \tag{4.16}$$

To use Eqs. (4.15) and (4.16), the quantities E_n and M_n must be computed at each sample of the input speech signal, even though a much lower sampling rate would suffice. Sometimes this is required anyway, as in the case of some waveform coding schemes discussed in Chapter 5, and this recursive method is then very attractive. However, when a lower sampling rate will suffice, the nonrecursive method may require less arithmetic. (See Problem 4.4.) Another factor of interest is the delay inherent in the lowpass filtering operation. The windows of Eqs. (4.8) and (4.9) have been defined so that they correspond to causal filters. Because they are symmetric, they have exactly linear phase with a delay of $(N-1)/2$ samples. Since they have linear phase, the origin of the energy function can be redefined to take this delay into account. For recursive implementations, the phase is nonlinear and therefore the delay cannot be exactly compensated.

126

4.3 Short-Time Average Zero-Crossing Rate

In the context of discrete-time signals, a zero-crossing is said to occur if successive samples have different algebraic signs. The rate at which zero crossings occur is a simple measure of the frequency content of a signal. This is particularly true of narrowband signals. For example, a sinusoidal signal of frequency F_0, sampled at a rate F_s, has F_s/F_0 samples per cycle of the sine wave. Each cycle has two zero crossings so that the "long-time average rate" of zero-crossings is

$$Z = 2F_0/F_s \quad \text{crossings/sample} \tag{4.17}$$

Thus, the "average zero-crossing rate" gives a reasonable way to estimate the frequency of a sine wave.

Speech signals are broadband signals and the interpretation of average zero-crossing rate is therefore much less precise. However, rough estimates of spectral properties can be obtained using a representation based on the "short-time average zero-crossing rate." Before discussing the interpretation of zero-crossing rate for speech, let us first define and discuss the required computations. An appropriate definition is

$$Z_n = \sum_{m=-\infty}^{\infty} |sgn[x(m)] - sgn[x(m-1)]| \, w(n-m) \tag{4.18}$$

where

$$sgn[x(n)] = 1 \quad x(n) \geqslant 0$$
$$= -1 \quad x(n) < 0 \tag{4.19}$$

and

$$w(n) = \frac{1}{2N} \quad 0 \leqslant n \leqslant N-1$$
$$= 0 \quad \textit{otherwise} \tag{4.20}$$

The operations involved in Eq. (4.18) are represented in block diagram form in Fig. 4.10. This representation shows that the short-time average zero-crossing rate has the same general properties as the short-time energy and the short-time average magnitude. However, Eq. (4.18) and Fig. 4.10 make the computation of Z_n appear more complex than it really is. All that is required is to check samples in pairs to determine where the zero-crossings occur and then the average is computed over N consecutive samples. (The division by N is

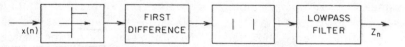

Fig. 4.10 Block diagram representation of short-time average zero-crossings.

127

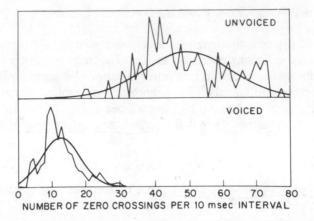

NUMBER OF ZERO CROSSINGS PER 10 msec INTERVAL

Fig. 4.11 Distribution of zero-crossings for unvoiced and voiced speech.

obviously unnecessary as well.) As before, this can be a weighted average, and if a symmetrical finite length window is used, the delay can be exactly compensated. Alternatively a recursive implementation similar to Eqs. (4.15) and (4.16) can be obtained. (See Problem 4.5.)

Now let us see how the short-time average zero-crossing rate applies to speech signals. The model for speech production suggests that the energy of voiced speech is concentrated below about 3 kHz because of the spectrum fall-off introduced by the glottal wave, whereas for unvoiced speech, most of the energy is found at higher frequencies. Since high frequencies imply high zero-crossing rates, and low frequencies imply low zero-crossing rates, there is a strong correlation between zero-crossing rate and energy distribution with frequency. A reasonable generalization is that if the zero-crossing rate is high, the speech signal is unvoiced, while if the zero-crossing rate is low, the speech signal is voiced. This, however, is a very imprecise statement because we have not said what is high and what is low, and, of course, it really is not possible to be precise. Figure 4.11 shows a histogram of average zero-crossing rates (averaged over 10 msec) for both voiced and unvoiced speech. Note that a Gaussian curve provides a reasonably good fit to each distribution. The mean short-time average zero-crossing rate is 49 per 10 msec for unvoiced and 14 per 10 msec for voiced. Clearly the two distributions overlap so that an unequivocal voiced/unvoiced decision is not possible based on short-time average zero-crossing rate alone. Nevertheless, such a representation is quite useful in making this distinction.

Some examples of average zero-crossing rate measurements are shown in Fig. 4.12. In these examples, the duration of the averaging window is 15 msec (150 samples at 10 kHz sampling rate) and the output is computed 100 times/sec (window moved in steps of 100 samples). Note that just as in the case of short-time energy and average magnitude, the short-time average zero-crossing rate can be sampled at a very low rate. Although the zero-crossing rate varies considerably, the voiced and unvoiced regions are quite prominent in Fig. 4.12.

128

There are a number of practical considerations in implementing a representation based on the short-time average zero-crossing rate. Although the basic algorithm for computing a zero-crossing requires only a comparison of signs of pairs of successive samples, special care must be taken in the sampling process. Clearly, the zero-crossing rate is strongly affected by dc offset in the analog-to-digital converter, 60 Hz hum in the signal, and any noise that may be present in the digitizing system. Therefore, extreme care must be taken in the analog processing prior to sampling to minimize these effects. For example, it is often preferable to use a bandpass filter(rather than a lowpass filter) as the anti-aliasing filter to eliminate dc and 60 Hz components in the signal. Additional considerations in the zero-crossing measurement are the sampling period, T, and the averaging interval, N. The sampling period determines the time (and frequency) resolution of the zero-crossing representation; i.e., fine resolution requires a high sampling rate. However, to preserve the zero-crossing information only 1-bit quantization (i.e., preserving the sign of the signal) is all that is required.

Because of the practical limitations, a variety of similar representations have been proposed. All of these variants introduce some feature intended to make the estimate less sensitive to noise, but each has its own set of limitations. Notable among these is the up-crossing representation studied by Baker [3]. This representation is based upon the time intervals between zero-

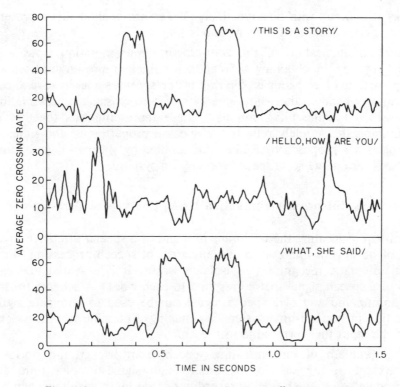

Fig. 4.12 Average zero-crossing rate for three different utterances.

129

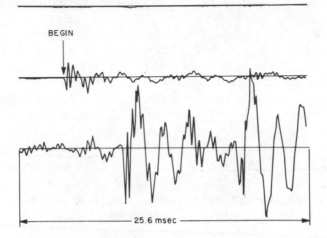

Fig. 4.13 Waveform for the beginning of the utterance /*eight*/. (After Rabiner and Sambur [6].)

crossings that occur with positive slope. Baker has applied this representation in phonetic classification of speech sounds [3].

Another application of the zero-crossing representation is as a simple intermediate step in obtaining a frequency domain representation of speech. The approach involves bandpass filtering of the speech signal in several contiguous frequency bands. Short-time energy and zero-crossing representations are then obtained for the filter outputs. These representations together give a representation that crudely reflects the spectral properties of the signal. Such an approach was proposed by Reddy, and studied by Vicens [4] and Erman [5] as the basis for a large-scale speech recognition system.

4.4 Speech vs. Silence Discrimination Using Energy and Zero-Crossings

The problem of locating the beginning and end of a speech utterance in a background of noise is of importance in many areas of speech processing. In particular, in automatic recognition of isolated words, it is essential to locate the regions of a speech signal that correspond to each word. A scheme for locating the beginning and end of a speech signal can be used to eliminate significant computation in nonreal-time systems by making it possible to process only the parts of the input that correspond to speech.

The problem of discriminating speech from background noise is not trivial, except in the case of extremely high signal-to-noise ratio acoustic environments — e.g., high fidelity recordings made in an anechoic chamber or a soundproof room. For such high signal-to-noise ratio environments, the energy of the lowest level speech sounds (e.g., weak fricatives) exceeds the background noise energy, and thus a simple energy measurement suffices. However, such ideal recording conditions are not practical for most applications.

The algorithm to be discussed in this section is based on two simple time-domain measurements — energy, and zero-crossing rate. Several simple examples will illustrate some difficulties encountered in locating the beginning and end of a speech utterance. Figure 4.13 shows an example (the beginning of the word /eight/) for which the background noise is easily distinguished from the speech, as denoted in the figure. In this case a radical change in the waveform energy between the background noise and the speech is the cue to the beginning of the utterance. Figure 4.14 shows another example (the beginning of the word /six/) for which it is easy to locate the beginning of the speech. In this case, the frequency content of the speech is radically different from the background noise, as seen by the sharp increase in zero crossing rate of the waveform. It should be noted that, in this case, the speech energy at the beginning of the utterance is comparable to the background noise energy.

Figure 4.15 gives an example of a case in which it is extremely difficult to locate the beginning of the speech signal. This figure shows the waveform for the beginning of the utterance /four/. Since /four/ begins with the weak (low energy) fricative /f/, it is very difficult to precisely identify the beginning point. Although the point marked B in this figure is a good candidate for the begin-

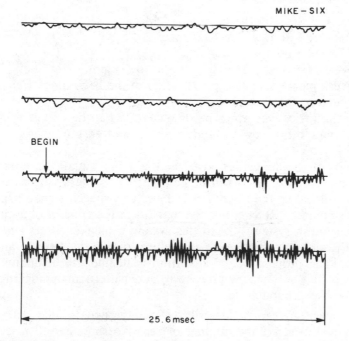

Fig. 4.14 Waveform for the beginning of the utterance /six/. (After Rabiner and Sambur [6].)

131

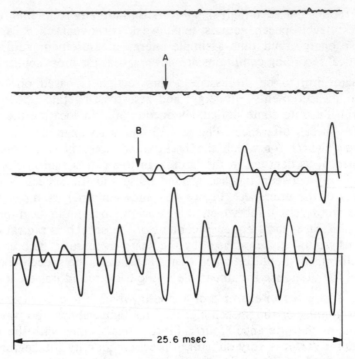

Fig. 4.15 Waveform for the beginning of the utterance /*four*/. (After Rabiner and Sambur [6].)

ning, point A is actually the beginning. In general, it is difficult to locate the beginning and end of an utterance if there are:

1. Weak fricatives (/f/, /th/, /h/) at the beginning or end.
2. Weak plosive bursts (/p/, /t/, /k/) at the beginning or end.
3. Nasals at the end.
4. Voiced fricatives which become devoiced at the end of words.
5. Trailing off of vowel sounds at the end of an utterance.

In spite of the difficulties posed by the above situations, energy and zero-crossing rate representations can be combined to serve as the basis of a useful algorithm for locating the beginning and end of a speech signal. One such algorithm was studied by Rabiner and Sambur [6] in the context of an isolated-word speech recognition system [7]. In this system a speaker utters a word during a prescribed recording interval, and the entire interval is sampled and stored for processing. The purpose of the algorithm is to find the beginning and end of the word so that subsequent processing and pattern matching can ignore the surrounding background noise.

The algorithm can be described by reference to Fig. 4.16. The basic representations used are the number of zero-crossings per 10 msec frame (Eq. (4.18)) and the average magnitude (Eq. (4.12)) computed with a 10 msec window. Both functions are computed for the entire recording interval at a rate of

100 times/sec. It is assumed that the first 100 msec of the interval contains no speech. The mean and standard deviation of the average magnitude and zero-crossing rate are computed for this interval to give a statistical characterization of the background noise. Using this statistical characterization and the maximum average magnitude in the interval, zero-crossing rate and energy thresholds are computed. (Details are given in [6].) The average magnitude profile is searched to find the interval in which it always exceeds a very conservative threshold (ITU in Fig. 4.16). It is assumed that the beginning and ending points lie outside this interval. Then working backwards from the point at which M_n first exceeded the threshold ITU, the point (labelled N_1 in Fig. 4.16) where M_n first falls below a lower threshold ITL is tentatively selected as the beginning point. A similar procedure is followed to find the tentative endpoint N_2. This double threshold procedure ensures that dips in the average magnitude function do not falsely signal the endpoint. At this stage it is reasonably safe to assume that the beginning and ending points are not within the interval N_1 to N_2. The next step is to move backwards from N_1 (forward from N_2) comparing the zero-crossing rate to a threshold (IZCT in Fig. 4.16) determined from the statistics of the zero-crossing rate for the background noise. This is limited to the 25 frames preceding N_1 (following N_2). If the zero-crossing rate exceeds the threshold 3 or more times, the beginning point N_1 is moved back to the first point at which the zero-crossing threshold was exceeded. Otherwise N_1 is defined as the beginning. A similar procedure is followed at the end.

Figure 4.17 shows examples of how the algorithm works on typical isolated words. In this figure there are 8 plots of the average magnitude function for 8 different words (for 2 different speakers). Some of the words were recorded in a noisy computer room (marked Mike) and others were recorded on analog tape (marked Tape) from a soundproof booth. The markers on each plot show the beginning point and ending point of each word, as determined by

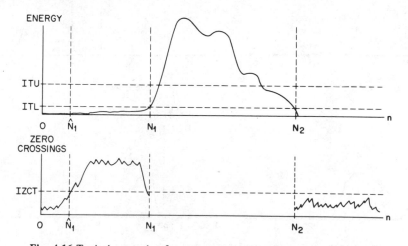

Fig. 4.16 Typical example of average magnitude and zero-crossing measurements for a word with a strong fricative at the beginning. (After Rabiner and Sambur [6].)

133

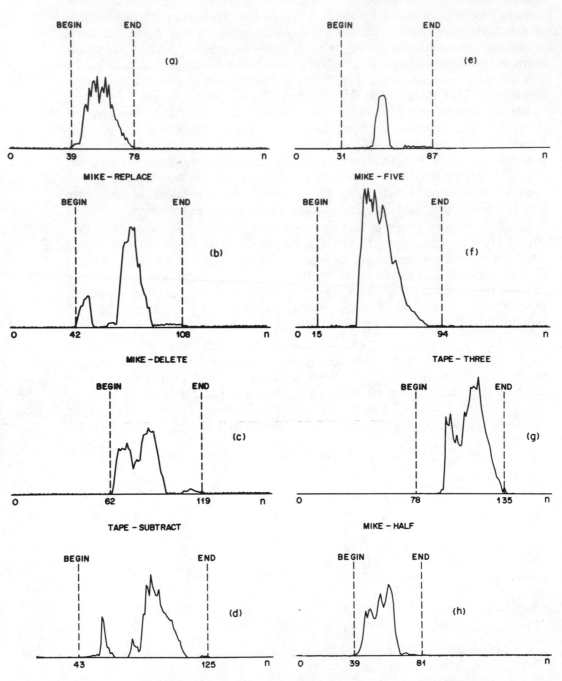

Fig. 4.17 Sequence of average magnitude plots showing how the end-point algorithm performed over a variety of words. (After Rabiner and Sambur [6].)

the automatic algorithm. For the example of Fig. 4.17a (the word nine), the average magnitude thresholds were sufficient to locate the boundary points. For the example of Fig. 4.17b (the word *replace*), the zero-crossing algorithm was used to determine the ending point due to the final fricative /s/. It should be noted that even though the final /s/ has fairly large average magnitude, since the average magnitude thresholds were set conservatively, this criterion was not able to find the actual endpoint of the word. Instead, the zero-crossing algorithm was relied upon in this case. In Fig. 4.17c the final /t/ in the word *delete* was correctly located because of the significant zero-crossing rate over the 70 msec burst when the /t/ was released. Thus, even though the average magnitude and zero crossing rate were very small for about 50 msec in the stop gap, the algorithm was able to correctly identify the endpoint because of the strength of the burst. On the other hand, if the burst had been weak, the ending point would have been located at the beginning of the stop gap.

In Fig. 4.17d there is an example where the average magnitude of the noise was significant in two places prior to the beginning of the word *subtract*, yet the algorithm successfully eliminated these places from consideration because of the low zero-crossing rates. In this example a relatively weak burst in the final /t/ was correctly labelled as the endpoint.

Figures 4.17e-4.17h show examples of words with fricatives at either the beginning or end of the word. In all cases the algorithm was able to correctly place the appropriate endpoint so that a reasonable amount of unvoiced duration was included within the boundaries of the word.

This application of zero-crossings and average magnitude illustrates the utility of these simple representations in a practical setting. These representations are particularly attractive since very little arithmetic is required for their implementation. We shall see ideas similar to those just discussed arising in later discussions in this chapter.

4.5 Pitch Period Estimation Using a Parallel Processing Approach

Pitch period estimation (or equivalently, fundamental frequency estimation) is one of the most important problems in speech processing. Pitch detectors are used in vocoders [8], speaker identification and verification systems [9,10], and aids-to-the handicapped [11]. Because of its importance, many solutions to this problem have been proposed [12-19]. All of the proposed schemes have their limitations, and it is safe to say that no presently available pitch detection scheme can be expected to give perfectly satisfactory results across a wide range of speakers, applications, and operating environments.

In this section we discuss a particular pitch detection scheme first proposed by Gold [13] and later modified by Gold and Rabiner [14]. Our reasons for discussing this particular pitch detector in this chapter are: (1) it has been used successfully in a wide variety of applications, (2) it is based on purely time domain processing, (3) it can be implemented to operate very quickly on a general purpose computer or it can be easily constructed in digital hardware, and (4) it illustrates the use of the basic principle of parallel processing in speech processing.

The basic principles of this scheme are as follows:

1. The speech signal is processed so as to create a number of impulse trains which retain the periodicity of the original signal and discard features which are irrelevant to the pitch detection process.
2. This processing permits very simple pitch detectors to be used to estimate the period of each impulse train.
3. The estimates of several of these simple pitch detectors are logically combined to infer the period of the speech waveform.

The particular scheme proposed by Gold and Rabiner [14] is depicted in Fig. 4.18. The speech waveform is sampled at a rate sufficient to give adequate time resolution; e.g., sampling at 10 kHz allows the period to be determined to within $T = 10^{-4}$ sec. The speech is lowpass filtered with a cutoff of about 900 Hz to produce a relatively smooth waveform. A bandpass filter passing frequencies between 100 Hz and 900 Hz may be necessary to remove 60 Hz noise in some applications. (This filtering can be done either with an analog filter before sampling or with a digital filter after sampling.)

Following the filtering, the "peaks and valleys" (local maxima and minima) are located, and from their locations and amplitudes, several impulse trains (6 in fig. 4.18) are derived from the filtered signal. Each impulse train consists of positive impulses occurring at the location of either the peaks or the valleys. The 6 cases used by Gold and Rabiner [14] are:

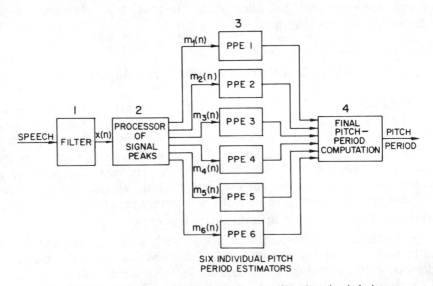

Fig. 4.18 Block diagram of a parallel processing time domain pitch detector.

136

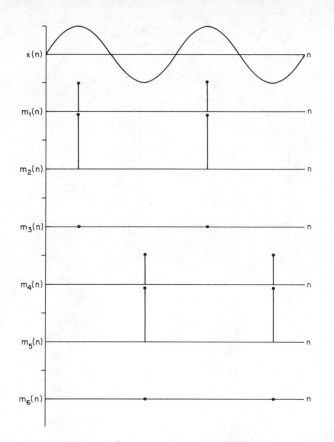

Fig. 4.19 Input (sinusoid) and corresponding impulse trains generated from the peaks and valleys.

1. $m_1(n)$: An impulse equal to the peak amplitude occurs at the location of each peak.
2. $m_2(n)$: An impulse equal to the difference between the peak amplitude and the preceding valley amplitude occurs at each peak.
3. $m_3(n)$: An impulse equal to the difference between the peak amplitude and the preceding peak amplitude occurs at each peak. (If this difference is negative the impulse is set to zero.)
4. $m_4(n)$: An impulse equal to the negative of the amplitude at a valley occurs at each valley.
5. $m_5(n)$: An impulse equal to the negative of the amplitude at a valley plus the amplitude at the preceding peak occurs at each valley.
6. $m_6(n)$: An impulse equal to the negative of the amplitude at a valley plus the amplitude at the preceding local minimum occurs at each valley. (If this difference is negative the impulse is set to zero.)

Figures 4.19 and 4.20 show two examples − a pure sine wave and a weak fundamental plus a strong second harmonic − together with the resulting impulse trains as defined above. Clearly the impulse trains have the same fundamental period as the original input signals, although $m_5(n)$ of Fig. 4.20 is close to

137

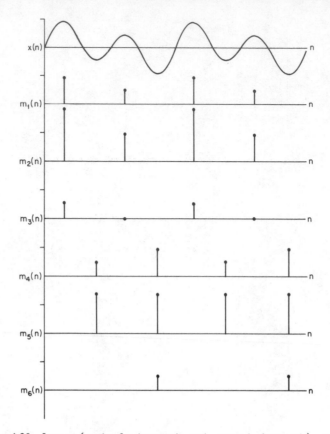

Fig. 4.20 Input (weak fundamental and second harmonic) and corresponding impulse trains generated from the peaks and valleys.

being periodic with half the fundamental period. The purpose of generating these impulse trains is to make it simple to estimate the period on a short-time basis. The operation of the simple pitch period estimators is depicted in Fig. 4.21. Each impulse train is processed by a time varying nonlinear system (called a peak detecting exponential window circuit in [13]). When an impulse of sufficient amplitude is detected in the input, the output is reset to the value of that impulse and then held for a blanking interval, $\tau(n)$ — during which no pulse can be detected. At the end of the blanking interval, the output begins to decay exponentially. When an impulse exceeds the level of the exponentially decaying output, the process is repeated. The rate of decay and the blanking interval are dependent upon the most recent estimates of pitch period [14]. The result is a kind of smoothing of the impulse train, producing a quasi-periodic sequence of pulses as shown in Fig. 4.21. The length of each pulse is an estimate of the pitch period. The pitch period is estimated periodically (e.g., 100 times/sec) by measuring the length of the pulse spanning the sampling interval.

This technique is applied to each of the six impulse trains thereby obtaining six estimates of the pitch period. These six estimates are combined with two of the most recent estimates for each of the six pitch detectors. These estimates are then compared and the value with the most occurrences (within some tolerance) is declared the pitch period at that time. This procedure produces very good estimates of the period of voiced speech. For unvoiced speech there is a distinct lack of consistency among the estimates. When this lack of consistency is detected, the speech is classified as unvoiced. The entire process is repeated periodically to produce an estimate of the pitch period and voiced/unvoiced classification as a function of time.

Although the above description may appear very involved, this scheme for pitch detection can be efficiently implemented either in special purpose hardware or on a general purpose computer. Indeed, near real-time operation (within a factor of 2 times real-time) is possible on present computers.

The performance of this pitch detection scheme is illustrated by Fig. 4.22 which shows the output for a sample of synthetic speech. The advantage of using synthetic speech is that the true pitch periods are known exactly (since they were artificially generated) and thus a measure of the accuracy of the algorithm can be obtained. The disadvantage of synthetic speech is that it is generated according to a simple model, and therefore may not show any of the unusual properties of natural speech. In any case, testing with synthetic speech has shown that most of the time, the method tracks the pitch period to within 2 samples. Furthermore it has been observed that at the initiation of voicing, (i.e., the first 10-30 msec of voicing) the speech is often classified as unvoiced. This result is due to the decision algorithm which requires about 3 pitch periods before a reliable pitch decision can be made — thus a delay of about 2 pitch periods is inherently built into the method. In a recent comparative study of pitch detection algorithms carried out under a wide range of conditions with natural speech this method compared well with other pitch estimation methods that have been proposed [12].

In summary, the details of this particular method are not so important as the basic principles that are introduced. (The details are available in Ref. [14].) First, note that the speech signal was processed to obtain a set of impulse trains which retain only the essential feature of periodicity (or lack of periodicity). Because of this simplification in the structure of the signal, a very simple pitch estimator suffices to produce good estimates of the pitch period. Finally, several estimates are combined to increase the overall reliability of the estimate. Thus, signal processing simplicity is achieved at the expense of increased logical complexity in estimating the desired feature of the speech signal. Because the

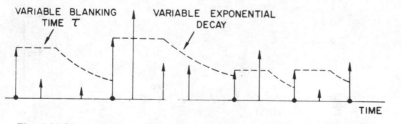

VARIABLE BLANKING TIME τ   VARIABLE EXPONENTIAL DECAY

TIME

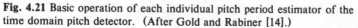

Fig. 4.21 Basic operation of each individual pitch period estimator of the time domain pitch detector. (After Gold and Rabiner [14].)

139

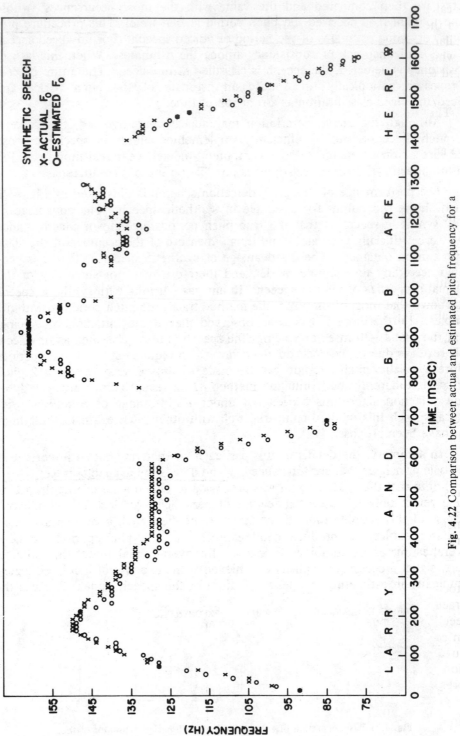

Fig. 4.22 Comparison between actual and estimated pitch frequency for a synthetic utterance. (After Gold and Rabiner [14].)

logical operations are carried out at a much lower rate (e.g., 100 times/sec) than the signal processing, this results in an overall speed-up in processing. A similar approach was used by Barnwell et al. [15] in designing a pitch detector in which the outputs of four simple zero-crossing pitch detectors were combined to produce a reliable estimate of pitch.

4.6 The Short-Time Autocorrelation Function

The autocorrelation function of a discrete-time deterministic signal is defined as

$$\phi(k) = \sum_{m=-\infty}^{\infty} x(m)x(m+k) \qquad (4.21)$$

If the signal is random or periodic the appropriate definition is

$$\phi(k) = \lim_{N\to\infty} \frac{1}{(2N+1)} \sum_{m=-N}^{N} x(m)x(m+k) \qquad (4.22)$$

In either case, the autocorrelation function representation of the signal is a convenient way of displaying certain properties of the signal. For example, if the signal is periodic with period P samples, then it is easily shown that

$$\phi(k) = \phi(k+P) \qquad (4.23)$$

i.e., the autocorrelation function of a periodic signal is also periodic with the same period. Other important properties of the autocorrelation function are:

1. It is an even function; i.e., $\phi(k) = \phi(-k)$.
2. It attains its maximum value at $k = 0$; i.e., $|\phi(k)| \leqslant \phi(0)$ for all k.
3. The quantity $\phi(0)$ is equal to the energy (Eq. (4.2)) for deterministic signals or the average power for random or periodic signals.

Thus, the autocorrelation function contains the energy as a special case. Even more important is the convenient way in which periodicity is displayed. If we consider Eq. (4.23) together with properties (1) and (2), we see that for periodic signals, the autocorrelation function attains a maximum at samples $0, \pm P, \pm 2P, \ldots$ That is, regardless of the time origin of the signal, the period can be estimated by finding the location of the first maximum in the autocorrelation function. This property makes the autocorrelation function an attractive basis for estimating periodicities in all sorts of signals, including speech. Furthermore, we shall see in Chapter 8 that the autocorrelation function contains much more information about the detailed structure of the signal. Thus it is extremely important to consider how the definition of the autocorrelation function can be adapted to obtain a "short-time" autocorrelation function representation of speech.

Using the same approach that was used to define the short-time representations that we have just discussed, we define the short-time autocorrelation function as

$$R_n(k) = \sum_{m=-\infty}^{\infty} x(m)w(n-m)x(m+k)w(n-k-m) \qquad (4.24)$$

This equation can be interpreted as follows: first a segment of speech is selected by multiplication by the window; then the deterministic autocorrelation definition Eq. (4.21) is applied to the windowed segment of speech. It is easily verified that

$$R_n(-k) = R_n(k) \qquad (4.25)$$

Using this expression, we can express $R_n(k)$ in the form of Eq. (4.10). First note that

$$R_n(k) = R_n(-k)$$

$$= \sum_{m=-\infty}^{\infty} x(m)x(m-k)[w(n-m)w(n+k-m)] \qquad (4.26)$$

If we define

$$h_k(n) = w(n)w(n+k) \qquad (4.27)$$

Then Eq. (4.26) can be written as

$$R_n(k) = \sum_{m=-\infty}^{\infty} x(m)x(m-k)h_k(n-m) \qquad (4.28)$$

Thus the value at time n of the kth autocorrelation "lag" is obtained by filtering the sequence $x(n)x(n-k)$ with a filter with impulse response, $h_k(n)$. This is depicted in Fig. 4.23.

The computation of the short-time autocorrelation function is usually carried out using Eq. (4.24) after rewriting it in the form

$$R_n(k) = \sum_{m=-\infty}^{\infty} [x(n+m)w'(m)][x(n+m+k)w'(k+m)] \qquad (4.29)$$

where $w'(n) = w(-n)$. Equation (4.29) states that the time origin of the input sequence is effectively shifted to sample n, whereupon it is multiplied by a window w' to select a short segment of speech. If the window w' is of finite duration as in Eqs. (4.8) and (4.9) then the resulting sequence, $x(n+m)w'(n)$ will

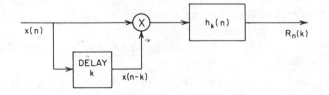

Fig. 4.23 Block diagram representation of the short-time autocorrelation.

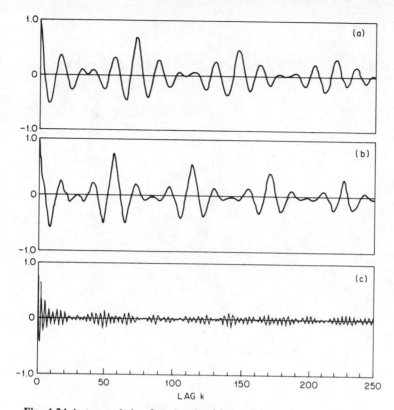

Fig. 4.24 Autocorrelation function for (a) and (b) voiced speech; and (c) unvoiced speech, using a rectangular window with $N = 401$.

be of finite duration and Eq. (4.29) becomes

$$R_n(k) = \sum_{m=0}^{N-1-k} [x(n+m)\,w'(m)][x(n+m+k)\,w'(k+m)] \qquad (4.30)$$

Note that when the rectangular or Hamming windows of Eqs. (4.8) and (4.9) are used for w' in Eq. (4.30) they correspond to a noncausal system in Eq. (4.28). For finite length windows, this poses no problem since suitable delay can be introduced into the processing, even in real-time applications.

The calculation of the kth autocorrelation lag using Eq. (4.30) would appear to require N multiplications for computing $x(n+m)\,w'(m)$, and $(N-k)$ multiplications and additions for computing the sum of lagged products. The computation of many lags as required in estimating periodicity thus requires a great deal of arithmetic. This can be reduced by taking advantage of some special properties of Eq. (4.30). Several techniques are given in the Appendix.

An alternative to Eq. (4.30) that is useful if only a few lags are required is provided by Eq. (4.28). If the window $w(n)$ is properly chosen, $R_n(k)$ can be computed recursively. (See Problem 4.7.)

Figure 4.24 shows three examples of autocorrelation functions computed for speech sampled at 10 kHz using Eq. (4.30) with $N = 401$. As shown, the

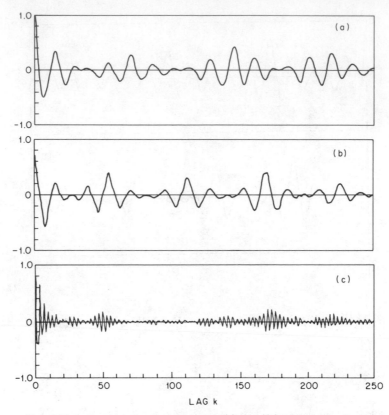

Fig. 4.25 Autocorrelation functions for (a) and (b) voiced speech; and (c) unvoiced speech, using a Hamming window with $N = 401$.

autocorrelation was evaluated for lags $0 \leqslant k \leqslant 250$. The first two cases are for voiced speech segments and the third is for an unvoiced segment.[4] For the first segment, peaks occur approximately at multiples of 72 indicating a period of 7.2 msec or a fundamental frequency of approximately 140 Hz. Note that even a very short segment of speech differs from a segment of a truly periodic signal. The "period" of the signal changes across a 401 sample interval and also the wave shape varies somewhat from period to period. This is part of the reason that the peaks get smaller for large lags. For the second voiced section (taken at a totally distinct place in the utterance) similar periodicity effects are seen, only now the local peaks in the autocorrelation are at multiples of 58 samples indicating an average pitch period of about 5.8 msec. Finally for the unvoiced section of speech there are no strong autocorrelation periodicity peaks thus indicating a *lack* of periodicity in the waveform. The autocorrelation function for unvoiced speech is seen to be a high frequency noise-like waveform, somewhat like the speech itself.

Figure 4.25 shows the same examples using a Hamming window. By comparing these results to those in Fig. 4.24, it can be seen that the rectangular window gives a much stronger indication of periodicity than the Hamming win-

---

[4]In this and subsequent plots, the autocorrelation function is normalized so that $R_n(0) = 1$.

dow. This is not surprising in view of the tapering of the speech segment introduced by the Hamming window.

The examples of Figures 4.24 and 4.25 were for a value of $N = 401$. An important issue is how N should be chosen to give a good indication of periodicity. Again we face conflicting requirements. Because of the changing properties of the speech signal, N should be as small as possible. On the other hand, it should be clear that to get any indication of periodicity in the autocorrelation function, the window must have a duration of at least two periods of the waveform. In fact, because of the finite length of the windowed speech segment involved in the computation of $R_n(k)$, there is less and less data involved in the computation as k increases. (Note the upper limit of summation in Eq. (4.30).) This leads to a reduction in amplitude of the correlation peaks as k increases. This can easily be verified for the case of a periodic impulse train (see Problem 4.8), and it is easily demonstrated for speech by examples. Figure 4.26 illustrates the effect for rectangular windows of different lengths. The dotted lines are plots of the equation

$$R(k) = 1 - k/N, \qquad |k| < N \qquad (4.31)$$

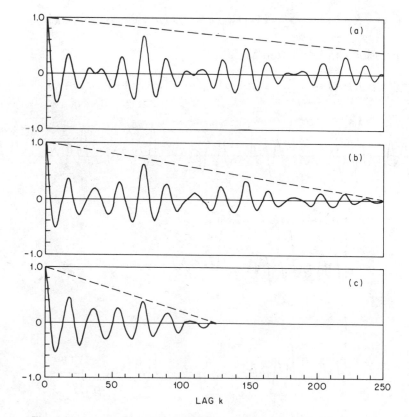

Fig. 4.26 Autocorrelation function for voiced speech with (a) $N = 401$; (b) $N = 251$; and (c) $N = 125$. Rectangular window used in all cases.

145

which is the autocorrelation function of the rectangular window. Clearly, this is a good bound on the amplitude of the correlation peaks. In Problem 4.8 it is shown that for a periodic impulse train, the peaks will lie exactly on such a straight line. For the present example, the peaks are further away from the line for $N = 401$ than for the other two cases. This is due to the fact that the pitch period and wave shape change more across an interval of 401 samples than across the shorter intervals. The effects combine to cause a greater reduction.

Figure 4.26c corresponds to a window length of 125 samples. Since the period for this example is about 72 samples, not even two complete pitch periods are included in the window. This is clearly a situation to be avoided, but avoiding it is difficult because of the wide range of pitch periods that may be encountered. One approach is to simply make the window long enough to accommodate the longest pitch period, but this leads to undesirable averaging of many periods when the pitch period is short. Another approach is to allow the window length to adapt to match the expected pitch period. Still another approach that allows the use of shorter windows is to modify the definition of the autocorrelation function.

The modified short-time autocorrelation function is defined as

$$\hat{R}_n(k) = \sum_{m=-\infty}^{\infty} x(m) w_1(n-m) x(m+k) w_2(n-m-k) \qquad (4.32)$$

This expression can be written as

$$\hat{R}_n(k) = \sum_{m=-\infty}^{\infty} x(n+m) \hat{w}_1(m) x(n+m+k) \hat{w}_2(m+k) \qquad (4.33)$$

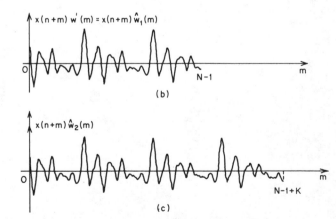

Fig. 4.27 Illustration of the samples involved in the computation of the short-time autocorrelation function.

146

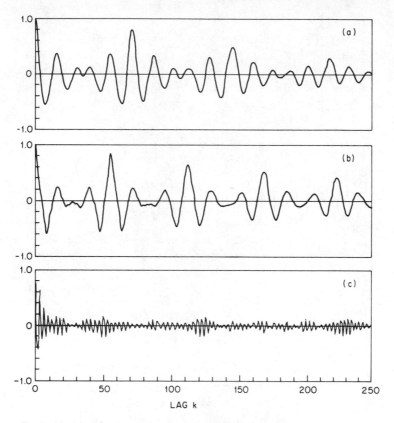

Fig. 4.28 Modified autocorrelation function for speech segments of Fig. 4.24 with $N = 401$.

where

$$\hat{w}_1(m) = w_1(-m) \qquad (4.34a)$$

and

$$\hat{w}_2(m) = w_2(-m) \qquad (4.34b)$$

To accomplish our goal of eliminating the fall-off due to the variable upper limit in Eq. (4.30) we can <u>choose the window $\hat{w}_2$ to include samples outside the nonzero interval of window $\hat{w}_1$.</u> That is, we define

$$\hat{w}_1(m) = 1 \qquad 0 \leqslant m \leqslant N - 1$$
$$= 0 \qquad otherwise \qquad (4.35a)$$

and

$$\hat{w}_2(m) = 1 \qquad 0 \leqslant m \leqslant N - 1 + K$$
$$= 0 \qquad otherwise \qquad (4.35b)$$

where K is the greatest lag of interest. Thus, Eq. (4.33) can be written as

$$\hat{R}_n(k) = \sum_{m=0}^{N-1} x(n+m)x(n+m+k) \qquad 0 \leqslant k \leqslant K \qquad (4.36)$$

147

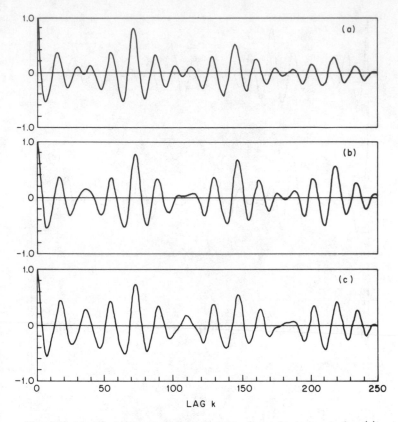

Fig. 4.29 Modified autocorrelation function for voiced speech for (a) $N = 401$; (b) $N = 251$; and (c) $N = 125$. Example corresponds to Fig. 4.26.

i.e., the average is always over N samples, and samples from outside the interval n to $n + N - 1$ are involved in the computation. The differences in the data involved in the computations of Eqs. (4.30) and (4.36) are depicted in Figure 4.27. Figure 4.27a shows a speech waveform and Figure 4.27b shows a segment of N samples selected by a rectangular window. For a rectangular window, this segment would be used for both terms in Eq. (4.30) and it would be the term $x(n+m)\hat{w}_1(m)$ in Eq. (4.36). Figure 4.27c shows the other term in Eq. (4.36). Note that K additional samples are included.

Equation (4.36) will be referred to as the *modified* short-time autocorrelation function. Strictly speaking, however, it is the *cross-correlation* function for the two different finite length segments of speech, $x(n+m)\hat{w}_1(m)$ and $x(n+m)\hat{w}_2(m)$. Thus $\hat{R}_n(k)$ has the properties of a cross-correlation function, not an autocorrelation function. For example, $\hat{R}_n(-k) \neq \hat{R}_n(k)$. Nevertheless, $\hat{R}_n(k)$ will display peaks at multiples of the period of a periodic signal and it will not display a fall-off in amplitude at large values of k. Figure 4.28 shows the modified autocorrelation functions corresponding to the examples of Figure 4.24. Because for $N = 401$ the effects of waveform variation dominate the tapering effect in Figure 4.24, the two figures look much alike. A comparison of Figure 4.29 with 4.26 shows that the difference is more apparent for smaller values of N. It is clear that the peaks in Figure 4.29 are less than the $k = 0$ peak only because of deviations from periodicity over the interval n to $n + N - 1 + K$ that is involved in the evaluation of Eq. (4.36). Problem 4.8 shows that for a perfectly periodic impulse train, all the peaks will be the same amplitude.

4.7 The Short-Time Average Magnitude Difference Function

As we have pointed out, the computation of the autocorrelation function involves considerable arithmetic, even using the simplifications discussed in the Appendix. A technique that eliminates the need for multiplications is based upon the idea that for a truly periodic input of period P, the sequence

$$d(n) = x(n) - x(n-k) \tag{4.37}$$

would be zero for $k = 0, \pm P, \pm 2P, \ldots$. For short-segments of voiced speech it is reasonable to expect that $d(n)$ will be small at multiples of the period, but not identically zero. The short-time average magnitude of $d(n)$ as a function of k should be small whenever k is close to the period. The short-time average magnitude difference function (AMDF) [16] is thus defined as

$$\gamma_n(k) = \sum_{m=-\infty}^{\infty} |x(n+m)w_1(m) - x(n+m-k)w_2(m-k)| \tag{4.38}$$

Clearly, if $x(n)$ is close to being periodic in the interval spanned by the window, then $\gamma_n(k)$ should dip sharply for $k = P, 2P, \ldots$. Note that it is most reasonable to choose the windows to be rectangular. If both have the same length, we obtain a function similar to the autocorrelation function of Eq. (4.30). If $w_2(n)$ is longer than $w_1(n)$ then we have a situation similar to the modified autocorrelation of Eq. (4.36). It can be shown [16] that

$$\gamma_n(k) \approx \sqrt{2}\, \beta(k) [\hat{R}_n(0) - \hat{R}_n(k)]^{1/2} \tag{4.39}$$

It is reported that $\beta(k)$ in Eq. (4.39) varies between 0.6 and 1.0 with different segments of speech, but does not change rapidly with k for a particular speech segment [16].

Figure 4.30 shows the AMDF function for the speech segments of Figures 4.24 and 4.28 for the same length window. It can be seen that the AMDF function does indeed have the shape suggested by Eq. (4.39) and thus, $\gamma_n(k)$ dips sharply at the pitch period of voiced speech and shows no comparable dips for unvoiced speech.

The AMDF function is implemented with subtraction, addition, and absolute value operations, in contrast to addition and multiplication operations for the autocorrelation function. With floating point arithmetic, where multiplies and adds take approximately the same time, about the same time is required for either method with the same window length. However, for special purpose hardware, or with fixed point arithmetic, the AMDF appears to have the advantage. In this case multiplies usually are more time consuming and furthermore

149

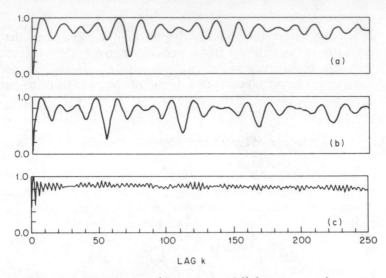

Fig. 4.30 AMDF function (normalized to 1.0) for same speech segments as in Figs. 4.24 and 4.28.

either scaling or a double precision accumulator is required to hold the sum of lagged products. For this reason the AMDF function has been used in numerous real-time speech processing systems.

4.8 Pitch Period Estimation Using the Autocorrelation Function

As demonstrated in Section 4.6, the short-time autocorrelation function provides a convenient representation upon which to base a scheme for determining the pitch period as a function of time. In this section we shall discuss several details of implementation of autocorrelation pitch detectors.

One of the major limitations of the autocorrelation representation is that in a sense it retains too much of the information in the speech signal. (We shall see in Chapter 8 that the low-time autocorrelation values ($0 \leqslant k \leqslant 10$ or 12) are sufficient to accurately estimate the vocal tract transfer function.) As a result, we note in Figure 4.26, for example, that the autocorrelation function has many peaks. Most of these peaks can be attributed to the damped oscillations of the vocal tract response which are responsible for the shape of each period of the speech wave. In Figures 4.26a and 4.26b, the peak at the pitch period has the greatest amplitude; however, in Figure 4.26c, the peak at about $k = 15$ is actually greater than the peak at $k = 72$. This occurs in this case because the window is short compared to the pitch period, but rapidly changing formant frequencies can also create such a situation. Clearly, in cases when the autocorrelation peaks due to the vocal tract response are bigger than those due to the periodicity of the vocal excitation, the simple procedure of picking the largest peak in the autocorrelation function will fail.

To avoid this problem it is again useful to process the speech signal so as to make the periodicity more prominent while suppressing other distracting features of the signal. This was the approach followed in Section 4.5 to permit the use of a very simple pitch detector. Techniques which perform this type of operation on a signal are sometimes called "spectrum flatteners" since their objective is to remove the effects of the vocal tract transfer function, thereby bringing each harmonic to the same amplitude level as in the case of a periodic impulse train. Numerous spectrum flattening techniques have been proposed [17]; however, a technique called "center clipping" [17] appears to be advantageous in the present context.

In the scheme proposed by Sondhi [17], the center clipped speech signal is obtained by a nonlinear transformation

$$y(n) = C[x(n)] \qquad (4.40)$$

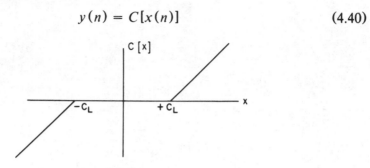

Fig. 4.31 Center clipping function.

where $C[\ ]$ is shown in Figure 4.31. The operation of the center clipper is depicted in Fig. 4.32. A segment of speech to be used in computing an autocorrelation function is shown in the upper plot. For this segment, the maximum amplitude, A_{max}, is found and the clipping level, C_L, is set equal to a fixed percentage of A_{max}. (Sondhi [17] used 30%.) From Figure 4.31, it can be seen that for samples above C_L, the output of the center clipper is equal to the input minus the clipping level. For samples below the clipping level the output

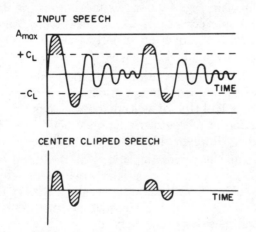

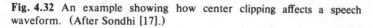

Fig. 4.32 An example showing how center clipping affects a speech waveform. (After Sondhi [17].)

151

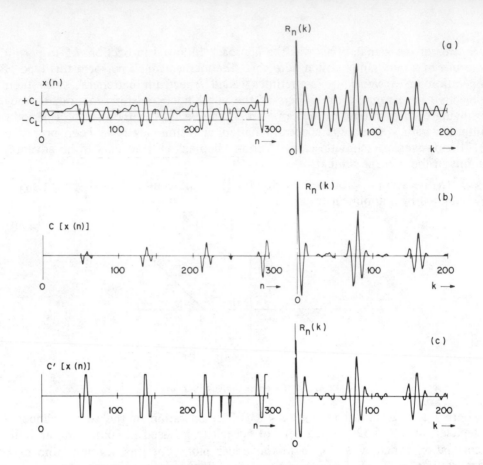

Fig. 4.33 Example of waveforms and correlation function; (a) no clipping; (b) center clipped; (c) 3-level center clipped. (All correlation functions normalized to 1.0.) (After Rabiner [18].)

is zero. The lower plot in Fig. 4.32 shows the output for the above input. In contrast to the scheme of Section 4.5, where peaks were converted to impulses, in this case, the peaks are converted to pulses consisting of the part of each peak that exceeds the clipping level.

Figure 4.33 [18] illustrates the effect of the center clipping operation on the computation of the autocorrelation function. Figure 4.33a shows a 300 sample segment ($F_s = 10$ kHz) of voiced speech. Note that there is a strong peak at the pitch period in the autocorrelation function for this segment which is shown on the right. However, it is also clear that there are many peaks that can be attributed to the damped oscillations of the vocal tract. Figure 4.33b shows the corresponding center clipped signal where the clipping level was set as shown in Figure 4.33a. (In this case it was set at 68% of the maximum magnitude in the first 100 samples.) Note that all that remains in the clipped waveform are several pulses spaced at the original pitch period. The resulting autocorrelation function has considerably fewer extraneous peaks to create confusion.

152

We shall return to part c of Fig. 4.33 soon. However, first let us examine the effect of the clipping level. Clearly, for high clipping levels, fewer peaks will exceed the clipping level and thus fewer pulses will appear in the output, and therefore, fewer extraneous peaks will appear in the autocorrelation function. This is illustrated by Fig. 4.34 which shows the autocorrelation functions for the segment of speech corresponding to Fig. 4.26a, for decreasing clipping levels. Clearly, as the clipping level is decreased, more peaks pass through the clipper and thus the autocorrelation function becomes more complex. (Note that a clipping level of zero corresponds to Fig. 4.26a.) The implication of this example is that the clearest indication of periodicity is obtained for the highest possible clipping level. There is a difficulty with using too high a clipping level. It is possible that the amplitude of the signal may vary appreciably across the duration of the speech segment (e.g., at the beginning or end of voicing) so that if the clipping level is set at a high percentage of the maximum amplitude across the whole segment, there is a possibility that much of the waveform will fall below the clipping level and be lost. For this reason Sondhi's original proposal was to set the clipping level at 30% of the maximum amplitude. A procedure which permits a greater percentage (60-80%) to be used is to find the

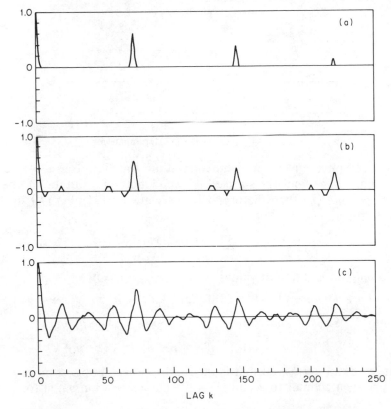

Fig. 4.34 Autocorrelation functions of center clipped speech using $N = 401$; (a) C_L set at 80% of maximum; (b) 64%; (c) 48%. (Speech segment same as for Fig. 4.26a.)

153

peak amplitude in both the first third and last third of the segment and set the clipping level at a fixed percentage of the minimum of these two maximum levels. (This was the procedure followed in Fig. 4.33b.)

The problem of extraneous peaks in the autocorrelation function can be greatly alleviated by center clipping prior to computing the autocorrelation function. However, another difficulty with the autocorrelation representation (that remains even with center clipping) is the large amount of computation that is required. A simple modification of the center clipping function leads to a great simplification in computation of the autocorrelation function with essentially no degradation in utility for pitch detection [19]. This modification is shown in Fig. 4.35. As indicated there, the output of the clipper is +1 if $x(n) > C_L$ and -1 if $x(n) < -C_L$. Otherwise the output is zero. This function will be called a 3-level center clipper. Figure 4.33c shows the output of the 3-level center clipper for the input segment of Fig. 4.33a. Note that although this operation tends to emphasize the importance of peaks that just exceed the clipping level, the autocorrelation function is very similar to that of the center clipper of Fig. 4.33b. That is, most of the extraneous peaks are eliminated, and a clear indication of periodicity is retained.

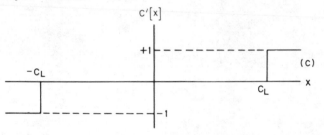

Fig. 4.35 3-level center clipping function.

The computation of the autocorrelation function for a 3-level center clipped signal is particularly simple. If we denote the output of the 3-level center clipper as $y(n)$ then the product terms $y(n+m)y(n+m+k)$ in the autocorrelation function

$$R_n(k) = \sum_{m=0}^{N-k-1} y(n+m)y(n+m+k) \tag{4.41}$$

can have only three different values

$$y(n+m)y(n+m+k) = 0 \qquad \text{if} \quad y(n+m) = 0 \ \ or \ \ y(n+m+k) = 0$$
$$= +1 \qquad \text{if} \quad y(n+m) = y(n+m+k)$$
$$= -1 \qquad \text{if} \quad y(n+m) \neq y(n+m+k) \tag{4.42}$$

Thus, in hardware terms, all that is required is some simple combinatorial logic and an up-down counter to accumulate the autocorrelation value for each value of k.

As a further comment on details of implementation we note that the modified autocorrelation definition of Eq. (4.36) could be used with either

154

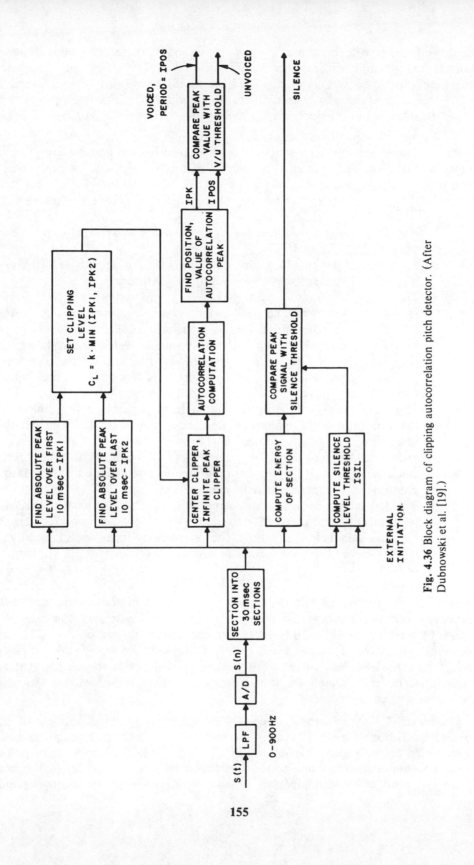

Fig. 4.36 Block diagram of clipping autocorrelation pitch detector. (After Dubnowski et al. [19].)

center clipped or 3-level center clipped speech input, therefore giving an auto-correlation function whose peaks do not fall off with increasing lag. Likewise, the input to the AMDF computation of Eq. (4.38) could also be some version of center clipped signal. Indeed, there are a multitude of combinations that could be used, some of which may have advantages in specific situations [18].

Numerous algorithms for estimating the pitch period from the short-time autocorrelation function representation have been proposed and no doubt many more will be proposed. We shall conclude this section with one example which has been implemented in digital hardware [19]. The details of the algorithm are depicted in Fig. 4.36 and a summary of the algorithm is given below:

1. The speech signal is filtered with a 900 Hz lowpass analog filter and sampled at a rate of 10 kHz.
2. Segments of length 30 msec (300 samples) are selected at 10 msec intervals. Thus, the segments overlap by 20 msec.
3. The average magnitude, Eq. (4.12), is computed with a 100 sample rectangular window. The peak signal level in each frame is compared to a threshold determined by measuring the peak signal level for 50 msec of background noise. If the peak signal level is above threshold, signifying that the segment is speech, not noise, then the algorithm procedes as follows; otherwise the segment is classed as silence and no further action is taken.
4. The clipping level is determined as a fixed percentage (e.g., 68%) of the minimum of the maximum absolute values in the first and last 100 samples of the speech segment.
5. Using this clipping level, the speech signal is processed by a 3-level center clipper and the correlation function is computed over a range spanning the expected range of pitch periods.
6. The largest peak of the autocorrelation function is located and the peak value is compared to a fixed threshold (e.g., 30% of $R_n(0)$). If the peak falls below threshold, the segment is classed as unvoiced and if it is above, the pitch period is defined as the location of the largest peak.

This is essentially the algorithm that was implemented in digital hardware [19]; however there is considerable latitude for variation in the details. For example, steps (4) and (5) could be altered to use the center clipper of Fig. 4.31 and standard arithmetic for the autocorrelation computation, or center clipping could be completely eliminated. Still another possibility is to use the AMDF function (and thus search for dips instead of peaks) either with or without some form of center clipping.

Figure 4.37 shows the outputs (pitch contours) of three variants of the above algorithm. Figure 4.37(a) is the pitch contour obtained using the auto-correlation of the speech signal without clipping. Note the scattering of points that are obviously errors due to the fact that a peak at a short lag was greater than the peak at the pitch period. Also note that the pitch period averages

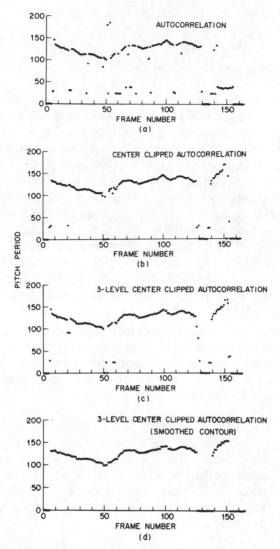

WE WERE AWAY A YEAR AGO

AUTOCORRELATION

(a)

CENTER CLIPPED AUTOCORRELATION

(b)

3-LEVEL CENTER CLIPPED AUTOCORRELATION

(c)

3-LEVEL CENTER CLIPPED AUTOCORRELATION
(SMOOTHED CONTOUR)

(d)

Fig. 4.37 Autocorrelation pitch detector outputs; (a) no clipping; (b) center clipping (Fig. 4.31); (c) 3-level center clipping (Fig. 4.35); (d) nonlinearly smoothed output from (c). (After Rabiner [18].)

between 100 and 150 samples so that the inherent fall-off of the autocorrelation function causes significant attenuation of the peak at the pitch period. Thus, peaks in the autocorrelation function due to the vocal tract response are likely to be greater than those due to periodicity. Figures 4.37b and 4.37c are respectively for the cases when center clipping and 3-level center clipping are used with the autocorrelation function. Clearly, most of the errors have been eliminated by the inclusion of clipping and furthermore there is no significant difference between the two results. A few obvious errors remain in both pitch contours. These errors can be effectively removed by a nonlinear smoothing

157

method to be discussed in the next section. An example is shown in Fig. 4.37d.

4.9 Median Smoothing and Speech Processing

In most signal processing applications a linear smoother (or a linear filter) is generally used to eliminate the noise-like components of a signal. For some speech processing applications, however, linear smoothers are not completely adequate due to the type of data being smoothed. An example is the pitch contour of Fig. 4.37c, which has obvious errors that must be brought back into line with the rest of the data. An ordinary linear lowpass filter would not only fail to bring the errant points back into line but would severely distort the contour at the transition between voiced and unvoiced speech (shown as zero period). For such cases some type of nonlinear smoothing algorithm which can preserve signal discontinuities yet still filter out large errors is required. Although an ideal nonlinear smoothing algorithm with these properties does not exist, a nonlinear smoother using a combination of running medians and linear smoothing (originally proposed by Tukey [20]) can be shown to have approximately the desired properties [21].

The basic concept of a linear smoother is the separation of signals based on their (approximately) nonoverlapping frequency content. For nonlinear smoothers it is more appropriate to consider separating signals based on whether they can be considered smooth or rough (noise-like). Thus a signal $x(n)$ can be considered to be of the form:

$$x(n) = S[x(n)] + R[x(n)] \qquad (4.43)$$

where $S[x]$ is the smooth part of the signal x, and $R[x]$ is the rough part of the signal x. A nonlinearity which is capable of separating $S[x(n)]$ from $R[x(n)]$ is the running median of $x(n)$. The output of the running median smoother, $M_L[x(n)]$, is simply the median of the L numbers, $x(n), \ldots, x(n-L+1)$. Running medians of length L have the following desirable properties for a smoother:

1. $M_L[\alpha x(n)] = \alpha M_L[x(n)]$
2. Medians will not smear out discontinuities in the signal if the signal has no other discontinuities within $L/2$ samples.
3. Medians will approximately follow low order polynomial trends in the signal. It should be reemphasized that running medians, like other nonlinear processing algorithms, do not obey the superposition property, i.e.,

$$M_L[\alpha x_1(n) + \beta x_2(n)] \neq \alpha M_L[x_1(n)] + \beta M_L[x_2(n)] \qquad (4.44)$$

Although running medians generally preserve sharp discontinuities in a signal, they often fail to provide sufficient smoothing of the undesirable noise-like components of a signal. A good compromise is to use a smoothing algo-

158

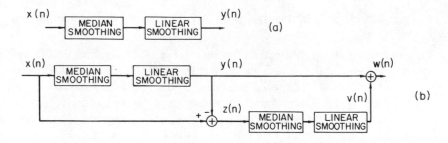

Fig. 4.38 Block diagram of a nonlinear smoothing system. (After Rabiner et al. [21].)

rithm based on a combination of running medians and linear smoothing. Since the running medians provide some smoothing, the linear smoother can be a low order system. Usually the linear filter is a symmetrical FIR filter so that delays can be exactly compensated. For example a hanning filter with impulse response

$$h(n) = 1/4 \quad n = 0$$
$$= 1/2 \quad n = 1$$
$$= 1/4 \quad n = 2 \quad (4.45)$$

is generally adequate [20]. Figure 4.38a shows a block diagram of a combination smoother based on running medians and linear smoothing. The signal $y(n)$ at the output of the smoother is an approximation to the signal $S[x(n)]$. Since the approximation is not ideal, a second pass of nonlinear smoothing is incorporated into the smoothing algorithm as shown in Fig. 4.38b. Since

$$y(n) = S[x(n)] \quad (4.46)$$

then

$$z(n) = x(n) - y(n) = R[x(n)] \quad (4.47)$$

The second pass of nonlinear smoothing of $z(n)$ yields a correction signal which is added to $y(n)$ to give $w(n)$, a refined approximation to $S[x(n)]$. The signal $w(n)$ satisfies the relation

$$w(n) = S[x(n)] + S[R[x(n)]] \quad (4.48)$$

If $z(n) = R[x(n)]$ exactly, i.e., the nonlinear smoother were ideal, then

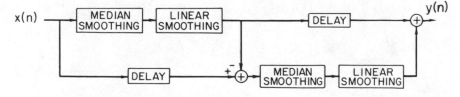

Fig. 4.39 Nonlinear smoothing system with delay compensation. (After Rabiner et al. [21].)

159

$S[R[x(n)]]$ would be identically zero and the correction term would be unnecessary.

To implement the nonlinear smoother of Fig. 4.38 in a realizable system requires accounting for the delays in each path of the smoother. Each median smoother has a delay of $(L-1)/2$ samples, and each linear smoother has a delay corresponding to the impulse response used. For example, a running median of 5 smoother has a delay of 2 samples, and a 3 point Hamming window linear filter has a delay of 1 sample. Figure 4.39 shows a block diagram of a realizable version of the smoother of Fig. 4.38b.

The final issue concerned with the implementation of the nonlinear smoother of Fig. 4.39 is the question of how the running median of the signal is defined at the beginning and end of the signal to be smoothed. Although a variety of approaches are possible, for speech applications extrapolating the signal backwards and forwards by assuming the signal stays constant is generally a reasonable solution.

Figure 4.40 shows the results of using several different smoothers on a zero crossing representation of a speech signal. The input signal (Fig. 4.40a) is rough due to the use of a short averaging time. It can be seen in Fig. 4.40d that the output of the median smoother alone (a 5 point median followed by a

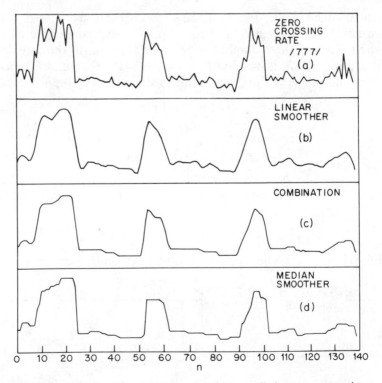

Fig. 4.40 Example of nonlinear smoothing applied to zero-crossing representation. (After Rabiner et al. [21].)

160

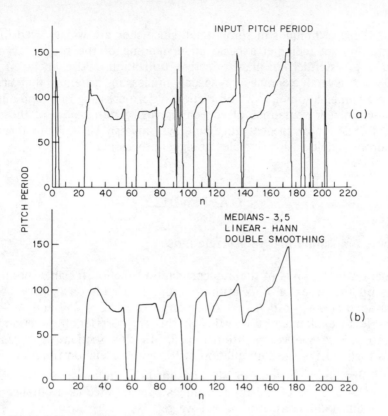

Fig. 4.41 Example of nonlinear smoothing of a pitch contour. (After Rabiner et al. [21].)

3 point median) has a block-like effect due to the presence of high frequency components in the smoothed output. The output of the linear smoother (a 19 point FIR lowpass filter), shown in Fig. 4.40b, is smeared whenever rapid changes occurred in the input signal. The output of the combination smoother (a median of 5 followed by a median of 3 followed by a 3-point hanning window), shown in Fig. 4.40c, is seen to follow the changes in the input signal quite well while eliminating most of the noise in the signal.

Figure 4.41 shows an example in which the combination smoother was used to smooth a pitch period contour for which several obvious errors were made in the estimate. An important property of median smoothing is that it can correct isolated errors in the data while combining this operation with the desired smoothing. As seen in Fig. 4.41 the combination smoother was able to eliminate the gross measurement errors and adequately smooth the signal, while leaving the voiced/unvoiced transitions intact.

4.10 Summary

In this chapter we have discussed several representations of speech that are based primarily on processing carried out directly in the time domain. We have

161

discussed these techniques in some detail since they are widely used in speech processing and we feel that a basic understanding of the properties of these techniques is essential to their effective utilization. Also included in this chapter were several examples of speech processing schemes that are based upon some combination of short-time energy, zero-crossing rate, and autocorrelation function representations of speech signals. The purpose of these examples was to show how speech processing systems can be built up through the application of simple basic principles of signal processing.

APPENDIX

Short-cuts for Computing Autocorrelations

Computation of K points of the autocorrelation function for an N point window requires on the order of $K \cdot N$ multiplications and additions. Since, for many practical applications, both K and N are large (e.g., $K = 250$ and $N = 401$), it is desirable to exploit any properties of the autocorrelation function that are known to reduce the computational load. In this section we discuss three methods for reducing the computation of the autocorrelation function.

The first short-cut, due to Blankenship [22], follows from the realization that for $m \neq 0$, most of the input samples appear twice as multiplicands, e.g., for the modified autocorrelation $k = 1$ we get

$$\hat{R}_n(1) = \sum_{m=0}^{N-1} x(m+n)x(m+n+1)$$

$$= x(n)x(n+1) + x(n+1)x(n+2) + \cdots + x(n+N-1)x(n+N)$$

$$= x(n+1)[x(n)+x(n+2)] + x(n+3)[x(n+2)+x(n+4)] + \cdots \text{(A1)}$$

Thus when $k \neq 0$, by using an expression of the form of Eq. (4.36) the number of multiplications can be effectively halved without increasing the number of additions. Formally the autocorrelation computation can be expressed in the form

$$\hat{R}(k) = B(k) + C(k) \tag{A2}$$

(suppressing the subscript n from this point on, for convenience) where N is expressed as

$$N = \underset{\substack{even \\ component}}{2qk} + \underset{\substack{optional \\ component}}{ak} + b \tag{A3}$$

with $a = 0$, or 1, and b in the range

$$0 \leqslant b < k \tag{A4}$$

In Eq. (A2),

$$B(k) = \sum_{j=0}^{q-1} \sum_{i=1}^{k} x(2jk+i+k)[x(2jk+i) + x(2jk+i+2k)] \qquad (A5)$$

and if $a = 0$

$$C(k) = \sum_{i=1}^{b} x(2qk+i)x(2qk+i+k) \qquad (A6)$$

or if $a = 1$

$$C(k) = \sum_{i=1}^{b} x(2qk+i+k)[x(2qk+i) + x(2qk+i+2k)]$$

$$+ \sum_{i=b+1}^{k} x(2qk+i)x(2qk+i+k) \qquad (A7)$$

For example, consider $N = 60$ with $k = 6$, 7, and 8. The values for q, a, and b of Eq. (A3) are

$$\begin{array}{lllll}
N = 60, & k = 6, & q = 5, & a = 0, & b = 0 \\
N = 60, & k = 7, & q = 4, & a = 0, & b = 4 \\
N = 60, & k = 8, & q = 3, & a = 1, & b = 4
\end{array}$$

Once values of q, a, and b are obtained, Eqs. (A2) and (A5)-(A7) are straightforwardly applied to give $\hat{R}(k)$. It is readily shown that the number of multiplications required to compute $\hat{R}(k)$ satisfies the relation

$$N_M < \frac{1}{2}(N+k) \qquad (A8)$$

Thus for $k \ll N$, this procedure leads to approximately a 2:1 reduction in the number of multiplications. When one is interested in only a few autocorrelation values (e.g., in linear prediction methods of Chapter 8) this method is quite a useful one. For example, Blankenship shows that if $K = 12$, and $N = 128$, a total of 1664 multiplications ($N \cdot (K+1)$) would be required for a direct evaluation of the autocorrelation, whereas only 912 multiplications are required for this modified procedure — a savings of a factor of 1.825 to 1 in the time required for multiplications. For computations where a large number of autocorrelation values are desired, e.g., the examples of Section 4.6, this procedure leads to very small savings.

A variation on the above procedure was given by Kendall [23] for the autocorrelation defined by Eq. (4.30). Denoting the weighted speech sample $x(n)w(n)$ as $\hat{x}(n)$, and again suppressing the subscript n, Eq. (4.30) becomes

$$R(k) = \sum_{m=0}^{N-1-k} \hat{x}(m)\hat{x}(m+k) \qquad (A9)$$

which can be expressed as (assuming N even)

$$R(k) = \sum_{m=0}^{(N-k)/2-1} [\hat{x}(2m) + \hat{x}(2m+k+1)][\hat{x}(2m+1) + \hat{x}(2m+k)]$$

$$- A(k) - B(k) \qquad k \ even \qquad (A10)$$

163

$$R(k) = \sum_{m=0}^{(N-k-1)/2-1} [\hat{x}(2m) + \hat{x}(2m+k+1)][\hat{x}(2m+1) + \hat{x}(2m+k)]$$

$$- A(k) - B(k) + \hat{x}(N-1-k)\hat{x}(N-1) \qquad k \; odd \qquad \text{(A11)}$$

where $A(k)$ and $B(k)$ are obtained via the recursion relations

$$A(k) = A(k+2) + \hat{x}(N-2-k)\hat{x}(N-1-k), \qquad k \; even \qquad \text{(A12)}$$

with initial condition $A(N) = 0$,

$$A(k) = A(k+1), \qquad k \; odd \qquad \text{(A13)}$$

and

$$B(k) = B(k+2) + \hat{x}(k)\hat{x}(k+1), \qquad k \; even \qquad \text{(A14)}$$

with initial condition $B(N) = 0$, and

$$B(k) = B(k+2) + \hat{x}(k)\hat{x}(k+1), \qquad k \; odd \qquad \text{(A15)}$$

with initial condition $B(N-1) = 0$. Equations (A10) and (A11) show that the number of multiplications required to compute $R(k)$ is approximately $(N-k-1)/2$ — i.e., half that normally required in a direct evaluation, whereas the number of additions is increased by about 50%. Furthermore it is seen that the reduction in the number of multiplications is valid for all k, not just when $k \ll N$ as in the preceding method.

The final method for speeding up the autocorrelation computation is the well known FFT method in which the autocorrelation is determined as the inverse DFT of the power density spectrum (the squared magnitude of the DFT) of the sequence [24,1,2]. For this method two DFT's and a squared magnitude are required. To avoid aliasing in the autocorrelation computation, a $2N$ point DFT (computed using an FFT algorithm) is required in which the N point data sequence is padded with N zero valued samples. The process of forming a squared magnitude requires about $2N$ multiplications, and a $2N$ point FFT requires $2N \log_2(2N)$ multiplications to give all N points of the autocorrelation function. Thus for the FFT method, the total number of multiplications required is:

$$N_F = 2 \cdot 2N \log_2(2N) + 2N \qquad \text{(A16)}$$

Kendall [23] has shown that the modified direct autocorrelation measurement is more efficient than the FFT method for values of $N \leqslant 256$, in terms of the number of multiplications required. If one includes additions in the computation, the modified direct method is more efficient for N in the vicinity of 128.

REFERENCES

1.   A. V. Oppenheim and R. W. Schafer, *Digital Signal Processing*, Prentice-Hall, Inc., Englewood Cliffs, N.J., 1975.

2. L. R. Rabiner and B. Gold, *Theory and Application of Digital Signal Processing,* Prentice-Hall, Inc., Englewood Cliffs, N.J., 1975.

3. J. M. Baker, "A New Time-Domain Analysis of Human Speech and Other Complex Waveforms," Ph.D Dissertation, Carnegie-Mellon Univ., Pittsburgh, PA., 1975.

4. P. J. Vicens, "Aspects of Speech Recognition by Computer," Ph.D. Thesis, Stanford Univ., AI Memo No. 85, Comp. Sci. Dept., Stanford Univ., 1969.

5. L. D. Erman, "An Environment and System for Machine Understanding of Connected Speech," Ph.D. Dissertation, Carnegie-Mellon Univ., Pittsburgh, PA, 1975.

6. L. R. Rabiner and M. R. Sambur, "An Algorithm for Determining the Endpoints of Isolated Utterances," *Bell Syst. Tech. J.,* Vol. 54, No. 2, pp. 297-315, February 1975.

7. M. R. Sambur and L. R. Rabiner, "A Speaker Independent Digit-Recognition System," *Bell Syst. Tech. J.,* Vol. 54, No. 1, pp. 81-102, January 1975.

8. J. L. Flanagan, *Speech Analysis, Synthesis and Perception,* 2nd Ed., Springer Verlag, N.Y., 1972.

9. B. S. Atal, "Automatic Speaker Recognition Based on Pitch Contours," *J. Acoust. Soc. Am.,* Vol. 52, pp. 1687-1697, December 1972.

10. A. E. Rosenberg and M. R. Sambur, "New Techniques for Automatic Speaker Verification," *IEEE Trans. Acoust., Speech, and Signal Proc.,* Vol. ASSP-23, pp. 169-176, April 1975.

11. H. Levitt, "Speech Processing Aids for the Deaf: An Overview," *IEEE Trans. Audio and Electroacoustics,* Vol. AU-21, pp. 269-273, June 1973.

12. L. R. Rabiner, M. J. Cheng, A. E. Rosenberg, and C. A. McGonegal, "A Comparative Performance Study of Several Pitch Detection Algorithms," *IEEE Trans. Acoust., Speech, and Signal Proc.,* Vol. ASSP-24, No. 5, pp. 399-418, October 1976.

13. B. Gold, "Computer Program for Pitch Extraction," *J. Acoust. Soc. Am.,* Vol. 34, No. 7, pp. 916-921, 1962.

14. B. Gold and L. R. Rabiner, "Parallel Processing Techniques for Estimating Pitch Periods of Speech in the Time Domain," *J. Acoust. Soc. Am.,* Vol. 46, No. 2, Pt. 2, pp. 442-448, August 1969.

15. T. P. Barnwell, J. E. Brown, A. M. Bush, and C. R. Patisaul, "Pitch and Voicing in Speech Digitization," Res. Rept. No. E-21-620-74-B4-1, Georgia Inst. of Tech., August 1974.

16. M. J. Ross, H. L. Shaffer, A. Cohen, R. Freudberg, and H. J. Manley, "Average Magnitude Difference Function Pitch Extractor," *IEEE Trans. Acoust., Speech and Signal Proc.*, Vol. ASSP-22, pp. 353-362, October 1974.

17. M. M. Sondhi, "New Methods of Pitch Extraction," *IEEE Trans. Audio and Electroacoustics*, Vol. AU-16, No. 2, pp. 262-266, June 1968.

18. L. R. Rabiner, "On the Use of Autocorrelation Analysis for Pitch Detection," *IEEE Trans. Acoust., Speech and Signal Proc.*, Vol. ASSP-25, No. 1, pp. 24-33, February 1977.

19. J. J. Dubnowski, R. W. Schafer, and L. R. Rabiner, "Real-Time Digital Hardware Pitch Detector," *IEEE Trans. Acoust., Speech, and Signal Proc.*, Vol. ASSP-24, No. 1, pp. 2-8, February 1976.

20. J. W. Tukey, "Nonlinear (Nonsuperposable) Methods for Smoothing Data," *Congress Record, 1974 EASCON*, p. 673, 1974.

21. L. R. Rabiner, M. R. Sambur, and C. E. Schmidt, "Applications of a Nonlinear Smoothing Algorithm to Speech Processing," *IEEE Trans. Acoust., Speech, and Signal Proc.*, Vol. ASSP-23, No. 6, pp. 552-557, December 1975.

22. W. A. Blankenship, "Note on Computing Autocorrelation," *IEEE Trans. Acoust., Speech, and Signal Proc.*, Vol. ASSP-22, No. 1, pp. 76-77, February 1974.

23. W. B. Kendall, "A New Algorithm for Computing Autocorrelations," *IEEE Trans. Computers*, Vol. C-23, No. 1, pp. 90-93, January 1974.

24. T. G. Stockham, Jr., "High-Speed Convolution and Correlation," 1966 Spring Joint Computer Conf., AFIPS Conf. Proc., Vol. 28, pp. 229-233, 1966.

PROBLEMS

4.1 The rectangular window is defined as

$$w_R(n) = 1 \quad 0 \leqslant n \leqslant N - 1$$
$$= 0 \quad otherwise$$

The Hamming window is defined as

$$w_H(n) = .54 - .46 \cos[2\pi n/(N-1)] \quad 0 \leqslant n \leqslant N - 1$$

$$= 0 \qquad otherwise$$

(a) Show that the Fourier transform of the rectangular window is

$$W_R(e^{j\omega}) = \frac{\sin(\omega N/2)}{\sin(\omega/2)} e^{-j\omega(N-1)/2}$$

(b) Sketch $W_R(e^{j\omega})$ as a function of ω. (Disregard the linear phase factor $e^{-j\omega(N-1)/2}$.)

(c) Express $w_H(n)$ in terms of $w_R(n)$ and thereby obtain an expression for $W_H(e^{j\omega})$ in terms of $W_R(e^{j\omega})$.

(d) Sketch the individual terms in $W_H(e^{j\omega})$. (Disregard the linear phase factor $e^{-j\omega(N-1)/2}$ that is common to each term.) Your sketch should illustrate how the Hamming window trades frequency resolution for increased suppression of higher frequencies.

4.2 The short-time energy of a sequence $x(n)$ is defined as

$$E_n = \sum_{m=-\infty}^{\infty} [x(m)w(n-m)]^2$$

For the particular choice

$$w(m) = a^m \quad m \geqslant 0$$
$$= 0 \quad m < 0$$

it is possible to find a recurrence formula for E_n.

(a) Find a difference equation that expresses E_n in terms of E_{n-1} and the input $x(n)$.

(b) Draw a digital network diagram of this equation.

(c) What general property must

$$h(m) = w^2(m)$$

have in order that it be possible to find a recursive implementation?

4.3 The short-time energy is defined as

$$E_n = \sum_{m=-N}^{N} h(m)x^2(n-m)$$

Suppose we wish to compute E_n at each sample of the input.

(a) Let $h(m)$ be

$$h(m) = a^{|m|} \quad |m| \leqslant N$$
$$= 0 \quad otherwise$$

Find a recurrence relation (i.e., a difference equation) for E_n.

(b) What is the savings in number of multiplications obtained by using the recurrence relation rather than directly computing E_n?

(c) Draw a digital network diagram of the recurrence formula for E_n. (As defined, $h(m)$ is noncausal. Therefore an appropriate delay must be inserted.)

4.4 Suppose that the average magnitude is to be estimated every L samples at the input sampling rate. One possibility is to use a finite length window as in

$$M_n = \sum_{m=n-N+1}^{n} |x(m)|w(n-m)$$

In this case, M_n is only computed once for each L samples of the input. Another approach is to use a window for which a recurrence formula can be obtained, e.g.,

$$M_n = aM_{n-1} + |x(n)| .$$

In this case M_n must be computed at each input sample, even though we may only want it every L samples.

(a)   How many multiplies and adds are required to compute M_n once for each L samples with the finite length window?

(b)   Repeat (a) for the recursive definition of M_n.

(c)   Under what conditions will the finite length window be more efficient?

4.5   The short-time average zero-crossing rate was defined in Eqs. (4.18)-(4.20) as

$$Z_n = \frac{1}{2N} \sum_{m=n-N+1}^{n} |sgn[x(m)] - sgn[x(m-1)]|$$

Show that Z_n can be expressed as

$$Z_n = Z_{n-1} + \frac{1}{2N} \left\{ |sgn[x(n)] - sgn[x(n-1)]| - |sgn[x(n-N)] - sgn[x(n-N-1)]| \right\}$$

4.6   To demonstrate how a parallel processing pitch detector can combine several independent pitch detectors, each with a fairly high error probability, and give a highly reliable result, consider the following idealized situation. Assume there are 7 *independent* pitch detectors, each having a probability p of correctly estimating pitch period, and probability $1 - p$ of incorrectly estimating pitch period. The decision logic is to combine the 7 pitch estimates in such a way that an overall error is made only if 4 or more of the individual pitch detectors make an error.

(a)   Derive an explicit expression for the probability of error of the parallel processing pitch detector in terms of p. (Hint: consider the result of each pitch detector a Bernoulli trial with probability $(1-p)$ of making an error and probability p of no error.)

(b)   Sketch a curve showing the overall error probability as a function of p.

(c)   For what value of p is the overall error probability less than 0.05?

4.7   As given by Eq. (4.24) the short-time autocorrelation function is defined as

$$R_n(k) = \sum_{m=-\infty}^{\infty} x(m) w(n-m) x(m+k) w(n-k-m)$$

(a)   Show that

$$R_n(k) = R_n(-k)$$

i.e., show that $R_n(k)$ is an even function of k.

(b) Show that $R_n(k)$ can be expressed as

$$R_n(k) = \sum_{m=-\infty}^{\infty} x(m)x(m-k)h_k(n-m)$$

where

$$h_k(n) = w(n)w(n+k)$$

(c) Suppose that

$$w(n) = a^n \qquad n \geqslant 0$$
$$= 0 \qquad n < 0$$

Find the impulse response, $h_k(n)$, for computing the kth lag.
(d) Find the z-transform of $h_k(n)$ in (c) and from it obtain a recursive implementation for $R_n(k)$. Draw a digital network implementation for computing $R_n(k)$ as a function of n for the window of (c).
(e) Repeat parts (c) and (d) for

$$w(n) = na^n \qquad n \geqslant 0$$
$$= 0 \qquad n < 0$$

4.8 Consider the periodic impulse train

$$x(m) = \sum_{r=-\infty}^{\infty} \delta(m-rP)$$

(a) Using Eq. (4.30), with $w'(m)$ a rectangular window whose length, N, satisfies

$$QP < N - 1 < (Q+1)P$$

where Q is an integer, find and sketch $R_n(k)$ for $0 \leqslant k \leqslant N - 1$.
(b) How would the result of (a) change if the window is a Hamming window of the same length?
(c) Find and sketch the modified short-time autocorrelation function, $\hat{R}_n(k)$, given by Eq. (4.36) for the same value of N.

4.9 The long-time autocorrelation function of a random signal or a periodic signal is defined as

$$\phi(k) = \lim_{N \to \infty} \frac{1}{2N+1} \sum_{m=-N}^{N} x(m)x(m+k)$$

The short-time autocorrelation function is defined as

$$R_n(k) = \sum_{m=0}^{N-|k|-1} x(n+m)w'(m)x(n+m+k)w'(m+k)$$

and the modified short-time autocorrelation function is defined as

$$\hat{R}_n(k) = \sum_{m=0}^{N-1} x(n+m)x(n+m+k)$$

Show whether or not the following statements are true or false.

(a) If $x(n) = x(n+P)$, $-\infty < n < \infty$, then

    (i) $\phi(k) = \phi(k+P)$ $-\infty < k < \infty$
    (ii) $R_n(k) = R_n(k+P)$ $-(N-1) \leqslant k \leqslant N-1$
    (iii) $\hat{R}_n(k) = \hat{R}_n(k+P)$ $-(N-1) \leqslant k \leqslant N-1$

(b)

    (i) $\phi(-k) = \phi(k)$ $-\infty < k < \infty$
    (ii) $R_n(-k) = R_n(k)$ $-(N-1) \leqslant k \leqslant N-1$
    (iii) $\hat{R}_n(-k) = \hat{R}_n(k)$ $-(N-1) \leqslant k \leqslant N-1$

(c)

    (i) $\phi(k) \leqslant \phi(0)$ $-\infty < k < \infty$
    (ii) $R_n(k) \leqslant R_n(0)$ $-(N-1) \leqslant k \leqslant N-1$
    (iii) $\hat{R}_n(k) \leqslant \hat{R}_n(0)$ $-(N-1) \leqslant k \leqslant N-1$

(d)

    (i) $\phi(0)$ is equal to the power in the signal.
    (ii) $R_n(0)$ is the short-time energy.
    (iii) $\hat{R}_n(0)$ is the short-time energy.

4.10 Consider the signal

$$x(n) = \cos \omega_0 n \quad -\infty < n < \infty$$

(a) Find the long-time autocorrelation function, $\phi(k)$, for $x(n)$ (Eq. (4.21)).
(b) Sketch $\phi(k)$ as a function of k.
(c) Find and sketch the long-time autocorrelation function of the signal

$$\begin{aligned} y(n) &= 1 \quad \text{if} \quad x(n) \geqslant 0 \\ &= 0 \quad \text{if} \quad x(n) < 0 \end{aligned}$$

4.11 The short-time average magnitude difference function (AMDF) of the signal $x(n)$ is defined as (see Eq. (4.38))

$$\gamma_n(k) = \frac{1}{N} \sum_{m=0}^{N-1} |x(n+m) - x(n+m-k)|$$

(a) Using the inequality [16]

$$\frac{1}{N} \sum_{m=0}^{N-1} |x(m)| \leqslant \left[\frac{1}{N} \sum_{m=0}^{N-1} |x(m)|^2 \right]^{1/2}$$

show that

$$\gamma_n(k) \leqslant \left[2(\hat{R}_n(0) - \hat{R}_n(k)) \right]^{1/2}$$

170

This result leads to Eq. (4.39)

(b) Sketch $\gamma_n(k)$ and the quantity $[2(\hat{R}_n(0)-\hat{R}_n(k))]^{1/2}$ for $0 \leqslant k \leqslant 200$ for the signal

$$x(n) = \cos(\omega_0 n)$$

with $N = 200$, $\omega_0 = 200\pi/(10000)$.

4.12 Consider the signal

$$x(n) = A\cos(\omega_0 n)$$

as input to a three-level center clipper which produces an output

$$\begin{aligned} y(n) &= 1 & x(n) &> C_L \\ &= 0 & |x(n)| &\leqslant C_L \\ &= -1 & x(n) &< -C_L \end{aligned}$$

(a) Sketch $y(n)$ as a function of n for $C_L = 0.5A$, $C_L = 0.75A$, and $C_L = A$.
(b) Sketch the autocorrelation function for $y(n)$ for the values of C_L in (a).
(c) Discuss the effect of the setting of C_L as it approaches A. Suppose that A varies with time such that

$$0 < A(n) \leqslant A_{max}$$

Discuss problems that this can cause if C_L is close to A_{max}.

5

Digital Representations
of the Speech Waveform

5.0 Introduction

"Watson, if I can get a mechanism which will make a current of electricity vary its intensity as the air varies in density when sound is passing through it, I can telegraph any sound, even the sound of speech" — A. G. Bell [1].

This simple idea, so important in the history of human communication, seems commonplace today. The basic principle embodied in Bell's great invention is fundamental to a multitude of devices and systems for recording, transmission or processing of speech signals in which the speech signal is represented by reproducing the amplitude fluctuations of the speech waveform. This is also the case for digital systems, where the speech waveform is often represented by a sequence of numbers which specifies the pattern of amplitude fluctuations.

The general nature of digital speech waveform representations is depicted in Fig. 5.1. As illustrated there, the speech waveform, which can be thought of as a continuous function of a continuous time variable, is sampled, generally periodically in time, to produce a sequence of samples, $x_a(nT)$. These samples generally would take on a continuum of values. Therefore, it is necessary to quantize them to a finite set of values in order to obtain a digital representation; i.e., one that is discrete in both time and amplitude.

As we shall see in this chapter, Fig. 5.1 is a convenient conceptualization of the process of obtaining a digital representation of the speech waveform. It

172

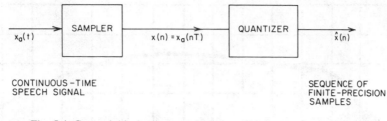

CONTINUOUS–TIME
SPEECH SIGNAL

SEQUENCE OF
FINITE–PRECISION
SAMPLES

Fig. 5.1 General block diagram depicting digital waveform representations.

may not always be possible to separate a given representation into two distinct stages; however, the two basic features, sampling and quantization, are inherent in all the schemes that we shall discuss in this chapter.

We shall begin by discussing the sampling process as applied to speech signals. Then we shall discuss a variety of schemes for quantizing the samples of speech signals.

5.1 Sampling Speech Signals

We have already discussed the sampling theorem in Chapter 2. There it was shown that samples of an analog signal are a unique representation if the analog signal is bandlimited and if the sampling rate is at least twice the Nyquist frequency. Since we are concerned with digital representations of speech signals, we need to consider the spectral properties of speech. We recall from the discussion of Chapter 3 that according to the steady state models for the produc-

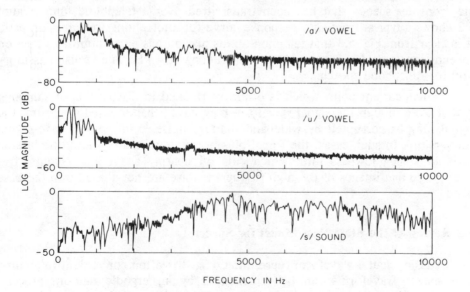

Fig. 5.2 Spectra of voiced sounds (/a/ vowel and /u/ vowel) and an unvoiced sound (/s/) for a 20 kHz sampling rate.

173

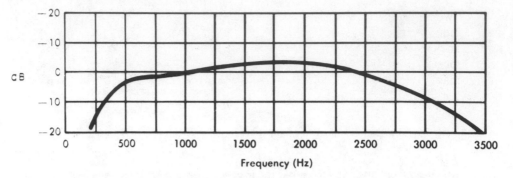

Fig. 5.3 Frequency response of a typical telephone transmission path. (After BTL, Transmission Systems for Communication, p. 73.)

tion of vowel and fricative sounds, speech signals are not inherently bandlimited, although the spectrum does tend to fall off rapidly at high frequencies. Figure 5.2 shows spectra of some typical speech sounds. It is observed that for of the voiced sounds, the high frequencies are more than 40 dB below the peak of the spectrum for frequencies above 4 kHz. On the other hand, for unvoiced sounds, the spectrum has not fallen off appreciably even above 8 kHz. Thus, to accurately represent all speech sounds would require a sampling rate greater than 20 kHz. In most applications, however, this sampling rate is not required. For example, if the sampling operation is a prelude to the process of estimating the first three formant frequencies of voiced speech, we are only interested in the portion of the spectrum up to abut 3.5 kHz. Therefore, if the speech is filtered by a sharp cutoff analog filter prior to sampling, so that the Nyquist frequency is 4 kHz, then a sampling rate of 8 kHz is possible. As another example, consider speech that has been transmitted over a telephone line. Figure 5.3 shows a typical frequency response curve for a telephone transmission path. It is clear from Fig. 5.3 that telephone transmission has a bandlimiting effect on speech signals, and indeed, a Nyquist frequency of 4 kHz is a realistic assumption for "telephone speech."

An important point which is often overlooked in discussions of sampling is that even though the signal waveform may have a bandlimited spectrum, the signal may be corrupted by wideband random noise, prior to analog-to-digital conversion. In such cases, the signal plus noise combination should be filtered with an analog lowpass filter which cuts off sharply above the Nyquist frequency, so that images of the high frequency noise are not aliased into the base band.

5.2 Review of the Statistical Model for Speech

In discussing digital waveform representations, it is often convenient to assume that speech waveforms can be represented by an ergodic random process. Although this is a gross simplification, we will see that a statistical point of view yields useful results thereby justifying the use of such a model.

If we assume that the signal $x_a(t)$ is a sample function of a continuous-time random process, then the sequence of samples derived by periodic sampling can likewise be thought of as a sample sequence of a discrete-time random process. For many purposes in communication system analysis, an adequate characterization of the analog signal consists of a first order probability density, $p(x)$, and the autocorrelation function of the random process, which is defined as

$$\phi_a(\tau) = E[x_a(t)x_a(t+\tau)] \tag{5.1}$$

where $E[\ ]$ denotes the expectation of the quantity within the brackets. The analog power spectrum is the Fourier transform of $\phi_a(\tau)$; i.e.,

$$\Phi_a(\Omega) = \int_{-\infty}^{\infty} \phi_a(\tau)e^{-j\Omega\tau}d\tau \tag{5.2}$$

The discrete-time signal obtained by sampling the random signal $x_a(t)$ has an autocorrelation function

$$\phi(m) = E[x(n)x(n+m)]$$
$$= E[x_a(nT)x_a(nT+mT)] = \phi_a(mT) \tag{5.3}$$

Thus, since $\phi(m)$ is just a sampled version of $\phi_a(\tau)$, then the power spectrum of $\phi(m)$ is given by

$$\Phi(e^{j\Omega T}) = \sum_{m=-\infty}^{\infty} \phi(m)e^{-j\Omega Tm}$$
$$= \frac{1}{T} \sum_{k=-\infty}^{\infty} \Phi_a\left(\Omega + \frac{2\pi}{T} k\right) \tag{5.4}$$

Eq. (5.4) shows that for the random process model of speech, the power spectrum of the sampled signal is an aliased version of the power spectrum of the original analog signal.

The probability density function for the amplitudes, $x(n)$, is the same as for the amplitudes, $x_a(t)$, since $x(n) = x_a(nT)$. Thus averages such as mean and variance are the same for the samples as for the original analog signal.

When applying statistical notions to speech signals, it is necessary to estimate the probability density and correlation function (or the power spectrum) from speech waveforms. The probability density is estimated by determining a histogram of amplitudes for a large number of samples; i.e., over a long time. Davenport [2] made extensive measurements of this kind, and more recently, Paez and Glisson [3], using similar measurements, have shown that a good approximation to measured speech amplitude densities is a gamma distribution of the form

$$p(x) = \left[\frac{\sqrt{3}}{8\pi\sigma_x|x|}\right]^{1/2} e^{-\frac{\sqrt{3}|x|}{2\sigma_x}} \tag{5.5}$$

175

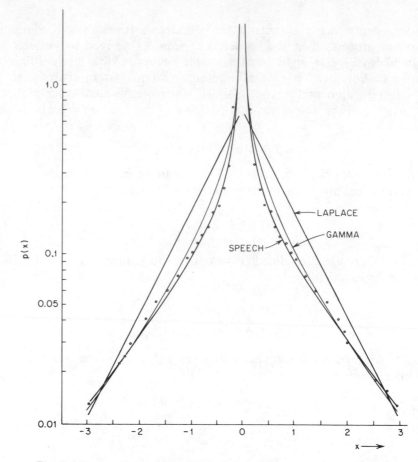

Fig. 5.4 Real speech and theoretical Gamma and Laplace probability densities. (After Paez and Glisson [3].)

A somewhat simpler approximation is the Laplacian density

$$p(x) = \frac{1}{\sqrt{2}\sigma_x} e^{-\frac{\sqrt{2}|x|}{\sigma_x}} \qquad (5.6)$$

Figure 5.4 shows a measured amplitude density for speech along with gamma and Laplacian densities, all of which have been normalized so that the mean is zero and the variance (σ_x^2) is unity. The gamma density is clearly a better approximation than the Laplacian density, but both are reasonably close.

The autocorrelation function and power spectrum of speech signals can be estimated by standard time-series analysis techniques. An estimate of the autocorrelation function of an ergodic random process can be obtained by estimating the time-average autocorrelation function from a long (but finite) segment of the signal. For example, the definition of the short-time autocorrelation function (Eq. (4.30) of Chapter 4) can be slightly modified to give the estimate of the long-time average autocorrelation function

176

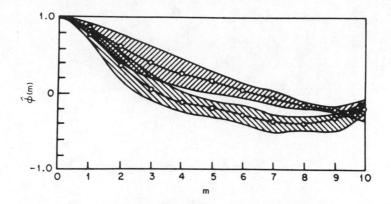

Fig. 5.5 Autocorrelation functions of speech signals; upper curves for lowpass speech, lower curves for bandpass speech. (After Noll [4].)

$$\hat{\phi}(m) = \frac{1}{L} \sum_{n=0}^{L-1-m} x(n)x(n+m), \quad 0 \leqslant |m| \leqslant L - 1 \tag{5.7}$$

where L is a large integer, An example of such an estimate is shown in Fig. 5.5 for an 8 kHz sampling rate [4]. The upper curve shows the correlation for lowpass filtered speech and the lower curve is for bandpass filtered speech. The shaded region around each curve shows the variation of the estimate due to

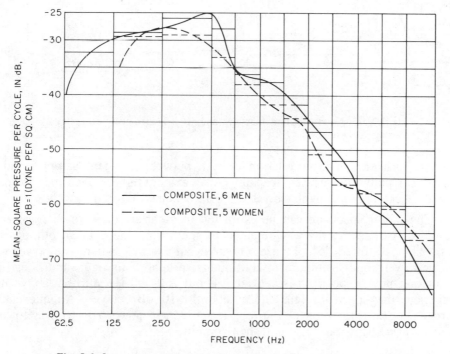

Fig. 5.6 Long-time power density spectrum for continuous speech. (After Dunn and White [5].)

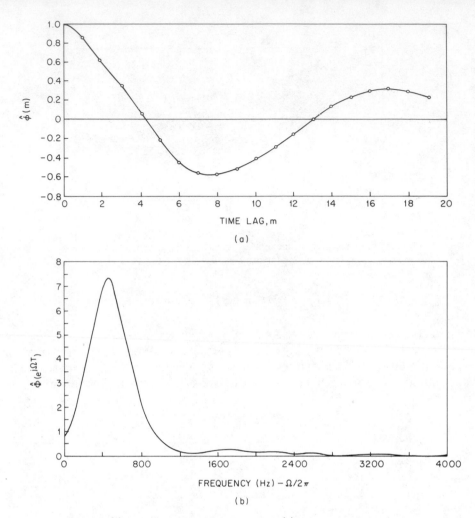

Fig. 5.7 (a) Autocorrelation function; and (b) power density estimates for speech. (After Noll [7].)

different speakers. The correlation is high between adjacent samples and it decreases rapidly for greater spacing. Also evident is the fact that lowpass filtered speech is more highly correlated than bandpass filtered speech.

The power spectrum can be estimated in a variety of ways. For speech, one of the earliest results was obtained by measuring the average output of a set of bandpass filters [5]. Figure 5.6 shows an example where the power was averaged over about a minute of continuous speech. This figure indicates that the average power spectrum is peaked at about 250-500 Hz and that above this frequency, the spectrum falls off at about 8-10 dB/octave. An alternative approach to the estimation of the long-term average power spectrum is to first estimate $\hat{\phi}(m)$ as in Eq. (5.7) and then compute

$$\hat{\Phi}(e^{j\Omega T}) = \sum_{m=-M}^{M} w(m)\hat{\phi}(m)e^{-j\Omega m T} \qquad (5.8)$$

178

for a discrete set of frequencies, $\Omega_k = 2\pi k/T$, for $k = 0, 1, \ldots, N - 1$, using the discrete Fourier transform [6] where $w(m)$ is a window (weighting) function on the autocorrelation function. An example of this method of spectrum estimation as applied to speech is shown in Fig. 5.7 for $w(m)$, a Hamming window [7]. Still another approach is to compute the power transfer function of a recursive digital filter whose output when excited by white noise has the same spectral properties as the given signal. (See Chapter 8.)

5.3 Instantaneous Quantization

As we have already pointed out, it is convenient to consider the processes of sampling and quantization separately even though it is often impossible to distinguish the two separate functions in actual implementations of digital waveform representations. Thus, let us assume that a speech waveform has been lowpass filtered and sampled at a suitable rate giving a sequence, $\{x(n)\}$, which is known with infinite precision. In most of our discussion in this chapter we will view this sequence of samples as a discrete-time random process. In order to transmit this sequence of samples over a digital communication channel, or to store them in digital memory or to use them as the input to a digital signal processing algorithm, the sample values must be quantized to a finite set of amplitudes so that they can be represented by a finite set of symbols. This process of quantization and coding is depicted in Fig. 5.8. Just as it is conceptually useful to separate sampling and quantization into two distinct steps, it is likewise useful to separate the process of representing the samples, $\{x(n)\}$, by a finite set of symbols, $\{c(n)\}$, into two stages; a quantization stage which produces a sequence of quantized amplitudes $\{\hat{x}(n)\} = \{Q[x(n)]\}$ and an encoding stage which represents each quantized sample by a code word, $c(n)$. This is depicted in Fig. 5.8a. (The quantity Δ in Fig. 5.8a represents the quantization step size for the quantizer.) Likewise it is convenient to define a decoder which takes a sequence of code words, $\{c'(n)\}$, and transforms it back into a sequence of quantized samples, $\{\hat{x}'(n)\}$, as depicted in Fig. 5.8b. If the code words $c'(n)$ are the same as the code words $c(n)$, i.e., no errors have been introduced, then the output of the ideal decoder is identical to the quantized samples; i.e., $\hat{x}'(n) = \hat{x}(n)$.

In most cases it is convenient to use binary numbers to represent the quantized samples. With B-bit binary code words it is possible to represent 2^B

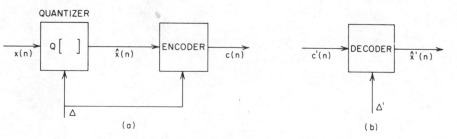

Fig. 5.8 Process of quantization and coding; (a) coder; (b) decoder.

179

different quantization levels. The information capacity required to transmit or store the digital representation is therefore:

$$I = B \cdot F_s = \textit{Bit rate in bits per second} \qquad (5.9)$$

where F_s is the sampling rate (i.e., samples/second) and B is the number of bits/sample. It is generally desirable to maintain the bit rate as low as possible while maintaining a required level of quality. For a given speech bandwidth, the minimum sampling rate is fixed by the sampling theorem. Therefore the only way to reduce the bit rate is to reduce the number of bits/sample. For this reason we shall now turn to a discussion of a variety of techniques for quantizing a signal.

In general, it is reasonable to assume that the samples $\{x(n)\}$ will fall in a finite range of amplitudes such that

$$|x(n)| \leqslant X_{\max} \qquad (5.10)$$

For convenience, it may be desirable to assume that $X_{\max}$ is infinite, as for example when we assume a particular form for the probability density function of amplitudes of $x(n)$ such as the gamma or Laplacian distribution. However, we should bear in mind that the assumption of a finite range of amplitudes is more realistic. Even if we assume a Laplacian density, it is easy to show (see Problem 5.2) that only 0.35% of the speech samples would fall outside the range

$$-4\sigma_x \leqslant x(n) \leqslant 4\sigma_x \qquad (5.11)$$

Thus, it is convenient to assume that the peak-to-peak range of the speech signal is proportional to the standard deviation of the signal.

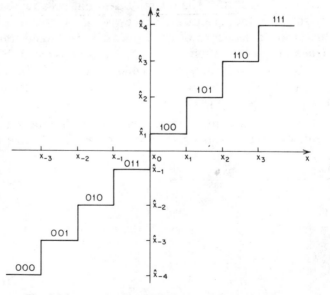

Fig. 5.9 Input-output characteristic of a 3-bit quantizer.

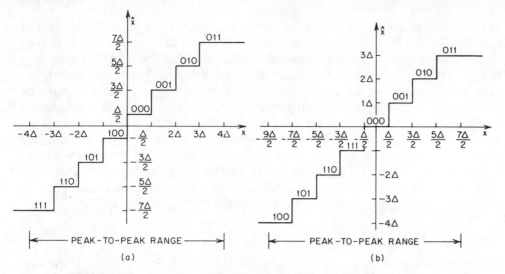

Fig. 5.10 Two common uniform quantizer characteristics; (a) mid-riser; (b) mid-tread.

The amplitudes of the samples are quantized by dividing the entire amplitude range into a finite set of amplitude ranges and assigning the same amplitude value to all samples falling in a given range. This is shown in Fig. 5.9 for an 8-level quantizer. For example, we see that for all values of $x(n)$ between x_1 and x_2, the output of the quantizer is $\hat{x}(n) = Q[x(n)] = \hat{x}_2$. Each of the eight quantizer levels is labelled with a 3-bit binary code word which serves as a symbolic representation of that amplitude level. For example, in Fig. 5.9 the coded output for a sample whose amplitude is between x_1 and x_2 would be the binary number 101. The particular labelling scheme in Fig. 5.9 is arbitrary. Any of the 8 factorial possible labelling schemes is a possibility; however, there are often good reasons for the choice of a particular scheme.

5.3.1 Uniform quantization

The quantization ranges and levels may be chosen in a variety of ways depending on the intended application of the digital representation. When the digital representation is to be processed by a digital system, the quantization levels and ranges are generally distributed uniformly. Thus to define a uniform quantizer using the example of Figure 5.9, we set

$$x_i - x_{i-1} = \Delta \tag{5.12}$$

and

$$\hat{x}_i - \hat{x}_{i-1} = \Delta \tag{5.13}$$

where Δ is the quantization step-size. Two common uniform quantizer characteristics are shown in Fig. 5.10 for the case of eight quantization levels. Figure 5.10a shows the case where the origin appears to be in the middle of a rising part of the staircase-like function. This class of quantizers is called the "mid-

181

riser" class. Likewise Fig. 5.10b shows an example of the "mid-tread" class of quantizers. For the case where the number of levels is a power of 2, as is convenient for a binary coding scheme, it can be seen that the mid-riser quantizer has the same number of positive and negative levels, and these are symmetrically positioned about the origin. In contrast, the mid-tread quantizer has one more negative level than positive; however, in this case one of the quantization levels is zero while there is no zero level in the mid-riser case. Code word assignments are shown in Fig. 5.10 in the manner of Fig. 5.9. In this case, the code words have been assigned so as to have a direct numerical significance. For example, in Fig. 5.10a, if we interpret the binary code words as a sign-magnitude representation with the left most bit being the sign bit, then the quantized samples are related to the code words by the relationship

$$\hat{x}(n) = \frac{\Delta}{2} \text{ sign } (c(n)) + \Delta c(n) \tag{5.14}$$

where $\text{sign}(c(n))$ is equal to $+1$ if the first bit of $c(n)$ is 0, and -1 if the first bit of $c(n)$ is 1. Similarly, we can interpret the binary code words in Fig. 5.10b as a 3-bit two's-complement representation, in which case the quantized samples are related to the code words by the relationship

$$\hat{x}(n) = \Delta c(n) \tag{5.15}$$

This latter method of assignment of code words to quantization levels is most commonly used when the sequence of samples is to be processed by a signal processing algorithm which is implemented with two's-complement arithmetic (as on most minicomputers), since the code words can serve as a direct numerical representation of the sample values.

For uniform quantizers (as shown in Fig. 5.10) there are only two parameters: the number of levels and the quantization step size, Δ. The number of levels is generally chosen to be of the form 2^B so as to make the most efficient use of B-bit binary code words. Together, Δ and B must be chosen so as to cover the range of input samples. If we assume that $|x(n)| \leqslant X_{\text{max}}$, then (assuming a symmetrical probability density function for $x(n)$) we should set

$$2X_{\text{max}} = \Delta 2^B \tag{5.16}$$

In discussing the effect of quantization it is helpful to represent the quantized samples $\hat{x}(n)$ as

$$\hat{x}(n) = x(n) + e(n) \tag{5.17}$$

where $x(n)$ is the unquantized sample and $e(n)$ is the quantization error or noise. It can be seen from both Figs. 5.10a and 5.10b that if Δ and B are chosen as in Eq. (5.16), then

$$-\frac{\Delta}{2} \leqslant e(n) \leqslant \frac{\Delta}{2} \tag{5.18}$$

By way of example, if we choose the peak-to-peak range of $x(n)$ to be $8\sigma_x$ and if we assume a Laplacian probability density function (as discussed in Section

5.2), then only 0.35% of the samples will fall outside the range of the quantizer. The clipped samples will incur a quantization error in excess of $\pm \Delta/2$; however, their number is so small that it is common to assume a range on the order $8\sigma_x$ and neglect the infrequent large errors in theoretical calculations [8].

It is clear that we do not know either $x(n)$ or $e(n)$, but only the quantized value $\hat{x}(n)$. To study quantization effects, it is convenient and useful to assume a simple statistical model for the quantization noise. This model is based on the following assumptions.

1. The quantization noise is a stationary white noise process; i.e.,

$$E[e(n)e(n+m)] = \sigma_e^2, \quad m = 0 \qquad (5.19)$$
$$= 0, \qquad otherwise$$

2. The quantization noise is uncorrelated with the input signal; i.e.,

$$E[x(n)e(n+m)] = 0, \quad for \; all \; m \qquad (5.20)$$

3. The distribution of quantization errors is uniform over each quantization interval and since the intervals are all of the same length,

$$p_e(e) = \frac{1}{\Delta}, \quad -\frac{\Delta}{2} \leqslant e \leqslant \frac{\Delta}{2} \qquad (5.21)$$
$$= 0, \qquad otherwise$$

These assumptions are clearly unrealistic for some types of signals. For example, if the input is a constant for all n, the above assumptions are not appropriate. Speech, however, is a complicated signal which fluctuates rapidly among all the quantization levels, and if Δ is small enough, the amplitude of the signal is likely to traverse many quantization steps in going from sample to sample. In this case, experiments have shown [9] that the above assumptions hold quite well.

An example which illustrates the validity of the above statistical assumptions is depicted in Fig. 5.11 [6]. Figure 5.11a shows 400 consecutive samples of a speech signal. This signal was quantized by 3- and 8-bit uniform quantizers of the form shown in Fig. 5.10b. The resulting quantization error signals are shown in Figs. 5.11b and 5.11c respectively. It is seen that the error sequence shows some correlation to the input signal in the case of 3-bit quantization; however, no correlation is readily apparent for the 8-bit quantizer. To verify this observation, the correlation functions for the 3- and 8-bit quantization error sequences are given in Figs. 5.12a and 5.12c respectively. Clearly, the correlation function shown in Fig. 5.12c is quite consistent with the assumption that $\phi(m) = \sigma_e^2\delta(m)$; however, Fig. 5.12a shows a significant correlation for $m > 0$. This is reflected in the resulting power spectrum estimates (given in Figures 5.12b and 5.12d respectively) which show that the 3-bit quantization noise spectrum tends to fall off at high frequencies (i.e., as the speech spectrum) while the 8-bit quantization noise spectrum is very flat.

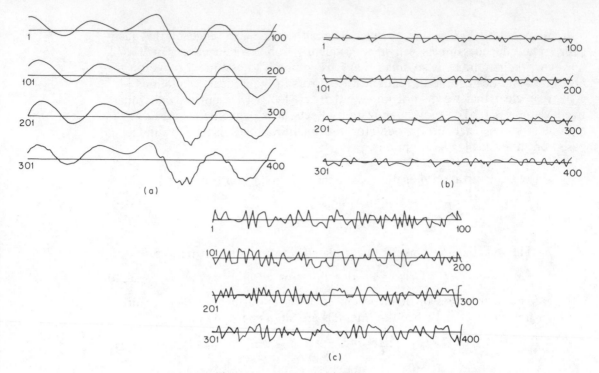

Fig. 5.11 (a) Speech waveform; (b) quantization error for 3-bit quantization (same scale as in (a)); (c) quantization error for 8-bit quantization (magnified 66 times with respect to (a)).

Note, however, that even for 3-bit quantization, the noise spectrum varies only 6 dB across the entire base band.

With this statistical model for the quantization noise, it is possible to relate the strength of the noise to the signal strength and the parameters of the quantizer. For this purpose it is convenient to compute the signal-to-quantization noise ratio defined as[1]

$$SNR = \frac{\sigma_x^2}{\sigma_e^2} = \frac{E[x^2(n)]}{E[e^2(n)]} = \frac{\sum_n x^2(n)}{\sum_n e^2(n)} \tag{5.22}$$

If the peak-to-peak quantizer range is assumed to be $2X_{max}$, then, for a B-bit quantizer, we get

$$\Delta = \frac{2X_{max}}{2^B} \tag{5.23}$$

If we assume a uniform amplitude distribution for the noise, we obtain (see Problem 5.1)

---

[1]Note we are assuming that $x(n)$ has zero mean. If this is not the case, the mean value of $x(n)$ should be subtracted out prior to *SNR* calculations.

$$\sigma_e^2 = \frac{\Delta^2}{12} = \frac{X_{max}^2}{(3)2^{2B}} \tag{5.24}$$

Substituting Eq. (5.24) into Eq. (5.22) gives

$$SNR = \frac{(3)2^{2B}}{\left[\dfrac{X_{max}}{\sigma_x}\right]^2} \tag{5.25}$$

or expressing the signal-to-quantizing noise in dB units,

$$SNR\,(dB) = 10\,\log_{10}\left[\frac{\sigma_x^2}{\sigma_e^2}\right]$$

$$= 6B + 4.77 - 20\,\log_{10}\left[\frac{X_{max}}{\sigma_x}\right] \tag{5.26}$$

if we assume that the quantizer range is such that $X_{max} = 4\sigma_x$, then Eq. (5.26) becomes [8]

$$SNR\,(dB) = 6B - 7.2 \tag{5.27}$$

Equation (5.27), which states that each bit in the code word contributes 6 dB to the signal-to-noise ratio, is valid subject to the assumptions

1. The input signal fluctuates in a complicated manner so that a statistical model for the noise sequence is valid.

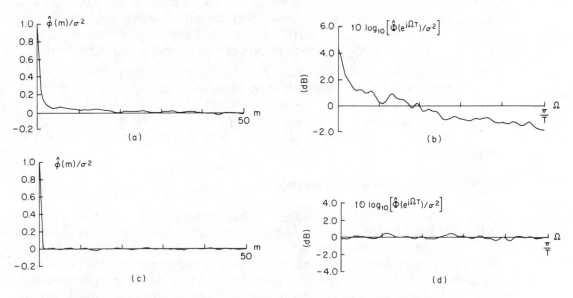

Fig. 5.12 (a) Normalized autocorrelation estimate for 3-bit quantization; (b) power spectrum for 3-bit quantization; (c) normalized autocorrelation estimate for 8-bit quantization; (d) power spectrum for 8-bit quantization.

2. The quantization step size is small enough to remove any possibility of signal correlated patterns in the noise waveform.
3. The range of the quantizer is set so as to match the peak-to-peak range of the signal; i.e., so that very few samples are clipped but yet the full range is utilized.

For speech signals, these first two assumptions hold up very well when the number of quantizer levels is reasonably large, say greater than 2^6. However, the third assumption is less valid for speech signals, since the signal energy may vary as much as 40 dB among speakers and with transmission environment. Also for a given speaking environment, the amplitude of the speech signal varies considerably from voiced speech to unvoiced speech and even within voiced sounds. Since Eq. (5.27) assumes a given range of amplitudes, if the signal fails to achieve that range, it is as if fewer quantization levels are available to represent the signal; i.e., as if fewer bits were used. For example, it is evident from Eq. (5.26) that if the input variance is actually only one half the range for which the quantizer was designed, the signal-to-noise ratio is reduced by 6 dB. Likewise, on a short-time basis, the variance of an unvoiced segment may be 20 to 30 dB below the variance for voiced speech. Thus, the "short-time" signal-to-noise ratio may be much less during unvoiced segments than during voiced segments.

In order to maintain a fidelity of representation with uniform quantization that is acceptable perceptually, it is necessary to use many more bits than might be implied by the previous analysis in which we have assumed that the signal is stationary. For example, whereas Eq. (5.27) suggests that $B = 7$ would provide about 36 dB *SNR* which would most likely provide adequate quality in a communications system, it is generally accepted that about 11 bits are required to provide high quality representation of speech signals with a uniform quantizer.

For all of the above reasons, it would be very desirable to have a quantizing system for which the signal-to-noise ratio was independent of signal level. That is, rather than the error being of constant variance independent of signal amplitude as for uniform quantization, it would be desirable to have a constant percentage error. This can be achieved using a nonuniform distribution of quantization levels.

5.3.2 Instantaneous companding

In order that the percentage error be a constant, the quantization levels must be logarithmically spaced. Alternatively, the logarithm of the input can be quantized rather than the input itself. This is depicted in Fig. 5.13 which shows the input amplitudes being compressed by the logarithm function prior to quantization and being expanded by the exponential function after decoding. To see that this leads to the desired insensitivity to signal amplitude, assume that

$$y(n) = ln|x(n)| \qquad (5.28)$$

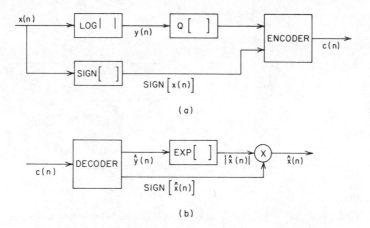

(a)

(b)

Fig. 5.13 Block diagram of a logarithmic encoder-decoder.

The inverse transformation is

$$x(n) = \exp[y(n)]\text{sign}[x(n)] \qquad (5.29)$$

where $\text{sign}[x(n)]$ is $+1$ if $x(n)$ is positive, and -1 if $x(n)$ is negative. Now the quantized log magnitude is

$$\hat{y}(n) = Q[\log|x(n)|]$$
$$= \log|x(n)| + \epsilon(n) \qquad (5.30)$$

where we have assumed as before that $\epsilon(n)$ is independent of $\log|x(n)|$. The inverse of the quantized log magnitude is

$$\hat{x}(n) = \exp[\hat{y}(n)]\text{sign}[x(n)]$$
$$= |x(n)|\text{sign}[x(n)]\exp[\epsilon(n)]$$
$$= x(n)\exp[\epsilon(n)] \qquad (5.31)$$

If $\epsilon(n)$ is small, we can approximate the above equation by

$$\hat{x}(n) \approx x(n)[1+\epsilon(n)] = x(n) + \epsilon(n)x(n) = x(n) + f(n) \qquad (5.32)$$

where $f(n) = x(n)\epsilon(n)$. Thus, since $x(n)$ and $\epsilon(n)$ are assumed independent,

$$\sigma_f^2 = \sigma_x^2 \cdot \sigma_\epsilon^2 \qquad (5.33)$$

and

$$SNR = \frac{\sigma_x^2}{\sigma_f^2} = \frac{1}{\sigma_\epsilon^2} \qquad (5.34)$$

That is, the signal-to-noise ratio is independent of signal variance. It depends only upon the step size.

This type of quantization is not practical since the dynamic range (ratio between largest and smallest values) is infinite and thus an infinite number of

187

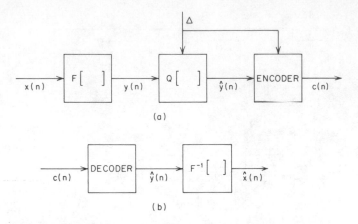

(a)

(b)

Fig. 5.14 Block diagram of a compressor/expander system for quantization.

quantization levels would be required. The above analysis, however impractical it may be, suggests that some approximation to a logarithmic compression characteristic would be desirable. The use of a compressor/expandor system for quantization is shown in Fig. 5.14. Smith [10] has investigated a compression characteristic that is called the μ-law. In this case

$$y(n) = F[x(n)]$$

$$= X_{max} \frac{\log\left[1 + \mu \dfrac{|x(n)|}{X_{max}}\right]}{\log[1+\mu]} \cdot \text{sign}[x(n)]. \qquad (5.35)$$

Figure 5.15 shows a family of curves of $y(n)$ versus $x(n)$ for different values of μ. It is clear that using the function of Eq. (5.35) avoids the problem of

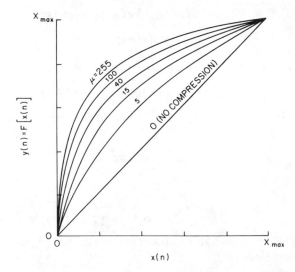

Fig. 5.15 Input-output relations for a μ-law characteristic. (After Smith [10].)

188

small input amplitudes since $y(n) = 0$ when $|x(n)| = 0$. If $\mu = 0$, Eq. (5.35) reduces to

$$y(n) = x(n); \qquad (5.36)$$

i.e., the quantization levels are uniformly spaced. However, for large μ, and for large $|x(n)|$,

$$|y(n)| \approx X_{\text{max}} \log \left| \frac{x(n)}{X_{\text{max}}} \right| \qquad (5.37)$$

Thus except for very low amplitudes, the μ-law curve gives a good approximation to constant percentage error. Figure 5.16 shows the distribution of quantization levels for the case $\mu = 40$ and 8 quantization levels. (The quantizer characteristic is antisymmetric about the origin.)

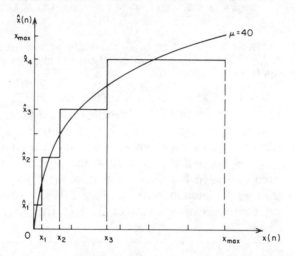

Fig. 5.16 Distribution of quantization levels for a μ-law 3-bit quantizer with $\mu = 40$.

Employing the same kind of assumptions that were used to analyze the uniform quantization case, Smith [10] derived a formula for the signal-to-quantizing noise ratio for a μ-law quantizer of the form

$$SNR\,(\text{dB}) = 6B + 4.77 - 20 \log_{10}[\ln(1+\mu)]$$

$$-10 \log_{10}\left[1 + \left[\frac{X_{\text{max}}}{\mu \sigma_x} \right]^2 + \sqrt{2} \left[\frac{X_{\text{max}}}{\mu \sigma_x} \right] \right] \qquad (5.38)$$

This equation, when compared to Eq. (5.26), indicates a much less severe dependence of SNR upon the quantity $(X_{\text{max}}/\sigma_x)$. It can be seen that as μ increases, the SNR becomes less and less sensitive to changes in $(X_{\text{max}}/\sigma_x)$; i.e., although the term $-20 \log_{10}[\ln(1+\mu)]$ reduces the SNR, the range of $(X_{\text{max}}/\sigma_x)$ for which the SNR is constant increases with μ. Figures 5.17 and 5.18 show Eqs. (5.26) and (5.38) plotted as a function of the quantity

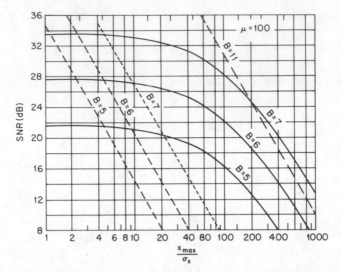

Fig. 5.17 *SNR* for μ-law and uniform quantizers as a function of X_{max}/σ_x for $\mu = 100$ and different numbers of bits (B) of the quantizer. (After Smith [10].)

(X_{max}/σ_x) for $\mu = 100$ and 500 respectively. The quantity X_{max} is a parameter of the quantizing system. It specifies the "overload" amplitude; i.e., the amplitude beyond which all samples are clipped. The quantity σ_x is a parameter of the signal. It specifies, in an average sense, the amplitude of the signal. The quantity (X_{max}/σ_x) gives an indication of how the signal is matched to the quantizer. The dotted curves in Fig. 5.17 show how the *SNR* (in dB) of a uniform quantizer varies as a function of (X_{max}/σ_x). We note that for a fixed value of X_{max}, if σ_x decreases by a factor of 2, then the *SNR* decreases by 6

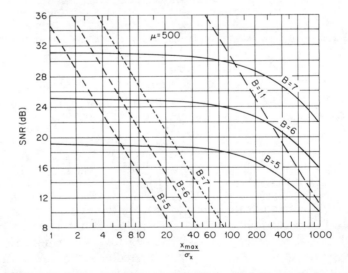

Fig. 5.18 *SNR* for μ-law and uniform quantizers for $\mu = 500$, $B = 5, 6, 7, 11$ bits. (After Smith [10].)

190

dB. We also note that for a fixed value of (X_{max}/σ_x) the SNR increases 6 dB for each added bit. This is true for both the uniform and μ-law quantizers, which are shown as solid lines.

A comment regarding the validity of Eqs. (5.26) and (5.38) and, thus, the curves in Figs. 5.17 and 5.18, is in order. One of the assumptions made in deriving these equations was that the quantizer overload was negligible; i.e., the probability of a sample exceeding the value X_{max} was very small. When the variance σ_x becomes on the order of X_{max}, i.e., $X_{max}/\sigma_x \approx 1$, this assumption is clearly violated. Measured curves of SNR vs. (X_{max}/σ_x) thus show a dramatic reduction in SNR as $(X_{max}/\sigma_x) \to 1$. Equations (5.26) and (5.38) do yield a good description of SNR for values of $(X_{max}/\sigma_x) > 8$ [10].

The important point that these curves show very clearly is that μ-law quantization can maintain roughly the same SNR over a rather wide range. For example from Fig. 5.17 it is clear that for the case $\mu = 100$, the SNR remains within 2 dB of the maximum attainable for

$$8 < \frac{X_{max}}{\sigma_x} < 30 \qquad (5.39)$$

From Fig. 5.18 we see that for $\mu = 500$ the SNR is within 2 dB of maximum for

$$8 < \frac{X_{max}}{\sigma_x} < 150 \qquad (5.40)$$

However, a comparison of Figs. 5.17 and 5.18 indicates that the maximum attainable SNR is about 2.6 dB greater for $\mu = 100$ than for $\mu = 500$. Thus by using large amounts of compression we achieve greater dynamic range with a rather small sacrifice in SNR.

It can be seen from Figs. 5.17 and 5.18 that with $B = 7$, a signal-to-noise ratio of about 34 dB can be maintained over a wide range of input signal levels. Indeed, 7-bit μ-law PCM is generally taken as a standard for "toll quality" representation of the speech waveform. As we have already noted, a uniform quantizer would require about 11 bits to obtain the same dynamic range as 7-bit log PCM. Figure 5.18, for example, shows that the curve for 11-bit uniform quantization exceeds the 7-bit $\mu = 500$ curve for $\sigma_x > .01 X_{max}$. Thus, we could say that 11-bit uniform quantization should be as good as or better than $\mu = 500$ log PCM for signal levels that are at least 1% of the quantizer maximum.

5.3.3 Quantization for optimum SNR

The μ-law quantizer strives to achieve constant SNR over a wide range of signal variances. As we have just seen, this is achieved at some sacrifice over the SNR performance that can be achieved if the quantizer step-size is matched to the variance of the signal. In cases where the signal variance is known, it is possible to choose the quantizer levels so as to minimize the quantization error variance, and thus, maximize the *SNR*. The problem was studied by Max [11] and later by Paez and Glisson [3].

The variance of the quantization noise is

$$\sigma_e^2 = E[e^2(n)] = E[(\hat{x}(n) - x(n))^2]$$ (5.41)

where $\hat{x}(n) = Q[x(n)]$. Generalizing from the example of Fig. 5.9, we observe that in general we have M quantization levels which can be labelled $\{\hat{x}_{-M/2}, \hat{x}_{-M/2+1}, \ldots, \hat{x}_{-1}, \hat{x}_1, \ldots, \hat{x}_{M/2}\}$ assuming M is an even number. The quantization level associated with the interval x_{j-1} to x_j is denoted $\hat{x}_j$. For a symmetric, zero mean amplitude distribution it is sensible to define the central boundary point $x_0 = 0$ and if the density function is nonzero for large amplitudes, such as the Laplacian or gamma densities, then the extremes of the outer range are set to $\pm\infty$; i.e., $x_{\pm M/2} = \pm\infty$. With this assumption, we can write

$$\sigma_e^2 = \int e^2 p_e(e)\, de$$ (5.42)

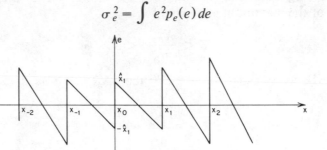

Fig. 5.19 Quantization error *e* versus signal level *x* for a nonuniform quantizer.

Figure 5.19 shows a plot of *e* versus *x*. It can be seen that contributions to the probability density function of *e* come from each of the quantization intervals of *x* in a fairly complicated manner. Since

$$e = \hat{x} - x$$ (5.43)

we can make a transformation of variables in Eq. (5.42), of the form

$$p_e(e) = p_e(\hat{x} - x) = p_{x/\hat{x}}(x/\hat{x}) \triangleq p_x(x)$$ (5.44)

giving

$$\sigma_e^2 = \sum_{i=-\frac{M}{2}+1}^{\frac{M}{2}} \int_{x_{i-1}}^{x_i} (\hat{x}_i - x)^2 p(x)\, dx$$ (5.45)

(Note that this formulation of the noise variance includes the errors due to clipping or "overload.") If $p(x) = p(-x)$ then the optimum quantizer characteristic will be antisymmetric so that $\hat{x}_i = -\hat{x}_{-i}$ and $x_i = -x_{-i}$. Thus

$$\sigma_e^2 = 2 \sum_{i=1}^{\frac{M}{2}} \int_{x_{i-1}}^{x_i} (\hat{x}_i - x)^2 p(x)\, dx$$ (5.46)

192

Now we wish to choose the sets of parameters $\{x_i\}$ and $\{\hat{x}_i\}$ so as to minimize σ_e^2. To do this we differentiate σ_e^2 with respect to each parameter and set the derivative equal to zero obtaining the equations [11]

$$\int_{x_{i-1}}^{x_i} (\hat{x}_i - x)p(x)\,dx = 0, \qquad\qquad i = 1, 2, \ldots, \frac{M}{2} \qquad (5.47a)$$

$$x_i = \frac{1}{2}(\hat{x}_i + \hat{x}_{i+1}), \qquad\qquad i = 1, 2, \ldots, \frac{M}{2} - 1 \quad (5.47b)$$

and by assumption,

$$x_0 = 0 \qquad\qquad\qquad (5.48a)$$

$$x_{\pm\frac{M}{2}} = \pm\infty \qquad\qquad\qquad (5.48b)$$

Equation (5.47b) shows that the optimum boundary points lie halfway between the $M/2$ quantizer levels. Equation (5.47a) shows that the optimum location of the quantization level $\hat{x}_i$ is at the centroid of the probability density over the interval x_{i-1} to x_i. These two sets of equations must be solved simultaneously for the $M - 1$ unknown parameters of the quantizer. Since these equations are generally nonlinear, closed form solutions can only be obtained in some special cases. Otherwise an iterative procedure must be used. Such an iterative procedure is given by Max [11]. Paez and Glisson [3] have used this procedure to solve for optimum boundary points for the Laplace and gamma probability density functions.

In general, the solution of Eq. (5.47) will result in a nonuniform distribution of quantization levels. Only in the special case of a uniform amplitude density will the optimum solution be uniform; i.e.,

$$\hat{x}_i - \hat{x}_{i-1} = x_i - x_{i-1} = \Delta \qquad\qquad (5.49)$$

We can, however, constrain the quantizer to be uniform, and solve for the value of the step size, Δ, which gives minimum quantization error variance and, therefore, maximum SNR. In this case,

$$x_i = \Delta \cdot i \qquad\qquad\qquad (5.50)$$

$$\hat{x}_i = \frac{(2i-1)\Delta}{2} \qquad\qquad\qquad (5.51)$$

and Δ satisfies the equation

$$\sum_{i=1}^{\frac{M}{2}-1} (2i-1) \int_{(i-1)\Delta}^{i\Delta} \left(\left[\frac{2i-1}{2}\right]\Delta - x\right)p(x)\,dx$$

$$+ (M-1) \int_{\left[\frac{M}{2}-1\right]\Delta}^{\infty} \left(\left[\frac{M-1}{2}\right]\Delta - x\right)p(x)\,dx = 0 \qquad (5.52)$$

193

Table 5.1 Optimum Quantizers for Signals with Laplace Density ($m_x=0$, $\sigma_x^2=1$). (After Paez and Glisson [3].)

N	2		4		8		16		32	
i	x_i	$\hat{x}_i$	x_i	$\hat{x}_i$	x_i	$\hat{x}_i$	x_i	$\hat{x}_i$	x_i	$\hat{x}_i$
1	∞	0.707	1.102	0.395	0.504	0.222	0.266	0.126	0.147	0.072
2			∞	1.810	1.181	0.785	0.566	0.407	0.302	0.222
3					2.285	1.576	0.910	0.726	0.467	0.382
4					∞	2.994	1.317	1.095	0.642	0.551
5							1.821	1.540	0.829	0.732
6							2.499	2.103	1.031	0.926
7							3.605	2.895	1.250	1.136
8							∞	4.316	1.490	1.365
9									1.756	1.616
10									2.055	1.896
11									2.398	2.214
12									2.804	2.583
13									3.305	3.025
14									3.978	3.586
15									5.069	4.371
16									∞	5.768
MSE	0.5		0.1765		0.0548		0.0154		0.00414	
SNR dB	3.01		7.53		12.61		18.12		23.83	

If $p(x)$ is known or assumed (e.g., Laplacian) then the integrations can be performed to yield a single equation which can be solved on a computer using iterative techniques by varying Δ until the optimum value is obtained.

Tables 5.1 and 5.2 show optimum quantizer parameters for Laplacian and gamma densities [3]. (Note that these numbers are derived assuming unit variance. If the variance of the input is σ_x^2, then the numbers in the tables should be multiplied by σ_x.) Figure 5.20 shows a 3-bit quantizer for a Laplacian density. It is clear from this figure that the quantization levels get further apart as the probability density decreases. This is consistent with intuition, which would suggest that the largest quantization errors should be reserved for the least frequently occurring samples. A comparison of Figs. 5.16 and 5.20 shows a similarity between the μ-law quantizer and the optimum nonuniform quantizer. Thus, the optimum nonuniform quantizers might be expected to have improved dynamic range. This is in fact true as discussed in [3].

Figure 5.21 shows the optimum step size for uniform quantizers for gamma and Laplacian densities [3] and a Gaussian density [11]. It is clear that, as expected, the step size decreases roughly exponentially with increasing numbers of bits. The details of the curves are, of course, attributable to the differences in the shape of the density functions.

Although optimum quantizers yield minimum mean squared error when matched to the variance and amplitude distribution of the signal, the nonstationary nature of the speech communication process leads to less than satisfac-

Table 5.2 Optimum Quantizers for Signals with Gamma Density ($m_x=0$, $\sigma_x^2=1$). (After Paez and Glisson [3].)

N	2		4		8		16		32	
i	x_i	$\hat{x}_i$	x_i	$\hat{x}_i$	x_i	$\hat{x}_i$	x_i	$\hat{x}_i$	x_i	$\hat{x}_i$
1	∞	0.577	1.205	0.302	0.504	0.149	0.229	0.072	0.101	0.033
2			∞	2.108	1.401	0.859	0.588	0.386	0.252	0.169
3					2.872	1.944	1.045	0.791	0.429	0.334
4					∞	3.799	1.623	1.300	0.630	0.523
5							2.372	1.945	0.857	0.737
6							3.407	3.798	1.111	0.976
7							5.050	4.015	1.397	1.245
8							∞	6.085	1.720	1.548
9									2.089	1.892
10									2.517	2.287
11									3.022	2.747
12									3.633	3.296
13									4.404	3.970
14									5.444	4.838
15									7.046	6.050
16									∞	8.043
MSE	0.6680		0.2326		0.0712		0.0196		0.0052	
SNR	1.77		6.33		11.47		17.07		22.83	

tory results. The simplest manifestation of this occurs in transmission systems during periods when no one is talking; i.e., the idle channel condition. In this case the input to the quantizer is very small (assuming low noise) so that the output of the quantizer will jump back and forth between the lowest magnitude quantization levels. For a symmetric quantizer such as Fig. 5.10a, if the lowest quantization levels are greater than the amplitude of the background noise, the output noise of the quantizer will be greater than the input noise. For this reason, optimum quantizers of the minimum mean squared error type are not practical when the number of quantization levels is small. Table 5.3 [3] shows a comparison of the smallest quantizer levels for several uniform and nonuniform optimum quantizers as compared to a μ-law quantizer with $\mu = 100$. It can be seen that the μ-law quantizer would produce much lower idle channel noise than any of the optimum quantizers. For larger values of μ the smallest quantization level would be even smaller. (If $\mu = 255$, the minimum quantization level is 0.031.) For this reason, μ-law quantizers are used in practice even though they provide somewhat lower SNR than optimum designs.

5.4 Adaptive Quantization

As we have seen in the previous section, we are confronted with a dilemma in quantizing speech signals. On the one hand we wish to choose the quantization step size large enough to accommodate the maximum peak-to-peak range of the

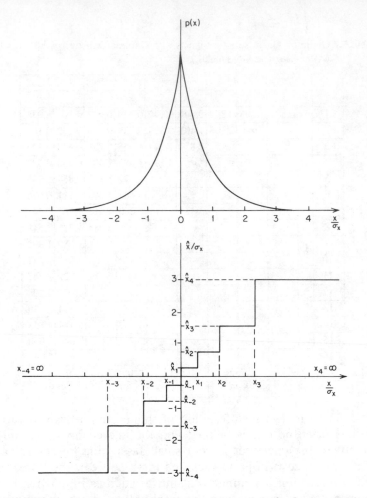

Fig. 5.20 Density function and quantizer characteristic for Laplace density function and a 3-bit quantizer.

signal. On the other hand we would like to make the quantization step small so as to minimize the quantization noise. This is compounded by the nonstationary nature of the speech signal and the speech communication process. The amplitude of the speech signal can vary over a wide range depending on the speaker, the communication environment, and within a given utterance, from voiced to unvoiced segments. As we have seen, one approach to accommodating these amplitude fluctuations is to use a nonuniform quantizer. An alternate approach is to adapt the properties of the quantizer to the level of the input signal. In this section, we shall discuss some general principles of adaptive quantization, and in later sections we will show examples of adaptive quantization schemes in conjunction with linear prediction. When adaptive quantization is used directly on samples of the input system it is called adaptive PCM or simply, APCM.

The basic idea of adaptive quantization is to let the step size Δ (or in general the quantizer levels and ranges) vary so as to match the variance of the

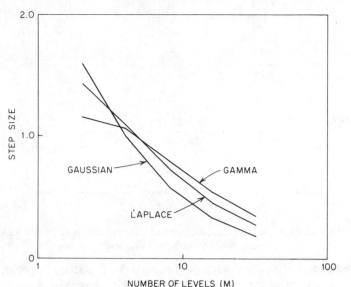

Fig. 5.21 Optimum step sizes for a uniform quantizer for Laplace, Gamma, and Gaussian density functions. (Data from [11].)

input signal. This is depicted schematically in Fig. 5.22a. An alternative point of view, depicted in Fig. 5.22b, is to consider a fixed quantizer characteristic preceded by a time varying gain which tends to keep the variance constant. In the first case the step size should increase and decrease with increases and decreases of the variance of the input. In the case of a nonuniform quantizer, this would imply that the quantization levels and ranges would be scaled linearly to match the variance of the signal. In the second point of view, which applies without modification to both uniform and nonuniform quantizers, the gain changes inversely with changes in the variance of the input so as to keep the variance of the quantizer input relatively constant. In either case, it is necessary to obtain an estimate of the time varying amplitude properties of the input signal.

Table 5.3 Signal-to-Noise Ratios for 3-bit Quantizers. (After Noll [12]).

Nonuniform Quantizers	SNR (dB)	Smallest Level $(\sigma_x=1)$
μ-law $(x_{max} = 8\sigma_x,\ \mu=100)$	9.5	0.062
Gaussian	14.6	0.245
Laplace	12.6	0.222
Gamma	11.5	0.149
Speech	12.1	0.124

Uniform Quantizers	SNR	Smallest Level $(\sigma_x=1)$
Gaussian	14.3	0.293
Laplace	11.4	0.366
Gamma	11.5	0.398
Speech	8.4	0.398

197

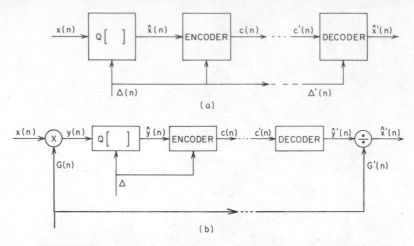

Fig. 5.22 Block diagram representation of adaptive quantization; (a) variable stepsize representation; (b) variable gain representation.

In discussing the time varying properties of speech signals, it is convenient to consider the time scale over which changes take place. In the case of amplitude changes, we will refer to changes from sample-to-sample or rapid changes within a few samples as *instantaneous* changes. General trends in the amplitude properties, as for example the peak amplitude in an unvoiced interval or in a voiced interval, remain essentially unchanged for relatively long time intervals. Such slowly varying trends are referred to as *syllabic* variations, implying that they occur at a rate comparable to the syllable rate in speaking. In discussing adaptive quantizing schemes, it will likewise be convenient to classify them according to whether they are slowly adapting or rapidly adapting; i.e., syllabic or instantaneous.

A dichotomy has evolved among the existing schemes for adaptive quantization. In one class of schemes, the amplitude or variance of the input is

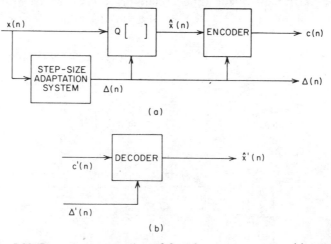

Fig. 5.23 General representation of feed-forward quantizers; (a) coder; (b) decoder.

198

estimated from the input itself. Such schemes are called feed-forward adaptive quantizers. In the other class of adaptive quantizers, the step size is adapted on the basis of the *output* of the quantizer, $\hat{x}(n)$, or equivalently, on the basis of the output code words, $c(n)$. These are called feedback quantizers. In general, the adaptation time of either class of quantizers can be either syllabic or instantaneous.

5.4.1 Feed-forward adaptation

Figure 5.23 depicts a general representation of the class of feed-forward quantizers. We assume for convenience in the discussion that the quantizer is uniform so that it is sufficient to vary a single step size parameter. It is straightforward to generalize this discussion to the case of nonuniform quantiz-

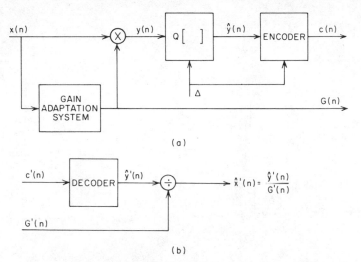

Fig. 5.24 General feed-forward adaptive quantizer with a time-varying gain; (a) coder; (b) decoder.

ers. The step size $\Delta(n)$, used to quantize the sample $x(n)$ in Fig. 5.23a, must be available at the receiver in Fig. 5.23b. Thus, the code words $c(n)$ and the step size $\Delta(n)$ together represent the sample $x(n)$. If $c'(n) = c(n)$ and $\Delta'(n) = \Delta(n)$, then $\hat{x}'(n) = \hat{x}(n)$; however, if $c'(n) \neq c(n)$ or $\Delta'(n) \neq \Delta(n)$, e.g., if there are errors in transmission, then $\hat{x}'(n) \neq \hat{x}(n)$. The effect of errors will depend upon the details of the adaptation scheme. Figure 5.24 shows the general feed-forward adaptive quantizer represented in terms of a time varying gain. In this case, the code words $c(n)$ and the gain $G(n)$ together represent the quantized samples.

To see how feed-forward schemes work, it is helpful to consider some examples. Most systems of this type attempt to obtain an estimate of the time-varying variance. Then, the step size or quantization levels are made proportional to the standard deviation, or the gain applied to the input can be made inversely proportional to the standard deviation.

199

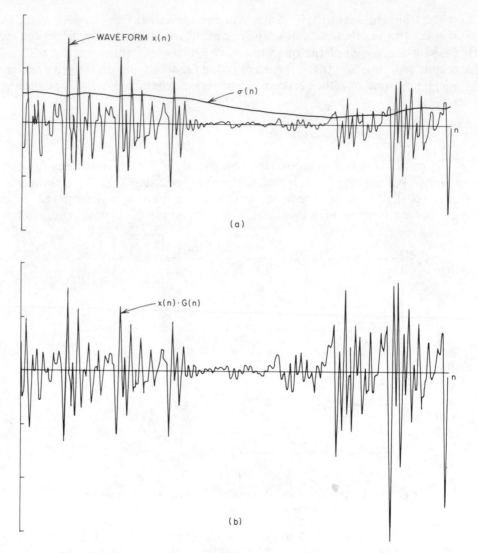

WAVEFORM x(n)

$\sigma(n)$

n

(a)

x(n)·G(n)

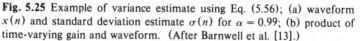

n

(b)

Fig. 5.25 Example of variance estimate using Eq. (5.56); (a) waveform $x(n)$ and standard deviation estimate $\sigma(n)$ for $\alpha = 0.99$; (b) product of time-varying gain and waveform. (After Barnwell et al. [13].)

A common approach is to assume that the variance is proportional to the short-time energy, which, as we have seen, is defined as the output of a lowpass filter with input, $x^2(n)$. That is,

$$\sigma^2(n) = \sum_{m=-\infty}^{\infty} x^2(m) h(n-m) \qquad (5.53)$$

where $h(n)$ is the impulse response of the lowpass filter. (For a stationary input signal, it can be easily shown that the expected value of $\sigma^2(n)$ is proportional to the variance, σ_x^2 [see Problem 5.7].)

A simple example is

$$h(n) = \alpha^{n-1}, \quad n \geq 1 \tag{5.54}$$

$$= 0, \qquad \textit{otherwise}$$

Using this in Eq. (5.53) gives

$$\sigma^2(n) = \sum_{m=-\infty}^{n-1} x^2(m)\alpha^{n-m-1} \tag{5.55}$$

It can be shown that $\sigma^2(n)$ in Eq. (5.55) also satisfies the difference equation

$$\sigma^2(n) = \alpha\sigma^2(n-1) + x^2(n-1) \tag{5.56}$$

(For stability we require $0 < \alpha < 1$.) The step size in Fig. 5.23a would therefore be of the form

$$\Delta(n) = \Delta_0\sigma(n) \tag{5.57}$$

or the time-varying gain in Fig. 5.24a would be of the form[2]

$$G(n) = \frac{G_0}{\sigma(n)} \tag{5.58}$$

The choice of the parameter α controls the effective interval that contributes to the variance estimate. Figure 5.25 shows an example of quantizing in a differential PCM system [13]. Figure 5.25a shows the standard deviation estimate superimposed upon the waveform for the case $\alpha = 0.99$. Figure 5.25b shows the product $y(n) = x(n)G(n)$. For this choice of α, the dip in amplitude of $x(n)$ is clearly not entirely compensated by the time varying gain. Figure 5.26 shows the same waveforms for the case $\alpha = 0.9$. In this case the system reacts much more quickly to changes in the input amplitude. Thus the variance of $y(n) = G(n)x(n)$ remains relatively constant even through the rather abrupt dip in amplitude in $x(n)$. In the first case, with $\alpha = 0.99$, the time constant (time for weighting sequence to decay to (e^{-1}) is about 100 samples (or 12.5 msec at an 8 kHz sampling rate). In the second case, with $\alpha = 0.9$, the time constant is only 9 samples, or about 1 msec at 8 kHz. Thus, it would be reasonable to classify the system with $\alpha = 0.99$ as syllabic and the system with $\alpha = 0.9$ as instantaneous.

As is evident from Figs. 5.25a and 5.26a, the standard deviation estimate and its reciprocal, $G(n)$, are slowly varying functions as compared to the original speech signal. The rate at which the gain (or step size) control signal must be sampled depends upon the bandwidth of the lowpass filter. For the cases shown in Figs. 5.25 and 5.26, the frequencies at which the filter gain is down by 3 dB are about 13 Hz and 135 Hz respectively for a sampling rate of 8 kHz. It is important to consider the lowest possible sampling rate for the gain, since the information rate of the digital representation is the sum of the information rate of the quantizer output and the information rate of the gain function. The gain

---

[2]The constants Δ_0 and G_0 would account for the gain of the filter.

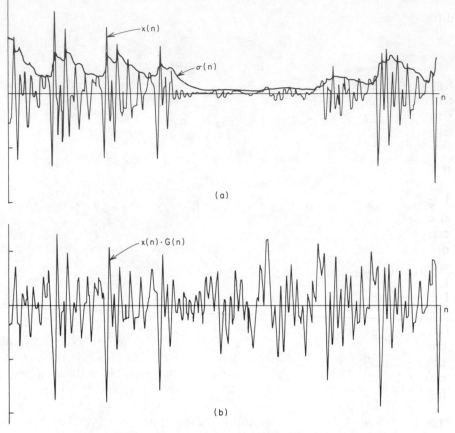

Fig. 5.26 Variance estimate using Eq. (5.56); (a) $x(n)$ and $\sigma(n)$ for $\alpha = 0.9$; (b) $x(n)\,G(n)$.

function (or step size) as used in Figs. 5.24 or 5.23 must be sampled and quantized before transmission.

To permit quantizing and because of constraints of physical implementations, it is common to limit the variation of the gain function or the step size. That is we define limits on $G(n)$ and $\Delta(n)$ of the form

$$G_{\min} \leqslant G(n) \leqslant G_{\max} \tag{5.59}$$

or

$$\Delta_{\min} \leqslant \Delta(n) \leqslant \Delta_{\max} \tag{5.60}$$

It is the ratio of these limits that determines the dynamic range of the system. Thus to obtain a relatively constant *SNR* over a range of 40 dB, requires $G_{\max}/G_{\min} = 100$ or $\Delta_{\max}/\Delta_{\min} = 100$.

An example of the improvement in *SNR* that can be achieved by adaptive quantization is given in a comparative study by Noll [12].[3] He considered a

---

[3]This technique was also studied by Croisier [14]. He used the term block companding to describe the process of evaluating the gain (or step size) every *M* samples.

feed-forward scheme in which the variance estimate was

$$\sigma^2(n) = \frac{1}{M} \sum_{m=n}^{n+M-1} x^2(m) \qquad (5.61)$$

The gain or step size is evaluated and transmitted every M samples. In this case the system requires a buffer of M samples to permit the quantizer gain or step size to be determined in terms of the samples that are to be quantized rather than in terms of past samples as in the previous example.

Table 5.4 shows a comparison of various 3-bit quantizers with a speech input of known variance.[4] The first column lists the various quantizer types. The second column gives the signal-to-noise ratios with no adaptation. The third and fourth columns give the signal-to-noise ratios for step size adaptation based upon the variance estimate of Eq. (5.61) with $M = 128$ and $M = 1024$ respectively. It can be readily seen that the adaptive quantizer achieves up to 8.0 dB better *SNR* for this particular speech material. Similar results can be expected with other speech utterances, with slight variations in all the numbers. Thus, it is evident that adaptive quantization achieves a definite advantage over fixed nonuniform quantizers. An additional advantage which is not placed in evidence by the numbers in Table 5.4 is that by appropriately choosing $\Delta_{\min}$ and $\Delta_{\max}$ it is possible to achieve the improvement in *SNR*, while maintaining low idle channel noise and wide dynamic range. This is true in general for most well-designed adaptive quantization systems. The combination of all these factors makes adaptive quantization an attractive alternative to instantaneous companding or minimum mean squared error quantization.

Table 5.4 Adaptive 3-bit Quantization with Feed-forward Adaptation. (After Noll [12].)

Nonuniform Quantizers	Nonadaptive SNR (dB)	Adaptive (M=128) SNR (dB)	Adaptive (M=1024) SNR (dB)
μ-law ($\mu=100$, $X_{\max}=8\sigma_x$)	9.5	–	–
Gaussian	7.3	15.0	12.1
Laplace	9.9	13.3	12.8
Uniform Quantizers			
Gaussian	6.7	14.7	11.3
Laplace	7.4	13.4	11.5

5.4.2 Feedback adaptation

The second class of adaptive quantizer systems is depicted in Figs. 5.27 and 5.28, where it is noted that the variance of the input is estimated from the quantizer output or equivalently from the code words. As in the case of feed-forward systems, the step size and gain are proportional and inversely proportional respectively to an estimate of the standard deviation of the input as in Eqs. (5.57) and (5.58). Such schemes have the distinct advantage that the step size or gain need not be explicitly retained or transmitted since they can be

---

[4]The results in this table are for quantization of actual speech signals.

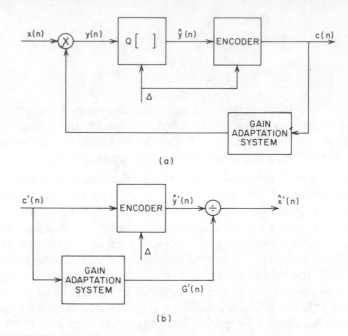

Fig. 5.27 General feedback adaptation of the time-varying gains; (a) coder; (b) decoder.

derived from the sequence of code words. The disadvantage of such systems is increased sensitivity to errors in the code words, since such errors imply not only an error in the quantizer level but also in the step size.

One simple approach is to apply Eq. (5.53) directly to the quantizer output; i.e.,

$$\sigma^2(n) = \sum_{m=-\infty}^{\infty} \hat{x}^2(m)h(n-m) \tag{5.62}$$

In this case, however, it will not be possible to use buffering to implement a noncausal filter. That is, the variance estimate must be based only on past values of $\hat{x}(n)$ since the present value of $\hat{x}(n)$ will not be available until the quantization has occured, which in turn must be after the variance has been estimated. For example, we could use a filter whose impulse response is

$$h(n) = \alpha^{n-1}, \quad n \geqslant 1$$
$$= 0 \qquad otherwise \tag{5.63}$$

as in Eq. (5.55). Alternatively the filter might have an impulse response

$$h(n) = 1/M, \quad 1 \leqslant n \leqslant M$$
$$= 0, \qquad otherwise \tag{5.64}$$

so that

$$\sigma^2(n) = \frac{1}{M} \sum_{m=n-M}^{n-1} x^2(m) \tag{5.65}$$

204

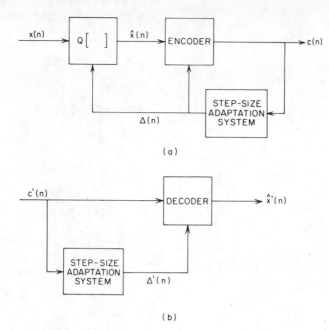

Fig. 5.28 General feedback adaptation of the stepsize; (a) coder; (b) decoder.

This system was studied by Noll [12], who found that with suitable adjustment of the constants Δ_0 or G_0 in Eqs. (5.57) or (5.58) a signal-to-noise ratio on the order of 12 dB could be obtained for a 3-bit quantizer with a window length of only 2 samples. Larger values of M produced only slightly better results.

A somewhat different approach, based on Fig. 5.28 has been studied extensively by Jayant [15]. In this method the step size of a uniform quantizer is adapted at each sample time by the formula

$$\Delta(n) = P\Delta(n-1) \tag{5.66}$$

where the step size multipler, P, is a function only of the magnitude of the previous code word, $|c(n-1)|$. This is depicted in Fig. 5.29 for a 3-bit uniform quantizer. With the choice of code words in Fig. 5.29, if we assume the most significant bit is the sign bit and the rest of the word is the magnitude, then

$$\hat{x}(n) = \frac{\Delta(n)\text{sign}(c(n))}{2} + \Delta(n)c(n) \tag{5.67}$$

where $\Delta(n)$ satisfies Eq. (5.66). Note that since $\Delta(n)$ depends upon the previous step size and the previous code word, the sequence of code words is all that is required to represent the signal. As a practical consideration, it is necessary to impose the limits

$$\Delta_{\min} \leqslant \Delta(n) \leqslant \Delta_{\max} \tag{5.68}$$

As mentioned before, the ratio $\Delta_{\max}/\Delta_{\min}$ controls the dynamic range of the quantizer.

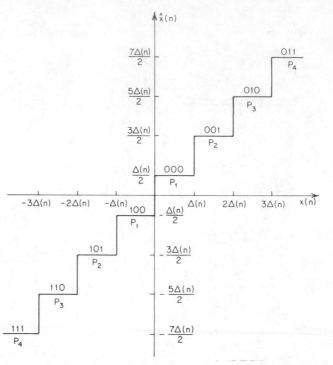

Fig. 5.29 Input-output characteristic of a 3-bit adaptive quantizer.

The manner in which the multiplier in Eq. (5.66) should vary with $|c(n-1)|$ is intuitively clear. If the previous code word corresponds to either the largest positive or largest negative quantizer level, then it is reasonable to assume that the quantizer is overloaded and, thus, that the quantizer step size is too small. In this case, then, the multiplier should be greater than one. Alternatively, if the previous code word corresponds to either the smallest positive or smallest negative level, then it is reasonable to decrease the step size by using a multiplier less than one. The design of such a quantizer involves the choice of multipliers to correspond to each of the 2^B code words for a B-bit quantizer. Jayant [15] has approached this problem by finding a set of step size multipliers that minimizes the mean squared quantization error. He was able to obtain theoretical results for Gaussian signals, and using a search procedure, he obtained empirical results for speech. The general conclusions of Jayant's study are summarized in Fig. 5.30, which shows the approximate way in which the step size multipliers should depend upon the quantity Q, defined as

$$Q = \frac{1 + 2\,|c(n-1)|}{2^B - 1} \qquad (5.69)$$

The shaded region in this figure represents the variation in the multipliers to be expected as the input statistics change or as B changes. The specific multiplier values should follow the general trend of Fig. 5.30, but the specific values are not overly critical. It is important, however, that the multipliers be such that step size increases occur more vigorously than step size decreases. Table 5.5 shows the sets of multipliers for $B = 2, 3, 4$ and 5.

The improvement in signal-to-noise ratio to be gained using this mode of adaptive quantization is shown in Table 5.6. The multipliers of Table 5.5 were used and $\Delta_{max}/\Delta_{min} = 100$. Table 5.6 shows a 4-7 dB improvement over μ-law quantization. A 2-4 dB improvement was also noted over nonadaptive optimum quantizers. In another study, Noll [7] noted signal-to-noise ratios for 3-bit μ-law and adaptive quantizers of 9.4 dB and 14.1 dB respectively. In this experiment the multipliers were $\{.8, .8, 1.3, 1.9\}$ in contrast to those used by Jayant which are seen from Table 5.5 to be $\{.85, 1, 1, 1.5\}$. The fact that such different multipliers can produce comparable results lends support to the contention that the values of the multipliers are not particularly critical.

5.4.3 General comments on adaptive quantization

As the discussion of this section clearly indicates, there are almost unlimited possibilities for adaptive quantization schemes. Most reasonable schemes will yield signal-to-noise ratios that exceed the signal-to-noise ratio of μ-law quantization, and with a suitable ratio $\Delta_{max}/\Delta_{min}$, the dynamic range of an adaptive quantizer can be fully comparable to that of μ-law quantization. Also, by choosing Δ_{min} to be small, the idle channel noise can be made very small. Thus adaptive quantization has many attractive features. However, it is unreasonable to expect that further sophistication of quantizer adaptation alone will yield dramatic savings in bit-rate since such techniques simply exploit our

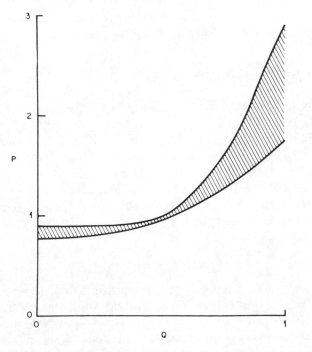

Fig. 5.30 General shape of optimal multiplier function in speech quantization for $B > 2$. (After Jayant [15].)

Table 5.5 Step Size Multipliers For Adaptive Quantization Methods. (After Jayant [15].)

B	Coder Type	
	PCM	DPCM
2	0.6, 2.2	0.8, 1.6
3	0.85, 1, 1, 1.5	0.9, 0.9, 1.25 1.75
4	0.8, 0.8, 0.8, 0.8, 1.2, 1.6, 2.0, 2.4	0.9, 0.9, 0.9, 0.9, 1.2, 1.6, 2.0, 2.4
5	0.85, 0.85, 0.85, 0.85, 0.85, 0.85, 0.85, 0.85, 1.2, 1.4, 1.6, 1.8, 2.0, 2.2, 2.4, 2.6	0.9, 0.9, 0.9, 0.9, 0.95, 0.95, 0.95, 0.95, 1.2, 1.5, 1.8, 2.1, 2,4, 2.7, 3.0, 3.3

Table 5.6 Improvements in Signal-to-Noise Ratio Using Optimum Step Size Multipliers for Adaptive Quantization. (After Jayant [15].)

B	Logarithmic PCM with μ-law ($\mu=100$) Quantization	Adaptive PCM with Uniform Quantization
2	3 db	9 db
3	8 db	15 db
4	15 db	19 db

knowledge of the amplitude distribution of the speech signal. Thus, we turn our attention in the next section to exploiting the sample-to-sample correlation through the techniques of differential quantization.

5.5 General Theory of Differential Quantization

Figure 5.7a shows that there is considerable correlation between adjacent speech samples, and indeed the correlation is significant even between samples that are several sampling intervals apart. The meaning of this high correlation is that, in an average sense, the signal does not change rapidly from sample to sample so that the difference between adjacent samples should have a lower variance than the variance of the signal itself. That this is so can be easily verified (see Problem 5.10). This fact provides the motivation for the general differential quantization scheme depicted in Fig. 5.31 [16,17]. In this system the input to the quantizer is a signal

$$d(n) = x(n) - \tilde{x}(n) \tag{5.70}$$

which is the difference between the unquantized input sample, $x(n)$, and an estimate, or prediction, of the input sample which is denoted $\tilde{x}(n)$. This predicted value is the output of a predictor system P, whose input is, as we will see, a quantized version of the input signal, $x(n)$. The difference signal may also be called the prediction error signal, since it is the amount by which the predictor fails to exactly predict the input. Temporarily leaving aside the ques-

208

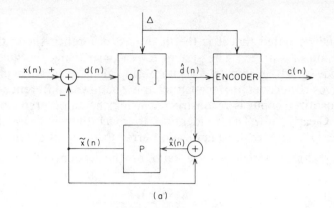

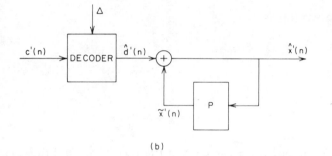

Fig. 5.31 General differential quantization scheme; (a) coder; (b) decoder.

tion of how the estimate, $\tilde{x}(n)$, is obtained, we note that it is the difference signal that is quantized rather than the input. The quantizer could be either fixed or adaptive, uniform or nonuniform, but in any case, its parameters should be adjusted to match the variance of $d(n)$. The quantized difference signal can be represented as

$$\hat{d}(n) = d(n) + e(n) \tag{5.71}$$

where $e(n)$ is the quantization error. According to Fig. 5.31a, the quantized difference signal is added to the predicted value $\tilde{x}(n)$ to produce a quantized version of the input; i.e.,

$$\hat{x}(n) = \tilde{x}(n) + \hat{d}(n) \tag{5.72}$$

Substituting Eqs. (5.70) and (5.71) into Eq. (5.72) we see that

$$\hat{x}(n) = x(n) + e(n) \tag{5.73}$$

That is, independent of the properties of the system labeled P, the quantized speech sample differs from the input only by the quantization error of the difference signal. Thus, if the prediction is good, the variance of $d(n)$ will be smaller than the variance of $x(n)$ so that a quantizer with a given number of levels can be adjusted to give a smaller quantization error than would be possible when quantizing the input directly.

It should be noted that it is the quantized difference signal that is coded for transmission or storage. The system for reconstructing the quantized input from the code words is implicit in Fig. 5.31a. This system, depicted in Fig. 5.31b, involves a decoder to reconstruct the quantized difference signal from which the quantized input is reconstructed using the same predictor as used in Fig. 5.31a. Clearly, if $c'(n)$ is identical to $c(n)$ then $\hat{x}'(n) = \hat{x}(n)$, which differs from $x(n)$ only by the quantization error incurred in quantizing $d(n)$.

The signal-to-quantizing noise ratio of the system of Fig. 5.31 is, by definition,

$$SNR = \frac{E[x^2(n)]}{E[e^2(n)]} = \frac{\sigma_x^2}{\sigma_e^2} \tag{5.74}$$

which can be written as

$$SNR = \frac{\sigma_x^2}{\sigma_d^2} \cdot \frac{\sigma_d^2}{\sigma_e^2} = G_P \cdot SNR_Q \tag{5.75}$$

where

$$SNR_Q = \frac{\sigma_d^2}{\sigma_e^2} \tag{5.76}$$

is the signal-to-quantizing noise ratio of the quantizer, and the quantity

$$G_P = \frac{\sigma_x^2}{\sigma_d^2} \tag{5.77}$$

is defined as the gain due to the differential configuration.

The quantity SNR_Q is dependent upon the particular quantizer that is used, and, given knowledge of the properties of $d(n)$, SNR_Q can be maximized by using the techniques of the previous sections. The quantity G_P, if greater than unity, represents the gain in SNR that is due to the differential scheme. Clearly, our objective should be to maximize G_P by appropriate choice of the predictor system, P. For a given signal, σ_x^2 is a fixed quantity so that G_P can be maximized by minimizing the denominator of Eq. (5.77); i.e., by minimizing the variance of the prediction error.

To proceed, we need to specify the nature of the predictor, P. One approach that is well motivated by our previous discussion of the model for speech production and by the fact that it leads to tractable mathematics is to use a linear predictor. That is, $\tilde{x}(n)$ is a linear combination of past quantized values

$$\tilde{x}(n) = \sum_{k=1}^{p} \alpha_k \hat{x}(n-k) \tag{5.78}$$

The predicted value is thus the output of a finite impulse response filter whose system function is

$$P(z) = \sum_{k=1}^{p} \alpha_k z^{-k} \tag{5.79}$$

and whose input is the reconstructed (quantized) signal $\hat{x}(n)$. We also note that the reconstructed signal is the output of a system whose system function is

$$H(z) = \frac{1}{1 - \sum\limits_{k=1}^{p} \alpha_k z^{-k}} \tag{5.80}$$

and whose input is the quantized difference signal. The variance of the prediction error in Fig. 5.31 is

$$\sigma_d^2 \doteq E[d^2(n)] = E[(x(n) - \tilde{x}(n))^2]$$

$$= E\left[[x(n) - \sum_{k=1}^{p} \alpha_k \hat{x}(n-k)]^2 \right]$$

$$= E\left[[x(n) - \sum_{k=1}^{p} \alpha_k x(n-k) - \sum_{k=1}^{p} \alpha_k e(n-k)]^2 \right] \tag{5.81}$$

In order to choose a set of predictor coefficients $\{\alpha_j\}$, $1 \leqslant j \leqslant p$, that minimize σ_d^2, we must differentiate σ_d^2 with respect to each parameter and set the derivatives equal to zero, thereby obtaining the set of p equations

$$\frac{\partial \sigma_d^2}{\partial \alpha_j} = -2E\left[[x(n) - \sum_{k=1}^{p} \alpha_k (x(n-k) + e(n-k))] \cdot [x(n-j) + e(n-j)] \right]$$

$$= 0, \quad 1 \leqslant j \leqslant p \tag{5.82}$$

Equation (5.82) can be written in the more compact form

$$E[(x(n) - \tilde{x}(n))\hat{x}(n-j)] = E[d(n)\hat{x}(n-j)] = 0 \quad 1 \leqslant j \leqslant p \tag{5.83}$$

from which we make the important observation that if the predictor coefficients are such as to minimize σ_d^2, then the difference signal (prediction error) is uncorrelated with (i.e., orthogonal to) the past values of the predictor input $\hat{x}(n-j)$, $1 \leqslant j \leqslant p$.

Equations (5.82) can be expanded into the set of p equations

$$E[x(n-j)x(n)] + E[e(n-j)x(n)] = \sum_{k=1}^{p} \alpha_k E[x(n-j)x(n-k)]$$

$$+ \sum_{k=1}^{p} \alpha_k E[e(n-j)x(n-k)]$$

$$+ \sum_{k=1}^{p} \alpha_k E[x(n-j)e(n-k)]$$

$$+ \sum_{k=1}^{p} \alpha_k E[e(n-j)e(n-k)] \tag{5.84}$$

where $1 \leqslant j \leqslant p$. Now, if we assume that the quantization is reasonably fine, it can be assumed that $e(n)$ is uncorrelated with $x(n)$ and that $e(n)$ is a stationary white noise sequence; i.e.,

$$E[x(n-j)e(n-k)] = 0, \quad \text{for all } n, j, \text{ and } k \tag{5.85}$$

211

and

$$E[e(n-j)e(n-k)] = \sigma_e^2 \delta(j-k). \tag{5.86}$$

Using these assumptions Eq. (5.84) can be simplified to

$$\phi(j) = \sum_{k=1}^{p} \alpha_k [\phi(j-k) + \sigma_e^2 \delta(j-k)] \quad 1 \leqslant j \leqslant p \tag{5.87}$$

where $\phi(j)$ is the autocorrelation function of $x(n)$. If we divide both sides of the above equations by σ_x^2 and define the normalized autocorrelation as

$$\rho(j) = \frac{\phi(j)}{\sigma_x^2} \tag{5.88}$$

then we can express Eq. (5.87) in matrix form as

$$\boldsymbol{\rho} = \mathbf{C} \, \boldsymbol{\alpha} \tag{5.89a}$$

where

$$\boldsymbol{\rho} = \begin{bmatrix} \rho(1) \\ \rho(2) \\ \cdot \\ \cdot \\ \cdot \\ \rho(p) \end{bmatrix} \tag{5.89b}$$

and

$$\mathbf{C} = \begin{bmatrix} (1 + \dfrac{1}{SNR}) & \rho(1) & \cdots & \rho(p-1) \\[2mm] \rho(1) & (1 + \dfrac{1}{SNR}) & \cdots & \rho(p-2) \\[2mm] \cdot & \cdot & & \cdot \\ \cdot & \cdot & & \cdot \\ \cdot & \cdot & & \cdot \\ \rho(p-1) & \rho(p-2) & \cdots & (1 + \dfrac{1}{SNR}) \end{bmatrix} \tag{5.89c}$$

and

$$\boldsymbol{\alpha} = \begin{bmatrix} \alpha_1 \\ \alpha_2 \\ \cdot \\ \cdot \\ \cdot \\ \alpha_p \end{bmatrix} \tag{5.89d}$$

and $SNR = \sigma_x^2/\sigma_e^2$. Thus the vector of optimum predictor coefficients is obtained as the solution of the matrix equation (5.89a); i.e.,

$$\boldsymbol{\alpha} = \mathbf{C}^{-1} \boldsymbol{\rho}. \tag{5.90}$$

In general, the matrix $\mathbf{C}^{-1}$ can be computed by a variety of numerical methods,

212

including methods that take advantage of the fact that $\mathbf{C}$ is a Toeplitz matrix (see Chapter 8). However, Eq. (5.89a) cannot be solved in the most general case since the matrix $\mathbf{C}$ contains terms which depend on the signal-to-noise ratio, $SNR = \sigma_x^2/\sigma_e^2$ (see Eq. (5.89c)); but SNR depends on the coefficients of the linear predictor, which in turn depend upon SNR through Eq. (5.89a). One possibility is to neglect the term $1/SNR$ in Eq. (5.89) in order to obtain a solution. For the case $p = 1$, however, such an assumption is unnecessary since Eq. (5.90) can be directly solved to give

$$\alpha_1 = \frac{\rho(1)}{1 + \dfrac{1}{SNR}} \tag{5.91}$$

Eq. (5.91) shows that $\alpha_1 < \rho(1)$.

Inspite of the difficulties in solving explicitly for the predictor coefficients, it is possible to obtain an expression for the optimum G_P in terms of the α_i's. To do this we solve for σ_d^2 by rewriting Eq. (5.81) in the form

$$\sigma_d^2 = E[(x(n) - \tilde{x}(n))(x(n) - \tilde{x}(n))]$$

$$= E[(x(n) - \tilde{x}(n))x(n)] - E[(x(n) - \tilde{x}(n))\tilde{x}(n)] \tag{5.92}$$

Using Eq. (5.83) it is straightforward to show that for the optimum predictor coefficients, the second term in the above equation is zero; i.e., the predicted value is also uncorrelated with the prediction error (see Problem 5.12). Thus we can write,

$$\sigma_d^2 = E[(x(n) - \tilde{x}(n))x(n)]$$

$$= E[x^2(n)] - E\left[\sum_{k=1}^{p} \alpha_k(x(n-k) + e(n-k))x(n)\right] \tag{5.93}$$

Using the assumptions of uncorrelated signal and noise, we obtain

$$\sigma_d^2 = \sigma_x^2 - \sum_{k=1}^{p} \alpha_k\phi(k) = \sigma_x^2\left[1 - \sum_{k=1}^{p} \alpha_k\rho(k)\right] \tag{5.94}$$

Thus, from Eq. (5.77),

$$(G_P)_{opt} = \frac{1}{1 - \displaystyle\sum_{k=1}^{p} \alpha_k\rho(k)} \tag{5.95}$$

where the α_k's satisfy Eq. (5.89a).

For the case $p = 1$ we can examine the effects of using a suboptimum value of α_1 on the quantity $G_P = \sigma_x^2/\sigma_d^2$. From Eq. (5.95) we get

$$(G_P)_{opt} = \frac{1}{1 - \alpha_1\rho(1)} \tag{5.96}$$

If we choose an arbitrary value for α_1, then by repeating the derivation leading to Eq. (5.94) we get

$$\sigma_d^2 = \sigma_x^2[1 - 2\alpha_1\rho(1) + \alpha_1^2] + \alpha_1^2\sigma_e^2 \tag{5.97}$$

213

or

$$(G_P)_{arb} = \cfrac{1}{1 - 2\alpha_1 \rho(1) + \alpha_1^2(1 + \cfrac{1}{SNR})} \qquad (5.98)$$

The term α_1^2/SNR represents the increase in variance of $d(n)$ due to the feedback of the error signal $e(n)$. It is readily shown (see Problem 5.13) that Eq. (5.98) can be rewritten in the form:

$$(G_P)_{arb} = \cfrac{1 - \cfrac{\alpha_1^2}{SNR_Q}}{1 - 2\alpha_1 \rho(1) + \alpha_1^2} \qquad (5.99)$$

for any value of α_1 (including the optimum value). Thus, for example, if $\alpha_1 = \rho(1)$ (which is suboptimum according to Eq. (5.91)) then

$$(G_P)_{subopt} = \cfrac{1 - \cfrac{\rho^2(1)}{SNR_Q}}{1 - \rho^2(1)} = \left[\cfrac{1}{1 - \rho^2(1)} \right] \left[1 - \cfrac{\rho^2(1)}{SNR_Q} \right] \qquad (5.100)$$

Thus the gain in prediction obtained without the quantizer, $1/(1 - \rho^2(1))$, is reduced by the second factor in Eq. (5.100) due to feedback of the error signal.

To obtain the optimum gain, Eq. (5.99) can be differentiated with respect to α_1 to give

$$\frac{d(G_P)}{d\alpha_1} = 0 \qquad (5.101)$$

which can be solved directly for the optimum value of α_1.[5]

For illustrative purposes we make the assumption that we can neglect the term $1/SNR$ in Eq. (5.89). Thus for a first order predictor, Eq. (5.91) becomes $\alpha_1 = \rho(1)$, and the gain due to prediction is

$$(G_P)_{opt} = \frac{1}{1 - \rho^2(1)} \qquad (5.102)$$

Thus so long as $\rho(1) \neq 0$ there will be some improvement due to prediction. We have already seen (Fig. 5.5) typical correlation functions for lowpass and bandpass filtered speech sampled at 8 kHz [4]. The shaded regions in this figure indicate the range of variation of $\rho(n)$ over four speakers, and the central curve is the average for these four speakers. We see from these curves that a reasonable assumption is that for lowpass filtered speech sampled at the Nyquist rate,

$$\rho(1) > .80 \qquad (5.103)$$

which implies that

$$(G_P)_{opt} > 2.77 \ (or \ 4.43 \ dB) \qquad (5.104)$$

[5]We are indebted to Professor Peter Noll for helpful comments on this analysis.

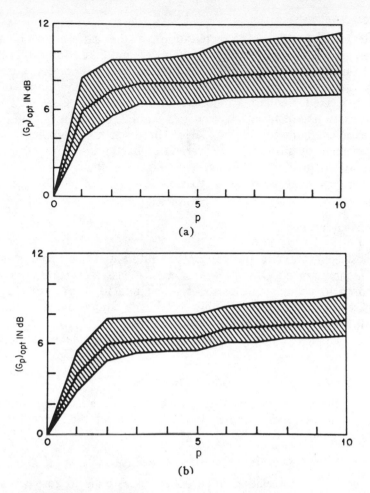

Fig. 5.32 Optimum *SNR* gain *G* versus number of predictor coefficients; (a) lowpass filtered speech; (b) bandpass filtered speech. (After Noll [7].)

Noll [4] has used the data shown in Fig. 5.5 to compute $(G_P)_{opt}$ as a function of p for a 55 second segment of speech that was both lowpass and bandpass filtered. The results are depicted in Fig. 5.32.[6] The shaded region shows the amount of variation obtained for four speakers, with the central curve representing the average over four speakers. It is clear that, even with the simplest predictor, it is possible to realize about a 6 dB improvement in *SNR*. This is of course equivalent to adding an extra bit to the quantizer; however since this bit is not actually added, the bit rate remains the same. Note also that in no case does the gain reach 12 dB, which would be required to achieve the effect of adding 2 bits. An alternative point of view is that differential quantization permits a reduction in bit rate while keeping the *SNR* the same. The price paid, of course, is increased complexity in the quantization system.

---

[6]Again the sampling rate was 8 kHz; thus *n* is a multiple of 125 μsec.

Some basic principles of application of the differential quantization scheme emerge from a consideration of Figs. 5.5 and 5.32. First, it is clear that differential quantization can yield improvements over direct quantization. Second, the amount of improvement is dependent upon the amount of correlation. Third, a fixed predictor cannot be optimum for all speakers and for all speech material. These facts have led to a variety of schemes that are based upon the basic configuration of Fig. 5.31. These schemes combine a variety of fixed and adaptive quantizers with a variety of fixed and adaptive predictors to achieve improved quality or lowered bit rate. We shall now discuss several examples that represent the range of possibilities.

5.6 Delta Modulation

The simplest application of the concept of differential quantization is in delta modulation (abbreviated DM) [18-24]. In this class of systems, the sampling rate is chosen to be many times the Nyquist rate for the input signal. As a result, adjacent samples become highly correlated. This is evident from the discussion of Section 5.2, where we showed that the autocorrelation of the sequence of samples is just a sampled version of the analog autocorrelation; i.e.,

$$\phi(m) = \phi_a(mT) \tag{5.105}$$

Given the properties of autocorrelation functions, it is reasonable to expect the correlation to increase as $T \to 0$. Indeed we expect that except for strictly uncorrelated signals,

$$\phi(1) \to \sigma_x^2 \quad as \quad T \to 0 \tag{5.106}$$

This high degree of correlation implies that as T tends to zero we should be better able to predict the input from past samples and as a result, the variance of the prediction error should be low. Therefore, because of the high gain due to the differential configuration, a rather crude quantizer can provide acceptable performance. Indeed, delta modulation systems employ a simple 1-bit (2 level) quantizer. Thus, the bit-rate of a delta modulation is simply equal to the sampling rate.

5.6.1 Linear delta modulation

The simplest delta modulation system is depicted in Fig. 5.33. In this case the quantizer has only two levels and the step size is fixed. The positive quantization level is represented by $c(n) = 0$ and the negative by $c(n) = 1$. Thus, $\hat{d}(n)$ is

$$\hat{d}(n) = \Delta \quad \text{if} \quad c(n) = 0$$
$$= -\Delta \quad \text{if} \quad c(n) = 1 \tag{5.107}$$

Figure 5.33 also incorporates a simple first order fixed predictor for which the

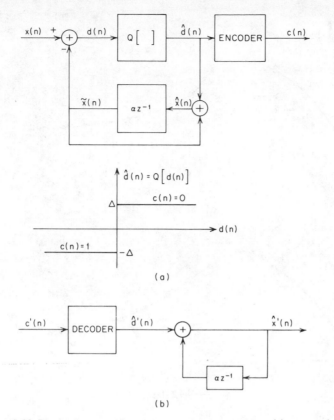

Fig. 5.33 Block diagram of a delta modulation system; (a) coder; (b) decoder.

optimum prediction gain is

$$(G_P)_{opt} = \frac{1}{1 - \rho^2(1)} \tag{5.108}$$

Thus as $\rho(1) \to 1$, $(G_P)_{opt} \to \infty$. This result can be viewed in only qualitative terms, however, since the assumptions under which the expression for $(G_P)_{opt}$ was derived tend to break down for such crude quantization.

The effect of quantization error can be observed from Fig. 5.34a, which shows an analog waveform $x_a(t)$ and resulting samples $x(n)$, $\tilde{x}(n)$ and $\hat{x}(n)$ for a given sampling period, T, and assuming α (the feedback multiplier) is set to 1.0. It can be seen from Fig. 5.33a that in general, $\hat{x}(n)$ satisfies the difference equation

$$\hat{x}(n) = \alpha\hat{x}(n-1) + \hat{d}(n) \tag{5.109}$$

With $\alpha \approx 1$, this equation is the digital equivalent of integration, in the sense that it represents the accumulation of positive and negative increments of magnitude Δ. We also note that the input to the quantizer is

$$d(n) = x(n) - \hat{x}(n-1) = x(n) - x(n-1) - e(n-1) \tag{5.110}$$

Thus except for the quantization error in $\hat{x}(n-1)$, $d(n)$ is a first backward

217

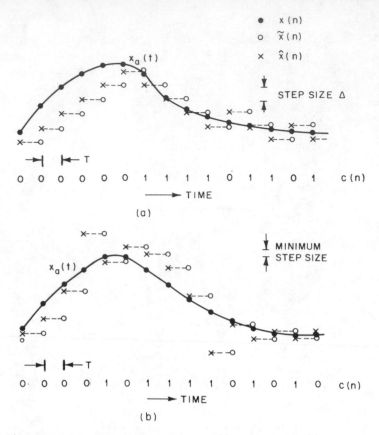

Fig. 5.34 Illustration of delta modulation; (a) fixed stepsize; (b) adaptive stepsize.

difference of $x(n)$, which can be viewed as a digital approximation to the derivative of the input and the inverse of the digital integration process. If we consider the maximum slope of the waveform, it is clear that in order for the sequence of samples $\{\hat{x}(n)\}$ to increase as fast as the sequence of samples $\{x(n)\}$ in a region of maximum slope of $x_a(t)$, we require

$$\frac{\Delta}{T} \geqslant \max\left|\frac{dx_a(t)}{dt}\right| \tag{5.111}$$

Otherwise, the reconstructed signal will fall behind as shown in the left side of Fig. 5.34a. This condition is called "slope overload" and the resulting quantization error is called *slope overload distortion* (noise). Note that since the maximum slope of $\hat{x}(n)$ is fixed by the step size, increases and decreases in the sequence $\hat{x}(n)$ tend to occur along straight lines. For this reason, fixed (nonadaptive) delta modulation is often called linear delta modulation (abbreviated LDM).

The step size, Δ, also determines the peak error when the slope is very small. For example, it is easily verified that when the input is zero (idle channel condition), the output of the quantizer will be an alternating sequence of

0's and 1's, in which case the reconstructed signal $\hat{x}(n)$ will alternate about zero (or some constant level) with a peak-to-peak variation of Δ. This latter type of quantization error, depicted on the right in Fig. 5.34a, is called *granular noise*.

As we have seen before, there is a need to have a large step size to accommodate wide dynamic range, while a small step size is required for accurate representation of low level signals. In this case, however, we are concerned with the dynamic range and amplitude of the difference signal (or derivative of the analog signal). It is intuitively clear that the choice of step size that minimizes the mean squared quantization error will represent a compromise between slope overload and granular noise.

Figure 5.35, which is from a detailed study of delta modulation by Abate [21], shows signal-to-noise ratio as a function of the normalized step size variable $\Delta/(E[(x(n) - x(n-1))^2])^{1/2}$, with oversampling index $F_0 = F_s/2F_N$ as a parameter where F_s is the sampling rate of the delta modulator, and F_N is the Nyquist frequency of the signal. Note that the bit rate is

$$Bit\ rate = F_s(1\ bit) = F_s = 2F_N \cdot F_0 \tag{5.112}$$

Thus the oversampling index plays the role of the number of bits/sample for a multi-bit quantizer with sampling at the Nyquist rate. These curves are for flat spectrum bandlimited Gaussian noise. Somewhat higher *SNR* values are obtained for speech since there is greater correlation; however, the shape of the curves is much the same. It can be seen from Fig. 5.35 that for a given value of F_0, the *SNR* curve has a rather sharp peak with values of Δ above the location of the peak corresponding to granular noise and values below the peak location corresponding to slope overload. Abate [21] gives the empirical formula

$$\Delta_{opt} = \{E[(x(n) - x(n-1))^2]\}^{1/2} \ln(2F_0) \tag{5.113}$$

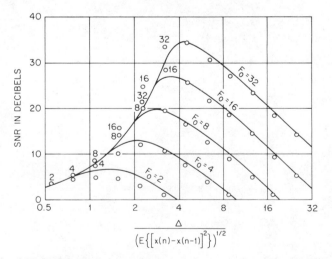

Fig. 5.35 *SNR* for delta modulators as a function of the normalized step size. (After Abate [21].)

219

for the optimum step size; i.e, for the location of the peak of the *SNR* curve for a given value of F_0. It can also be seen from Fig. 5.35 that the optimum *SNR* increases at the rate of about 9 dB for each doubling of F_0. Since doubling F_0 is equivalent to doubling F_s, we note that doubling the bit rate increases the *SNR* by 9 dB. This is in contrast to PCM where if we double the bit rate by doubling the number of bits/sample, we achieve a 6 dB increase for each *added* bit; thus the increase of *SNR* with bit rate is much more dramatic for PCM than for LDM.

Another important feature of the curves of Fig. 5.35 is the sharpness of the peak of the *SNR* curve, which implies that the *SNR* is very sensitive to the input level. [Note that $E[(x(n) - x(n-1))^2] = 2\sigma_x^2(1-\rho(1))$.] Thus, it can be seen from Fig. 5.35 that to obtain an *SNR* of 35 dB for a Nyquist frequency of 3 kHz would require a bit rate of about 200 Kb/sec. Even at this rate, however, this quality can only be maintained over a rather narrow range of input levels if the step size is fixed. To achieve toll quality, i.e., quality comparable to 7-bit log PCM, for speech requires much higher bit rates.

The main advantage of LDM is its simplicity. The system can be implemented with simple analog and digital integrated circuits and since only a one-

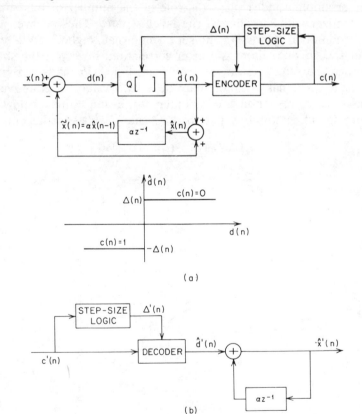

Fig. 5.36 Delta modulator with adaptive stepsize; (a) coder; (b) decoder.

bit code is required, no synchronization of bit patterns is required between transmitter and receiver. The limitations on the performance of linear delta modulation systems stem mainly from the crude quantization of the difference signal. In view of our previous discussion of adaptive quantization, it is natural to suppose that adaptive quantization schemes would greatly improve the performance of a delta modulator. Of greatest interest are simple adaptive quantization schemes which improve performance but do not greatly increase the complexity of the system.

5.6.2 Adaptive delta modulation

A large variety of adaptive delta modulation (ADM) schemes has been proposed. Most of these schemes are of the feedback type in which the step size for the two-level quantizer is adapted from the output code words. The general form of such systems is shown in Fig. 5.36. Such schemes maintain the advantage that no synchronization of bit-patterns is required since, in the absence of errors, the step size information can be derived from the code word sequence at both the transmitter and the receiver.

In this section we shall illustrate the use of adaptive quantization in delta modulation through the discussion of two specific adaptation algorithms. There are many other possibilities which can be found in the literature [20-24].

The first system that we shall discuss has been studied extensively by N. S. Jayant [22]. Jayant's algorithm for adaptive delta modulation is a modification of the quantization scheme discussed in Section 5.4.2. As for the case of multi-bit quantizer, the step size obeys the rule

$$\Delta(n) = M\Delta(n-1) \tag{5.114a}$$

$$\Delta_{min} \leqslant \Delta(n) \leqslant \Delta_{max} \tag{5.114b}$$

In this case, the multiplier is a function of the present and the previous code words, $c(n)$ and $c(n-1)$. This is possible since $c(n)$ depends only on the sign of $d(n)$ which is given by

$$d(n) = x(n) - \alpha\hat{x}(n-1). \tag{5.115}$$

Thus the sign of $d(n)$ can be determined before the determination of the actual quantized value $\hat{d}(n)$ which must await the determination of $\Delta(n)$ from Eq. (5.114). The algorithm for choosing the step size multiplier in Eq. (5.114a) is

$$M = P > 1 \quad \text{if} \quad c(n) = c(n-1)$$

$$M = Q < 1 \quad \text{if} \quad c(n) \neq c(n-1) \tag{5.116}$$

This adaptation strategy is motivated by the bit patterns observed in linear delta modulation. For example, in Fig. 5.34a we note that periods of slope overload are signaled by runs of consecutive 0's or 1's. Periods of granularity are signaled by alternating sequences of the form ...0 1 0 1 0 1.... Figure 5.34b shows how the waveform in Fig. 5.34a would be quantized by an adaptive delta modulator of the type described by Eqs. (5.114) and (5.116). For convenience, the

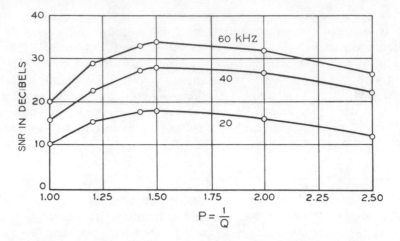

Fig. 5.37 Signal-to-noise ratios of an adaptive delta modulator as functions of P. (After Jayant [22].)

parameters of the system are set at $P = 2$, $Q = 1/2$, $a = 1$, and the minimum step size is shown in the figure. It can be seen that the region of large positive slope still causes a run of 0's but in this case the step size increases exponentially so as to follow the increase in slope of the waveform. The region of granularity to the right in the figure is again signaled by an alternating sequence of 0's of 1's but in this case the step size falls rapidly to the minimum (Δ_{min}) and remains there as long as the slope is small. Since the minimum step size can be made much smaller than that required for optimum performance of a linear delta modulator, granular noise can be greatly reduced. Likewise the maximum step size can be made larger than the maximum slope of the input signal so as to reduce slope overload noise.

The parameters of this adaptive delta modulation system are P, Q, Δ_{min} and Δ_{max}. The step size limits should be chosen to provide the desired dynamic range for the input signal. The ratio $\Delta_{max}/\Delta_{min}$ should be large enough to maintain a high *SNR* over a desired range of input signal levels. The minimum step size should be as small as is practical so as to minimize the idle channel noise. Jayant [22] has shown that P and Q should satisfy the relation

$$PQ \leqslant 1 \tag{5.117}$$

for stability; i.e., to maintain the step size at values appropriate for the level of the input signal. Figure 5.37 shows the results of a simulation for speech signals with $PQ = 1$ for three different sampling rates. It is evident that the maximum *SNR* is obtained for $P = 1.5$; however, the peak of all three curves is very broad with *SNR* being within a few dB of the maximum for

$$1.25 < P < 2 \tag{5.118}$$

The results of Fig. 5.37 are replotted in Fig. 5.38 to compare ADM to LDM and log PCM. It is noted that with $P = 1/Q$ the condition $P = 1 = 1/Q$ implies no adaptation at all; i.e., LDM. The *SNR* values for this condition and for $P = 1.5$ are plotted as a function of bit rate in Fig. 5.38. Also shown there

222

is maximum *SNR* for $\mu = 100$ (log PCM) as a function of bit rate as computed from Eq. (5.38) assuming sampling at the Nyquist rate ($2F_N = 6.6$ kHz).

Figure 5.38 shows that ADM is superior to LDM by 8 dB at 20 kb/s and the *SNR* advantage increases to 14 dB at 60 kb/s. For LDM we observe about a 6 dB increase with a doubling of sampling rate (and bit rate), whereas with ADM the corresponding increase is 10 dB. Comparing ADM and log PCM, we note that for bit rates below 40 kb/s, ADM out-performs log PCM. For higher bit rates log PCM has a higher *SNR*. For example, Fig. 5.38 shows that the ADM system requires about 60 kb/s to achieve the same quality as 7-bit log PCM, having a bit rate of about 46 kb/s.

The improved quality of the ADM system is achieved with only a slight increase in complexity. Since the step size adaptation is done using the output bit stream, the ADM system retains the basic simplicity of delta modulation systems; i.e., no code word framing is required. Thus, for many applications ADM may be preferred to log PCM even at the expense of slightly higher information rate.

Another example of adaptive quantization in delta modulation is known as continuously variable slope delta modulation (CVSD). (A system of this type was first proposed by Greefkes [23].) This system is again based upon Fig. 5.36, with the step size logic being defined by the equations

$$\Delta(n) = \beta\Delta(n-1) + D_2 \quad \text{if} \quad c(n) = c(n-1) = c(n-2) \qquad (5.119a)$$

$$= \beta\Delta(n-1) + D_1 \quad \textit{otherwise} \qquad (5.119b)$$

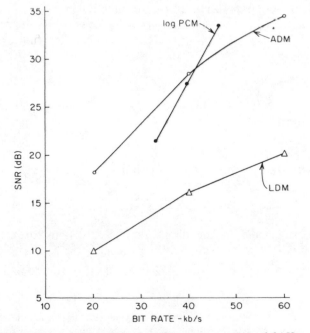

Fig. 5.38 *SNR* versus bit rate for 3 coding schemes using a 6.6 kHz sampling rate.

223

where $0 < \beta < 1$ and $D_2 >> D_1 > 0$. In this case the minimum and maximum step sizes are inherent in the recurrence formula for $\Delta(n)$. (See Problem 5.14.)

The basic principle is, as before, to increase the step size in response to patterns in the bit stream that indicate slope overload. In this case, a run of three consecutive 1's or three consecutive 0's causes an increment D_2 to be added to the step size. In the absence of such patterns the step size decays (because $\beta < 1$) until it reaches Δ_{min}. Thus the step size will increase during slope overload conditions and will decrease otherwise. Again, Δ_{min} and Δ_{max} can be chosen to provide desired dynamic range and low granular noise during idle channel conditions. The parameter β controls the speed of adaptation. If β is close to 1, the rate of build-up and decay of $\Delta(n)$ is slow, whereas if β is much less than one, the adaptation is much faster. Thus, this basic adaptation scheme can be adjusted to be either syllabic or instantaneous.

This system has been used in situations requiring low sensitivity to channel errors with speech quality requirements below those required for commercial communication channels. In this situation the parameters of the system are adjusted to provide syllabic adaptation. Also the predictor coefficient, α, is set at a value considerably less than one so that the effect of channel errors dies out quickly. The price paid for insensitivity to errors is, of course, decreased quality when no errors occur. A major advantage of the ADM system in this situation is that it has sufficient flexibility to provide effective tradeoffs between quality and robustness.

5.6.3 Higher-order predictors in delta modulation

For simplicity, most LDM and ADM systems use a first order fixed predictor of the form

$$\tilde{x}(n) = \alpha \hat{x}(n-1) \tag{5.120}$$

as shown in Fig. 5.36. In this case, the reconstructed signal satisfies the difference equation

$$\hat{x}(n) = \alpha \hat{x}(n-1) + \hat{d}(n) \tag{5.121}$$

which is characterized by the system function

$$H_1(z) = \frac{1}{1 - \alpha z^{-1}} \tag{5.122}$$

This, we have suggested, is the digital equivalent of an integrator (if $\alpha = 1$). When $\alpha < 1$ it is sometimes called a "leaky integrator."

The results shown in Figure 5.32 suggest[7] that for delta modulation sys-

---

[7]To be more specific one would have to know exact values of the speech autocorrelation function for lags less than 125 μsec (corresponding to the higher sampling rates of delta modulation systems) to calculate high order prediction gains.

tems, a greater *SNR* is possible with a second order predictor; i.e., with

$$\tilde{x}(n) = \alpha_1 \hat{x}(n-1) + \alpha_2 \hat{x}(n-2) \qquad (5.123)$$

In this case

$$\hat{x}(n) = \alpha_1 \hat{x}(n-1) + \alpha_2 \hat{x}(n-2) + d(n) \qquad (5.124)$$

which is characterized by

$$H_2(z) = \frac{1}{1 - \alpha_1 z^{-1} - \alpha_2 z^{-2}} \qquad (5.125)$$

It has been shown empirically [25] that second order prediction gives improved performance over first order prediction when the poles of $H_2(z)$ are both real; i.e.,

$$H_2(z) = \frac{1}{(1-az^{-1})(1-bz^{-1})}, \qquad 0 < a,b < 1 \qquad (5.126)$$

This is often called "double integration." Improvements over first order prediction may be as high as 4 dB depending on speaker and speech material [25].

Unfortunately, the use of higher order prediction in ADM systems is not just a simple matter of replacing the first order predictor with a second order predictor since the adaptive quantization algorithm interacts with the prediction algorithm. For example, the idle channel condition will be signaled by different bit patterns depending on the order of the predictor. For a second order predictor, the bit pattern for the idle channel condition might be ...010101... or ...00110011... depending upon the choice of α_1 and α_2 and the past state of the system before the input became zero. This clearly calls for an adaptation algorithm based upon more than two consecutive bits if the step size is to fall to its minimum value for idle channel conditions.

The design of ADM systems with high order predictors has not been extensively studied. Whether the added complexity in both the predictor and the quantizer could be justified would depend upon the amount of improvement in quality that could be obtained. The use of multi-bit quantizers of the type discussed in Section 5.4 simplifies the design somewhat at the expense of the need for framing the bit stream. We now turn to a discussion of differential quantization using multi bit quantizers.

5.7 Differential PCM (DPCM)

Any system of the form shown in Fig. 5.31 could be called a differential PCM (DPCM) system. Delta modulators, as discussed in the previous section, for example, could also be called 1-bit DPCM systems. Generally, however, the term differential PCM is reserved for differential quantization systems in which the quantizer has more than two levels.

As is clear from Fig. 5.32, DPCM systems with fixed predictors can provide from 4 to 11 dB improvement over direct quantization (PCM). The

225

greatest improvement occurs in going from no prediction to first order prediction with somewhat smaller additional gains resulting from increasing the predictor order up to 4 or 5, after which little additional gain results. As pointed out in Section 5.5, this gain in *SNR* implies that a DPCM system can achieve a given *SNR* using one less bit than would be required when using the same quantizer directly on the speech waveform. Thus, the results of Sections 5.3 and 5.4 can be applied to obtain a reasonable estimate of the performance that can be obtained for a particular quantizer used in a differential configuration. For example, for a differential PCM system with a uniform fixed quantizer, the *SNR* would be approximately 6 dB greater than the *SNR* for a quantizer with the same number of levels acting directly on the input. The differential scheme would behave in much the same manner as the direct PCM scheme; i.e., the *SNR* would increase by 6 dB for each bit added to the code words, and the *SNR* would show the same dependence upon signal level. Similarly, the *SNR* of a μ-law quantizer would be improved by about 6 dB by use in a differential configuration and at the same time its characteristic insensitivity to input signal level would be maintained.

Figure 5.32 displays a wide variation of prediction gain with speaker and with bandwidth. Similar wide variations are observed among different speech utterances. All of these effects are a result, of course, of the nonstationarity of the speech signal. No single set of predictor coefficients can be optimum for a wide variety of speech material or a wide range of speakers.

This variation of performance with speaker and speech material, together with variations in signal level inherent in the speech communication process, make adaptive prediction and adaptive quantization necessary to achieve best performance over a wide range of speakers and speaking situations. Such systems are called adaptive differential PCM systems (ADPCM). We shall first discuss the use of adaptive quantization with fixed prediction, and conclude with a discussion of adaptive prediction.

5.7.1 DPCM with adaptive quantization

The discussion of adaptive quantization in Section 5.4 can be applied directly to the case of DPCM. As Section 5.4 pointed out, there are two basic approaches to the control of adaptive quantizers.

Figure 5.39 shows how a feed-forward-type adaptive quantizer is used in an ADPCM system [7]. In schemes of this type, the quantizer step size is proportional to the variance of the input to the quantizer. However, since the difference signal $d(n)$ will be proportional to the input, it is reasonable to control the step size either from $d(n)$ or, as depicted in Fig. 5.39, from the input, $x(n)$. Several algorithms for adjusting the step size are given in Section 5.4.1. The discussion of Section 5.4.1 indicates that such adaptation procedures can provide about 5 dB improvement in *SNR* over standard μ-law nonadaptive PCM. This improvement coupled with the 6 dB that can be obtained from the differential configuration with fixed prediction means that ADPCM with feed-

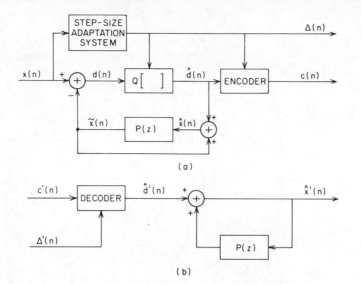

Fig. 5.39 ADPCM system with feed-forward adaptive quantization; (a) coder; (b) decoder.

forward adaptive prediction should achieve an *SNR* that is 10-11 dB greater than could be obtained with a fixed quantizer with the same number of levels.

Figure 5.40 shows how a feedback-type adaptive quantizer can be used in an ADPCM system [26]. If for example the adaptation strategy described by Eqs. (5.66)-(5.68) is used, we can again expect an improvement of 4-6 dB over a fixed μ-law quantizer with the same number of bits. Thus, both the feed-forward and feedback adaptive quantizers can be expected to achieve about 10-

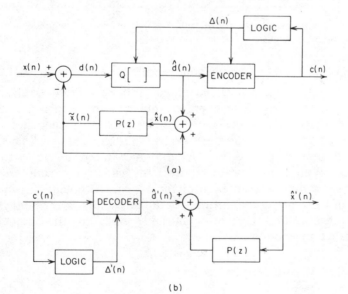

Fig. 5.40 ADPCM system with feedback adaptive quantization; (a) coder; (b) decoder.

227

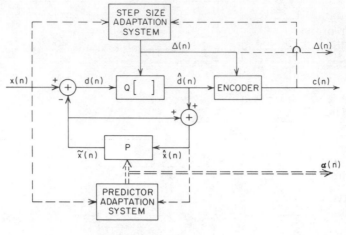

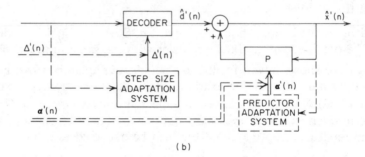

(b)

Fig. 5.41 ADPCM system with both adaptive quantization and adaptive prediction; (a) coder; (b) decoder.

12 dB improvement over a fixed quantizer with the same number of levels.

In either case the quantizer adaptation provides improved dynamic range as well as improved *SNR*. The main advantage of the feedback control is that the step size information is derived from the code word sequence, so that no additional step size information need be transmitted or stored. This, however, makes the quality of the reconstructed output more sensitive to errors in transmission. With feed-forward control, the code words and the step size together serve as the representation of the signal. Although this increases the complexity of the representation, there is the possibility of transmitting the step size with error protection, thereby significantly improving the output quality for high error rate transmission [27,28].

5.7.2 DPCM with adaptive prediction

So far we have considered only fixed predictors, and have found that even with higher order predictors, we can expect that differential quantization will provide, under the best circumstances, about 10-12 dB improvement. Further-more the amount of improvement is a function of speaker and of speech material. In order to effectively cope with the nonstationarity of the speech communication process, it is natural to consider adapting the predictor as well as the quantizer to match the temporal variations of the speech signal [29]. A general adaptive DPCM system with both adaptive quantization and adaptive prediction is depicted in Fig. 5.41. The dotted lines indicate that both the quantizer adaptation and the predictor adaptation algorithms can be either of the feed-forward or the feedback type. If feed-forward control is used for the quantizer or the predictor, then $\Delta(n)$ or the predictor coefficients, $\alpha(n) = \{\alpha_k(n)\}$, (or both) are also required in addition to the code words, $c(n)$, to complete the representation of the speech signal.

The predictor coefficients are assumed to be time dependent so that the predicted value is

$$\tilde{x}(n) = \sum_{k=1}^{p} \alpha_k(n)\hat{x}(n-k) \tag{5.127}$$

In adapting the predictor coefficients $\alpha(n)$ it is common to assume that the pro-perties of the speech signal remain fixed over short time intervals. The predic-tor coefficients are therefore chosen to minimize the average squared prediction error over a short time interval. For feed-forward control, the predictor adapta-tion is based upon measurements on the input signal. (This can be seen to be equivalent to neglecting the term $1/SNR$ in the analysis of Section 5.5.) Using the same type of manipulations that were used to derive Eqs. (5.87) and (5.89), and neglecting the effect of quantization error, it can be shown that the optimum predictor coefficients satisfy the equations,

$$R_n(j) = \sum_{k=1}^{p} \alpha_k(n) R_n(j-k), \quad j = 1, 2, \ldots, p \tag{5.128}$$

where $R_n(j)$ is the short-time autocorrelation function (Eq. (4.24))

$$R_n(j) = \sum_{m=-\infty}^{\infty} x(m) w(n-m) x(j+m) w(n-m-j), \quad 0 \leqslant j \leqslant p \tag{5.129}$$

and $w(n-m)$ is a window function that is positioned at sample n of the input sequence. A rectangular window, or one with much less abrupt tapering of the data (e.g., a Hamming window of length N) can be used. Since the parameters of speech vary rather slowly, it is reasonable to adjust the predictor parameters $\alpha(n)$ infrequently. For example, a new estimate may be computed every 10-20 msec, with the values being held fixed between estimates. The window dura-tion may be equal to the interval between estimates or it may be somewhat larger. In the latter case, successive segments of speech would overlap. As defined by Eq. (5.129), the computation of the correlation estimates required in Eq. (5.128) would require the accumulation of N samples of $x(n)$ in a buffer before computing $R_n(j)$. The set of coefficients $\alpha(n)$ satisfying Eq. (5.128) are used in the configuration of Fig. 5.41a to quantize the input during the interval of N samples beginning at sample n. Thus, to reconstruct the input from the

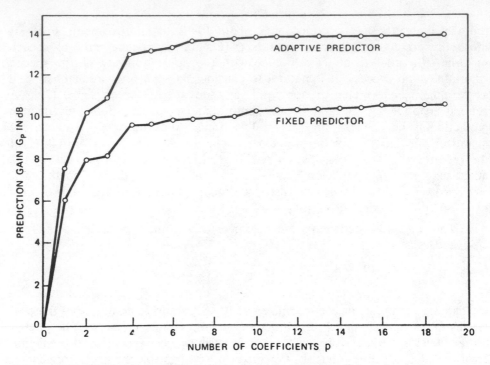

Fig. 5.42 Prediction gains versus number of predictor coefficients for one female speaker (band from 0-3200 Hz). (After Noll [7].)

quantizer code words we also need the predictor coefficients (and possibly the quantizer step size) as depicted in Fig. 5.41b. The details of computing the time-varying predictor parameters are discussed in Chapter 8.

In order to quantitatively express the benefits of adaptive prediction, Noll [7] has examined the dependence of the prediction gain, G_P, upon predictor order for both fixed and adaptive predictor. Figure 5.42[8] shows the quantity

$$10 \log_{10}[G_P] = 10 \log_{10}\left[\frac{E[x^2(n)]}{E[d^2(n)]}\right] \qquad (5.130)$$

as a function of predictor order, p, for both fixed and adaptive prediction. The lower curve, obtained by computing a long term estimate of the autocorrelation for a given speech utterance and solving for the set of predictor coefficients satisfying Eq. (5.89), shows a maximum gain of about 10.5 dB. The upper curve was obtained by finding the value of window length, L, and the predictor coefficients $\alpha(n)$ that maximized G_P across the entire utterance for a fixed value of p. This maximum value is plotted for each value of p. In this case, the maximum gain is about 14 dB. Thus, Noll [7] suggests that reasonable upper bounds on the performance of DPCM systems with fixed and adaptive prediction are 10.5 and 14 dB respectively. Not evident in the curves of Fig.

[8]The results in this figure are for a single speaker. In addition no error feedback was included in the system which was studied.

5.42 is the fact that the optimum fixed predictor is likely to be very sensitive to speaker and speech material, whereas the adaptive prediction scheme is inherently less sensitive.

The adaptive predictor tends to remove redundancy from the speech signal. Indeed, if perfect prediction were possible, the prediction error signal, $d(n)$, would be completely uncorrelated (white noise). We might say that the redundancy is removed by incorporating a model for the speech wave. It can be seen from Fig. 5.42 that little gain results from increasing the order of prediction beyond 4 or 5. Such prediction, however, ignores an important source of redundancy in speech; namely, the correlation due to the quasi-periodic nature of voiced speech. One approach to exploitation of this correlation was considered by Atal and Schroeder [29], who used a two stage predictor of the form

$$\tilde{x}(n) = \beta\hat{x}(n-M) + \sum_{k=1}^{p} \alpha_k[\hat{x}(n-k) - \beta\hat{x}(n-k-M)] \qquad (5.131)$$

where the predictor parameters β, M and $\{\alpha_k\}$ are all adapted at intervals of N samples. Neglecting the effect of quantization error in $\hat{x}(n)$, we can express the prediction error as

$$d(n) = x(n) - \tilde{x}(n) \qquad (5.132)$$

$$= x(n) - \beta x(n-M) - \sum_{k=1}^{p} \alpha_k[x(n-k) - \beta x(n-k-M)]$$

which can be expressed as

$$d(n) = v(n) - \sum_{k=1}^{p} \alpha_k v(n-k) \qquad (5.133)$$

where

$$v(n) = x(n) - \beta x(n-M) \qquad (5.134)$$

The computation of the values of β, M and $\{\alpha_k\}$ that minimize the variance of $d(n)$ is not straightforward. For this reason, Atal and Schroeder [29] consider a sub-optimum solution in which the variance of $v(n)$ was first minimized, and then the variance of $d(n)$ was minimized subject to fixed values of β and M. Thus, the autocorrelation of the input was computed as before, but in this case $R_n(j)$ was determined over a range which would encompass typical pitch periods of speech. The predictor coefficient β was chosen to be a value of the peak of the normalized autocorrelation over the entire range of lags and M was set at the position of the peak of $R_n(j)$. Thus, β accounts for the variability of amplitude between consecutive periods, while M is the pitch period (in samples). Given M and β, the sequence $v(n)$ can be determined and its autocorrelation computed for $j = 0, 1, \ldots ,p$, from which the prediction coefficients α_k can be obtained from Eq. (5.128) with $R_n(j)$ being the short-time autocorrelation of the sequence $v(n)$.

In order to represent speech using such a system, it is necessary to transmit or store the quantized difference signal, the quantizer step size (if feed forward control is used), and the (quantized) predictor coefficients. In the original work of Atal and Schroeder, a one-bit quantizer was used for the difference signal and the step size was adapted every 5 msec (33 samples at a 6.67 kHz sampling rate), so as to minimize the quantization error. Likewise, the predictor parameters were also estimated every 5 msec. Although no explicit *SNR* data was given, it was suggested that high quality reproduction of the speech signal could be achieved at bit rates on the order of 10 kb/s. Jayant [8] asserts that using reasonable quantization of the parameters, gains of 20 dB should be possible over PCM.

Unfortunately, no careful study of the limits of performance of adaptive prediction including pitch parameters has been done. However, it is apparent that schemes such as this represent the extreme of complexity of digital waveform coding systems. On the other end of the scale would be linear delta modulation with its simple quantization process and unstructured stream of one-bit binary code words. The choice of quantization scheme depends on a variety of factors including the desired bit rate, desired quality, coder complexity, and complexity of the digital representation. In the next section we will summarize some comparative studies that help to place the wide variety of quantization schemes in perspective. But first, let us consider briefly the question of feedback control of the adaptive predictor.

One approach is to base the computation of the correlation function upon the quantized signal $\hat{x}(n)$ rather than on the input. Thus, in Eq. (5.128) $R_n(j)$ would be replaced by

$$R_n(j) = \sum_{m=-\infty}^{\infty} \hat{x}(m)\,w(n-m)\,\hat{x}(m+j)\,w(n-m-j), \quad 0 \leqslant j \leqslant p \quad (5.135)$$

In this case, the window must be of the form

$$w(m) = 1, \quad 0 \leqslant m \leqslant N-1 \quad (5.136)$$
$$= 0, \quad otherwise$$

That is, the estimate of predictor coefficients must be based upon *past* quantized values rather than *future* values which cannot be obtained until the predictor coefficients are available. As in the case of adaptive quantizer control, the feedback mode has the advantage that only the quantizer code words need be transmitted. Feedback control of adaptive predictors, however, has not been widely used due to the inherent sensitivity to errors and the inferior performance that results from basing the control upon a noisy input. An interesting approach to feedback control was considered by Stroh [30] who studied a gradient scheme for adjusting the predictor coefficients.

5.8 Comparison of Systems

In comparing digital waveform coding systems, it is convenient to use signal-to-quantization noise as a criterion. However, the ultimate criterion for systems that are to be used for voice communication is a perceptual one. The question of how well the coded speech sounds in comparison to the original unquantized speech is often of paramount importance. Unfortunately this perceptual criterion is often the most difficult to quantify and there is no unified set of results that we can refer to. Thus, in this section we shall briefly summarize the results of objective *SNR* measurements for a variety of speech coding systems and then summarize a few perceptual results that appear to be particularly illuminating.

Noll [7] has performed a very illuminating comparative study of digital waveform coding schemes. He considered the following systems:

1. $\mu = 100$ log PCM with $X_{max} = 8\sigma_x$. (PCM)
2. Adaptive PCM (optimum Gaussian quantizer) with feed-forward control. (PCM-AQF)
3. Differential PCM with first order fixed prediction and adaptive Gaussian quantizer with feedback control. (DPCM1-AQB)
4. Adaptive DPCM with first order adaptive predictor and adaptive Gaussian quantizer with feed-forward control of both the quantizer and the predictor (window length 32). (ADPCM1-AQF)
5. Adaptive DPCM with fourth order adaptive predictor and adaptive Laplacian quantizer, both with feed-forward control (window length 128). (ADPCM4-AQF)
6. Adaptive DPCM with twelfth order adaptive predictor and adaptive Gamma quantizer, both with feed-forward control (window length 256). (ADPCM12-AQF)

In all these systems the sampling rate was 8 kHz and the quantizer word length ranged from 2 bits/sample to 5 bits/sample. Thus the bit rate ranges from 16 kb/s to 40 kb/s. Signal-to-quantizing noise ratios for all the systems are plotted in Fig. 5.43. The curves of Fig. 5.43 display a number of interesting features. First it is seen that the lowest curve corresponds to the use of a 2-bit quantizer and moving upward from one curve to the next corresponds to adding one bit to the quantizer word length. Note that the curves are displaced from one another by roughly 6 dB. Notice also the sharp increase in *SNR* with the addition of both fixed prediction and adaptive quantization, and note that almost no gain results from adapting a simple first order predictor. However, it also is clear that higher order adaptive prediction offers significant improvements.

For telephone transmission, it is generally accepted that acceptable speech quality is obtained with a μ-law quantizer with 6-7 bits/sample. From Eq. (5.38), it can be seen that 7-bit $\mu = 100$ PCM would have an *SNR* of about 33 dB. On the basis of Fig. 5.43 it would appear that comparable quality could be obtained using a 5-bit quantizer with adaptive quantization and adaptive prediction. In practice there is strong evidence to suggest that the perceived quality of ADPCM coded speech is better than a comparison of *SNR* values would suggest. In a study of an ADPCM system with fixed prediction and feedback con-

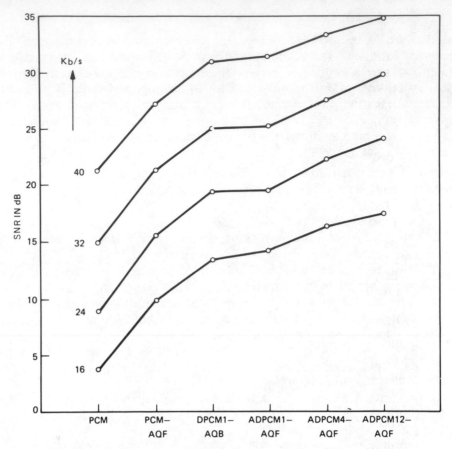

Fig. 5.43 Signal-to-noise ratio values for quantization with two bits per sample (16 kb/s) up to five bits per sample (40 kb/s). Code: AQF - Adaptive quantizer - feed forward; AQB - Adaptive quantizer - feed backward; ADPCM$_r$ - ADPCM system with r^{th} order predictor. (After Noll [7].)

trol of an adaptive quantizer, Cummiskey et al., [26] found that listeners preferred ADPCM coded speech to log-PCM coded speech with higher *SNR*. The results of a preference test are given in Table 5.7, where the PCM system is like system 1 of Noll's study and the ADPCM system is like system 3 of Noll's study. Table 5.7 shows that 4-bit ADPCM is preferred to 6-bit log PCM.

Table 5.7 Comparison of Objective and Subjective Performance of ADPCM and Log-PCM. (After Cummiskey, Jayant, and Flanagan [26].)

Objective Rating (SNR)	Subjective Rating (Preference)
7-bit PCM	7-bit PCM (High Preference)
6-bit PCM	4-bit ADPCM
4-bit ADPCM	6-bit PCM
5-bit PCM	3-bit ADPCM
3-bit ADPCM	5-bit PCM
4-bit PCM	4-bit PCM (Low Preference)

Recalling that the *SNR* improvement for ADPCM with fixed prediction and adaptive quantization is expected to be on the order of 10-12 dB, or roughly 2-bits, it is not surprising that the systems would be comparable, but in fact, the 4-bit ADPCM was preferred to the 6-bit log-PCM even though the *SNR* of 4-bit ADPCM was somewhat lower.

In their study of adaptive prediction Atal and Schroeder [29] found that their ADPCM system with a one-bit adaptive quantizer and complex adaptive predictor yielded coded speech whose quality was slightly inferior to 6-bit log PCM. The estimated bit rate for this system was about 10 kb/s in contrast to the 40 kb/s required for 6-bit PCM at a sampling rate of 6.67 kHz. Especially in this case, the subjective quality was greater than would be expected from consideration of the *SNR*.

A precise explanation of this phenomenon is difficult to obtain; however it is reasonable to conjecture that it is due to a combination of such factors as better idle channel performance of the adaptive quantizer and greater correlation between the quantization noise and the signal [7].

5.9 Direct Digital Code Conversion

It is abundantly clear from the discussion of this chapter that there are limitless possibilities for quantizing the waveform of speech signals. These schemes range in complexity from linear delta modulation which is extremely simple to implement but requires a high bit rate, to a wide variety of adaptive differential PCM systems which provide good quality at low bit rates but are rather complex signal processing algorithms.

As a result, a problem of major concern is that of direct conversion from one digital representation to another without intervening analog processing. This problem is important for a number of reasons. (1) In large communication systems there are likely to be situations such as local communication loops which call for low terminal cost with information transmission rate being of less concern. In other situations such as long distance transmission or storage of speech in digital memory, low bit rate may be the overriding consideration. At interfaces between parts of a communication system where different digital representations are used, it is very desirable to be able to convert from one digital representation to another in a way that avoids degradation of speech quality. (2) The implementation of low bit rate representations is often simplified by using entirely digital techniques. For example, it may be desirable to use a simple coding scheme such as linear delta modulation to perform the initial A/D conversion and then, though digital processing, convert to a lower bit rate representation such as PCM or ADPCM. (3) In situations requiring processing of the speech signal, it is necessary to have the signal represented in uniform PCM format so that arithmetic processing (e.g., digital filtering) can be performed on the samples of the speech signal.

For all these reasons, the problem of direct digital code conversion is very important. To illustrate how digital signal processing techniques can be applied in this important area, we shall discuss two examples.

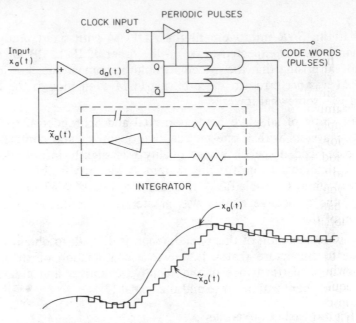

Fig. 5.44 Circuit implementation of a linear delta modulator.
(After Baldwin and Tewksbury [31]).

5.9.1 LDM-to-PCM conversion

In order to obtain a high quality representation of speech, a linear delta modulator employs a very high sampling rate and a simple 1-bit quantizer. Such systems are simple to implement using a combination of simple analog and digital components. Indeed, an entire LDM coder can easily be implemented as a single integrated circuit [31]. For example Fig. 5.44 shows a typical analog comparator circuit to create a difference signal, a flipflop to sense the polarity of the difference signal, and an integrator to reconstruct a predicted signal to compare to the input. This simple combination of analog and digital circuitry is all that is required to implement a delta modulator. The output of the flipflop is a train of pulses which correspond to the sequence of 1-bit binary code words that represent the input. A circuit of this type has been implemented as an integrated circuit which is capable of sampling rates up to 17 MHz [31].

The simplicity of such circuits with capabilities of very high data rates makes LDM an extremely attractive possibility for low cost, high quality analog-to-digital conversion. The price of this simplicity is of course the extremely high data rate required for high quality.

The bit rate can be reduced, however, by using digital signal processing techniques to convert the LDM code words (1's and 0's) into another, more efficient representation such as PCM or ADPCM. One of the most important conversions is from LDM to uniform PCM since uniform PCM is required whenever numerical processing of the samples of an analog waveform is desired.

236

The process of converting the LDM representation to a PCM representation involves first obtaining a PCM representation at the LDM sampling rate followed by a reduction in sampling rate to the Nyquist rate. The first step is accomplished simply by decoding the 1's and 0's into numbers with value $\pm\Delta$, and then accumulating the resulting positive and negative increments to obtain quantized samples of $x_a(t)$ at the LDM sampling rate. The resulting sequence has quantization noise throughout the band $|\Omega| \leqslant \pi/T$, where T is the LDM sampling period, even though the spectrum of the input speech may be bandlimited to a much lower frequency. Thus, before reducing the sampling rate to the Nyquist rate for the input, it is necessary to remove the quantization noise in the band from the Nyquist frequency up to one-half the LDM sampling frequency. As discussed in Section 2.4.2, this can be done very efficiently with an FIR digital lowpass filter [32], whose cutoff frequency is the Nyquist frequency of the speech signal. The output of the filter is computed only every M samples, where M is the ratio of the LDM sampling frequency to the PCM sampling frequency. Thus, an LDM-to-PCM converter is essentially an accumulator or up-down counter whose output is filtered and sampled. As noted in Fig. 5.45, an LDM-to-PCM converter in conjunction with an LDM system constitutes an analog-to-PCM converter whose implementation is almost completely digital.

5.9.2 PCM-to-ADPCM conversion

Another example of code conversion is from uniform PCM-to-ADPCM [33]. Interest in this conversion process stems from the desire to obtain a more efficient representation than uniform PCM. Figure 5.46 shows a diagram of the process. It is clear from Fig. 5.46 that the basic approach is to implement the particular ADPCM algorithm directly on the uniform PCM samples. This can be done with ordinary digital hardware for the arithmetic and logical operations depicted in Fig. 5.46. An important consideration in any code conversion process is the degradation introduced by further digital processing. Clearly the PCM input will have some quantization error which we can characterize by a

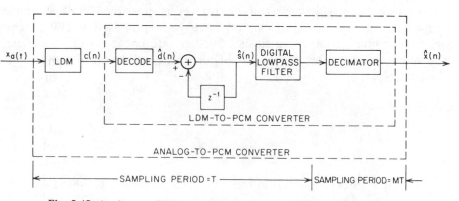

Fig. 5.45 Analog to PCM converter using an LDM system and an LDM-PCM converter.

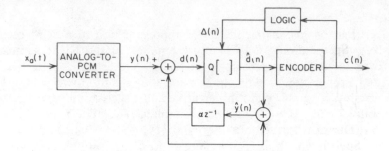

Fig. 5.46 PCM to ADPCM code converter.

signal-to-noise ratio SNR_1. Similarly, the ADPCM algorithm will introduce further quantization error. If we assume that this error is independent of the true signal and the PCM quantization error, we can approximate the overall SNR of the system as (see Problem 5.17)

$$SNR = \frac{SNR_1}{1 + \dfrac{SNR_1}{SNR_2}} \qquad (5.137)$$

where SNR_2 is the signal-to-noise ratio of the ADPCM system. We see from this equation that the overall SNR can be no better than SNR_1. However, the degradation is rather small if SNR_2 is on the order of SNR_1. Specifically, if $SNR_1 = SNR_2$, we note that SNR is only 3 dB worse than SNR_1.

Digital code conversion can be viewed as a means of accurate and efficient implementation of a particular waveform coding algorithm. If it is possible to obtain uniform PCM samples — for example as described in Section 5.9.1 — then digital processing can be applied to implementing a particular coding scheme.

5.10 Summary

This chapter has presented a detailed discussion of digital waveform coding. We have seen that a wide variety of approaches is possible. We have made no attempt to cover all the systems that have been proposed but have chosen instead to emphasize basic principles. Further references on this topic can be found in the review paper by Jayant [8], or in the collection of reprints edited by Jayant [34].

REFERENCES

1. Robert V. Bruce, *Bell,* Little Brown and Co., Boston, p. 144, 1973.

2. W. B. Davenport, "An Experimental Study of Speech-wave Probability Distributions," *J. Acoust. Soc. Am.,* Vol. 24, pp. 390-399, July 1952.

3. M. D. Paez and T. H. Glisson, "Minimum Mean Squared-Error Quantization in Speech," *IEEE Trans. Comm.,* Vol. Com-20, pp. 225-230, April 1972.

4. P. Noll, "Non-adaptive and Adaptive DPCM of Speech Signals," *Polytech. Tijdschr. Ed. Elektrotech/Elektron* (The Netherlands), No. 19, 1972.

5. H. K. Dunn and S. D. White, "Statistical Measurements on Conversational Speech," *J. Acoust. Soc. Am.,* Vol. 11, pp. 278-288, January 1940.

6. A. V. Oppenheim and R.W. Schafer, *Digital Signal Processing,* Prentice-Hall, Inc., Englewood Cliffs, N.J., 1975.

7. P. Noll, "A Comparative Study of Various Schemes for Speech Encoding," *Bell System Tech. J.,* Vol. 54, No. 9, pp. 1597-1614. November 1975.

8. N. S. Jayant, "Digital Coding of Speech Waveforms: PCM, DPCM, and DM Quantizers," *Proc. IEEE,* Vol. 62, pp. 611-632, May 1974.

9. W. R. Bennett, "Spectra of Quantized Signals," *Bell System Tech. J.,* Vol. 27, No. 3, pp. 446-472, July 1948.

10. B. Smith, "Instantaneous Companding of Quantized Signals, *Bell System Tech. J.,* Vol. 36, No. 3, pp. 653-709, May 1957.

11. J. Max, "Quantizing for Minimum Distortion," *IRE Trans. Inform. Theory,* Vol. IT-6, pp. 7-12, March 1960.

12. P. Noll, "Adaptive Quantizing in Speech Coding Systems," *Proc. 1974 Zurich Seminar on Digital Communications,* Zurich, March 1974.

13. T. P. Barnwell, A. M. Bush, J. B. O'Neal, and R. W. Stroh, "Adaptive Differential PCM Speech Transmission," *RADC-TR-74-177,* Rome Air Development Center, July 1974.

14. A. Croisier, "Progress in PCM and Delta Modulation: Block-Companded Coding of Speech Signals," *Proc. 1974 Zurich Seminar on Digital Communication,* March 1974.

15. N. S. Jayant, "Adaptive Quantization With a One Word Memory," *Bell System Tech. J.,* pp. 1119-1144, September 1973.

16. C. C. Cutler, "Differential Quantization of Communications," U. S. Patent 2, 605, 361, July 29, 1952.

17. R. A. McDonald, "Signal to Noise and Idle Channel Performance of DPCM Systems — Particular Applications to Voice Signals," *Bell System Tech. J.,* Vol. 45, No.7, pp. 1123-1151, September 1966.

18. J. S. Schouten, F. E. DeJager, and J. A. Greefkes, "Delta Modulation, a New Modulation System for Telecommunications," *Philips Tech. Rept.* pp. 237-245, March 1952.

19. F. E. DeJager, "Delta Modulation, a Method of PCM Transmission Using a 1-Unit Code," *Phillips Res. Rep.,* pp. 442-466, December 1952.

20. H. R. Schindler, "Delta Modulation," *IEEE Spectrum,* Vol. 7, pp. 69-78, October 1970.

21. J. E. Abate, "Linear and Adaptive Delta Modulation," *Proc. IEEE,* Vol. 55, pp. 298-308, March 1967.

22. N. S. Jayant, "Adaptive Delta Modulation With a One-Bit Memory," *Bell System Tech. J.,* pp. 321-342, March 1970.

23. J. A. Greefkes, "A Digitally Companded Delta Modulation Modem for Speech Transmission," *Proc. IEEE Int. Conf. Comm.,* pp. 7-33 to 7-48, June 1970.

24. R. Steele, *Delta Modulaton Systems,* Halsted Press, London, 1975.

25. P. Cummiskey, Unpublished work, Bell Laboratories.

26. P. Cummiskey, N. S. Jayant, and J. L. Flanagan, "Adaptive Quantization in Differential PCM Coding of Speech," *Bell System Tech. J.,* Vol. 52, No. 7, pp. 1105-1118, September 1973.

27. P. Noll, "Effect of Channel Errors on the Signal-to-Noise Performance of Speech Encoding Systems," *Bell System Tech. J.,* Vol. 54, No. 9, pp. 1615-1636, November 1975.

28. N. S. Jayant, "Step-Size Transmitting Differential Coders for Mobile Telephony," *Bell System Tech. J.,* Vol. 54, No. 9, pp 1557-1582, November 1975.

29. B. S. Atal and M. R. Schroeder, "Adaptive Predictive Coding of Speech Signals,"
Bell System Tech. J., Vol. 49, No. 8, pp. 1973-1986, October 1970.

30. R. W. Stroh, "Optimum and Adaptive Differential PCM," Ph.D. Dissertation, Polytechnic Inst. of Brooklyn, Farmingdale, N.Y., 1970.

31. G. L. Baldwin and S. K. Tewksbury, "Linear Delta Modulator Integrated Circuit With 17-Mbit/s Sampling Rate," *IEEE Trans. on Comm.,* Vol. COM-22, No. 7, pp. 977-985, July 1974.

32. D. J. Goodman, "The Application of Delta Modulation to Analog-to-PCM Encoding," *Bell System Tech. J.,* Vol. 48, No. 2, pp. 321-343, February 1969.

33. S. L. Bates, "A Hardware Realization of a PCM-ADPCM Code Converter," M. S. Thesis, MIT, Cambridge, Mass., January 1976.

34. *Waveform Quantization and Coding,* N. S. Jayant, Editor, IEEE Press, 1976.

PROBLEMS

5.1 The uniform probability density function is defined as

$$p(x) = \frac{1}{\Delta} \quad |x| < \Delta/2$$
$$= 0 \quad \text{otherwise}$$

Find the mean and variance of the uniform distribution.

5.2 Consider the Laplacian probability density function

$$p(x) = \frac{1}{\sqrt{2}\,\sigma_x}\, e^{-\sqrt{2}\,|x|/\sigma_x}$$

Find the probability that $|x| > 4\sigma_x$.

5.3 Let $x(n)$, the input to a linear shift-invariant system, be a stationary, zero mean, white noise process. Show that the autocorrelation function of the output is

$$\phi(m) = \sigma_x^2 \sum_{k=-\infty}^{\infty} h(k)h(k+m)$$

where σ_x^2 is the variance of the input and $h(n)$ is the impulse response of the linear system.

5.4 Consider the design of a high-quality digital audio system. The specifications are: 60 dB signal-to-noise ratio must be maintained over a range of peak signal levels of 100 to 1. The useful signal bandwidth must be at least 8 kHz.
(a) Draw a block diagram of the basic components needed for A/D and D/A conversion.
(b) How many bits are required in the A/D and D/A converter?
(c) What are the main considerations in choosing the sampling rate? What types of analog filters should be used prior to the A/D converter and following the D/A converter? Estimate the lowest sampling rate that would be possible in a practical system.
(d) How would the specifications and answers change if the objective was only to maintain telephone quality representation of speech?

5.5 A speech signal is bandlimited by an ideal lowpass filter, sampled at the Nyquist rate, quantized by a uniform B-bit quantizer, and converted back to an analog signal by an ideal D/A converter as shown in Fig. P5.5a. Define $y(n) = x(n) + e_1(n)$ where $e_1(n)$ is the quantization error. Assume that the quantization step is $\Delta = 8\sigma_x/2^B$ and that B is large

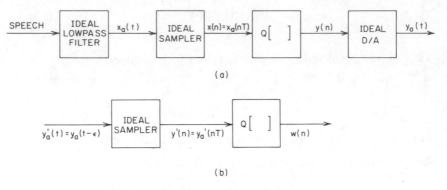

(a)

(b)

Fig. P5.5

enough so that we can assume:
1. $e_1(n)$ is stationary
2. $e_1(n)$ is uncorrelated with $x(n)$
3. $e_1(n)$ is a uniformly distributed white noise sequence
We have seen that under these conditions, the signal-to-quantizing noise ratio is

$$SNR_1 = \frac{\sigma_x^2}{\sigma_{e_1}^2} = \frac{12}{64} \cdot 2^{2B}$$

Now assume that the analog signal $y_a(t)$ is sampled again at the Nyquist rate and quantized by an identical B-bit quantizer as shown in Fig. P5.5b. (Assume that $0 < \epsilon < T$, i.e., the two sampling systems are not exactly synchronized in time.)
Assume that $w(n) = y'(n) + e_2(n)$ where $e_2(n)$ has identical properties to $e_1(n)$.
(a)   Show that the overall signal-to-noise ratio is

$$SNR_2 = \frac{SNR_1}{2}$$

(b)   Generalize the result of (a) to N stages of A/D and D/A conversion.

5.6   Although it is common to treat the quantization error as being independent of the signal $x(n)$, it can easily be shown that this assumption breaks down for a small number of quantization levels.
(a)   Show that $e(n) = x(n) - \hat{x}(n)$ is *not* statistically independent of $x(n)$. ($\hat{x}(n)$ is the quantized signal.) Hint: Represent $\hat{x}(n)$ as

$$\hat{x}(n) = \left[\frac{x(n)}{\Delta} \right] \cdot \Delta + \frac{\Delta}{2}$$

where [·] denotes the "greatest integer in," i.e., the greatest integer less than or equal to the quantity within the brackets.   Also represent $x(n)$ as

$$x(n) = \left[\frac{x(n)}{\Delta}\right] \cdot \Delta + x_f(n) = x_i(n) + x_f(n)$$

where $x_i(n)$ is the integer part of $x(n)$, and $x_f(n)$ is the fractional part of $x(n)$. Then $e(n)$ can be determined as a function of $x(n)$. Argue that they cannot be exactly statistically independent.

(b) Under what conditions is the approximation that $x(n)$ and $e(n)$ are statistically independent valid?

(c) Figure P5.6 shows a method which has been suggested for making $e(n)$ and $x(n)$ statistically independent, even for a small number of quantization levels. For this case $z(n)$ is a pseudorandom, uniformly distributed, white noise sequence with probability density function

$$p(z) = \frac{1}{\Delta} \qquad -\frac{\Delta}{2} \leqslant z \leqslant \frac{\Delta}{2}$$

Show that in this case the quantization error $e(n) = x(n) - \hat{y}(n)$ is statistically independent of $x(n)$ for all values of B. (The noise sequence, $z(n)$, being added to the signal is called dither noise.)

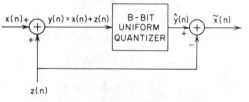

Fig. P5.6

Hint: Look at the range of values for $e(n)$ for ranges of $y(n)$.

(d) Show that the variance of the quantization error at the output of the B-bit quantizer is greater than the variance of the quantization error for the undithered case — i.e., show that

$$\sigma_{e_1}^2 > \sigma_e^2$$

where

$$e_1(n) = x(n) - \hat{y}(n)$$

and

$$e(n) = x(n) - \hat{x}(n)$$

(e) Show that by simply subtracting off the dither noise $z(n)$ from the quantizer output, the variance of the quantization error $e_2(n) = x(n) - (\hat{y}(n) - z(n))$ is the same as the variance of the undithered case; i.e., $\sigma_{e_2}^2 = \sigma_e^2$.

5.7 A common approach to estimating the signal variance is to assume that it is proportional to the short-time energy of the signal, defined as

$$\sigma^2(n) = \sum_{m=-\infty}^{\infty} x^2(m)h(n-m)$$

(a) Show that if $x(n)$ is stationary with zero mean and variance σ_x^2, then $E[\sigma^2(n)]$ is proportional to σ_x^2.

(b) For

$$h(n) = \alpha^n \quad n \geqslant 0 \quad (|\alpha| < 1)$$
$$= 0 \quad n < 0$$

and for

$$E[x^2(m)x^2(l)] = B \quad m = l$$
$$= 0 \quad m \neq l$$

determine the variance of $\sigma^2(n)$ as a function of B and α.

(c) Explain the behavior of the variance of $\sigma^2(n)$ of part (b) as α varies from 0 to 1.

5.8 Consider the adaptive quantization system shown in Fig. P5.8a. The 2-bit quantizer characteristic and code word assignment is shown in Fig. P5.8b. Suppose the step size is adapted according to the following rule:

$$\Delta(n) = M\Delta(n-1)$$

where M is a function of the previous codeword $c(n-1)$ and

$$\Delta_{min} \leqslant \Delta(n) \leqslant \Delta_{max}$$

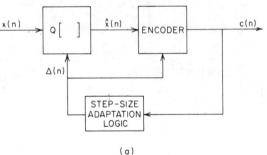

(a)

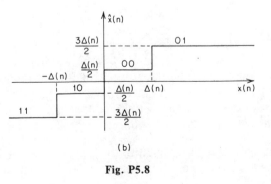

(b)

Fig. P5.8

Furthermore suppose that

$$M = P \qquad \text{if} \qquad c(n-1) = 01 \qquad \text{or} \qquad 11$$
$$= 1/P \qquad \text{if} \qquad c(n-1) = 00 \qquad \text{or} \qquad 10$$

(a) Draw a block diagram of the step-size adaptation system.
(b) Suppose that

$$
\begin{aligned}
x(n) &= 0 & n < 5 \\
&= 20 & 5 \leqslant n \leqslant 13 \\
&= 0 & 13 < n
\end{aligned}
$$

Assume that $\Delta_{min} = 2$ and $\Delta_{max} = 30$ and $P = 2$. Make a table of values of $x(n)$, $\Delta(n)$, $c(n)$ and $\hat{x}(n)$ for $0 \leqslant n \leqslant 25$. (Assume that at $n = 0$, $\Delta(n) = \Delta_{min} = 2$, and $c(n) = 00$.)
(c) Plot the samples $x(n)$ and $\hat{x}(n)$ on the same coordinate scale.

5.9 Consider the 2-bit adaptive quantizing system of Problem 5.8. In this case, however, the step size adaptation algorithm is:

$$\Delta(n) = \beta\Delta(n-1) + D \qquad \text{if} \qquad \sum_{k=1}^{M} LSB[c(n-k)] \geqslant 2$$

$$= \beta\Delta(n-1) \qquad\qquad\qquad otherwise$$

where $LSB[c(n-k)]$ means "least significant bit" of the codeword $c(n-k)$.
(a) Draw a block diagram of the step-size adaptation system.
(b) In this case the maximum step-size is built into the algorithm for adaptation. Find Δ_{max} in terms of β and D. (Hint: consider the step response of the first equation of this problem.)
(c) Again suppose that

$$
\begin{aligned}
x(n) &= 0 & n < 5 \\
&= 20 & 5 \leqslant n \leqslant 13 \\
&= 0 & 13 < n
\end{aligned}
$$

Also suppose that $M = 1$, $\beta = 0.8$, and $D = 6$. Make a table of values of $x(n)$, $\Delta(n)$, $c(n)$ and $\hat{x}(n)$ for $0 \leqslant n \leqslant 25$. (Assume that at $n=0$, $\Delta(n) = 0$, and $c(n)=00$.) Plot the samples $x(n)$ and $\hat{x}(n)$ on the same coordinate system.
(d) Find the value of β such that the time constant of the step-size adaptation system is 10 msec.

5.10 Consider the first order linear prediction

$$\tilde{x}(n) = \alpha x(n-1)$$

where $x(n)$ is a stationary, zero mean signal.
(a) Show that the prediction error

$$d(n) = x(n) - \tilde{x}(n)$$

245

has variance

$$\sigma_d^2 = \sigma_x^2(1 + \alpha^2 - 2\alpha\phi(1)/\sigma_x^2)$$

(b) Show that σ_d^2 is minimized for

$$\alpha = \phi(1)/\sigma_x^2 = \rho(1)$$

(c) Show that the minimum prediction error variance is

$$\sigma_d^2 = \sigma_x^2(1-\rho^2(1))$$

(d) Under what conditions will it be true that $\sigma_d^2 < \sigma_x^2$?

5.11 Given a sequence $x(n)$ with long time autocorrelation $\phi(m)$, show that the difference signal

$$d(n) = x(n) - x(n-n_0)$$

has lower variance than the original signal $x(n)$ as long as there is reasonable correlation between $x(n)$ and $x(n-n_0)$. (Assume $x(n)$ has zero mean value.)

(a) State the conditions on $\phi(n_0)$ such that

$$\sigma_d^2 \leqslant \sigma_x^2$$

(b) If $d(n)$ is formed as

$$d(n) = x(n) - \alpha x(n-n_0)$$

where

$$\alpha = \frac{\phi(n_0)}{\phi(0)}$$

state the conditions on $\phi(n_0)$ such that

$$\sigma_d^2 \leqslant \sigma_x^2$$

5.12 Using Eqs. (5.78) and (5.83), prove the assertion that for the optimum predictor coefficients

$$E[(x(n)-\tilde{x}(n))\tilde{x}(n)] = E[d(n)\tilde{x}(n)] = 0$$

That is, that the optimum prediction error is uncorrelated with the predicted signal.

5.13 Consider the difference signal

$$d(n) = x(n) - \alpha_1 \hat{x}(n-1)$$

where $\hat{x}(n)$ is the quantized signal in a differential coder.

(a) Show that

$$\sigma_d^2 = \sigma_x^2 \left[1 - 2\alpha_1\rho(1) + \alpha_1^2 \right] + \alpha_1^2\sigma_e^2$$

(b) Using the result of (a) show that

$$G_p = \frac{\sigma_x^2}{\sigma_d^2} = \frac{1 - \dfrac{\alpha_1^2}{SNR_Q}}{1 - 2\alpha_1\rho(1) + \alpha_1^2}$$

where

$$SNR_Q = \frac{\sigma_d^2}{\sigma_e^2}$$

5.14 In the CVSD adaptive delta modulator, the step-size adaptation algorithm is

$$\begin{aligned}
\Delta(n) &= \beta\Delta(n-1) + D_2 \qquad \text{if } c(n) = c(n-1) = c(n-2)\\
&= \beta\Delta(n-1) + D_1 \qquad \textit{otherwise}
\end{aligned}$$

where $0 < \beta < 1$ and $0 < D_1 \ll D_2$.

(a) The maximum step size is attained if the input to the step size filter is constant at D_2 as would occur in a prolonged period of slope-overload. Find $\Delta_{\max}$ in terms of D_2 and β.

(b) The minimum step size would be attained if the pattern $c(n) = c(n-1) = c(n-2)$ does not occur for a prolonged period as in the idle channel condition. Find $\Delta_{\min}$ in terms of D_1 and β.

5.15 Consider the adaptive delta modulator of Fig. P5.15a. The 1-bit quantizer is given in Fig. P5.15b. The step size is adapted according to the following rule

$$\Delta(n) = M\Delta(n-1)$$

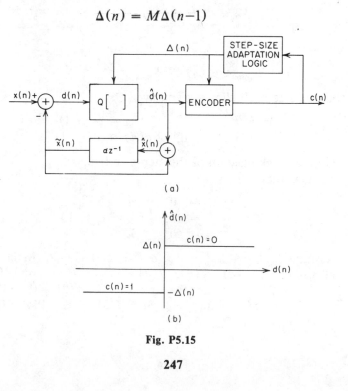

(a)

(b)

Fig. P5.15

247

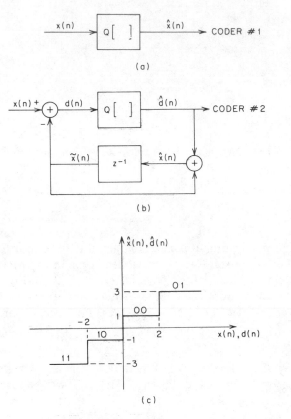

Fig. P5.16

where

$$\Delta_{min} \leqslant \Delta(n) \leqslant \Delta_{max}$$

and the step size multiplier is given by

$$M = P \quad \text{if} \quad c(n) = c(n-1)$$
$$ = 1/P \quad \text{if} \quad c(n) \neq c(n-1)$$

(a) Draw a block diagram of the step-size logic.

(b) Suppose that

$$x(n) = 0 \quad n < 5$$
$$ = 20 \quad 5 \leqslant n \leqslant 13$$
$$ = 0 \quad 13 < n$$

Assume $\Delta_{min} = 1$ $\Delta_{max} = 15$, $\alpha = 1$, and $P = 2$. Make a table of values of $x(n)$, $\tilde{x}(n)$, $d(n)$, $\Delta(n)$, $\hat{d}(n)$ and $\hat{x}(n)$ for $0 \leqslant n \leqslant 25$. Assume that at $n = 0$, $x(0) = 0$, $\tilde{x}(0) = 1$, $d(0) = -1$, $\Delta(0) = \Delta_{min} = 1$, and $\hat{d}(0) = -1$. Plot $x(n)$ and $\hat{x}(n)$ for $0 \leqslant n \leqslant 25$.

5.16 Consider the two coders shown in Fig. P5.16a and b. Each coder uses a 2-bit quantizer with input-output characteristic shown in Fig. P5.16c.

Consider the idle channel case, i.e. the case where $x(n)$ is a low level noise. For simplicity we assume $x(n)$ is of the form

$$x(n) = 0.1 \ cos(\pi n/4)$$

(a)   For $0 \leqslant n \leqslant 20$ make a plot of $\hat{x}(n)$ for both coders.

(b)   For which coder would the "idle channel noise" be more objectionable in a real communications system? Why?

5.17 Consider the PCM-to-ADPCM code conversion system of Fig. 5.46. The PCM coded signal, $y(n)$, can be represented as

$$y(n) = x(n) + e_1(n)$$

where $x(n) = x_a(nT)$, and $e_1(n)$ is the quantization error in the PCM representation. The quantized ADPCM signal, $\hat{y}(n)$, can be represented as

$$\hat{y}(n) = y(n) + e_2(n)$$

where $e_2(n)$ is the ADPCM quantization error.

(a)   Assuming that the quantization errors $e_1(n)$ and $e_2(n)$ are uncorrelated, show that the overall signal-to-noise ratio is

$$SNR = \frac{\sigma_x^2}{\sigma_{e_1}^2 + \sigma_{e_2}^2}$$

(b)   Show that SNR can be expressed as

$$SNR = \frac{SNR_1}{1 + \dfrac{1 + SNR_1}{SNR_2}}$$

where

$$SNR_1 = \frac{\sigma_x^2}{\sigma_{e_1}^2}$$

and

$$SNR_2 = \frac{\sigma_y^2}{\sigma_{e_2}^2}$$

6

Short-Time Fourier Analysis

6.0 Introduction

In many areas of science and engineering, the representation of signals or other functions by sums of sinusoids or complex exponentials leads to convenient solutions to problems and often to greater insight into physical phenomena than is available by other means. Such representations — Fourier representations as they are commonly called — are useful in signal processing for two basic reasons. The first is that for linear systems it is very convenient to determine the response to a superposition of sinusoids or complex exponentials. The second reason is that the Fourier representations often serve to place in evidence certain properties of the signal that may be obscure or at least less evident in the original signal.

Speech communication research and technology are areas where the concept of a Fourier representation has traditionally played a major role. To see why this is so, it is helpful to recall that the model for the production of a steady state speech sound such as a vowel or fricative simply consists of a linear system excited by a source which is either periodically or randomly varying with time. In general, the spectrum of the output of such a model would be the product of the frequency response of the vocal tract system and the spectrum of the excitation. Thus, it is to be expected that the spectrum of the output would reflect the properties of both the excitation and the vocal tract frequency response. We have seen, however, that speech waveforms are generally much more complicated than simply a sustained vowel or fricative sound. Thus the standard Fourier representations that are appropriate for periodic, transient, or stationary random signals are not directly applicable to the representation of speech signals whose properties change markedly as a function for time. However, we have already seen ample evidence that the short-time analysis principle is a valid approach to speech processing. We have seen, for example, that temporal properties such as energy, zero crossings and correlation can be assumed fixed over time intervals on the order of 10 to 30 msec. We shall demonstrate in this chapter that spectral properties of speech likewise can be assumed to change relatively slowly with time.

In order to study spectral properties of speech signals, we shall find it convenient to formally introduce the concept of a time-varying Fourier representation of a signal. We shall define a time-varying Fourier transform and the operation of synthesis from a time-varying Fourier transform. In so doing, it will be convenient to consider Fourier analysis in the context of a bank of filters. This will lead to both theoretical and practical (computational) insights into time-varying Fourier analysis. We shall also consider other computational techniques based upon fast computation algorithms for discrete Fourier analysis (FFT algorithms). Finally, having considered the theoretical and computational details of time-varying Fourier representations, we shall consider applications to the analysis/synthesis of speech (vocoders), spectrum displays, and such basic speech analysis problems as formant analysis and pitch detection.

6.1 Definitions and Properties

In defining a time dependent Fourier representation, we are motivated by the need for a spectral representation which reflects the time-varying properties of the speech waveform. A useful definition of the time dependent Fourier transform is

$$X_n(e^{j\omega}) = \sum_{m=-\infty}^{\infty} w(n-m)x(m)e^{-j\omega m} \qquad (6.1)$$

In Eq. (6.1), $w(n-m)$ is a real "window" sequence which determines the portion of the input signal that receives emphasis at a particular time index, n. The time dependent Fourier transform is clearly a function of two variables: the time index, n, which is discrete, and the frequency variable, ω, which is continuous. An alternative form of Eq. (6.1) is obtained by a change of summation index which yields the expression

$$X_n(e^{j\omega}) = \sum_{m=-\infty}^{\infty} w(m)x(n-m)e^{-j\omega(n-m)}$$

$$= e^{-j\omega n} \sum_{m=-\infty}^{\infty} x(n-m)w(m)e^{j\omega m} \qquad (6.2)$$

If we define

$$\tilde{X}_n(e^{j\omega}) = \sum_{m=-\infty}^{\infty} x(n-m)w(m)e^{j\omega m} \qquad (6.3)$$

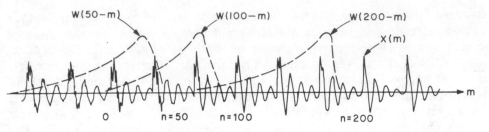

Fig. 6.1 Sketches of $x(m)$ and $w(n-m)$ for several values of n.

then $X_n(e^{j\omega})$ can be expressed as

$$X_n(e^{j\omega}) = e^{-j\omega n}\, \tilde{X}_n(e^{j\omega}) \tag{6.4}$$

These equations can be interpreted in two distinct ways. First, if we assume that n is fixed, we note that $X_n(e^{j\omega})$ is simply the normal Fourier transform of the sequence $w(n-m)x(m)$, $-\infty < m < \infty$. Therefore for fixed n, $X_n(e^{j\omega})$ has the same properties as a normal Fourier transform. The second interpretation follows by considering $X_n(e^{j\omega})$ as a function of the time index n with ω fixed. In this case we observe that both Eq. (6.1) and Eq. (6.3) are in the form of a convolution. This interpretation leads us naturally to consider the time dependent Fourier representation in terms of linear filtering. As we will see, both interpretations lead to useful insights and we shall find it worthwhile to examine the time dependent Fourier transform in detail from both viewpoints.

6.1.1 Fourier transform interpretation

To begin, we shall consider $X_n(e^{j\omega})$ as the normal Fourier transform of the sequence $w(n-m)x(m)$, $-\infty < m < \infty$, for fixed n. The time-dependent Fourier transform is a function of the time index, n, which takes on all integer values so as to "slide" the window, $w(n-m)$, along the sequence, $x(m)$. This is depicted in Fig. 6.1, which shows $x(m)$ and $w(n-m)$ as functions of m for several values of n. (Note that the signal and the window are plotted as continuous functions for convenience even though they are only defined for integer values of m and $n-m$.)

Conditions for the existence of the time-dependent Fourier transform representation are easily obtained if we recall that a sufficient condition for the existence of the conventional Fourier transform is that the sequence be absolutely summable. In this case, we require that the sequence $x(m)w(n-m)$ be absolutely summable for all values of n. If, as is often the case, $w(n-m)$ is of finite duration, then this condition is clearly satisfied.

As in the case of normal Fourier transforms of discrete-time signals, the time-varying Fourier transform is periodic in ω with period 2π. This is easily seen by substituting $\omega + 2\pi$ into Eq. (6.1). Also note that it is possible to express the time-varying Fourier transform in terms of a variety of frequency variables. For example, if $\omega = \Omega T$, where T is the sampling period used to obtain the sequence $x(m)$, then Ω is analog radian frequency. Also, by making the substitutions $\omega = 2\pi f$ or $\omega = 2\pi FT$, we can express the time-varying Fourier transform as a function of either normalized cyclic frequency (f) or conventional analog cyclic frequency ($F-$ in Hertz) respectively. We shall have occasion to use a variety of different frequency variables in equations and figures in this chapter and in the remainder of the book. However, there should be no confusion once the simple relationships become familiar.

The fact that, for a given value of n, $X_n(e^{j\omega})$ has the same properties as a normal Fourier transform leads to a simple proof that the input sequence $x(m)$ can be recovered *exactly* from the time-varying Fourier transform. Recalling our earlier observation that $X_n(e^{j\omega})$ is simply the normal Fourier transform of $w(n-m)x(m)$, we can write

$$w(n-m)x(m) = \frac{1}{2\pi} \int_{-\pi}^{\pi} X_n(e^{j\omega}) e^{j\omega m} d\omega \qquad (6.5)$$

Note that the integration in Eq. (6.5) could be over any interval of length 2π (e.g., 0 to 2π) since the entire integrand is periodic with period 2π. Now if $w(0) \neq 0$, Eq. (6.5) can be evaluated for $m = n$, thereby obtaining

$$x(n) = \frac{1}{2\pi w(0)} \int_{-\pi}^{\pi} X_n(e^{j\omega}) e^{j\omega n} d\omega \qquad (6.6)$$

Thus, with the rather mild requirement that $w(0)$ be nonzero, the sequence $x(n)$ can be exactly recovered from $X_n(e^{j\omega})$, if $X_n(e^{j\omega})$ is known for all values of ω over one complete period. This is an important theoretical result, which, we shall see, takes on practical significance with the imposition of a simple additional constraint on the window.

An important property of $X_n(e^{j\omega})$ concerns its relation to the short-time autocorrelation function as defined in Chapter 4. Given that $X_n(e^{j\omega})$ is the normal Fourier transform of $w(n-m)x(m)$ for each value of n, then it is easily seen that

$$S_n(e^{j\omega}) = |X_n(e^{j\omega})|^2 = X_n(e^{j\omega}) \cdot X_n^*(e^{j\omega}) \qquad (6.7)$$

is the Fourier transform of

$$R_n(k) = \sum_{m=-\infty}^{\infty} w(n-m)x(m)w(n-k-m)x(m+k) \qquad (6.8)$$

Equations (6.7) and (6.8) thus serve to relate the short-time spectrum representation to the short-time correlation as discussed in Chapter 4.

The short-time Fourier transform, $X_n(e^{j\omega})$, can be expressed in a variety of alternative forms. One particularly simple form is in terms of its real and imaginary parts,[1] i.e.

$$X_n(e^{j\omega}) = a_n(\omega) - jb_n(\omega) \qquad (6.9)$$

---

[1]Note that $a_n(\omega)$ is the real part and $b_n(\omega)$ is minus the imaginary part of $X_n(e^{j\omega})$. The negative sign is used for convenience in later discussion.

For the case when $x(m)$ and $w(n-m)$ are both real then $a_n(\omega)$ and $b_n(\omega)$ can be shown to satisfy certain symmetry and periodicity relations (see Problem 6.1). Another representation for $X_n(e^{j\omega})$ is in terms of magnitude and phase as

$$X_n(e^{j\omega}) = |X_n(e^{j\omega})| e^{j\theta_n(\omega)} \tag{6.10}$$

The quantities $|X_n(e^{j\omega})|$ and $\theta_n(\omega)$ can readily be related to $a_n(\omega)$ and $b_n(\omega)$ and vice versa. see Problem 6.3). Additional properties of $a_n(\omega)$, $b_n(\omega)$, and $X_n(e^{j\omega})$ are emphasized in other problems given at the end of this chapter.

So far we have not considered the role of the window, $w(n-m)$, beyond its obvious function of selecting the portion of the sequence $x(m)$ to be analyzed. The shape of the window sequence has an important effect on the nature of the time-dependent Fourier transform, and the present viewpoint provides a convenient way to interpret the role of the window sequence, $w(n-m)$. If $X_n(e^{j\omega})$ is thought of as the normal Fourier transform of the sequence $w(n-m)x(m)$, and if we assume that the normal Fourier transforms

$$X(e^{j\omega}) = \sum_{m=-\infty}^{\infty} x(m)e^{-j\omega m} \tag{6.11}$$

and

$$W(e^{j\omega}) = \sum_{m=-\infty}^{\infty} w(m)e^{-j\omega m} \tag{6.12}$$

exist, then the normal Fourier transform of $w(n-m)x(m)$ (for fixed n) is the convolution of the transforms of $w(n-m)$ and $x(m)$. Since, for fixed n, the Fourier transform of $w(n-m)$ is $W(e^{-j\omega})e^{-j\omega n}$, then

$$X_n(e^{j\omega}) = \frac{1}{2\pi} \int_{-\pi}^{\pi} W(e^{-j\theta})e^{-j\theta n}X(e^{j(\omega-\theta)}) d\theta \tag{6.13}$$

By changing θ into $-\theta$ in Eq. (6.14) we can also write

$$X_n(e^{j\omega}) = \frac{1}{2\pi} \int_{-\pi}^{\pi} W(e^{j\theta})e^{j\theta n}X(e^{j(\omega+\theta)}) d\theta \tag{6.14}$$

Thus, we observe that the Fourier transform of the sequence $x(m)$, $-\infty < m < \infty$ is convolved with the Fourier transform of the shifted window sequence. This result needs to be qualified by recognizing that strictly speaking the normal Fourier transform of a speech signal does not exist. However, Eq. (6.14) can be useful if we first recall that the purpose of the window is to emphasize a finite segment of the speech waveform in the vicinity of sample n, and to deemphasize the remainder of the waveform. Indeed, typical window sequences may be such that $w(n-m) = 0$ for m outside a finite interval around n. Insofar as the final result is concerned, then, it is entirely reasonable to assume that the properties of $x(m)$ inside the window persist outside the window. For example, if the speech signal within the window corresponds to a vowel or other voiced sound, we can just as well consider that the resulting sequence $x(m)w(n-m)$ arose from a periodic sustained voiced sound. Like-

254

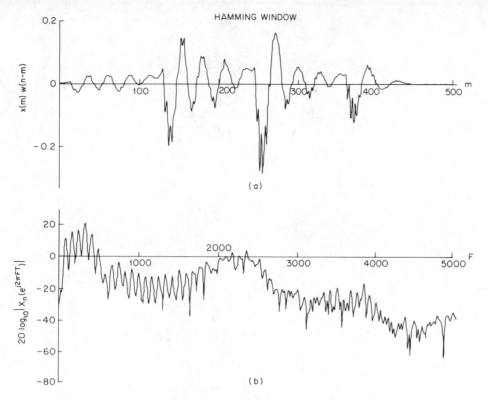

Fig. 6.2 Spectrum analysis for voiced speech using a 50 msec (a,b) Hamming window; (c,d) rectangular window. Parts (a) and (c) show time waveforms; parts (b) and (d) show corresponding spectra.

wise if the speech within the window is unvoiced, we can assume that the same unvoiced properties exist outside the window. An equally appropriate point of view is that the signal is simply zero outside the window. This would be appropriate for the analysis of transient sounds such as plosives.

Thus, Eq. (6.14) is meaningful if we assume that $X(e^{j\theta})$ stands for the Fourier transform of a signal whose basic properties either continue outside the window or which is zero outside the window. Thus the time dependent Fourier transform can be interpreted as a smoothed version of the Fourier transform of the part of the signal within the window.

With this point of view, the properties of the window Fourier transform, $W(e^{j\theta})$, become important. It is clear from Eq. (6.14) that for faithful reproduction of the properties of $X(e^{j\omega})$ in $X_n(e^{j\omega})$, the function $W(e^{j\theta})$ should appear as an impulse with respect to $X(e^{j\omega})$. In Chapter 4 we have already discussed the properties of the rectangular and Hamming windows. It was shown that the width of the main lobe of $W(e^{j\theta})$ is inversely proportional to the length of the window, whereas the levels of the side lobes are essentially independent of the window length.

The effects of using windows for speech spectral analysis are shown in Figures 6.2-6.5. Part a of each of these figures shows the windowed signal

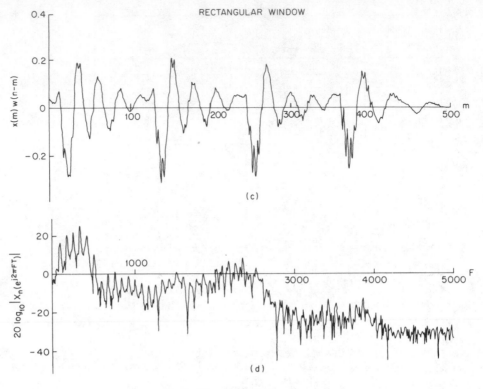

Fig. 6.2 (Continued)

$x(m)w(n-m)$ for a Hamming window; part b shows the log magnitude of $X_n(e^{j\omega})$ (in dB units); part c shows the windowed signal using a rectangular window, and part ·d shows the resulting log magnitude spectrum. Figure 6.2 shows results for a window duration of 500 samples (50 msec for 10 kHz sampling rate) for a section of voiced speech. The periodicity of the signal is clearly seen in Figure 6.2a (the time waveform) as well as in Figure 6.2b in which the fundamental frequency and its harmonics show up as narrow peaks at equally spaced frequencies in the short-time Fourier transform. The spectrum of Fig. 6.2b is also seen to consist of a strong first formant peak at about 300-400 Hz, and a broad peak at about 2200 Hz which corresponds to the second and third formants. A fourth formant peak at about 3800 Hz is also seen. Finally the spectrum shows a tendency to fall off at higher frequencies due to the lowpass nature of the glottal pulse spectrum.

A comparison of the spectra of Fig. 6.2b (Hamming window) and 6.2d (rectangular window) shows considerable overall similarity in terms of the pitch harmonics, formant structure, and gross spectral shape. Differences in the spectra can also be seen, the most notable being the increased sharpness of the pitch harmonics of Fig. 6.2d, due to the greater frequency resolution of the rectangular window relative to that of the same length Hamming window. Another difference in the spectra is that the relatively large side lobes of the rectangular window produce a "ragged" or noisy spectrum. This effect occurs because the side lobes due to adjacent harmonics interact in the space between the harmonics — sometimes reinforcing, sometimes cancelling — thereby pro-ducing a rather random appearing variation between harmonics. This undesir-able "leakage" between adjacent harmonics tends to offset the benefits of the narrower main lobe of the rectangular window. As a result, such windows are rarely used in speech spectrum analysis.

Figure 6.3 shows a similar set of comparisons for a 50 sample (5 msec) section of voiced speech. For such short windows the time sequences $x(m)w(n-m)$ (Figures 6.3a, c) do not show the signal periodicity, nor do the signal spectra. (Figures 6.3b, d). In contrast to Figure 6.2, the spectra of Fig. 6.3 show only a few rather broad peaks at about 400, 1400, and 2200 Hz, corresponding to the first three formants of the speech section within the win-dow. Comparison of the spectra of Figures 6.3b and 6.3d again shows the increased frequency resolution obtained with a rectangular window.

Figures 6.4 and 6.5 show the effects of windows for a section of unvoiced speech (corresponding to the fricative /sh/) for a 500 sample segment (Figure 6.4) and a 50 sample segment (Figure 6.5). From these figures it is seen that the spectra show a slowly varying trend with a series of sharp peaks and valleys superimposed. The ragged appearance of the spectrum (for both windows) is due to the random nature of unvoiced speech. Finally the use of a Hamming

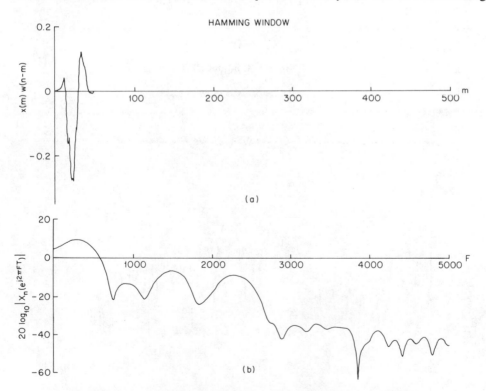

Fig. 6.3 Spectrum analysis of voiced speech using a 5 msec (a,b) Ham-ming window; (c,d) rectangular window. Parts (a) and (c) show time waveforms; parts (b) and (d) show corresponding spectra.

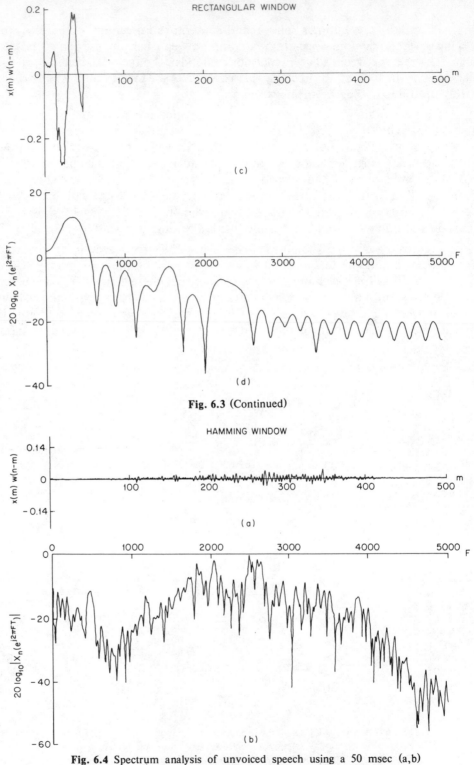

RECTANGULAR WINDOW

(c)

(d)

Fig. 6.3 (Continued)

HAMMING WINDOW

(a)

(b)

Fig. 6.4 Spectrum analysis of unvoiced speech using a 50 msec (a,b) Hamming window; (c,d) rectangular window. Parts (a) and (c) show time waveforms; parts (b) and (d) show corresponding spectra.

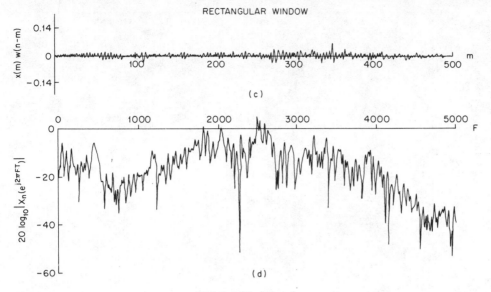

(c)

(d)

Fig. 6.4 (Continued)

window is seen to produce a somewhat smoother spectrum than a rectangular window.

The examples of Figs. 6.2-6.5 clearly illustrate the basic relationship between the time duration of the window and the properties of the short-time Fourier transform. That is, frequency resolution varies inversely with the length of the window. When we recall that the purpose of the window is to limit the time interval to be analyzed so that the properties of the waveform do not change appreciably, we see that a compromise is required. In Fig. 6.2c, for example, it can be seen that the formant frequencies are obviously changing across the 50 msec interval. A shorter analysis interval is required in order to display this temporal variation. Windows of 5 msec duration positioned at the beginning and end of the 50 msec interval would yield distinctly different

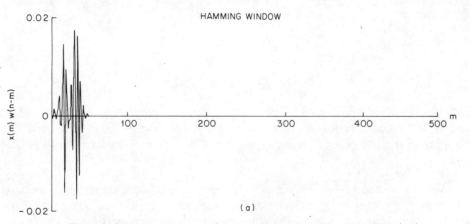

Fig. 6.5 Spectrum analysis of unvoiced speech using a 5 msec (a,b) Hamming window; (c,d) rectangular window. Parts (a) and (c) show time waveforms; parts (b) and (d) show corresponding spectra.

259

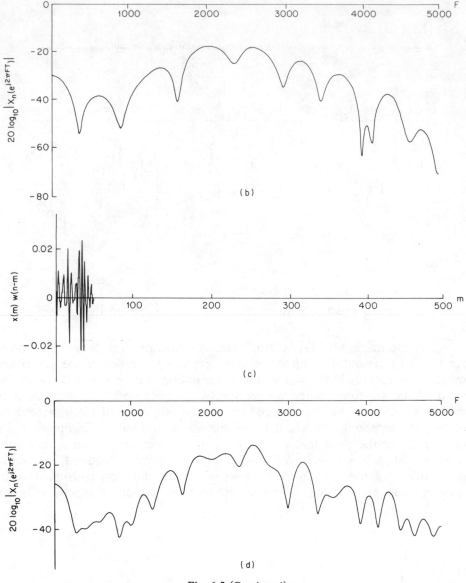

Fig. 6.5 (Continued)

short-time Fourier transforms. Thus, good temporal resolution requires a short window while good frequency resolution calls for a long window. We shall see examples of the use of both types of windows later in our discussion of applications.

We have seen that an interpretation of the time-dependent Fourier transform as the conventional Fourier transform of a windowed segment of the speech signal leads to useful insights into both the properties of the time-dependent Fourier representation and the role of the window. Further insight will result from the linear filtering interpretation.

6.1.2 Linear filtering interpretation

It is obvious from Eq. (6.1), that for each value of ω, $X_n(e^{j\omega})$ is the convolution of the sequence $w(n)$ with the sequence $x(n)e^{-j\omega n}$. Thus, for a particular value of ω, $X_n(e^{j\omega})$ can be thought of as the output of the system depicted in Fig. 6.6a where $w(n)$ plays the role of the impulse response of a linear shift invariant system. Note that in Fig. 6.6a, the input and output of the linear system are complex. Expressing $X_n(e^{j\omega})$ as

$$X_n(e^{j\omega}) = a_n(\omega) - jb_n(\omega) \tag{6.15}$$

then the operations required to obtain $a_n(\omega)$ and $b_n(\omega)$ are shown in Fig. 6.6b where all the sequences are real.

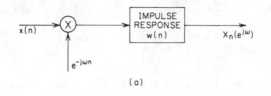

(a)

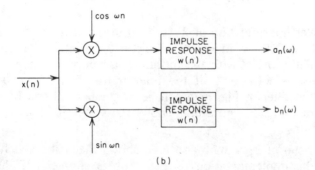

(b)

Fig. 6.6 Linear filtering interpretation of short-time spectrum analysis; (a) complex operations; (b) real operations only.

To see how the system of Fig. 6.6a operates to form the short-time Fourier transform at frequency ω, it is helpful to again assume that the normal Fourier transform of $x(n)$ exists. To avoid confusion of frequency variables, we denote the Fourier transform of $x(n)$ as $X(e^{j\theta})$. (Recall that we are now considering ω to be a particular value of radian frequency.) Then, as a result of the modulation process, the Fourier transform of the input to the linear filter is $X(e^{j(\theta+\omega)})$. Thus the spectrum of $x(n)$ at frequency ω is shifted to zero frequency. Since the Fourier transform of the output of the filter is $X(e^{j(\theta+\omega)}) W(e^{j\theta})$, then if the filter is a lowpass filter with a very narrow passband, the output of the filter will depend essentially upon $X(e^{j\omega})$. Thus, as in the previous interpretation, $W(e^{j\theta})$ should be nonzero over a very narrow band around zero frequency and as small as possible outside this band. It is interesting to note in passing that Eq. (6.14) is exactly the inverse Fourier transform of $W(e^{j\theta})X(e^{j(\theta+\omega)})$.

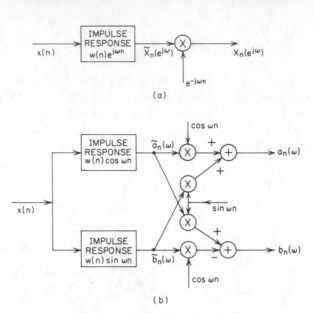

(a)

(b)

Fig. 6.7 Another interpretation of short-time spectral analysis in terms of linear filtering; (a) complex operations; (b) real operations only.

Still another interpretation of $X_n(e^{j\omega})$ in terms of linear filtering is evident from Eq. (6.2). As shown in Fig. 6.7a, $X_n(e^{j\omega})$ can also be thought of as the result of modulating $e^{-j\omega n}$ with the output of a complex bandpass filter whose impulse response is $w(n)e^{j\omega n}$. If the Fourier transform $W(e^{j\theta})$ is a lowpass function then the filter in Fig. 6.7a will be a bandpass filter whose passband is centered at frequency ω. Figure 6.7b shows the system of Fig. 6.7a in terms of only real quantities.

A comparison of Figs. 6.6b and 6.7b shows that if both $a_n(\omega)$ and $b_n(\omega)$ are required, the implementation of Fig. 6.6b is simpler. If, however, only $|X_n(e^{j\omega})|$ is required, implementation with bandpass filters is simpler. To see this note that from Eqs. (6.4) and (6.9),

$$|X_n(e^{j\omega})| = [a_n^2(\omega) + b_n^2(\omega)]^{1/2} \qquad (6.16a)$$

$$= |\tilde{X}_n(e^{j\omega})| = [\tilde{a}_n^2(n) + \tilde{b}_n^2(\omega)]^{1/2} \qquad (6.16b)$$

Figure 6.8a depicts Eq. (6.16a) and Fig. 6.8b depicts Eq. (6.16b). The system of Fig. 6.8b would in general be simpler.

With the point-of-view that $X_n(e^{j\omega})$ at a particular value of ω is the output of a system as depicted in Figs. 6.6 or 6.7, we can call on a knowledge of linear systems to help to understand the properties of the time-varying Fourier representation. For example, it is helpful to recall that the impulse response of a discrete-time, linear, shift-invariant system can be either of finite (FIR) or infinite (IIR) duration. Similarly we may define two classes of windows for time-varying Fourier analysis. Also, recall that a linear shift-invariant system can be either causal or noncausal depending upon whether or not its impulse response is zero for $n < 0$. In like manner we can classify windows as either causal or noncausal. A causal window is one for which

$$w(n) = 0, \quad n < 0 \tag{6.17a}$$

or equivalently,

$$w(n-m) = 0, \quad n < m. \tag{6.17b}$$

The Hamming window and the rectangular window are examples of finite duration windows. By appropriate choice of time origin they can also be defined as causal windows. Such windows are, as we shall see, appropriate for use in implementations based upon Figs. 6.6 and 6.7 as well as implementations based upon the discrete Fourier transform. Infinite duration windows are also useful, especially when $X_n(e^{j\omega})$ is computed using linear filtering as in Figs. 6.6 and 6.7. In such cases, we can obtain a recurrence formula that gives $X_n(e^{j\omega})$ in terms of values at previous times (see Problem 6.6).

6.1.3 Sampling rates of $X_n(e^{j\omega})$ in time and frequency[2]

The short-time Fourier transform is a two-dimensional representation of the one-dimensional signal $x(n)$. That is $X_n(e^{j\omega})$ is a function of both n which represents time, and radian frequency ω. A basic consideration in the digital implementation of systems for short-time Fourier analysis is the rate at which $X_n(e^{j\omega})$ should be sampled in both the time and frequency dimensions to provide an unaliased representation of $X_n(e^{j\omega})$ from which $x(n)$ can be exactly

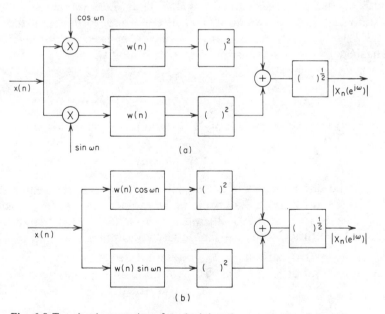

Fig. 6.8 Two implementations for obtaining the magnitude of the short-time spectrum; (a) using lowpass filters; (b) using bandpass filters.

[2]The material presented in the remainder of Section 6.1 is based on References 1-4.

recovered. This question is by no means a trivial one and requires careful consideration of the factors entering into the computation of $X_n(e^{j\omega})$ to arrive at the correct sampling rates in both time and frequency. As will be shown, a complicating factor in the discussion of the choice of the proper sampling rates for $X_n(e^{j\omega})$ is that sampling rates lower than the theoretically minimum rate can be used in either the time or frequency dimensions, and $x(n)$ can still be exactly recovered from the aliased (undersampled) short-time transform. Such undersampled representations are indeed quite useful for applications in which one is only interested in short-time Fourier analysis (e.g., spectral estimation, pitch and formant analysis, digital speech spectrograms etc.) and for vocoder applications in which minimization of overall bit rate of the system is of prime importance. For applications in which one is interested in obtaining a short-time Fourier transform of the signal, performing some modification on the signal (e.g., linear or nonlinear filtering) and then resynthesizing the modified signal, it is essential that no aliasing occur in either the time or frequency domains.

We begin by discussing the required sampling rate of $X_n(e^{j\omega})$ in the time dimension. In this case the linear filtering interpretation of the previous section provides the necessary insight. In that section it was shown that for a fixed value of ω, $X_n(e^{j\omega})$ could be shown to be the output of a filter with impulse response $w(n)$. If we denote the Fourier transform of $w(n)$ as $W(e^{j\omega})$ then we have already shown that for most reasonable windows $W(e^{j\omega})$ has the properties of of a lowpass filter frequency response. Let us denote the effective bandwidth of the analysis window as B Hz.[3] Thus the sequence $X_n(e^{j\omega})$ has the same bandwidth as the window, and therefore according to the sampling theorem, $X_n(e^{j\omega})$ must be sampled at a rate of at least $2B$ samples/second to avoid aliasing. For an example consider a Hamming window, i.e.,

$$w(n) = 0.54 - 0.46*\cos(2\pi n/(L-1)) \quad 0 \leqslant n \leqslant L - 1$$

$$= 0 \qquad\qquad\qquad\qquad otherwise \qquad (6.18)$$

Then the approximate bandwidth of $W(e^{j\omega})$ in terms of analog frequencies is

$$B = \frac{2F_s}{L} \quad \text{(Hz)} \qquad\qquad (6.19)$$

where F_s is the sampling rate of the signal $x(n)$ and thus the required sampling rate of $X_n(e^{j\omega})$ in the time dimension is $2B$ samples/sec $= 4F_s/L$ samples/sec. Thus for $L = 100$, $F_s = 10000$ Hz, we get $B = 200$ Hz, so that $X_n(e^{j\omega})$ must be evaluated 400 times/second — i.e., every 25 samples.

Since $X_n(e^{j\omega})$ is periodic in ω with period 2π, it is only necessary to sample over an interval of length 2π. To determine an appropriate finite set of frequencies $\omega_k = 2\pi k/N$, $k = 0, 1,...,N - 1$ at which $X_n(e^{j\omega})$ must be specified to

---

[3]Note that there is a possibility of confusion of frequency variables. Recall that when $X_n(e^{j\omega})$ is viewed as a function of time, ω is fixed. However, we also use the variable ω to denote the frequency variable associated with the time variation of $X_n(e^{j\omega})$.

exactly recover $x(n)$, we use the Fourier transform interpretation of $X_n(e^{j\omega})$. If the window is time-limited, then if $X_n(e^{j\omega})$ is viewed as a Fourier transform, its inverse transform is time-limited. In this case the sampling theorem requires that we sample $X_n(e^{j\omega})$ in the frequency dimension at a rate of at least twice its "time width." Since the inverse Fourier transform of $X_n(e^{j\omega})$ is the signal $x(m)w(n-m)$ and this signal is of duration L samples (again due to the finite duration window $w(n)$) then according to the sampling theorem, $X_n(e^{j\omega})$ must be sampled (in frequency) at the set of frequencies

$$\omega_k = \frac{2\pi k}{L} , \quad k = 0, 1, \ldots, L-1 \qquad (6.20)$$

in order to exactly recover $x(n)$ from $X_n(e^{j\omega_k})$. (See Problem 6.8.) Thus for the example of a Hamming window of duration $L = 100$ samples, we require $X_n(e^{j\omega})$ to be evaluated at at least 100 uniformly spaced frequencies around the unit circle.

Based on the above discussion we can determine the total number of samples of $X_n(e^{j\omega_k})$ that must be computed per second to give an unaliased representation of the original signal $x(n)$. The minimum sampling rate of $X_n(e^{j\omega_k})$ in the time dimension is $2B$ where B is the frequency bandwidth of the window, and the minimum number of samples in the frequency dimension is L, the time width of the window. Thus the total sampling rate (SR) of $X_n(e^{j\omega_k})$ is

$$SR = 2B \cdot L \quad \text{samples/sec} \qquad (6.21)$$

For most practical windows B can be represented as a multiple of (F_s/L) where F_s is the sampling frequency of $x(n)$, i.e.,

$$B = C \frac{F_s}{L} \quad \text{(Hz)} \qquad (6.22)$$

where C is the proportionality constant. Thus Eq. (6.21) can be written as

$$SR = 2CF_s \quad \text{samples/sec} \qquad (6.23)$$

The ratio of SR to F_s is therefore

$$\frac{SR}{F_s} = 2C \qquad (6.24)$$

The quantity $2C$ indicates the "oversampling" ratio of the short-time analysis as compared to a conventional sampling representation of $x(n)$.

By way of example, if $w(n)$ is a Hamming window, then $2C = 4$ whereas if $w(n)$ is a rectangular window (and if the bandwidth B is defined to be the frequency of the first zero of $W(e^{j\omega})$) then $2C = 2$. Thus the short-time spectrum representation of $x(n)$ is seen to require on the order of 2 to 4 times as many samples as required to represent the waveform; however, in return one obtains a very flexible representation of the signal from which extensive modifications in both the time and frequency domains can be made.

In this section we have discussed the required sampling rates of $X_n(e^{j\omega_k})$ in time and frequency to obtain an unaliased representation of $x(n)$. Although the sampling rates which were derived are theoretically the minimum rates for the signal, there exist special cases in which $X_n(e^{j\omega_k})$ can be undersampled in either the time or frequency dimensions, and for which $x(n)$ can be exactly recovered with no aliasing error. Such cases are of practical importance for the implementation of systems in which minimum storage (bit rate) of the representation is of importance, e.g., an analysis-synthesis system, a spectral display etc. We shall discuss how to design and implement such systems later in this chapter. First we shall show two distinct ways in which $x(n)$ can be recovered from a sampled version of $X_n(e^{j\omega_k})$ and then we will discuss the effects of modifications of $X_n(e^{j\omega_k})$ on the resulting reconstructed signal.

6.1.4 Filter bank summation method of short-time synthesis

The first synthesis method is related to the filter bank interpretation of the short-time spectrum, in which it was shown that for any frequency ω_k, $X_n(e^{j\omega_k})$ is a lowpass representation of the signal in a band centered at ω_k. From Eqs. (6.1) and (6.2), $X_n(e^{j\omega_k})$ can be expressed as either

$$X_n(e^{j\omega_k}) = \sum_{m=-\infty}^{\infty} w_k(n-m)x(m)e^{-j\omega_k m} \tag{6.25}$$

or

$$X_n(e^{j\omega_k}) = e^{-j\omega_k n}\sum_{m=-\infty}^{\infty} x(n-m)w_k(m)e^{j\omega_k m} \tag{6.26}$$

where $w_k(m)$ is the window used at frequency ω_k. If we define

$$h_k(n) = w_k(n)e^{j\omega_k n} \tag{6.27}$$

then Eq. (6.26) can be expressed as

$$X_n(e^{j\omega_k}) = e^{-j\omega_k n}\sum_{m=-\infty}^{\infty} x(n-m)h_k(m) \tag{6.28}$$

Since the window $w_k(n)$ has the properties of a lowpass filter, Eq. (6.28) can be interpreted as in Fig. 6.7 as a bandpass filter with impulse response $h_k(n)$ followed by modulation with a complex exponential $e^{-j\omega_k n}$. If we define

$$y_k(n) = X_n(e^{j\omega_k})e^{j\omega_k n} \tag{6.29}$$

we see from Eq. (6.28) that

$$y_k(n) = \sum_{m=-\infty}^{\infty} x(n-m)h_k(m) \tag{6.30}$$

Thus $y_k(n)$ is simply the output of a bandpass filter with impulse response $h_k(n)$ as given by Eq. (6.27). The operations of Eqs. (6.28) and (6.29) are depicted in Fig. 6.9a. Since Eqs. (6.25) and (6.28) are equivalent we also note

266

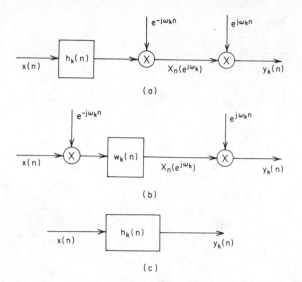

Fig. 6.9 Methods for implementing synthesis of a single channel in terms of linear filtering.

that either form for $X_n(e^{j\omega_k})$ can be used in Eq. (6.29), and in both cases the overall system relating $x(n)$ to $y_k(n)$ is a bandpass filter with impulse response $h_k(n)$. This is depicted in Fig. 6.9, where Fig. 6.9a depicts Eqs. (6.28) and (6.29) and Fig. 6.9b depicts Eqs. (6.25) and (6.29). Figure 6.9c shows the equivalent bandpass filter for both cases.

The result summarized in Fig. 6.9 provides the key to a practical method for reconstructing the input signal from its time dependent Fourier transform. Consider now a set of N frequencies $\{\omega_k\}$, $k = 0, 1, ..., N - 1$, and suppose that $X_n(e^{j\omega_k})$ is available for each frequency. Now consider N bandpass filters of the form Eq. (6.27). Suppose for example that $w_k(n)$ is the impulse response of an ideal lowpass filter with cutoff frequency ω_{pk}. The frequency response $W_k(e^{j\omega})$ of such a filter is shown in Fig. 6.10a.[4] The frequency response of the corresponding complex bandpass filter with impulse response $h_k(n) = w_k(n)e^{j\omega_k n}$ is

$$H_k(e^{j\omega}) = W_k(e^{j(\omega - \omega_k)})\tag{6.31}$$

as shown in Fig. 6.10b. Note that the center frequency is ω_k and the bandwidth is $2\omega_{pk}$.

We now consider a set of N bandpass filters chosen with their center frequencies uniformly spaced so that the entire base frequency band is covered. Thus

$$\omega_k = \frac{2\pi k}{N} \quad k = 0, 1, \ldots, N - 1\tag{6.32}$$

We also assume the window is the same for all channels, i.e.,

$$w_k(n) = w(n) \quad k = 0, 1, \ldots, N - 1\tag{6.33}$$

[4]Note that we are now using ω as the frequency variable.

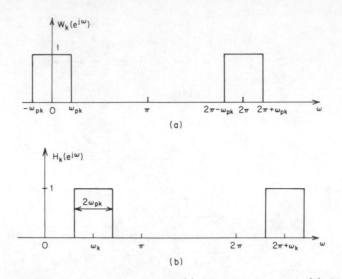

(a)

(b)

Fig. 6.10 Frequency responses of (a) ideal lowpass; and (b) ideal bandpass filters.

Then if we consider the entire collection of bandpass filters, each having the same input and their outputs added together as in Fig. 6.11, the composite frequency response relating $y(n)$ to $x(n)$ is

$$\tilde{H}(e^{j\omega}) = \sum_{k=0}^{N-1} H_k(e^{j\omega}) = \sum_{k=0}^{N-1} W(e^{j(\omega-\omega_k)}) \tag{6.34}$$

If $W(e^{j\omega_k})$ is properly sampled in frequency (i.e., if $N \geqslant L$ where L is the time duration of the window), then it can be shown that

$$\frac{1}{N} \sum_{k=0}^{N-1} W(e^{j(\omega-\omega_k)}) = w(0) \tag{6.35}$$

for all ω.

The derivation of Eq. (6.35) is as follows. The inverse Fourier transform of $W(e^{j\omega})$ is $w(n)$, the window. If $W(e^{j\omega})$ is sampled in frequency at N uni-

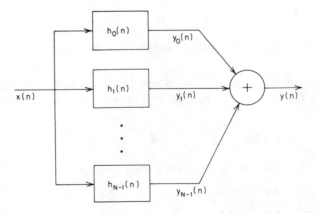

Fig. 6.11 Equivalent linear system relating $y_k(n)$ and $y(n)$ to $x(n)$.

formly spaced points, the inverse discrete Fourier transform of the sampled version of $W(e^{j\omega_k})$ is

$$\frac{1}{N} \sum_{k=0}^{N-1} W(e^{j\omega_k}) e^{j\omega_k n} = \sum_{r=-\infty}^{\infty} w(n+rN) \tag{6.36}$$

i.e., an aliased representation of $w(n)$ is obtained. (See Problem 6.8.) If $w(n)$ is of duration L samples, then

$$w(n) = 0 \quad n < 0, \ n \geqslant L \tag{6.37}$$

and no aliasing occurs due to sampling in frequency of $W(e^{j\omega})$. Thus, in this case, if Eq. (6.36) is evaluated for $n = 0$ we get

$$\frac{1}{N} \sum_{k=0}^{N-1} W(e^{j\omega_k}) = w(0) \tag{6.38}$$

Eq. (6.35) is readily obtained by noting that $W(e^{j(\omega-\omega_k)})$ is a uniformly sampled version of $W(e^{j\omega})$ evaluated at $\omega - \omega_k$ rather then ω_k. According to the sampling theorem, *any* set of N uniformly spaced samples is adequate. Thus, Eq. (6.35) follows from Eq. (6.38) and the sampling theorem.

From Eqs. (6.38) and (6.35) we see that the impulse response of the composite system

$$\tilde{h}(n) = \sum_{k=0}^{N-1} h_k(n) = \sum_{k=0}^{N-1} w(n) e^{j\omega_k n} = Nw(0)\delta(n) \tag{6.39}$$

is simply equal to a scaled unit sample $N\,w(0)\,\delta(n)$, and thus the composite output $y(n)$ will be $Nw(0)x(n)$.

Thus for the filter bank summation method, the reconstructed signal is formed as

$$y(n) = \sum_{k=0}^{N-1} y_k(n)$$

$$= \sum_{k=0}^{N-1} X_n(e^{j\omega_k}) e^{j\omega_k n} \tag{6.40}$$

and we have shown that if $X_n(e^{j\omega_k})$ is properly sampled in frequency, then $y(n) = Nw(0)x(n)$ independent of the specific shape of $w(n)$. The operations of analysis and synthesis implied by Eq. (6.40) are depicted in Fig. 6.12 where the filters are bandpass filters.

We have just shown a very important result; i.e. under the condition that $w(n)$ has finite duration, L, then the sequence $x(n)$ can be reconstructed exactly from the time-dependent Fourier transform sampled in both the time and frequency dimensions. It can also be shown, that if $W(e^{j\omega})$ is perfectly bandlimited in frequency, then $x(n)$ can likewise be reconstructed exactly from $X_n(e^{j\omega_k})$. Indeed we shall see that numerous possibilities exist for exact reconstruction of $x(n)$ from the short-time transform.

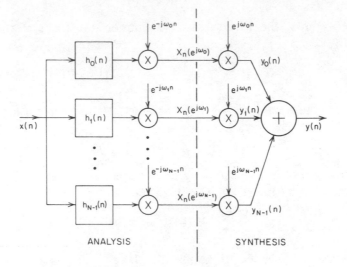

Fig. 6.12 Analysis and synthesis operations for short-time spectrum analysis.

We have shown that to avoid time aliasing, $X_n(e^{j\omega_k})$ must be evaluated at least at L uniformly spaced frequencies where L is the window duration. The bandwidth of a window of duration L samples is generally from $2\pi/L$ (for a rectangular window) to $4\pi/L$ (for a Hamming window). Since the analysis frequencies are $2\pi k/L$, the effective bandpass filters overlap in frequency. As mentioned earlier there is a way in which $X_n(e^{j\omega_k})$ can be evaluated in nonoverlapping bands, and for which $x(n)$ can still be exactly (at least theoretically) recovered.

To show this we assume that the window length for all bands is L samples, and that the same window is used for N equally spaced frequency bands with analysis frequencies

$$\omega_k = \frac{2\pi k}{N} \quad k = 0, 1, \ldots, N-1 \tag{6.41a}$$

where N may be less than L. We also assume that $w(n)$ is an ideal lowpass filter with cutoff frequency

$$\omega_p = \frac{\pi}{N} \tag{6.41b}$$

This situation is depicted in Fig. 6.13 which shows the composite response for $N = 6$ equally spaced ideal filters. For this case Eq. (6.39) becomes

$$\tilde{h}(n) = \sum_{k=0}^{N-1} w(n) e^{j\omega_k n} = w(n) \sum_{k=0}^{N-1} e^{j\omega_k n} \tag{6.42}$$

If we define

$$p(n) = \sum_{k=0}^{N-1} e^{j\omega_k n} = \sum_{k=0}^{N-1} e^{j\frac{2\pi}{N} kn} \tag{6.43}$$

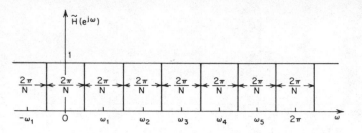

Fig. 6.13 Composite frequency response for $N = 6$ equally spaced ideal filters.

then $\tilde{h}(n)$ can be written

$$\tilde{h}(n) = w(n)p(n) \tag{6.44}$$

The sequence $p(n)$ is easily seen to be periodic with period N. Indeed, $p(n)$ can easily be shown (see Problem 6.7) to be a periodic train of impulses of amplitude N, i.e.,

$$p(n) = N \sum_{r=-\infty}^{\infty} \delta(n-rN) \tag{6.45}$$

Thus, $\tilde{h}(n)$ is

$$\tilde{h}(n) = N \sum_{r=-\infty}^{\infty} w(rN)\delta(n-rN) \tag{6.46}$$

Thus the composite impulse response is simply the window sequence sampled at intervals of N samples. This is depicted in Fig. 6.14. Figure 6.14a shows the sequence $p(n)$. Figure 6.14b shows the impulse response of an ideal lowpass filter with cutoff frequency π/N; i.e.,

$$w(n) = \frac{\sin \dfrac{\pi}{N} n}{\pi n} \tag{6.47}$$

It can be seen by comparing Figs. 6.14a and 6.14b that the product $\tilde{h}(n) = p(n)w(n)$ will be zero everywhere except at $n = 0$ where the product is unity. Thus the composite impulse response is

$$\tilde{h}(n) = \delta(n) \tag{6.48}$$

Although this assumes an ideal lowpass filter, the details of the way $p(n)$ and $w(n)$ interact to produce the composite response suggest a multitude of ways of choosing $w(n)$ so that the signal can be reconstructed from the sampled short-time transform. First of all, note that if $w(n)$ is of finite length $L < N$ and causal, then the composite impulse response will be as in Eq. (6.39) thus verifying the discussion of the previous section. Figure 6.14c shows an example of this case. Alternatively, a causal window with length greater than N can be used if $w(n)$ has the properties.

$$w(n) = \frac{1}{N}, \quad n = r_0 N$$

$$= 0 \quad n = rN \quad \begin{cases} r \neq r_0 \\ r = 0, \pm 1, \pm 2, \ldots \end{cases} \tag{6.49a}$$

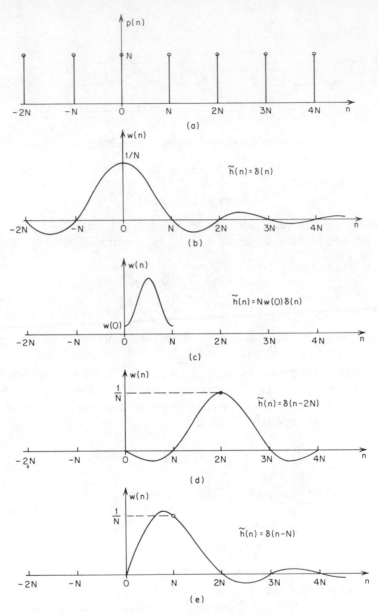

Fig. 6.14 Typical sequences for $p(n)$ and $w(n)$ for composite filter bank.

then

$$\tilde{h}(n) = p(n)w(n) = \delta(n - r_0 N) \qquad (6.49b)$$

Figure 6.14d shows an example of a finite duration window where $r_0 = 2$. In fact it is clear that $w(n)$ need be neither time limited or frequency limited in order that it be possible to exactly reconstruct a delayed replica of $x(n)$ from $X_n(e^{j\omega_k})$. All that is required is that Eq. (6.49a) hold for $w(n)$. There is no restriction that the window be of finite length as long as it has equally spaced zeros. Indeed Fig. 6.14e depicts an infinite duration window with the appropriate properties.

272

The implication of Eq. (6.49b) is that the composite frequency response of an analysis/synthesis system such as Fig. 6.12 has a flat magnitude response and linear phase corresponding to a delay of r_0N samples. That is,

$$\tilde{H}(e^{j\omega}) = e^{-j\omega r_0 N} \qquad (6.50)$$

This implies that the output of the analysis/synthesis system is

$$y(n) = x(n - r_0 N) \qquad (6.51)$$

Thus, except for a delay of r_0N samples, the output of the system for time-dependent Fourier analysis and synthesis is an *exact replica* of the input sequence. Therefore we have shown that exact reconstruction of the input is possible with a number of frequency channels less than that required by the sampling theorem and with a causal window which permits the realization of the analysis with causal bandpass or lowpass filters. Thus an important practical issue is how well we can design digital filters to approximate the behavior shown in Figure 6.14. We discuss this issue extensively in Section 6.2.

Before proceding to an alternative approach to synthesis from the short-time spectrum, we must discuss the way in which Eq. (6.40) (the synthesis equation) is practically implemented since we have shown that $X_n(e^{j\omega_k})$ need only be computed at a rate related to the bandwidth of the window. As such we assume that the kth channel is computed once every D_k samples of the input. (For uniformly spaced channels $D_k = D$, independent of k). Assuming that $X_n(e^{j\omega_k})$ is computed at the sampling rate of the input (although this may not be the case in practice) we can modify Figs. 6.9a and 6.9b to reflect the fact that $X_n(e^{j\omega_k})$ need only be sampled at a rate of F_s/D_k by including a decimator at the output of the analysis and an interpolator at the input to the synthesis as depicted in Fig. 6.15. As discussed in Chapter 2, the decimation is simply implemented by discarding $D_k - 1$ samples out of every D_k, or equivalently, by

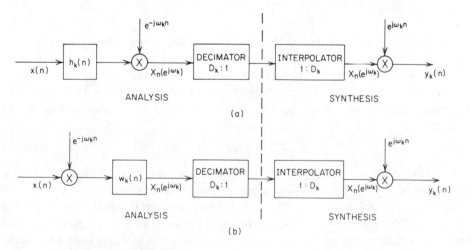

Fig. 6.15 Complete filter bank implementation for a single channel for short-time spectrum analysis.

273

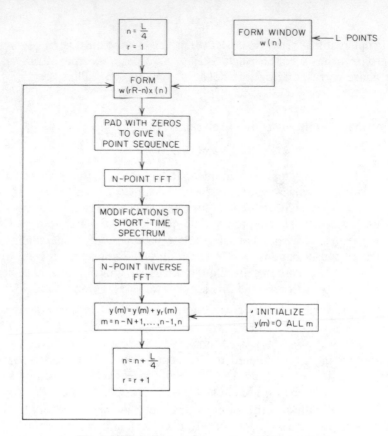

Fig. 6.16 Flow diagram for overlap addition method.

only computing $X_n(e^{j\omega_k})$ once every D_k samples of the input. The interpolation is implemented by filling in $D_k - 1$ zero valued samples between every sample of $X_n(e^{j\omega_k})$ at the reduced rate, and filtering the result with an appropriate lowpass filter.

6.1.5 Overlap addition method for short-time synthesis

An alternative method for reconstructing $x(n)$ from its short-time spectrum is based on the normal Fourier transform interpretation of the short-time spectrum. Since, as shown earlier, $X_n(e^{j\omega_k})$ can be considered the normal discrete Fourier transform of the sequence

$$y_n(m) = x(m)w(n-m) \tag{6.52}$$

then $x(m)$ can be reconstructed by computing the inverse discrete Fourier transform of $X_n(e^{j\omega_k})$ and dividing out the window (assuming it is nonzero for all values of m which are considered). In this manner L signal values of $x(m)$ can be reconstructed for for each window (where L is the window duration). Then the window can be moved by L samples and the process repeated. Based on the discussion of Section 6.1.3, it can be seen that this procedure uses an

274

"undersampled in time" representation of $X_n(e^{j\omega_k})$ and thus is highly suscepti-
ble to aliasing errors. Thus although such a procedure is valid, it has not been
found useful for many applications in which one is interested in reconstructing
the original signal (or a processed version of it). In this section we present a
more robust synthesis procedure similar to the overlap addition method for a
periodic convolution using discrete Fourier transforms.

Assume that the short-time transform is sampled with period R samples
in the time dimension; i.e. we have $Y_r(e^{j\omega_k}) = X_{rR}(e^{j\omega_k})$ were r is an integer
and $0 \leqslant k \leqslant N - 1$. The overlap addition method is based upon the equation

$$y(n) = \sum_{r=-\infty}^{\infty} \left[\frac{1}{N} \sum_{k=0}^{N-1} Y_r(e^{j\omega_k}) e^{j\omega_k n} \right] \tag{6.53}$$

That is, to reconstruct the signal, the inverse transform of $Y_r(e^{j\omega_k})$ is computed
for each value of r giving the sequences

$$y_r(m) = x(m) w(rR - m) \qquad -\infty < m < \infty \tag{6.54}$$

Then the signal at time n is obtained by summing the values at time n of all the
sequences $y_r(m)$ that overlap at time n. That is,

$$y(n) = \sum_{r=-\infty}^{\infty} y_r(n) = x(n) \sum_{r=-\infty}^{\infty} w(rR - n) \tag{6.55}$$

It is readily shown (see Problem 6.8) that if $w(n)$ has a bandlimited Fourier
transform and if $X_n(e^{j\omega_k})$ is properly sampled in time, i.e. R is small enough to
avoid time aliasing,[5] then

$$\sum_{r=-\infty}^{\infty} w(rR - n) \approx W(e^{j0})/R \tag{6.56}$$

regardless of the value of n.

Thus Eq. (6.55) becomes

$$y(n) = x(n) W(e^{j0})/R \tag{6.57}$$

showing that the synthesis rule of Eq. (6.53) can lead to exact reconstruction of
$x(n)$ (to within a constant multipler) by adding overlapping sections of the
waveform.

Figures 6.16 and 6.17 illustrate how the overlap addition method is implemented for $w(n)$, an L-point Hamming window with $R = L/4$. Figure 6.16
gives a flow chart of the method, assuming the signal $x(n)$ is 0 for $n < 0$.
Since a time overlap of 4 to 1 is required for a Hamming window, to obtain the
correct initial conditions the first analysis section is positioned to begin at
$n = L/4$ as shown in Fig. 6.17. The window (assumed to be nonzero for
$0 \leqslant n \leqslant L - 1$) is used to give the signal $y_r(m) = w(rR - m)x(m)$ which is
nonzero for $rR - L + 1 \leqslant m \leqslant rR$. This L-point sequence is padded with

---

[5] For an L-point Hamming window $R \leqslant L/4$.

275

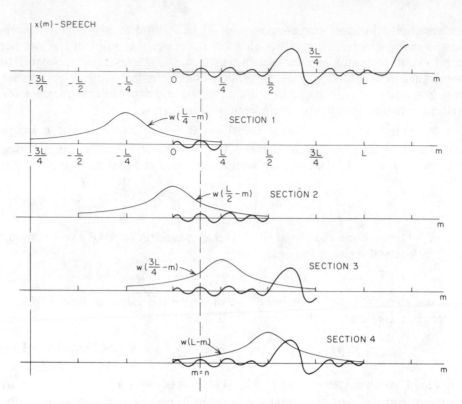

Fig. 6.17 Reconstruction procedure for $w(n)$ using an L-point Hamming window.

sufficient zeros to account for the effects of modifications of the short-time spectrum (as discussed in the next section), and then an N-point FFT of the resulting sequence is used to give $Y_r(e^{j\omega_k})$.

To reconstruct the signal at time n, we use Eq. (6.53). Figure 6.17 illustrates the operations implied by Eq. (6.53) for a value of n such that $0 \leqslant n \leqslant R - 1$. Note that $y(n)$ consists of the sum of 4 numbers; i.e.

$$y(n) = x(n)w(R-n) + x(n)w(2R-n) + x(n)w(3R-n)$$
$$+ x(n)w(4R-n) . \tag{6.58}$$

Clearly for values of n such that $R \leqslant n \leqslant 2R - 1$, the term $x(n)w(R-n)$ would be replaced by a term $x(n)w(5R-n)$, etc.

It is interesting to note that the filter bank summation method and the overlap addition method are essentially duals of one another; i.e. one depends on a sampling relation in frequency, and one depends on a sampling relation in time. The filter bank summation requires that the sampling in frequency be such that the window transform obeys the relation

$$\sum_{k=0}^{N-1} W(e^{j(\omega-\omega_k)}) = w(0) \tag{6.59a}$$

276

whereas the overlap addition method requires that the sampling in time be such that the window obeys the relation

$$\sum_{r=-\infty}^{\infty} w(rR-n) = W(e^{j0})/R \tag{6.59b}$$

The duality between Eqs. (6.59a) and (6.59b) is evident.

In order to compare and contrast these two methods of reconstructing a signal from the short-time transform, the next sections discuss the effects of modifications of the short-time spectrum on the resulting synthesis.

6.1.6 Effects of modifications to the short-time spectrum on the resulting synthesis

We have just shown that there are two distinctly different methods for reconstructing a signal from its short-time spectrum. Both methods have been shown capable of reconstructing the original signal exactly (to within a scale factor) in the case when the short-time spectrum is properly sampled in both time and frequency. For many applications, however, one is interested in making modifications to the short-time spectrum in order to perform fixed or dynamic (i.e., time-varying) filtering on the signal being analyzed. In this section we show the effects of fixed and time-varying modifications of the short-time spectrum on the resulting synthesis.

6.1.6a Filter Bank Summation (FBS) Method

We represent a fixed modification to the short-time spectrum as

$$\hat{X}_n(e^{j\omega_k}) = X_n(e^{j\omega_k})P(e^{j\omega_k}) \tag{6.60}$$

where $P(e^{j\omega_k})$ is a frequency weighting function on the short-time spectrum. We assume that the inverse discrete Fourier transform of $P(e^{j\omega_k})$ exists, and we call this sequence $p(n)$ where

$$p(n) = \frac{1}{N} \sum_{k=0}^{N-1} P(e^{j\omega_k})e^{j\omega_k n} \tag{6.61}$$

where N is the number of frequencies at which $P(e^{j\omega_k})$ is evaluated — i.e., the number of analysis frequencies. The reconstructed signal $\hat{y}(n)$ from the FBS method is obtained by substituting Eq. (6.60) into Eq. (6.40) giving

$$\hat{y}(n) = \sum_{k=0}^{N-1} X_n(e^{j\omega_k})P(e^{j\omega_k})e^{j\omega_k n}$$

$$= \sum_{k=0}^{N-1} [\sum_{m=-\infty}^{\infty} w(n-m)x(m)e^{-j\omega_k m}]P(e^{j\omega_k})e^{j\omega_k n}$$

$$= \sum_{m=-\infty}^{\infty} w(n-m)x(m) \sum_{k=0}^{N-1} P(e^{j\omega_k})e^{j\omega_k (n-m)}$$

$$= \sum_{m=-\infty}^{\infty} w(n-m)x(m)Np(n-m)$$

$$= Nx(n) * [w(n)p(n)] \qquad (6.62)$$

Thus, the effect of the fixed spectral modification $P(e^{j\omega_k})$ is to convolve the signal $x(n)$ with the product of the window $w(n)$ and the periodic sequence, $p(n)$. The motivation for making modifications of the form of Eq. (6.60) to the short-time Fourier transform is to effect a linear filtering operation on $x(n)$. That is, it is likely that we would desire that

$$w(n)p(n) = h_p(n) \qquad (6.63)$$

be the impulse response of a desired linear filter. We should point out that $p(n)$ is a periodic sequence, so that if $w(n)$ is longer than N, there will be a kind of repetitive structure in $h_p(n)$. In the next section we shall see that modifications of this form arise in the design of filter banks using IIR filters. Thus for the filter bank summation method fixed spectral modifications are strongly affected by the window, and only in the case when $p(n)$ is highly concentrated or when a rectangular window is used is it even approximately true that

$$h_p(n) \approx p(n) \qquad 0 \leqslant n \leqslant N-1 \qquad (6.64)$$

as might be desired.

For time-varying modifications we model $\hat{X}_n(e^{j\omega_k})$ as

$$\hat{X}_n(e^{j\omega_k}) = X_n(e^{j\omega_k}) P_n(e^{j\omega_k}) \qquad (6.65)$$

and we define the time-varying impulse response due to the modification, $p_n(m)$, as

$$p_n(m) = \frac{1}{N} \sum_{k=0}^{N-1} P_n(e^{j\omega_k}) e^{j\omega_k m} \qquad (6.66)$$

Proceeding as before we solve for $\hat{y}(n)$, due to the modification, as

$$\hat{y}(n) = \sum_{k=0}^{N-1} X_n(e^{j\omega_k}) P_n(e^{j\omega_k}) e^{j\omega_k n}$$

$$= \sum_{k=0}^{N-1} e^{-j\omega_k n} \sum_{m=-\infty}^{\infty} x(n-m) w(m) e^{j\omega_k m} P_n(e^{j\omega_k}) e^{j\omega_k n}$$

$$= \sum_{m=-\infty}^{\infty} x(n-m) w(m) \sum_{k=0}^{N-1} P_n(e^{j\omega_k}) e^{j\omega_k m}$$

$$= \sum_{m=-\infty}^{\infty} x(n-m) w(m) N p_n(m)$$

$$= N \sum_{m=-\infty}^{\infty} x(n-m) [p_n(m) w(m)] \qquad (6.67)$$

Eq. (6.57) again shows that for the FBS method the time response of the spectral modification is weighted by the window before being convolved with $x(n)$.

In summary, for the filter bank summation method, the effect of a spectral modification (either fixed or time-varying) is to convolve the original signal with a time-limited, window weighted version of the time response due to the modification.

6.1.6b Overlap Addition (OLA) Method

Using the representation of Eq. (6.60) for the modification we can solve for the reconstructed signal by using Eq. (6.53), giving

$$
\begin{aligned}
\hat{y}(n) &= \sum_{r=-\infty}^{\infty} \frac{1}{N} \sum_{k=0}^{N-1} Y_r(e^{j\omega_k}) P(e^{j\omega_k}) e^{j\omega_k n} \\
&= \frac{1}{N} \sum_{r=-\infty}^{\infty} \sum_{k=0}^{N-1} \sum_{l=-\infty}^{\infty} x(l) w(rR-l) e^{-j\omega_k l} P(e^{j\omega_k}) e^{j\omega_k n} \\
&= \frac{1}{N} \sum_{l=-\infty}^{\infty} x(l) \left[\sum_{k=0}^{N-1} P(e^{j\omega_k}) e^{j\omega_k (n-l)} \right] \left[\sum_{r=-\infty}^{\infty} w(rR-l) \right] \\
&= \sum_{l=-\infty}^{\infty} x(l) p(n-l) W(e^{j0})/R
\end{aligned}
\tag{6.68}
$$

or

$$
\hat{y}(n) = (1/R) W(e^{j0}) [x(n) * p(n)] \tag{6.69}
$$

Eq. (6.69) shows that $\hat{y}(n)$ is the convolution of the original signal with the time response of the spectral modification — i.e., no window modifications on $p(n)$ are obtained with this method.[6] (The reader should realize that appropriate modifications must be made to the analysis as shown in Fig. 6.16 — i.e., padding the signal with a sufficient number of zeros — to prevent aliasing when implementing the analysis and synthesis operations with FFT's.)

For the case of a time-varying modification we obtain

$$
\hat{y}(n) = \sum_{r=-\infty}^{\infty} \frac{1}{N} \left[\sum_{k=0}^{N-1} Y_r(e^{j\omega_k}) P_r(e^{j\omega_k}) \right] e^{j\omega_k n} \tag{6.70}
$$

which can be manipulated into the form

$$
\hat{y}(n) = \frac{1}{N} \sum_{l=-\infty}^{\infty} x(l) \sum_{r=-\infty}^{\infty} w(rR-l) \left[\sum_{k=0}^{N-1} P_r(e^{j\omega_k}) e^{j\omega_k (n-l)} \right] \tag{6.71}
$$

Using Eq. (6.66) we get

$$
\hat{y}(n) = \sum_{l=-\infty}^{\infty} x(l) \sum_{r=-\infty}^{\infty} w(rR-l) p_r(n-l) \tag{6.72}
$$

---

[6]Because of the way Eq. (6.68) is implemented; i.e., by computing the output in blocks of N samples, $p(n)$ in Eq. (6.68) is not periodic but at most only N samples long.

If we let $q = n - l$, or $l = n - q$ then Eq. (6.72) becomes

$$\hat{y}(n) = \sum_{q=-\infty}^{\infty} x(n-q) \sum_{r=-\infty}^{\infty} p_r(q) w(rR-n+q) \qquad (6.73)$$

If we define $\hat{p}$ by

$$\hat{p}(n-q,q) = \hat{p}(m,q) = \sum_{r=-\infty}^{\infty} p_r(q) w(rR-m) \qquad (6.74)$$

then Eq. (6.72) becomes

$$\hat{y}(n) = \sum_{q=-\infty}^{\infty} x(n-q)\hat{p}(n-q,q) \qquad (6.75)$$

The interpretation of Eq. (6.74) is that for the qth value, $\hat{p}(m,q)$ is the convolution of $p_r(q)$ and $w(r)$. Thus, each coefficient of the time response due to the time-varying modification is smoothed (i.e., lowpass filtered) by the window. Thus for the overlap add method, any modification is *bandlimited by the window* but the modification acts as a true convolution on the input signal. This is in direct contrast to the filter bank summation method in which the modifications were *time limited* by the window, and could change instantaneously.

6.1.7 Additive modifications

We have been discussing the effects of nonrandom multiplicative modifications to the short-time spectrum. It is also important to understand the effects of additive, signal independent (random), modifications to the short-time spectrum as might be expected to occur when implementing the analysis with finite precision (i.e., roundoff noise), or when quantizing the short-time spectrum as for a vocoder.

We model such additive modifications to the short-time spectrum as

$$\hat{X}_n(e^{j\omega_k}) = X_n(e^{j\omega_k}) + E_n(e^{j\omega_k}) \qquad (6.76)$$

where we define the noise sequence corresponding to $E_n(e^{j\omega_k})$ as

$$e(n) = \sum_{k=0}^{N-1} E_n(e^{j\omega_k}) e^{j\omega_k n} \qquad (6.77)$$

(In the case where $e(n)$ is a random noise, then a statistical model for $e(n)$ and $E(e^{j\omega_k})$ is warranted. The results to be presented are not dependent on such a statistical model.)

For the FBS method the effect of the additive modification of Eq. (6.76) is

$$\hat{y}(n) = \sum_{k=0}^{N-1} [X_n(e^{j\omega_k}) + E_n(e^{j\omega_k})] e^{j\omega_k n} \qquad (6.78)$$

which, by linearity, can be put in the form

280

$$\hat{y}(n) = y(n) + \sum_{k=0}^{N-1} E_n(e^{j\omega_k}) e^{j\omega_k n} \qquad (6.79)$$

or

$$\hat{y}(n) = y(n) + e(n) \qquad (6.80)$$

Thus, an additive spectral modification results in an additive component in the reconstructed signal. The reader should notice that the analysis window has no direct effect on the additive term in the synthesis.

For the OLA method the effect of the additive modification of Eq. (6.76) is

$$\hat{y}(n) = \sum_{r=-\infty}^{\infty} \frac{1}{N} \sum_{k=0}^{N-1} (Y_r(e^{j\omega_k}) + E_r(e^{j\omega_k})) e^{j\omega_k n} \qquad (6.81)$$

which can be put in the form

$$\hat{y}(n) = y(n) + \sum_{r=-\infty}^{\infty} \left[\frac{1}{N} \sum_{k=0}^{N-1} E_r(e^{j\omega_k}) e^{j\omega_k n} \right]$$

$$= y(n) + \sum_{r=-\infty}^{\infty} e_r(n) \qquad (6.82)$$

Thus, for additive modifications the resulting synthesis contains a larger additive (noise) signal for the OLA method than for the FBS method due to the overlap between analysis frames. For a Hamming window with a 4-to-1 overlap, the additive term in the synthesis will be on the order of 4 times larger for the OLA method than for the FBS method. As such the OLA method tends to be more sensitive to noise than the FBS method, and thus would be less useful for vocoding applications, etc.

6.1.8 Summary of the method of short-time analysis and synthesis of speech

In this chapter we have shown that a reasonable definition of the short-time Fourier transform of a signal, $X(e^{j\omega})$, is

$$X_n(e^{j\omega}) = \sum_{m=-\infty}^{\infty} w(n-m) x(m) e^{-j\omega m}$$

where $w(n)$ is an analysis window which determines the portion of the input signal that receives emphasis at time index n. We have shown that $X_n(e^{j\omega})$ can be interpreted in terms of linear filtering as the output of a bandpass filter translated to 0 frequency, or equivalently it can be interpreted as the normal Fourier transform of the sequence $w(n-m) x(m)$.

It was then shown that sampling rates in both time and frequency can be defined for $X_n(e^{j\omega})$ based on application of the sampling theorem to both the time- and frequency domain representations of the window. The required rate of the properly sampled short-time spectrum representation was shown to be from 2 to 4 times higher than for the equivalent time-domain representation of the signal itself.

Based on the two interpretations of the short-time analysis, two distinct synthesis procedures were discussed. The first method, called filter bank summation, synthesizes the signal as

$$y(n) = \sum_{k=0}^{N-1} X_n(e^{j\omega_k}) e^{j\omega_k n}$$

i.e., the outut signal is a sum of the signals from each band of the filter bank, translated to the original center frequency of the band.

The second synthesis method, called the overlap addition method, synthesizes the signal as

$$y(n) = \sum_{r=-\infty}^{\infty} \frac{1}{N} \sum_{k=0}^{N-1} Y_r(e^{j\omega_k}) e^{j\omega_k n}$$

where $Y_r(e^{j\omega_k}) = X_{rR}(e^{j\omega_k})$; i.e. windowed segments spaced by R samples in time and overlapped and added to produce the reconstructed signal.

These two synthesis methods were shown to have certain dual properties with regard to the synthesis itself, and to the way in which the systems handle modifications to the short-time spectra.

This concludes our formal discussion of the general properties of methods for short-time analysis and synthesis of speech. In the following sections we concentrate our attention on design methods for digital filter banks for the "undersampled in frequency" case (for vocoder type applications), and we discuss some of the numerous applications of the theory of short-time analysis and synthesis to speech processing.

6.2 Design of Digital Filter Banks

In this section, we consider some practical methods for designing filter banks for use with the filter bank summation method of analysis/synthesis. The goal here is to design filter banks whose composite frequency response closely approximates the ideal of flat magnitude and linear phase. We shall begin with certain details that are common to the design of filter banks regardless of the type of filters used. Then we shall show examples of the use of both IIR and FIR digital filters.

6.2.1 Practical considerations

Our previous theoretical discussion showed that it is possible to achieve exact reproduction of the input signal at the output of a filter bank of equally spaced filters if the lowpass filter (analysis window), $w(n)$, has zeros equally spaced at N sample intervals. Therefore to design such a filter bank would seem to require first simply choosing the number of filters (and therefore the frequency spacing) and then designing a lowpass filter with appropriate frequency resolution and appropriate spacing of zeros in the time domain. Unfortunately, a number of practical considerations arise to complicate this simple procedure. First, it is often desirable to use a nonuniform spacing of filters; thus, our previous proof that a perfect composite response can be obtained does not provide direct guidance to a practical solution. Second, in practical situations, it is common to omit certain parts of the spectrum from analysis. It is therefore necessary to consider the effect of omission of filters upon the composite response. Finally, most of the design procedures for lowpass filters do not permit simultaneous constraints upon both the frequency response and the time response. Thus it may be impossible to specify both frequency resolution and the locations of zeros of the impulse response.

In order to simply represent some of these special considerations, and in anticipation of solutions to some of the above problems, it is helpful to consider a slightly modified filter bank structure[7] in which each complex channel signal in Fig. 6.11 (or equivalently 6.12) is multiplied by a complex constant, denoted $P_k = |P_k|e^{j\phi_k}$. This is depicted for a single channel in Fig. 6.18. Note that since each channel is a linear system the multiplier can be applied at either the input or the output. In Figs. 6.18b and 6.18c, the complex constant

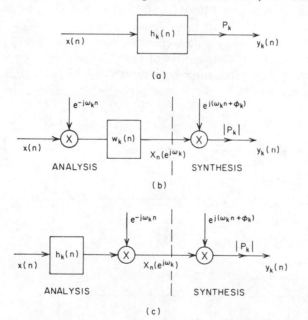

Fig. 6.18 Modified filter bank structure for a single channel (a) equivalent bandpass filters for analysis-synthesis; (b) implementation using modulators and lowpass filter; (c) implementation using modulators and bandpass filter.

[7]This modification is of the general type discussed in Section 6.1.6.

is incorporated into the synthesis part of the channel. Alternatively, if the complex constant were applied at the input, the output of the analysis stage would be $P_k X_n(e^{j\omega_k})$. In any case, if we consider the composite output of a system of N complex channels, we obtain

$$y(n) = \sum_{k=0}^{N-1} P_k y_k(n) = \sum_{k=0}^{N-1} P_k X_n(e^{j\omega_k}) e^{j\omega_k n} \qquad (6.83)$$

The composite impulse response of this system is

$$\tilde{h}(n) = \sum_{k=0}^{N-1} P_k h_k(n) = \sum_{k=0}^{N-1} |P_k| w_k(n) e^{j(\omega_k n + \phi_k)} \qquad (6.84)$$

Thus, in terms of time dependent Fourier analysis/synthesis, $X_n(e^{j\omega_k})$ is weighted by the complex sequence $\{P_k\}$, $k = 0, 1, \ldots, N-1$. In terms of a filter bank, the complex constant P_k provides for adjustment of the gain and phase of the individual filters in the bank.

The first step in design of a filter bank system (or a time-dependent Fourier analysis/synthesis system) is to choose the set of analysis frequencies $\{\omega_k\}$ for $0 \leqslant k \leqslant N-1$. In this choice, we are normally guided by a requirement on frequency resolution. For example, if we wish to resolve the voice fundamental and its harmonics, we need to use closely spaced analysis frequencies and lowpass or bandpass filters (equivalently analysis windows) with narrow enough bandwidths. In many cases, equally spaced analysis frequencies and equal bandwidth filters will be desired. Sometimes, however, it will be desirable to use a nonuniform distribution of analysis frequencies. This is the case, for example, in speech processing systems which attempt to take advantage of the decreasing sensitivity of the ear at higher frequencies. In any case, it is

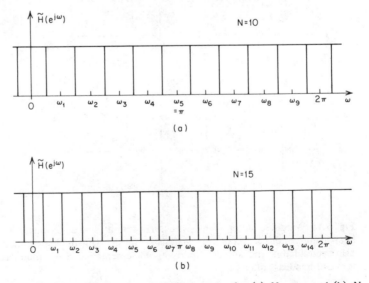

(a)

(b)

Fig. 6.19 Distributions of analysis frequencies for (a) N even; and (b) N odd filter banks.

284

common to choose the set of frequencies symmetrically across the band $0 \leqslant \omega < 2\pi$; i.e., so that $\omega_{N-k} = 2\pi - \omega_k$. This is depicted in Fig. 6.19 for N even and N odd. Note that for N even, a single channel is centered upon $\omega = \pi$. If in addition $w_k(n) = w_{N-k}(n)$, then we have seen that

$$X_n(e^{j\omega_k}) = X_n^*(e^{j(2\pi-\omega_k)}) = X_n^*(e^{-j\omega_k}) \tag{6.85}$$

This leads to a simplification of the filter bank so that only real bandpass filters are needed. Using the assumptions $\omega_k = 2\pi - \omega_{N-k}$, $P_k = P_{N-k}^*$ and $w_k(n) = w_{N-k}(n)$ in Eq. (6.84) it can easily be shown that for N even

$$\tilde{h}(n) = P_0 w_0(n) + \sum_{k=1}^{\frac{N}{2}-1} 2|P_k| w_k(n)\cos(\omega_k n + \phi_k)$$

$$+ P_{\frac{N}{2}} w_{\frac{N}{2}}(n)(-1)^n \tag{6.86}$$

and for N odd,

$$\tilde{h}(n) = P_0 w_0(n) + \sum_{k=1}^{\frac{N-1}{2}} 2|P_k| w_k(n)\cos(\omega_k n + \phi_k) \tag{6.87}$$

Thus for this case, we can consider the filter bank to consist of a lowpass filter, $P_0 w_0(n)$, and a set of bandpass filters with real impulse responses

$$h_k(n) = 2|P_k| w_k(n)\cos(\omega_k n + \phi_k) \tag{6.88}$$

When N is even, an additional highpass channel centered at $\omega = \pi$ is required to completely cover the entire frequency range $0 \leqslant \omega < 2\pi$. The impulse response of this filter is $P_{N/2} w_{N/2}(n)(-1)^n$.

Except for the constraint that analysis frequencies be placed symmetrically in the interval $0 \leqslant \omega < 2\pi$, there are no other constraints on the set of frequencies $\{\omega_k\}$. Once these frequencies are chosen, we must find a corresponding set of lowpass filters or analysis windows, $\{w_k(n)\}$, that have the desired frequency resolution and also provide the desired composite response.

Often it is only necessary to compute $X_n(e^{j\omega_k})$ at frequencies in a subband of the base frequency band $0 \leqslant \omega < 2\pi$. For example, analysis at $\omega_k = 0$ is often omitted since this part of the speech spectrum is not of interest in most speech processing schemes. Similarly, analysis at high frequencies (ω_k *close to* π) is often omitted since in the process of bandlimiting the input prior to sampling, this part of the spectrum is usually greatly attenuated and therefore it carries little reliable information.

To see the effect of omitting channels from the composite response let us return to the case of equally spaced analysis frequencies ($\omega_k = 2\pi k/N$). If we assume identical analysis windows ($w_k(n) = w(n)$) then from Eq. (6.84),

$$\tilde{h}(n) = w(n) \sum_{k=0}^{N-1} P_k e^{j\frac{2\pi}{N}kn} \tag{6.89}$$

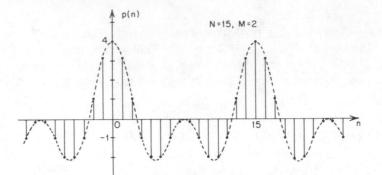

Fig. 6.20 Plot of $p(n)$ for a filter bank with $N = 15$, $M = 2$.

Thus, if we define

$$p(n) = \sum_{k=0}^{N-1} P_k e^{j \frac{2\pi}{N} kn} \tag{6.90}$$

then

$$\tilde{h}(n) = w(n)p(n) \tag{6.91}$$

as in Eq. (6.89). As before, $p(n)$ is seen to be periodic with period N; indeed the complex channel gains play the role of the coefficients in a discrete Fourier series. Note that if $P_k = 1$, $0 \leqslant k \leqslant N-1$, all channels are included and Eq. (6.89) becomes identical to Eq. (6.42). The effect of omitting channels is very conveniently observed by setting the appropriate P_k's equal to zero. For example, to exclude the zero frequency channel, $P_0 = 0$. To also exclude all channels above $\omega_M = 2\pi M/N$, $P_k = 0$ for $k > M$. In this case,

$$p(n) = \sum_{k=1}^{M} e^{j \frac{2\pi}{N} kn} + \sum_{k=N-M}^{N-1} e^{j \frac{2\pi}{N} kn}$$

$$= \sum_{k=1}^{M} \left[e^{j \frac{2\pi}{N} kn} + e^{-j \frac{2\pi}{N} kn} \right] \tag{6.92}$$

This expression can be placed in the more compact form

$$p(n) = \frac{\sin[\frac{\pi}{N} (2M+1)n]}{\sin[\frac{\pi}{N} n]} - 1 \tag{6.93}$$

As an example, Eq. (6.93) is plotted in Fig. 6.20 for the case $N = 15$ and $M = 2$. In this case, $p(n)$ is clearly periodic with period 15, but instead of single unit samples every 15 samples, $p(n)$ now consists of pulses whose amplitude and width depend upon N and M. It can be seen that if the channel at zero frequency is included the term -1 on the right hand side of Eq. (6.93) will disappear. Also if N is odd so that when all channels are included $M = (N-1)/2$, then $p(n)$ can also be expressed as

$$p(n) = \frac{\sin(\pi n)}{\sin(\frac{\pi}{N} n)} = N \sum_{r=-\infty}^{\infty} \delta(n-rN) \qquad (6.94)$$

That is, only in the case when all the channels are included is the sequence $p(n)$ such that a window $w(n)$ with zeros spaced at intervals of N samples will be sufficient to guarantee that

$$\tilde{h}(n) = \delta(n-r_0 N). \qquad (6.95)$$

This is of course reasonable since not all of the frequency spectrum is included. However, it is reasonable to suppose that over the range of frequencies included, the composite magnitude response will be flat and the composite phase will be linear. Indeed it should be clear that leaving out the zero frequency channel and a high frequency set of channels is equivalent to bandpass filtering. We shall see evidence of this in subsequent examples.

A final general consideration stems from the fact that many of the standard filter design methods do not permit simultaneous constraints on the frequency response and the time response. Thus, it may not be possible to obtain a lowpass filter whose impulse response has zeros every N samples. To see the effect of this, consider Fig. 6.21. Here, we assume for convenience that all channels are included so that $p(n)$ is a train of unit samples with period N. The envelope of the sequence $w(n)$ is shown as a continuous curve. The pro-

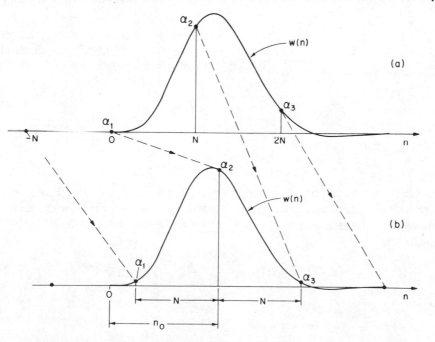

Fig. 6.21 Illustration of how to adjust the parameter n_0; (a) composite impulse response for $n_0 = 0$; (b) n_0 chosen to minimize magnitude and phase ripple (dotted lines indicate movement of individual pulses). (After Schafer and Rabiner [1].)

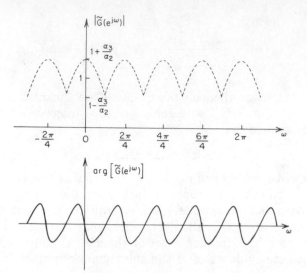

Fig. 6.22 Magnitude and phase of a composite filter bank.

duct of $p(n)$ and $w(n)$ is shown as the samples labelled α_1, α_2, α_3, etc. It is clear in this case that the composite impulse response is approximately

$$\tilde{h}(n) = \alpha_2\delta(n-N) + \alpha_3\delta(n-2N) \tag{6.96}$$

if we neglect the impulses at $3N$, $4N$, etc. In this case, the composite frequency response is of the form

$$\tilde{H}(e^{j\omega}) = \alpha_2 e^{-j\omega N} + \alpha_3 e^{-j\omega 2N}$$

$$= \alpha_2 e^{-j\omega N}(1 + \frac{\alpha_3}{\alpha_2} e^{-j\omega N})$$

$$= \alpha_2 e^{-j\omega N} \cdot \tilde{G}(e^{j\omega}) \tag{6.97}$$

where

$$\tilde{G}(e^{j\omega}) = 1 + \frac{\alpha_3}{\alpha_2} e^{-j\omega N} \tag{6.98}$$

represents the deviation of the composite frequency response from $\alpha_2 e^{-j\omega N}$, which would correspond to perfect reproduction of the input with delay of N samples and a scale factor of α_2. The magnitude and phase of $\tilde{G}(e^{j\omega})$ are respectively

$$|\tilde{G}(e^{j\omega})| = [1 + (\frac{\alpha_3}{\alpha_2})^2 + 2(\frac{\alpha_3}{\alpha_2})\cos \omega N]^{1/2} \tag{6.99}$$

and

$$\arg[\tilde{G}(e^{j\omega})] = \tan^{-1}\left[\frac{-\dfrac{\alpha_3}{\alpha_2}\sin \omega N}{1 + \dfrac{\alpha_3}{\alpha_2}\cos \omega N}\right] \tag{6.100}$$

288

These functions are sketched in Fig. 6.22 for the case $N = 4$. It can be seen that the magnitude of the composite response has a multiplicative error which in general oscillates with period $2\pi/N$. This is of course the spacing between filters. Indeed the peaks of $|\tilde{G}(e^{j\omega})|$ occur at the analysis frequencies $\omega_k = (2\pi/N)k$, and the valleys occur half way between; i.e., at the "crossover" points of the individual bandpass filters. The size of the ripple in the composite response depends upon the relative sizes of α_2 and α_3. Likewise the phase exhibits a deviation from linear that is also periodic in frequency with period $2\pi/N$. Notice that both the magnitude and phase error disappear when $\alpha_3 = 0$, and the errors are maximum when $\alpha_3 = \alpha_2$.

Several approaches to reducing this error in the composite response are suggested by Fig. 6.21. One possibility is to use a somewhat shorter window, $w(n)$; i.e., make α_3 smaller by making $w(n)$ smaller at $n = 2N$. This unfortunately will have the effect of broadening the Fourier transform of $w(n)$, and thus the frequency resolution of the individual channels will be sacrificed. A second similar possibility is to make N larger while keeping $w(n)$ the same. A third possibility is suggested by Fig. 6.21b. If we simply shift the infinite sequence of impulses relative to $w(n)$, we obtain a sequence of three impulses instead of two; i.e.,

$$\tilde{h}(n) = \alpha_1' \delta(n-n_0) + \alpha_2' \delta(n-N-n_0) + \alpha_3' \delta(n-2N-n_0) \qquad (6.101)$$

The composite frequency response is

$$\tilde{H}(e^{j\omega}) = e^{-j\omega(n_0+N)}[\alpha_1' e^{j\omega N} + \alpha_2' + \alpha_3' e^{-j\omega N}] \qquad (6.102)$$

If we define as before

$$\tilde{G}'(e^{j\omega}) = \frac{\alpha_1'}{\alpha_2'} e^{j\omega N} + 1 + \frac{\alpha_3'}{\alpha_2'} e^{-j\omega N} \qquad (6.103)$$

then

$$\tilde{H}(e^{j\omega}) = \alpha_2' e^{-j\omega(n_0+N)} \tilde{G}'(e^{j\omega}) . \qquad (6.104)$$

It can be seen that if $\alpha_1' = \alpha_3'$, then

$$\tilde{G}'(e^{j\omega}) = \frac{2\alpha_3'}{\alpha_2'} \cos \omega N + 1 \qquad (6.105)$$

that is, there is no phase error. The amplitude error is plotted in Fig. 6.23 for $N = 4$. In the first case the peak-to-peak error is $2\alpha_3/\alpha_2$. In the second case, the peak-to-peak error is $4\alpha_3/\alpha_2$. Thus, in addition to the fact that the phase error is zero if $\alpha_1' = \alpha_3'$, if

$$\frac{2\alpha_3'}{\alpha_2'} < \frac{\alpha_3}{\alpha_2} , \qquad (6.106)$$

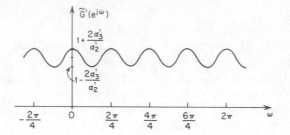

Fig. 6.23 Amplitude for composite filter banks when $p(n)$ is shifted relative to $w(n)$.

then the magnitude error will also be smaller.

The mechanism for shifting $p(n)$ relative to $w(n)$ is available in Eq. (6.90). To obtain an infinite sequence of impulses shifted by n_0, we simply let

$$P_k = e^{-j \frac{2\pi}{N} k n_0}, \quad 0 \leqslant k \leqslant N-1 \tag{6.107}$$

in which case,

$$p(n) = \sum_{k=0}^{N-1} e^{j \frac{2\pi}{N} k(n-n_0)} = N \sum_{r=-\infty}^{\infty} \delta(n-n_0-rN) \tag{6.108}$$

If some of the channels are omitted by setting some of the P_k's to zero, we obtain a shifted pulse train rather than impulses, but in any case by adjusting the phase of the channels according to Eq. (6.107) it is possible to shift $p(n)$ relative to $w(n)$. The phase adjustment can be accomplished as depicted in Fig. 6.18.

So far we have discussed some general principles of filter bank design. We now consider some examples of IIR and FIR filter banks.

6.2.2 Filter bank design using IIR filters

The design of filter banks using IIR filters is in Ref. [1] which shows that the general principle of adjusting the phases of the individual channels according to Eq. (6.107) can be useful in optimizing the design of IIR filter banks.

The following example is taken from Ref. [1]. It is assumed that the input sequence is obtained by sampling at a rate of 10000 samples/sec. A filter bank is to be designed with a uniform spacing of 100 Hz (in analog terms) between filters. This implies that $N = 10,000/100 = 100$ and that the analysis frequencies are $\omega_k = 2\pi k/100$, $k = 0, 1, \ldots, M$. The range of analysis is to be between 100 Hz and 3000 Hz. This means that 30 channels will be required; i.e., $M = 30$. The objective of flat composite amplitude response and linear phase is most easily achieved if the individual filters have these same properties. For this reason, Bessel (maximally flat delay) filters were used for the individual filters. For this example, a digital lowpass filter was derived from a 6th order Bessel analog filter by the impulse invariance method. Figure 6.24 shows the response properties of the basic analysis filter. The envelope of the lowpass impulse response is shown in Fig. 6.24a. Note that 30 msec

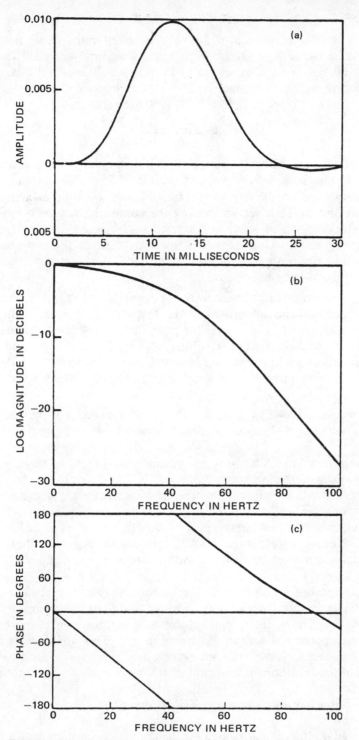

Fig. 6.24 Sixth-order Bessel filter characteristics; (a) impulse response; (b) magnitude response; (c) phase response. (After Schafer and Rabiner [1].)

291

corresponds to 300 samples at a 10 kHz sampling rate. Also note that the phase in Fig. 6.24c is quite linear. The nominal asymptotic cutoff frequency of the filter is 60 Hz. This filter was used to design a 30 channel filter bank with 100 Hz spacing between channels. The zero frequency channel was omitted. Using Eq. (6.93) with $M = 30$ and $N=100$, we see that

$$p(n) = \frac{\sin[0.61\pi n]}{\sin[0.01\pi n]} - 1 \qquad (6.109)$$

Note that $p(n)$ is periodic with period $N = 100$ samples. The resulting composite response characteristics are shown in Fig. 6.25. The composite impulse response shows the broadening of the impulse that is to be expected when all the channels are not included in forming the composite response. Also we see that since the duration of $w(n)$ is greater than $2N = 200$ samples, $\tilde{h}(n)$ exhibits two strong peaks. This leads to significant ripple in the composite magnitude and phase response; specifically, about 4 dB ripple in the magnitude and 25 degrees peak-to-peak phase error.

To improve the performance without changing $w(n)$, sufficient delay was included to equalize the amplitudes of the first and third peaks. This is shown in Fig. 6.26. As seen in Fig. 6.26a, the peak in $p(n)$ at $n = 0$ has been shifted 129 samples to the right. The resulting composite magnitude and phase are much improved as seen in Figs. 6.26b and 6.26c respectively. In this case the magnitude ripple is about 0.8 dB and the peak-to-peak phase error is only 0.6 degrees.

The above example suggests a trial and error procedure that will generally produce a filter bank with good composite response.

1. Determine the filter spacing and number of filters required to cover the desired analysis band.
2. Design a filter which gives the desired frequency selectivity for each channel. This yields $w(n)$.
3. Evaluate $w(n)$ and determine n_0 such that $\alpha_1' = \alpha_3'$ as in Fig. 6.21.
4. Evaluate the response. If it is not satisfactory, the filter spacing or bandwidth must be changed and the process repeated.

This procedure is appropriate for uniformly spaced analysis frequencies. Reference [1] also discusses an approach to the design of nonuniformly spaced filter banks in which the filter bank is broken down into several sub-banks, each of which is composed of uniformly spaced filters. This approach is reasonably successful; however, vastly superior results can be obtained using FIR filters. Thus we shall not belabor the question of IIR filter bank design further.

6.2.3 Filter bank design using FIR filters

FIR digital filters are attractive for design of speech filter banks for several reasons. First, such filters can be designed to have precisely linear phase simply by imposing the constraint

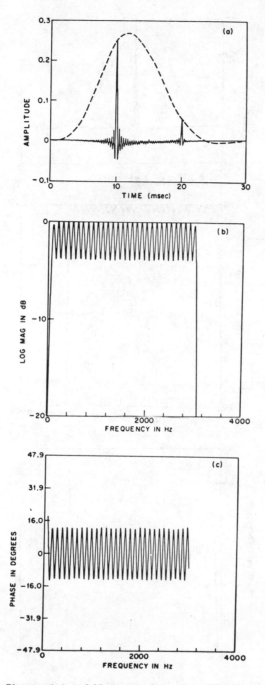

Fig. 6.25 Characteristics of 30 channel filter bank; (a) impulse response (dotted curve is the impulse response of the prototype lowpass filter); (b) composite magnitude response; (c) composite phase response after subtracting 10 msec delay. (After Schafer and Rabiner [1].)

293

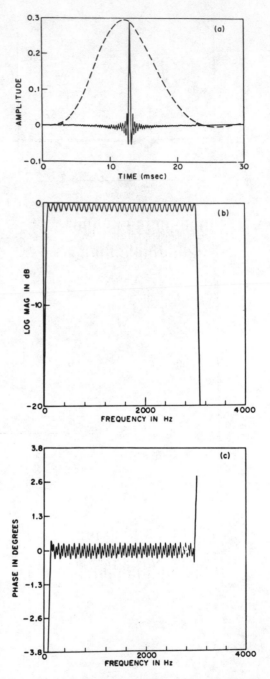

Fig. 6.26 Characteristics of 30-channel filter bank; (a) impulse response for $n_0 = 129$ (dotted curve is the impulse response); (b) composite magnitude response; (c) composite phase response after subtracting 12.9 msec delay. (After Schafer and Rabiner [1].)

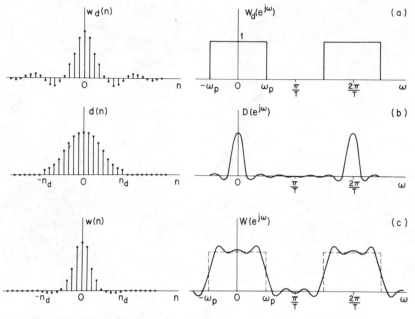

Fig. 6.27 Window design of an ideal lowpass filter. (After Schafer et al. [2].)

$$w(n) = w(L-1-n), \quad 0 \leqslant n \leqslant L-1 \qquad (6.110)$$

(on each individual filter band[8]) where $w(n)$ is the impulse response of the filter and L is its length in samples. This means that the criterion of linear phase for the composite filter bank response is trivially met if the individual filters have identical linear phase characteristics. Therefore, it is possible to focus attention on achieving arbitrary frequency selective properties for the individual filters and on obtaining the desired flat response for the composite filter bank. The second advantage of FIR filters is that there exist a variety of design methods ranging from the straightforward windowing method to iterative approximation methods that allow great flexibility in realizing complicated design specifications.

The window design method [2] appears to have a number of advantages for design of the lowpass or bandpass FIR filters for use in filter banks. This method is depicted in Fig. 6.27. Before considering the details of filter bank design let us review the window design method. First, a desired ideal lowpass filter of the form

$$W_d(e^{j\omega}) = e^{-j\omega n_d} \quad |\omega| \leqslant \omega_p$$

$$= 0 \qquad otherwise \qquad (6.111)$$

[8]It is assumed, for simplicity, that the impulse response of each bandpass filter is of duration L samples, although it is trivial to remove this restriction by adding appropriate delays for each channel.

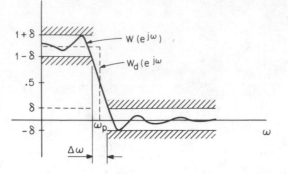

Fig. 6.28 Resulting lowpass design from windowing. (After Schafer et al. [2].)

is defined by choosing the cutoff frequency ω_p. Note that for simplicity we have omitted in the figure the linear phase term $e^{-j\omega n_d}$ corresponding to a delay of n_d samples. The value of n_d required is $n_d = (L-1)/2$. This means that if L is even, the delay corresponds to a noninteger number of samples. The ideal impulse response is therefore

$$w_d(n) = \frac{1}{2\pi} \int_{-\omega_p}^{\omega_p} e^{-j\omega n_d} e^{j\omega n} d\omega = \frac{\sin[\omega_p(n-n_d)]}{\pi(n-n_d)}, \quad all\ n \quad (6.112)$$

This impulse response is infinite in extent and must be truncated to obtain an FIR filter. This is done by defining

$$w(n) = d(n-n_d)w_d(n), \quad (6.113)$$

where $d(n)$ is a filter design window function, and $w(n)$ is the impulse response of the resulting lowpass filter.[9] The length of the window, denoted by L, can be either an even integer ($L=2q$) or an odd integer ($L=2q+1$). Figure 6.27 shows the case when L is odd.

The result of multiplying the ideal lowpass impulse response by the design window corresponds to a convolution in the frequency domain of the ideal frequency response and the Fourier transform, $D(e^{j\omega})$, of the design window, i.e.,

$$W(e^{j\omega}) = \frac{1}{2\pi} \int_{-\pi}^{\pi} W_d(e^{j\theta}) D(e^{j(\omega-\theta)}) d\theta \quad (6.114)$$

The result of this convolution is depicted in Fig. 6.27c. It can be seen that the main effects are the introduction of a smooth transition between the passband and the stopband and the introduction of ripples in the passband and the stopband regions. The properties of this approximation are depicted in Fig. 6.28. If ω_p is larger than the width of the "main lobe" of $D(e^{j\omega})$, then the following set of properties are generally true:

---

[9]It is important not to become confused with terminology here. We have called $d(n)$ the filter design window, and $w(n)$ the impulse response of a lowpass filter; $w(n)$ is also the window for time-dependent Fourier analysis.

296

1. The transition region, $\Delta\omega$, is inversely proportional to L.
2. The function $W(e^{j\omega})$ is very nearly antisymmetric about the point $(\omega_p=0.5)$.
3. The peak approximation errors in the passband and stopband are very nearly equal.
4. The approximation error is greatest in the vicinity of ω_p and it decreases for values of ω in both directions away from ω_p.

The above properties of the windowing design method are true of all the commonly used windows. However, Kaiser has proposed a family of window functions that are very flexible and nearly optimum for filter design purposes [5]. Specifically, the Kaiser design window is

$$d(n) = \frac{I_0[\alpha\sqrt{1-(n/n_d)^2}]}{I_0(\alpha)} \quad |n| \leqslant n_d$$

$$= 0 \qquad\qquad otherwise \qquad\qquad (6.115)$$

where $n_d = (L-1)/2$ and $I_0[x]$ is the modified zero order Bessel function of the first kind. By adjusting the parameter α, one can trade off between transition width and peak approximation error. Furthermore, Kaiser [5] has formalized the window design procedure by giving the empirical design formula

$$L = \frac{-20\log_{10}\delta - 7.95}{14.36\,\Delta f} + 1 \qquad (6.116a)$$

where L is the filter order, δ is the peak approximation error, and Δf is the normalized transition width

$$\Delta f = \frac{\Delta\omega}{2\pi} . \qquad (6.116b)$$

To use this formula δ and Δf are fixed at values which provide the desired frequency selectivity. Then Eq. (6.116a) can be used to compute L and the parameter α can be computed from the equation [5]

$$\alpha = 0.1102(-20\log_{10}\delta - 8.7), \qquad \text{for } -20\log_{10}\delta > 50$$

$$= 0.5842(-20\log_{10}\delta - 21)^{0.4}$$

$$+ 0.07886(-20\log_{10}\delta - 21), \qquad \text{for } 21 < -20\log_{10}\delta < 50 \quad (6.117)$$

In the present application of this design method, the choice of δ and Δf depends upon the specifications of the bandpass filters that constitute the filter bank.

Recall that the filter bank consists of a set of bandpass filters with impulse responses of the form

$$h_k(n) = P_k w_k(n) e^{j\omega_k n}, \quad 0 \leqslant k \leqslant N-1 \qquad (6.118)$$

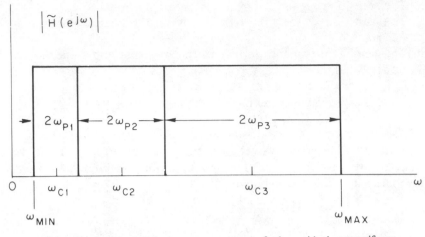

Fig. 6.29 Composite frequency response of three ideal nonuniform bandpass filters.

where $w_k(n)$ is a lowpass filter impulse response. Figure 6.29 shows three ideal bandpass filters whose composite response is perfectly flat over the range $\omega_{min} \leqslant \omega \leqslant \omega_{max}$. In designing the lowpass filters, we choose the set of analysis frequencies $\{\omega_k\}$ and the set of cutoff frequencies $\{\omega_{pk}\}$ so as to exactly cover the desired band as depicted in Fig. 6.29. The ideal bandpass filters are then approximated by the window design method.

The choice of peak approximation error for the filter depends upon how much stopband attenuation is deemed necessary in a given application. Typical values of $-20 \log_{10} \delta$ would most likely be between 40 and 60 dB. Using Eq. (6.117) the appropriate value of α can be computed. Finally, the normalized transition width Δf must be fixed in order to compute L from Eq. (6.116a). Again the choice of Δf (or $\Delta \omega$) is governed by consideration of the desired frequency selectivity for the individual filters. Clearly, the transition width $\Delta \omega_k$ should not be more than $2\omega_{pk}$.

In the filter bank context we shall require that $\Delta \omega$ be the same for all filters so that we can take advantage of property 2 discussed above. That is, if all the filters have identical transition regions and, furthermore, if these transitions are antisymmetric about the crossover points, then we can expect that the sum of the frequency responses will be very close to unity. This implies that $d(n)$ should be the same for each filter in the filter bank.

To see the full implications of choosing the filter design window, $d(n)$, to be the same for each frequency channel, let us consider the composite frequency response which as we see from Eq. (6.118) is

$$\tilde{H}(e^{j\omega}) = \sum_{k=0}^{N-1} P_k W_k(e^{j(\omega - \omega_k)}) \tag{6.119}$$

Now if the same design window is used for each analysis frequency, we can write

298

$$W_k(e^{j(\omega-\omega_k)}) = \frac{1}{2\pi} \int_{-\pi}^{\pi} W_{dk}(e^{j(\theta-\omega_k)}) D(e^{j(\omega-\theta)}) d\theta \qquad (6.120)$$

If we substitute Eq. (6.120) into Eq. (6.119) and interchange the order of summation and integration, we obtain

$$\tilde{H}(e^{j\omega}) = \frac{1}{2\pi} \int_{-\pi}^{\pi} \left[\sum_{k=0}^{N-1} P_k W_{dk}(e^{j(\theta-\omega_k)}) \right] D(e^{j(\omega-\theta)}) d\theta \qquad (6.121)$$

If we define

$$\tilde{H}_d(e^{j\omega}) = \sum_{k=0}^{N-1} P_k W_{dk}(e^{j(\omega-\omega_k)}) \qquad (6.122)$$

Then Eq. (6.121) can be written as

$$\tilde{H}(e^{j\omega}) = \frac{1}{2\pi} \int_{-\pi}^{\pi} \tilde{H}_d(e^{j\theta}) D(e^{j(\omega-\theta)}) d\theta \qquad (6.123)$$

The function $\tilde{H}_d(e^{j\omega})$ is seen to be simply the desired composite frequency response. If we assume, for example, that $P_k = 1$ for $0 \leqslant k \leqslant N-1$, and that the bandwidths and center frequencies of $W_{dk}(e^{j(\omega-\omega_k)})$ are such that the entire frequency range $-\pi \leqslant \omega \leqslant \pi$ is covered, then

$$\tilde{H}_d(e^{j\omega}) = e^{-j\omega n_d}, \quad -\pi \leqslant \omega \leqslant \pi \qquad (6.124)$$

Substituting this into Eq. (6.123) gives

$$\tilde{H}(e^{j\omega}) = \frac{1}{2\pi} \int_{-\pi}^{\pi} e^{-j\theta n_d} D(e^{j(\omega-\theta)}) d\theta$$

$$= d(n_d) e^{-j\omega n_d} \qquad (6.125)$$

This implies that the composite impulse response is

$$\tilde{h}(n) = d(n_d)\delta(n-n_d) \qquad (6.126)$$

Thus, if sufficient channels are included so that the composite desired frequency response is flat with linear phase, then the actual composite response of the filter bank which uses filters all designed with the same window is also ideal. That is, independent of how the center frequencies and bandwidths are distributed, the composite response is ideal no matter what filter design window is used as long as the same window is used for all the channels. Thus perfect reproduction of the input is theoretically possible using FIR filters with an arbitrary distribution of center frequencies and bandwidths.

The effect of failing to include portions of the frequency range $-\pi \leqslant \omega \leqslant \pi$ can be easily seen by noting that Eq. (6.123) holds regardless of how the P_k's are chosen in Eq. (6.122). Thus, if both the low and high frequency regions are omitted from analysis as in Fig. 6.29, then $P_0 = 0$ and $P_k = 0$ for $k > M$, where M is the number of channels included. In general the desired composite response will be of the form

$$\tilde{H}_d(e^{j\omega}) = e^{-j\omega n_d} \qquad \omega_{min} \leqslant |\omega| \leqslant \omega_{max} \qquad (6.127)$$

Thus, for a given design window, $d(n)$, and ω_{min} and ω_{max}, the composite response will be a bandpass filter with transition regions and passband and stopband ripples identical to those of the individual channel filters. This is because the same filter design window multiplies each individual ideal impulse response. Thus, the composite response is again independent of the number and distribution of the individual bandpass filters.

The design of filter banks according to the above principles is illustrated by the following examples.

6.2.3a Example 1

Suppose that the input sampling rate is 9.6 kHz and that we wish to design a bank of 15 equally spaced filters that covers the range 200 Hz to 3200 Hz. The cutoff frequency for all of the lowpass filters is

$$F_p = \frac{\omega_p}{2\pi T} = \frac{3200 - 200}{2(15)} = 100 \text{ Hz} \qquad (6.128)$$

and the center frequencies are

$$F_k = \frac{\omega_k}{2\pi T} = 200 \ k + 100 \qquad k = 1, 2, \ldots, 15 \qquad (6.129)$$

With this choice of center frequencies and bandwidths, the 15 ideal filters exactly cover the interval from 200 to 3200 Hz. If we assume that 60 dB attenuation is required outside the transition regions of each channel, we find from Eq. (6.117) that $\alpha = 5.65326$. Since the cutoff frequency is 100 Hz for all the prototype lowpass filters, the widest transition band that is reasonable is 200 Hz. Using this value and $-20 \log_{10}\delta = 60$ in Eq. (6.116a) we obtain $L = 175$ as the lowest reasonable value for L. Note that if lower attenuation is acceptable, then L can be smaller for the same Δf.

The filter bank designed with the above parameters is shown in Fig. 6.30. The top part of the figure shows the individual bandpass filters. Note how the falloff in the upper transition band of a given filter compliments the ascent of the next filter. Also note that adjacent channels cross at an amplitude value of 0.5. The lower half of the figure shows the composite magnitude response of the filter bank. The phase is of course linear with $n_d = (175-1)/2 = 87$ samples delay. It is clear that the filters merge together very well at the edges of the frequency bands. Indeed the deviation from unity is less than or equal to the peak approximation error $\delta = 0.001$ that was used in designing the prototype lowpass filters.

6.2.3b Example 2

A nonuniform spacing of the filters is often used to exploit the ear's decreasing frequency resolution with increasing frequency. Suppose that we

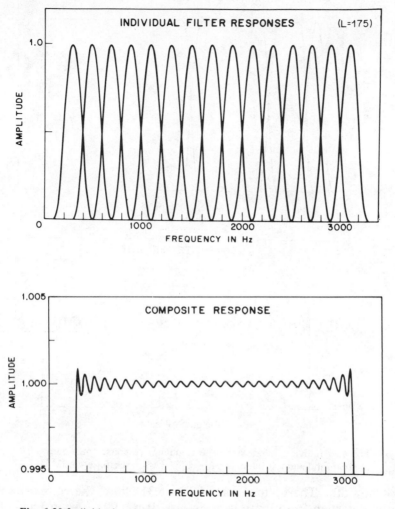

Fig. 6.30 Individual and composite frequency responses of a bank of 15 uniform bandpass filters for $L = 175$. (After Schafer et al. [2].)

wish to cover the same range 200 Hz to 3200 Hz as in Example 1; however, we wish to use only four octave band filters. That is, each successive filter will have a bandwidth twice the bandwidth of the previous filter. This implies that the frequency range from 200 Hz to 3200 Hz must be divided into four bands of width 200, 400, 800, and 1600 Hz with center frequencies 300, 600, 1200, and 2400 Hz, respectively. Again assuming that 60 dB attenuation is required, we note that the smallest lowpass cutoff frequency is 100 Hz so that the smallest reasonable transition width is 200 Hz. This leads again to a minimum value of $L = 175$. The filter bank corresponding to these design parameters is shown in Fig. 6.31. The complementary relationship between the ascending and descending transitions between adjacent filters is noted at the top part of the figure. It is seen that since L and α are the same for each of the prototype lowpass designs, the shape of the curves in the transition region is independent

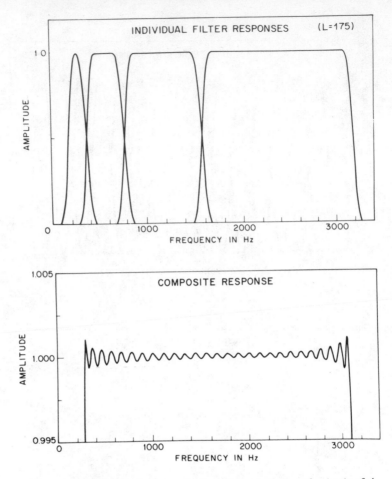

Fig. 6.31 Individual and composite frequency responses of a bank of 4 nonuniform bandpass filters for $L = 175$. (After Schafer et al. [2].)

of the bandwidth. The bottom part of Fig. 6.31 shows the composite response where the deviation from unity is again less than 0.001. As in Example 1, the phase is linear corresponding to 87 samples delay.

A comparison of Figs. 6.30 and 6.31 verifies the fact that the same composite response is obtained in both cases.

6.2.3c Example 3

Suppose that all the parameters remain the same as in Example 2 except that we require narrower transition regions. This means that a larger value of L is required. In fact Eq. (6.116a) shows that L and Δf are roughly inversely proportional. Figure 6.32 shows the filter bank corresponding to the parameters of Example 2 except that $L = 301$ and $\Delta f = 0.012082$ (transition width is 116 Hz). The sharper transitions are apparent in the top part of the figure and the lower part shows that the composite response remains very flat. In this case the delay is 150 samples.

302

6.3 Implementation of the Filter Bank Summation Method Using the Fast Fourier Transform

In the preceding section we have shown that it is possible, using causal filters, to design a filter bank whose composite output is identical to the input except for a delay and scale factor. In particular, we saw that finite impulse response filters are particularly well suited to this purpose. Since time-dependent Fourier analysis and synthesis is equivalent to such a filter bank, it is also true that finite duration analysis windows can be used effectively in the design of analysis/synthesis systems. One of the major disadvantages of FIR systems is the extensive computation required to implement them. Fortunately in the particular context of time-dependent Fourier analysis, there are several methods for reducing the computation over that required for straightforward implementation.

6.3.1 Analysis techniques

Consider a time-dependent Fourier analysis/synthesis system with equally spaced analysis frequencies $\omega_k = 2\pi k/N$, $0 \leqslant k \leqslant N-1$. In Section 6.1.3 it

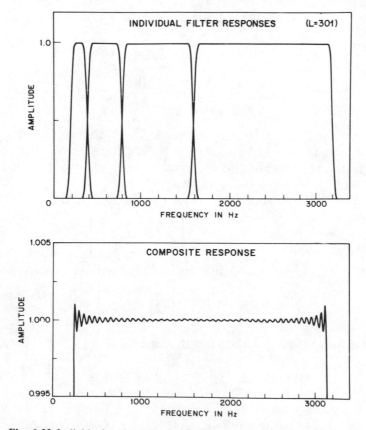

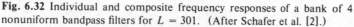

Fig. 6.32 Individual and composite frequency responses of a bank of 4 nonuniform bandpass filters for $L = 301$. (After Schafer et al. [2].)

303

was shown that the sequences $X_n(e^{j\omega_k})$ need not be computed at the same rate as the input sampling rate because each sequence $X_n(e^{j\omega_k})$ is effectively the output of a lowpass filter with digital cutoff frequency π/N. Thus the output can be computed only once for each N consecutive samples of the input. FIR systems are especially suited to this application because it is possible to compute only the desired output samples without computing the intervening $N-1$ samples. With IIR systems, the inherent recursive nature of the implementation requires that all values of the output be computed.

An additional improvement in computational efficiency can be obtained by the use of Fast Fourier Transform (FFT) techniques [6]. To see this, we express the time-dependent Fourier transform as

$$X_n\left(e^{j\frac{2\pi}{N}k}\right) = \sum_{m=-\infty}^{\infty} x(m)\,w(n-m)\,e^{-j\frac{2\pi}{N}km} \qquad 0 \leqslant k \leqslant N-1 \quad (6.130)$$

We observe that if the limits of summation were 0 to $N-1$, Eq. (6.130) would be in the form of a discrete Fourier transform. If $w(m)$ is of finite duration, Eq. (6.130) can be manipulated into the form of a DFT and thus an FFT algorithm can be employed in computing $X_n(e^{j\,2\pi k/N})$ for $0 \leqslant k \leqslant N-1$. By a substitution of variable of summation, Eq. (6.130) becomes

$$X_n\left(e^{j\frac{2\pi}{N}k}\right) = e^{-j\frac{2\pi}{N}kn} \sum_{m=-\infty}^{\infty} x_n(m)\,e^{-j\frac{2\pi}{N}km} \qquad (6.131)$$

where

$$x_n(m) = x(n+m)\,w(-m), \qquad -\infty < m < \infty \qquad (6.132)$$

That is, the sequence $x_n(m)$ is obtained by redefining the origin of the sequence $x(m)\,w(n-m)$ to be at sample n, thus focusing our attention on the sequence in the neighborhood of the time at which $X_n(e^{j\,2\pi k/N})$ is to be computed. Next, by a substitution $m = Nr + q$, $-\infty < r < \infty$ and $0 \leqslant q \leqslant N-1$, we can express Eq. (6.131) as the double sum

$$X_n\left(e^{j\frac{2\pi}{N}k}\right) = e^{-j\frac{2\pi}{N}kn} \sum_{r=-\infty}^{\infty} \left[\sum_{q=0}^{N-1} x_n(Nr+q)\right]e^{-j\frac{2\pi}{N}k(Nr+q)} \qquad (6.133)$$

Since $e^{-j2\pi kr} = 1$, we can interchange the order of summations and obtain

$$X_n\left(e^{j\frac{2\pi}{N}k}\right) = e^{-j\frac{2\pi}{N}kn} \sum_{q=0}^{N-1} \left[\sum_{r=-\infty}^{\infty} x_n(Nr+q)\right]e^{-j\frac{2\pi}{N}kq} \qquad (6.134)$$

If we define $u_n(q)$ to be the finite length sequence

$$u_n(q) = \sum_{r=-\infty}^{\infty} x_n(Nr+q), \qquad 0 \leqslant q \leqslant N-1 \qquad (6.135)$$

then we observe that $X_n(e^{j2\pi k/N})$ is simply $e^{-j2\pi kn/N}$ times the N-point DFT of the sequence $u_n(q)$. Alternatively, $X_n(e^{j2\pi k/N})$ is the N-point DFT of the sequence $u_n(q)$ after a circular shift of n modulo N. That is,

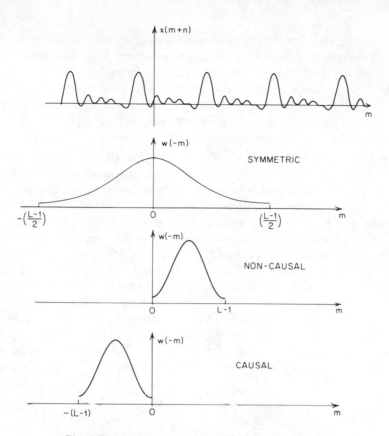

Fig. 6.33 Plots of the sequences $x(m+n)$ and $w(-m)$.

$$X_n\left(e^{j\frac{2\pi}{N}k}\right) = \sum_{m=0}^{N-1} u_n((m-n))_N e^{-j\frac{2\pi}{N}km} \tag{6.136}$$

where the notation $((\;))_N$ means that the integer inside the double set of parentheses is to be interpreted modulo N. Thus, we have succeeded in manipulating $X_n(e^{j2\pi k/N})$ into the form of an N-point DFT of a finite length sequence that is derived from the windowed input sequence. In summary, the procedure for computing $X_n(e^{j2\pi k/N})$ for $0 \leqslant k \leqslant N-1$ is:

1. Form the sequence $x_n(m)$ as in Eq. (6.132) by multiplying $x(m+n)$ by the reversed window sequence $w(-m)$. Figure 6.33 shows $x(m+n)$ and three special cases of $w(-m)$.
2. Break the resulting sequence up into segments of length N samples and add these segments together according to Eq. (6.135) to produce the finite length sequence $u_n(q)$, $0 \leqslant q \leqslant N-1$.
3. Circularly shift $u_n(q)$ by n modulo N to produce $u_n((m-n))_N$, $0 \leqslant m \leqslant N-1$.
4. Compute the N-point DFT of $u_n((m-n))_N$ to produce $X_n(e^{j2\pi k/N})$, $0 \leqslant k \leqslant N-1$.

This procedure must be repeated at each value of n at which $X_n(e^{j2\pi k/N})$ is desired; however, it is clear that n can be incremented in any desired manner. For example $X_n(e^{j2\pi k/N})$ can be computed for $n = 0, \pm R, \pm 2R, \ldots$; i.e., at intervals of R samples of the input signal. This, we recall, is justified since $X_n(e^{j2\pi k/N})$ is the output of a lowpass filter whose nominal cutoff frequency is π/N radians. Thus, as long as $R \leqslant N$, the "samples" of $X_n(e^{j2\pi k/N})$ will suffice to reconstruct the input signal.

Note that this method gives $X_n(e^{j2\pi k/N})$ for all values of k. In general, because of the conjugate symmetry of $X_n(e^{j\omega})$, at most only about half of the channels need be computed. Also, as we have seen, often the very low frequency and high frequency channels are not implemented. Thus, the question arises as to whether the FFT method is more efficient than direct implementation. To compare, let us assume that we only require $X_n(e^{j2\pi k/N})$ for $1 \leqslant k \leqslant M$. Further, assume that the window is of length L. Then to obtain the complete set of values of $X_n(e^{j2\pi k/N})$ would require $4LM$ real multiplications and about $2LM$ real additions using the method of Fig. 6.12. Assuming a rather straightforward complex FFT algorithm where N is a power of two,[10] it can be shown that the FFT method would require $L + 2N \log_2 N$ real multiplications and $L + 2N \log_2 N$ real additions to obtain all N values of $X_n(e^{j2\pi k/N})$. If we take the number of real multiplications as the basis for comparison, and assume that $L = N$, then it is easily shown that the FFT method requires less computation unless

$$M \leqslant \frac{\log_2 N}{2} \tag{6.137}$$

For example, if $N = 128 = 2^7$, we see that the FFT method is more efficient than the direct method unless $M \leqslant 3.5$; i.e., for fewer than 4 channels. Thus, in any application where fine frequency resolution is required, it is almost certain that the FFT method would be most efficient. (Note that if $L > N$, the comparison is even more favorable to the FFT method.)

6.3.2 Synthesis techniques

The previous discussion of analysis techniques showed that by using a fast Fourier transform algorithm we can compute all N equally spaced values of $X_n(e^{j2\pi k/N})$ with an amount of computation that is less than that required to compute M channels using a direct implementation: By rearranging the computations required for synthesis, a similar savings can occur in reconstructing $x(n)$ from values of $X_n(e^{j2\pi k/N})$ which are available at every R samples of $x(n)$, where $R \leqslant N$ [7].

From Eq. (6.83) with $\omega_k = 2\pi k/N$, the output of the synthesis system is

$$y(n) = \sum_{k=0}^{N-1} Y_n(k) e^{j \frac{2\pi}{N} kn} \tag{6.138}$$

[10]This does not take advantage of the fact that $u_n((m-n))_N$ is real. This could be used to reduce computation by another factor of 2.

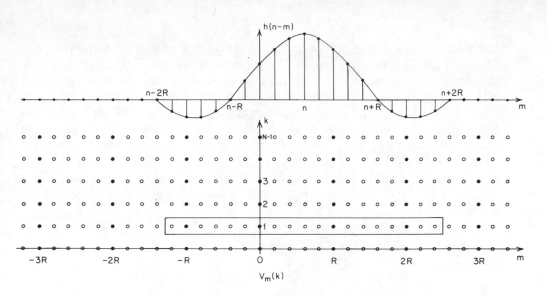

Fig. 6.34 Samples involved in computing $Y_n(k)$.

where

$$Y_n(k) = P_k X_n\left(e^{j\frac{2\pi}{N}k}\right), \quad 0 \leqslant k \leqslant N-1 \tag{6.139}$$

Recall that the complex weighting coefficients, P_k, permit adjustments of the magnitude and phase of the channels. If $X_n(e^{j2\pi k/N})$ is available only at integer multiples of R, then the intermediate values must be filled in by interpolation. To do this it is helpful to define the sequence

$$V_n(k) = P_k X_n\left(e^{j\frac{2\pi}{N}k}\right) \qquad n = 0, \pm R, \pm 2R, \ldots$$

$$= 0 \qquad\qquad\qquad otherwise \tag{6.140}$$

There is a sequence of the above form for each value of k. Now, for each value of k, the intermediate values are filled in by processing the sequence $V_n(k)$ with a lowpass filter with a cutoff frequency of π/N radians. If we denote the unit sample response of this filter by $h(n)$, and assume that it is symmetrical with total length $2RQ - 1$, then for each value of k, for $0 \leqslant k \leqslant N-1$,

$$Y_n(k) = \sum_{m=n-RQ+1}^{n+RQ-1} h(n-m) V_m(k), \quad -\infty < n < \infty \tag{6.141}$$

Equation (6.141) together with Eq. (6.138) describes the operations required to compute the output of the synthesis stage when the time dependent Fourier transform is available at intervals of R samples. This process is illustrated in Fig. 6.34 which depicts $V_m(k)$ as a function of m and k. Remember that m is the time index and k is the frequency index. The heavy dots denote points at which $V_m(k)$ is nonzero, i.e., the points at which $X_m(e^{j2\pi k/N})$ is known. The open circles denote points at which $V_m(k)$ is zero, and at which we desire to

307

interpolate the values of $Y_n(k)$. The impulse response of the interpolating filter (for $Q=2$) is shown positioned at a particular time n. Each channel signal is interpolated by convolution with the impulse response of the interpolation filter. As an example, the samples involved in the computation of $Y_3(1)$ are enclosed in a box. In general, the box indicating which samples are involved in computing $Y_n(k)$ will slide along the kth row of Fig. 6.34, with the center of the box being positioned at n. Notice that each interpolated value is dependent upon $2Q$ of the known values of $X_n(e^{j2\pi k/N})$. If we assume that M channels are available for synthesis, then it easily shown that $2(Q+1)M$ real multiplications and $2QM$ real additions are required to compute each value of the output sequence, $y(n)$.

To see how the output can be computed more efficiently, let us substitute Eq. (6.141) into Eq. (6.138), obtaining

$$y(n) = \sum_{k=0}^{N-1} \sum_{m=n-RQ+1}^{n+RQ-1} h(n-m) V_m(k) e^{j\frac{2\pi}{N}kn} \tag{6.142}$$

Interchanging the order of summation gives

$$y(n) = \sum_{m=n-RQ+1}^{n+RQ-1} h(n-m) v_m(n) \tag{6.143}$$

where

$$v_m(r) = \sum_{k=0}^{N-1} V_m(k) e^{j\frac{2\pi}{N}kr} \tag{6.144}$$

Using Eq. (6.140) it can be seen that

$$v_m(r) = \sum_{k=0}^{N-1} P_k X_m\left(e^{j\frac{2\pi}{N}k}\right) e^{j\frac{2\pi}{N}kr}, \qquad m = 0, \pm R, \pm 2R, \ldots$$

$$= 0 \qquad\qquad\qquad\qquad otherwise \tag{6.145}$$

Thus we see that rather than interpolate the time-dependent Fourier transform, and then evaluate Eq. (6.138), we can instead compute $v_m(r)$ at each time at which $X_n(e^{j2\pi k/N})$ is known (i.e., $m = 0, \pm R, \ldots$), and then interpolate $v_m(r)$ as in Eq. (6.143).

It can be seen that $v_m(r)$ is in the form of an inverse discrete Fourier transform, and thus $v_m(r)$ is periodic in r with period N. Thus, in Eq. (6.143), the index n in the term $v_m(n)$ must be interpreted modulo N. The interpolation process implied by this is depicted in Fig. 6.35. The heavy dots in the two-dimensional net of points represent the points at which $v_m(r)$ is nonzero. The remaining points in Fig. 6.35 can be interpolated in the same manner described for interpolation of $Y_n(k)$. However, we do not need to interpolate all these points since all that is required are the values of the one-dimensional sequence $y(n)$. From Eq. (6.143) and the periodic nature of $v_m(r)$, it can be

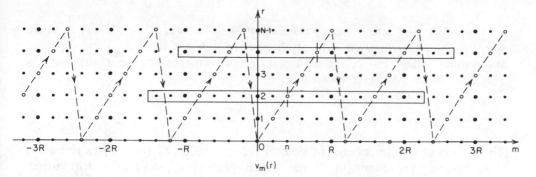

Fig. 6.35 The interpolation process for $v_m(r)$. (After Portnoff [7].)

seen that $y(n)$ is equal to the values of the interpolated sequence along the "saw tooth" pattern in Fig. 6.35.

In implementing the synthesis system in this manner, the N-point sequences $v_m(r)$, $0 \leqslant r \leqslant N-1$ can be computed for each value of m at which $X_m(e^{j2\pi k/N})$ is known by using a fast Fourier transform algorithm to perform the inverse DFT computation of Eq. (6.145). Note that a channel can be omitted simply by setting its value equal to zero before computing the inverse discrete Fourier transform. Likewise, if it is desirable to implement a linear phase shift by choosing $P = e^{-j2\pi kn_0/N}$, it easily shown that the effect is to simply circularly shift the sequences $v_m(r)$. Thus, multiplications by the factor $e^{-j2\pi kn_0/N}$ can be avoided by performing the inverse discrete Fourier transform operation directly upon $X_m(e^{j2\pi k/N})$ and then circularly shifting the result by the desired n_0 samples. Once the sequences $v_m(r)$ are obtained then the output can be computed by interpolating $v_m(r)$ as in Eq. (6.143). For each value of $y(n)$, $2Q$ values of $v_m(r)$ are involved. The samples involved for two different values of n are shown enclosed in a box in Fig. 6.35. It should be clear that for R consecutive values of $y(n)$, the values of $v_m(r)$ are obtained from the same $2Q$ columns. Thus it is convenient to compute the output in blocks of R samples.

The amount of computation required to implement time-dependent Fourier synthesis in the above manner can again be estimated by assuming that N is a power of 2 and that a straightforward complex FFT algorithm is used to compute the inverse transforms called for in Eq. (6.145). For this assumption, the synthesis requires $(2QR+2N\log_2 N)$ real multiplications and $(2QR-1+2N\log_2 N)$ real additions to compute a group of R consecutive values of the output, $y(n)$. The direct method of synthesis requires $2(Q+1)MR$ real multiplications and $2QMR$ real additions to compute R consecutive samples of the output. If we consider the situation when the direct method requires fewer multiplications than the FFT method, we find that

$$M < \frac{Q + \dfrac{N}{R}\log_2 N}{Q + 1} \tag{6.146}$$

For typical values of $N = 128$, $Q = 2$ (interpolation over 4 samples as in Fig.

309

6.35), and $R = N$ (the lowest possible sampling rate for $X_n(e^{j2\pi k/N})$) then we see that the direct method is most efficient only when $M < 3$. Thus for most applications, the FFT offers significant improvements in the computational efficiency of the synthesis operation.

6.4 Spectrographic Displays

The notion of a time-dependent Fourier representation of speech was prevalent long before the advent of digital signal processing techniques for speech analysis. Indeed, speech researchers have relied heavily upon spectrum analysis techniques since the 1930's. One of the earliest embodiments of the time-dependent Fourier representation was the sound spectrograph, a device that has become an essential tool in almost every phase of speech research. In this device, a short (2 second) speech utterance repeatedly modulates a variable frequency oscillator. The modulated signal is input to a bandpass filter. The average energy in the output of the bandpass filter at a given time and frequency is a crude measure of the time-dependent Fourier transform. This energy is recorded by an ingenious electromechanical system on teledeltos paper. The result, called a spectrogram, is a two-dimensional representation of the time-dependent spectrum in which the vertical dimension on the paper represents frequency and the horizontal dimension represents time. The spectrum magnitude is represented by the darkness of the marking on the paper. If the bandpass filter has a wide bandwidth (300 Hz) the spectrogram displays good temporal resolution and poor frequency resolution. On the other hand, if the bandpass filter has a narrow bandwidth (45 Hz), the spectrogram has good frequency resolution and poor time resolution. Examples are shown in Fig. 6.36.

Figure 6.36a shows a wideband spectrogram of the utterance "Every salt breeze comes from the sea." This example illustrates a number of characteristic features of wideband time-dependent spectra. First, we observe that at a particular time, the spectrum varies with frequency in a manner suggested by Figs. 6.3 and 6.5; i.e., the spectrum consists of a few broad peaks corresponding to the formant frequencies. The spectrogram clearly displays the variation of the formant frequencies with time. Another interesting feature of the wideband spectrogram is the vertical striations that appear in regions of voiced speech. These are due to the fact that the impulse response of the analyzing filter (ie., the spectrum analysis window) is of about the same duration as the pitch period. Thus, the energy in the filter output is maximum when the peak of the impulse response is aligned with the maximum of each individual pitch period. At other times, the output energy is significantly less. For unvoiced speech, which is not, of course, periodic, the vertical striations do not appear and the spectral pattern is much more ragged.

Figure 6.36b is a narrowband spectrogram of the same utterance. In this case, the bandwidth of the filter is such that individual harmonics are resolved in voiced regions. Thus, while formant frequencies are still in evidence, a

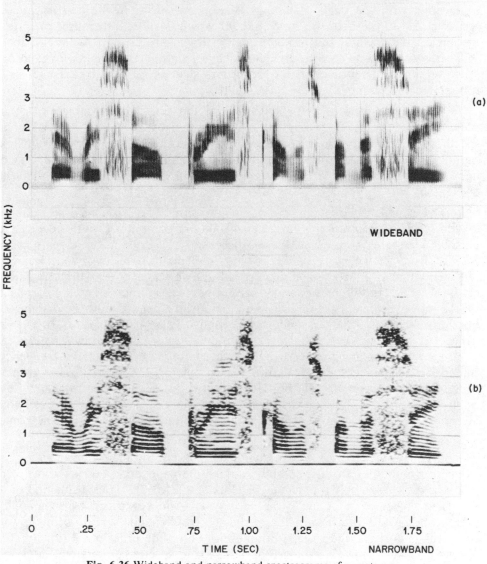

FREQUENCY (kHz)

(a)

WIDEBAND

(b)

TIME (SEC)   NARROWBAND

Fig. 6.36 Wideband and narrowband spectrograms of a sentence.

cross-section at a particular time is reminiscent of Figs. 6.2 and 6.4. No longer is the pattern striated in the voiced regions, since the narrowband impulse response spans several pitch periods; but, rather, the frequency dimension now clearly places in evidence the fundamental frequency and its harmonics. Unvoiced regions are distinguished by a lack of periodicity in the frequency dimension.

The wideband and the narrowband spectrograms display a great deal of information about the properties of a speech utterance. Indeed, when apparatus for displaying such time-dependent Fourier representations first became available, it was hoped that such displays could provide a new "language" for communicating with the deaf. Although this hope was not realized, subsequent research lead to the book *Visible Speech* [8] which is still a rich source of information on the spectral and temporal properties of speech. In the years since this early work, many speech researchers have made measurements by hand on spectrograms to determine speech parameters such as formant frequencies and fundamental frequency.

Another outgrowth of the invention of the sound spectrograph was the notion that a speaker's identity could be revealed by a detailed analysis of a spectrogram or "voiceprint" of a speech utterance. Although there remains significant question as to the reliability of voice identification techniques based upon spectrograms [9], these techniques have gained some acceptance in courts of law [10].

The sound spectrograph was for a long time the basic analysis tool in speech research. However, with the availability of computer facilities dedicated to speech research, this is no longer the case. The previous sections of this chapter have shown ways to design and implement time-dependent Fourier representations of much greater sophistication than was ever possible using analog hardware. These representations can, of course, be implemented either as a special purpose digital hardware or as a program in a general purpose computer. For example, using the techniques of Section 6.3, we can obtain $X_n(e^{j2\pi k/N})$ which is a complex two-dimensional representation of the speech signal which is discrete in time and frequency and furthermore is periodic in the frequency dimension. Thus, we are faced with the problem of how to display such a representation. Generally, all the information is not needed in a display. Often only $|X_n(e^{j2\pi k/N})|$ would be displayed. Also, since $|X_n(e^{j2\pi k/N})|$ is even and periodic in k with period N, it is only necessary to display values in the range $0 \leqslant k \leqslant N/2$.

When a device such as a graphics oscilloscope or incremental plotter is available for output from a computer, the time-dependent Fourier transform can be plotted as simply a sequence of plots of $|X_n(e^{j2\pi k/N})|$ as a function of k for fixed values of n. Usually the values of n will be spaced by an amount corresponding to Nyquist sampling of the spectrum channels. For example, for narrowband analysis, the spacing in time may be on the order of 10 to 20 msec. Figure 6.37 [11] shows a sequence of narrowband spectra computed at intervals of 20 msec. It is clear from Fig. 6.37 that the entire interval of speech is voiced.

An alternative to displaying the spectra as sections through the surface defined by $|X_n(e^{j2\pi k/N})|$ is to display that surface in a perspective drawing. An example of this type of display is shown in Fig. 6.38 [12]. Clearly, this plot is less useful for quantitative measurements, but has the virtue, like the spectrogram, of displaying the entire utterance in a compact form.

In view of the demonstrated usefulness and wide acceptance of the spectrogram as a basic tool, a digitally generated spectrogram is probably more use-

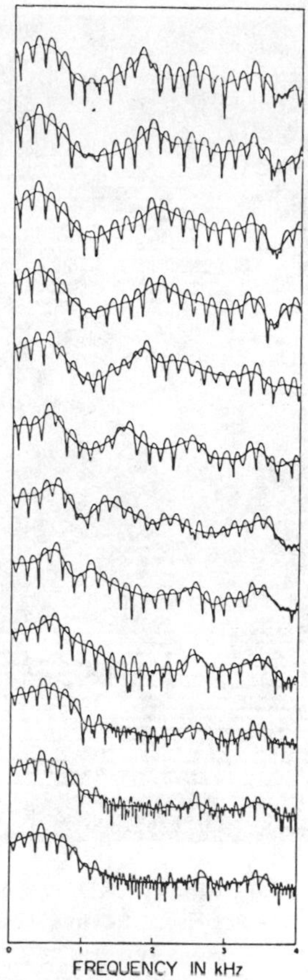

FREQUENCY IN kHz

Fig. 6.37 Sequence of narrowband spectra of a voiced section of speech.
(After Schafer and Rabiner [11].)

ful than either of the former displays. If a TV or CRT display is available to output sampled images,[11] then $|X_n(e^{j2\pi k/N})|$ for an appropriate sized interval can be thought of as just such a sampled image. A number of researchers have investigated such outputs and have found that it is possible to duplicate the appearance of analog spectrograms. Indeed, since the teledeltos paper only has a gray scale range of 12 dB [13], a rather crude quantization of the values of $|X_n(e^{j2\pi k/N})|$ can be used in the display if the objective is to duplicate spectrogram appearance. However, most digital image display systems have a much greater dynamic range so that more of the spectral information may be portrayed than with the analog system.

Another advantage of the digital spectrogram is that the spectrum can be conveniently shaped in sophisticated ways to enhance the usefulness of the display. An example is the use of high frequency emphasis to counteract the

---

[11]Usually this is achieved using a dedicated auxiliary memory from which the display is refreshed. However, L. R. Morris has discussed techniques for displaying spectrograms using only the memory and output capability of a standard minicomputer. (IEEE Trans. on Acoustics, Speech, and Signal Proc., June 1975.)

313

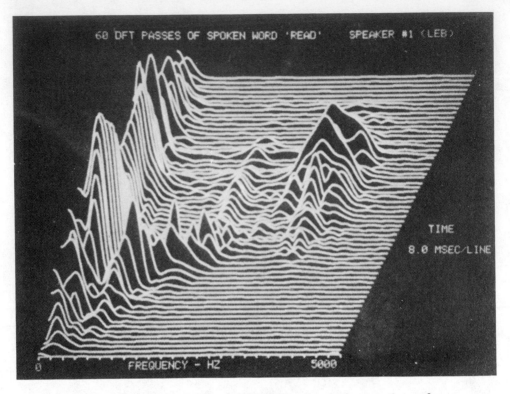

Fig. 6.38 Spectrogram of the word "READ" computed from contiguous 8 msec speech segments. (After Tufts, Levinson, and Rao [12].)

natural fall-off of the speech spectrum. (This is also used in analog spectrographs.) A simple way to introduce high frequency emphasis is to compute the spectrum of the first difference of the input signal. (See Problem 6.11.) Another more flexible way is to directly shape the spectrum prior to display. This latter approach was used by Oppenheim [14] in producing computer generated spectrograms similar to the one shown in Fig. 6.39. Oppenheim also points out that one has a great deal of flexibility in displaying the spectrum data. For example, the frequency and time dimensions can be expanded or contracted at will.

Still another approach to producing spectrograms by computer is required when no image output capability is available. If a printing device is available with strikeover capability, it is possible to obtain a gray scale range comparable to that of an analog spectrogram by representing each darkness level by a set of superimposed printer chatacters. An example of this type of output is shown in Fig. 6.40. The details of procedures for producing such plots are given in [15].

6.5 Pitch Detection

We have seen that in a narrowband time-dependent Fourier representation, the excitation for voiced speech is manifested in sharp peaks that occur at integer multiples of the fundamental frequency. This fact has served as the basis of a

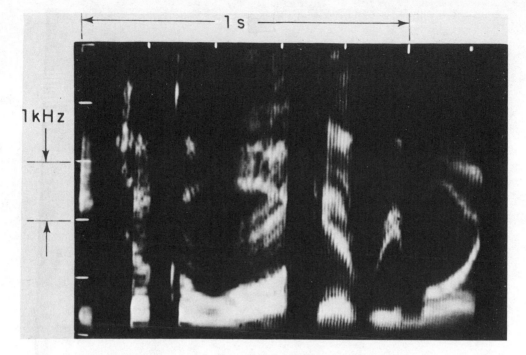

1kHz

Fig. 6.39 An example of a spectrogram produced using the short-time spectrum and a computer graphics system. (After Oppenheim [14].)

number of pitch detection schemes. In this section we shall briefly discuss an example of a pitch detector based upon time-dependent spectrum analysis. This example illustrates both the basic concepts of using the short-time spectrum for pitch detection and the flexibility afforded by digital processing methods. It will be clear to the thoughtful reader that many more possibilities exist for exploiting the time-dependent Fourier representation in determining excitation parameters. (Another example is suggested by Problem 6.14.)

One approach involves the computation of the *harmonic product spectrum* [16] which is defined as

$$P_n(e^{j\omega}) = \prod_{r=1}^{K} |X_n(e^{j\omega r})|^2 \qquad (6.147)$$

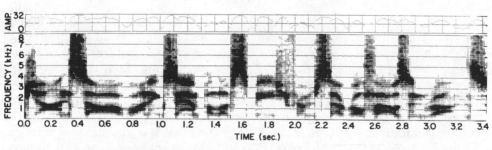

Fig. 6.40 800-point DFT spectrogram. (After Silverman and Dixon [15].)

315

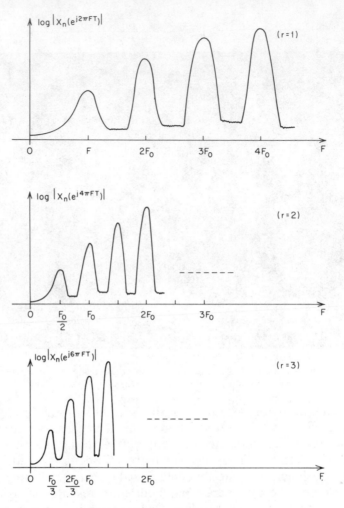

Fig. 6.41 Representations of terms in the log harmonic product spectrum.

Taking the logarithm gives the *log harmonic product spectrum*

$$\hat{P}_n(e^{j\omega}) = 2 \sum_{r=1}^{K} \log |X_n(e^{j\omega r})| \qquad (6.148)$$

The function $\hat{P}_n(e^{j\omega})$ is seen to be a sum of K frequency compressed replicas of $\log|X_n(e^{j\omega})|$. The motivation for using the function of Eq. (6.148) is that for voiced speech, compressing the frequency scale by integer factors should cause harmonics of the fundamental frequency to coincide at the fundamental frequency. At frequencies between harmonics, some of the frequency compressed harmonics will coincide, but only at the fundamental will there always be reinforcement. This is depicted schematically in Fig. 6.41. Note that for the continuous function $|X_n(e^{j2\pi FTr})|$, the peak at F_o becomes sharper as r increases; thus, the sum of Eq. (6.148) will have a sharp peak at F_o with possibly some lesser peaks elsewhere. This technique has been found to be especially resistant to additive noise, since the contributions of the noise to $X_n(e^{j\omega})$ have no coherent structure when viewed as a function of frequency. Therefore, in Eq. (6.148) the noise components in $X_n(e^{j\omega r})$ also tend to add incoherently. For the same reason, unvoiced speech will not exhibit a peak in $\hat{P}_n(e^{j\omega})$. Another important point is that a peak at the fundamental frequency need not be present in $|X_n(e^{j\omega})|$ for there to be a peak in $\hat{P}_n(e^{j\omega})$. Thus, this

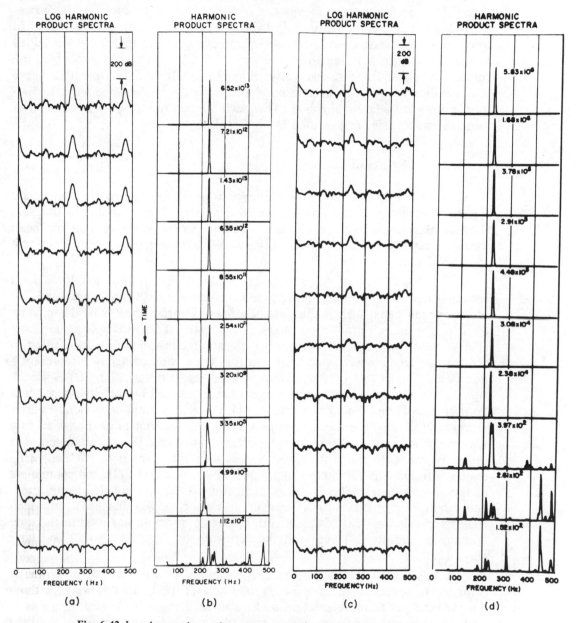

Fig. 6.42 Log harmonic product spectra and harmonic product spectra for (a) and (b) noise free; and (c) and (d) 0 dB *SNR*. (After Noll [17].)

method is attractive for operation on highpass filtered speech such as telephone speech.

An example of the use of this technique is shown in Fig. 6.42 [16]. The input speech was sampled at a 10 kHz sampling rate and every 10 msec the signal was multiplied by a 40 msec Hamming window (400 samples). Then values of $X_n(e^{j2\pi k/N})$ were computed using an FFT algorithm with $N = 2048$. Figures 6.42a and 6.42b show a sequence of log harmonic product spectra respectively for the case $K = 5$ in Eqs. (6.147) and (6.148). Figures 6.42c and 6.42d were computed using exactly the same parameters, except that noise was added to the input signal at a signal-to-noise ratio of 0 dB. The clarity with which the fundamental frequency stands out is remarkable. It is clear from this figure that a rather simple pitch estimation algorithm could be designed with the harmonic product spectrum as input. That such an algorithm should have superior noise resistance has been verified by Noll [17].

6.6 Analysis-by-Synthesis

In Sections 6.4 and 6.5 we have shown that the basic speech parameters are clearly in evidence in the time-dependent Fourier representation. In this section we shall consider a technique, called analysis-by-synthesis, that has been useful in estimating formant frequencies and in estimating the glottal waveform for voiced speech.

The basic idea of analysis-by-synthesis is the following. First it is assumed that we begin with the speech waveform or some other representation of the speech signal such as the time-dependent Fourier transform. Then some form of the speech production model is assumed. This model (e.g., terminal analog, vocal tract, etc.) has a number of parameters which can be adjusted to produce different speech sounds. From the model we can derive a representation of the model that is of the same form as the representation of the speech signal. For example, if the speech signal is represented by the time-dependent Fourier transform, then we would likewise obtain a time-dependent Fourier representation of the model. Then by varying the parameters of the model in a systematic way, we can, for example, attempt to find a set of parameters that cause the model to match the speech signal with minimum error. When such a match is found, the parameters of the model are assumed to be the parameters of the speech signal. This is a very general approach and not tied to the time-dependent Fourier transform. However, this principle was first used for speech analysis in this way [18]. Subsequently the same principle was used in the time domain by Pinson [19] and with the cepstrum by Oppenheim [20] and later Olive [21].

Some of the earliest reported applications of the analysis-by-synthesis principle to speech were done by a group at MIT [22]. In this work, a time-dependent Fourier representation was obtained using a bank of analog filters. The filter outputs were sampled and read into a digital computer. The resulting crude spectrum representation was then matched by an iterative procedure that adjusted the parameters of a model for speech production that included spectral components for the vocal tract transfer function, the glottal wave shape, and the radiation load. Although the algorithm for adjusting the model parameters was not completely automatic, this work showed the feasibility of the analysis-by-synthesis principle by producing excellent estimates of formant frequencies for voiced speech [22].

Possibly the most significant limitation of the scheme described by Bell et al. [18] was the use of an analog filter bank for the Fourier analysis. This limitation was not present in the scheme devised by Mathews, Miller, and David [23]. They began with samples of the speech signal and implemented the Fourier analysis using digital computations. Their approach introduced an additional new concept into the spectrum analysis of speech; namely the concept of pitch synchronous analysis of voiced speech. Although the work of Mathews et al. [23] occurred before many of the important advances in the implementation and understanding of discrete Fourier analysis, we shall take advantage of such knowledge in explaining their approach.

6.6.1 Pitch synchronous spectrum analysis

First let us recall from Chapter 3 that our digital model for voiced speech assumes that a short segment of voiced speech is identical to the same length segment from the periodic sequence

$$\tilde{x}(n) = \sum_{m=-\infty}^{\infty} h_v(n+mN_p) \tag{6.149}$$

where $h_v(n)$ represents the convolution of the vocal tract impulse response, $v(n)$, with the glottal pulse, $g(n)$, and the impulse response of the radiation load, $r(n)$. That is,

$$h_v(n) = r(n)*v(n)*g(n) \tag{6.150}$$

The quantity N_p is the pitch period in samples. The radiation effects, which basically appear as a differentiation at low frequencies, are adequately modelled for most purposes by a simple first difference, for which the z-transform representation is

$$R(z) = 1 - z^{-1} \tag{6.151}$$

The vocal tract is characterized by a transfer function of the form

$$V(z) = \frac{A}{\displaystyle\prod_{k=1}^{M} (1 - 2e^{-\sigma_k T}\cos(2\pi F_k T)z^{-1} + e^{-2\sigma_k T}z^{-2})} \tag{6.152}$$

where the number of poles included depends upon the sampling frequency of the input data. Finally, the glottal pulse is of finite duration, implying that the z-transform of $g(n)$ is a polynomial in z of the form

$$G(z) = \sum_{n=0}^{N_g} g(n)z^{-n} = B \prod_{n=1}^{N_g} (1-z_n z^{-1}) \tag{6.153}$$

319

where N_g is less than N_p. From Eq. (6.150) we observe that the z-transform of $h_v(n)$ is

$$H_v(z) = R(z) \cdot V(z) \cdot G(z) \qquad (6.154)$$

and the corresponding Fourier transform would be

$$H_v(e^{j\omega}) = R(e^{j\omega}) V(e^{j\omega}) G(e^{j\omega}) \qquad (6.155)$$

The Fourier transform of the periodic signal $\tilde{x}(n)$ will consist of very sharp spectral lines at multiples of the fundamental frequency.

The periodic signal $\tilde{x}(n)$ can be represented by a Fourier series of the form

$$\tilde{x}(n) = \frac{1}{N_p} \sum_{k=0}^{N_p-1} \tilde{X}(k) e^{j \frac{2\pi}{N_p} kn} \qquad (6.156)$$

where

$$\tilde{X}(k) = H_v(e^{j \frac{2\pi}{N_p} k}) \qquad (6.157)$$

By substituting Eqs. (6.156) and (6.157) into Eq. (6.1) it is easily shown that

$$\tilde{X}_n(e^{j\omega}) = \frac{1}{N_p} \sum_{k=0}^{N_p-1} H_v(e^{j \frac{2\pi}{N_p} k}) W_n(e^{j(\omega - \frac{2\pi k}{N_p})}), \qquad (6.158)$$

where $W_n(e^{j\omega})$ is the Fourier transform of the analysis window, $w(n-m)$. We have seen that the character of the time-dependent Fourier transform is strongly dependent on the length and shape of the analysis window. We recall from Figures 6.2 and 6.3 that from the point of view of estimating the parameters of the model (other than pitch), we have a dilemma. If we perform a narrowband analysis (i.e., long analysis window) the spectrum envelope information is obscured by the pitch peaks, while if we perform a wideband analysis (i.e., short analysis window), we find the formant peaks smeared out by convolution with the Fourier transform of the window. Furthermore, Eq. (6.158) suggests that even though $\tilde{x}(n)$ is periodic, $\tilde{X}_n(e^{j\omega})$ is a function of the window position. The approach suggested by Mathews et al. takes advantage of the fact that the Fourier series coefficients for a periodic signal such as Eq. (6.149) are simply as given by Eq. (6.157) and can also be computed directly from one period of $\tilde{x}(n)$. That is

$$\tilde{X}(k) = H_v(e^{j \frac{2\pi}{N_p} k}) = \sum_{n=0}^{N_p-1} \tilde{x}(n) e^{-j \frac{2\pi}{N_p} kn}, \quad 0 \leqslant k \leqslant N_p-1 \qquad (6.159)$$

Thus, by isolating one period of the periodic signal we can compute samples of $H_v(e^{j\omega})$ at N_p equally spaced values. When one uses one "period" of voiced speech in place of $\tilde{x}(n)$ in Eq. (6.159) the resulting time-dependent Fourier transform is termed a pitch synchronous time dependent Fourier transform. In general, this approach to voiced speech analysis is called pitch synchronous analysis.

320

This approach is completely consistent with our general discussion of time dependent Fourier analysis with only slight modifications. First, the times at which $X_n(e^{j\omega})$ is computed are dependent upon the voice pitch period. Since pitch varies with time, this requires a nonuniform sampling in the time dimension. Also, since the number of values of frequency that are obtained depends upon the pitch period, the sampling rate in the frequency dimension also is time dependent. The window used in this case is normally rectangular, i.e., a single "period" of the speech wave is isolated and then transformed using Eq. (6.159). As shown in Problem 6.15 this is consistent with Eq. (6.158) because for a rectangular window of length N_p, $W(e^{j\omega})$ has zeros spaced at multiples of $2\pi/N_p$. Thus, if Eq. (6.158) is evaluated at frequencies $\omega_k = 2\pi k/N_p$, then

$$\tilde{X}_n(e^{j\frac{2\pi}{N_p}k}) = H_v(e^{j\frac{2\pi}{N_p}k}) \qquad 0 \leqslant k \leqslant N_p - 1 \qquad (6.160)$$

This approach avoids the problems of pitch in the frequency domain by coping with it first in the time domain. Thus, the time ambiguity of the narrowband spectrum is avoided, and the frequency dimension smearing of the wideband analysis is avoided by being content with an accurate estimate of the spectrum at only N_p samples.

6.6.2 Pole-zero analysis using analysis-by-synthesis

With this pitch synchronous spectrum as a starting point, Mathews et al. [23] used an iterative procedure to estimate the parameters of the speech model. They used an analog model for the transfer function of the radiation load, the vocal tract, and the glottal pulse. This necessitated a higher pole correction factor which probably would not have been necessary had Eqs. (6.151) through (6.154) been used. The basic approach remains the same regardless of the particular functional form for the speech model, so we shall continue our exposition using the digital model. Details of the functions used are given in Ref. [23].

The parameters of $H_v(e^{j\omega})$ can be determined by an iterative approximation process. Mathews et al. guessed a set of parameters for $H_v(e^{j\omega})$, computed values at frequencies $2\pi k/N_p$, and then evaluated an error function of the form:

$$E = \sum_k Q(k)[\log|H_v(e^{j\frac{2\pi}{N_p}k})| - \log|X_n(e^{j\frac{2\pi}{N_p}k})|]^2 \qquad (6.161)$$

where $Q(k)$ is a weighting function on the spectrum and $X_n(e^{j2\pi k/N_p})$ is the pitch synchronous spectrum of the speech signal. The parameters were adjusted in a systematic way so as to minimize the error function. The rules for adjusting the poles of $V(z)$ and the zeros of $G(z)$ are discussed in Ref. [23]. Figure 6.43 shows some examples of spectrum matches obtained by Mathews et al. [23]. When the error is minimized, the resulting values of the poles of $V(z)$ are taken as estimates of the formant frequencies. The zero locations give information about the glottal wave.

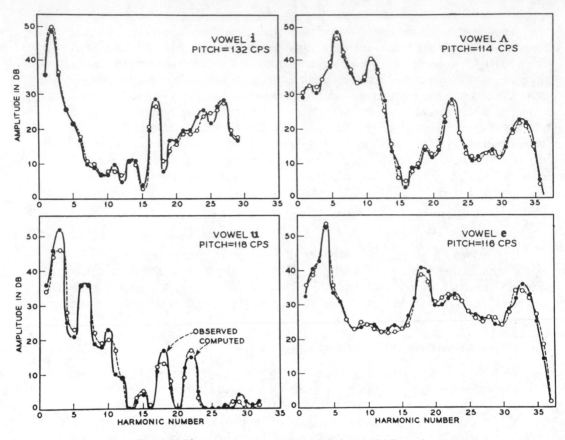

Fig. 6.43 Pitch synchronous spectral matches; solid lines show observed spectra; dotted lines show fitted spectra. (After Mathews, Miller and David [23].)

6.6.3 Pitch synchronous estimation of the glottal wave

The work reported by Mathews, Miller, and David was primarily concerned with the distribution of zeros of the approximations, and attempts were made to relate the spacing and arrangement of the zeros to the shape of the glottal pulse. Later work by Miller and Mathews (unpublished) modified the earlier technique to obtain estimates of the glottal pulse. In this case, the model was of the form

$$H_v(z) = R(z)\,G_f(z)\,V(z) \qquad (6.162)$$

where in this case the glottal wave contributions to the spectrum were initially modelled by a fixed transfer function whose equivalent digital form would be

$$G_f(z) = \frac{1}{(1-az^{-1})(1-bz^{-1})} \qquad (6.163)$$

Again, the parameters of $V(z)$ were varied systematically to minimize a similar error criterion. The resulting pole locations serve as an estimate of the formant

322

frequencies. To obtain the glottal wave shape for the particular period of speech being analyzed, Miller and Mathews computed the quantity

$$\tilde{G}(k) = \frac{X_n(e^{j\frac{2\pi}{N_p}k})}{R(e^{j\frac{2\pi}{N_p}k})V(e^{j\frac{2\pi}{N_p}k})} \qquad 0 \leqslant k \leqslant N_p-1 \qquad (6.164)$$

The values of $\tilde{G}(k)$ for $0 \leqslant k \leqslant N_p-1$ are used as the Fourier coefficients of the glottal pulse, $\tilde{g}(n)$, which is computed using the inverse DFT; i.e.,

$$\tilde{g}(n) = \frac{1}{N_p} \sum_{k=0}^{N_p-1} \tilde{G}(k)e^{j\frac{2\pi}{N_p}kn} \qquad (6.165)$$

Note that this is feasible since $\tilde{g}(n)$ is a finite duration pulse, even though the N_p samples of $H_v(e^{j\omega})$ that are obtained by pitch synchronous analysis would in general not be adequate to completely specify $h_v(n)$ which is in general longer than N_p. Thus with the aid of the model for speech production it is possible to extract the component of the convolution which is of finite duration. This technique was used with considerable success by Rosenberg [24] in a study of the effect of glottal pulse shape on vowel quality. Figure 6.44 shows an example of a speech wave and the corresponding glottal wave that was extracted by the above procedure.

This technique for estimating the glottal pulse is sensitive to the model that is assumed. In cases where the model fits the speech signal well, as in the steady state vowels, the results are excellent. In other situations a more complex model is required. Another factor that affected the results was the way in which the pitch period was isolated in the speech waveform. It was found that considerable care was needed in determining the beginning and end of the period. In fact, the speech signal was interpolated to a higher sampling rate to facilitate the exact location of the beginning of each cycle. It is not surprising that this is required in view of the fact that in general it is very unlikely that the exact point of glottal opening (and closing) would occur at a sampling instant.

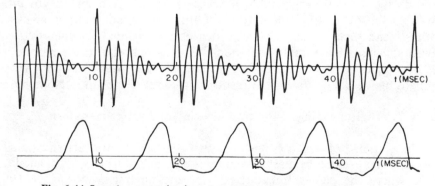

Fig. 6.44 Speech output (top) and analyzed excitation waveform (bottom) for vowel in HOD. (After Rosenberg [24].)

323

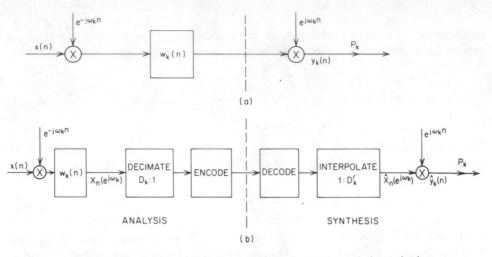

Fig. 6.45 Block diagram of the coding for a single analysis-synthesis channel.

6.7 Analysis-Synthesis Systems

So far in this chapter, we have discussed the basic theory of time-dependent Fourier analysis and synthesis, and have shown that the time-dependent Fourier transform is a useful basis for a variety of schemes for estimating the parameters of the model for speech production. However, applications of the most important result of this chapter have so far not been considered. That is, we have not discussed the practical implications of the fact that the speech signal can be exactly recovered from the time-dependent Fourier representation. This fact is the basis for an important class of speech coding schemes called vocoders. The main purpose of vocoder systems is to obtain a digital representation of speech at a much lower bit-rate than is possible with waveform coding schemes. Other applications of vocoders include the removal of additive noise or the effects of reverberation, and the modification of basic speech parameters to permit alteration of the time or frequency dimensions of a speech signal.

In this section we shall discuss several speech coding systems that are based on the theoretical principles of Sections 6.1-6.3. We shall begin with systems that are direct implementations of time-dependent Fourier analysis and synthesis and then discuss systems, such as the channel vocoder that can be viewed as approximate implementations. This is opposite to the chronology of the development of vocoders, however, this approach has the virtue of highlighting the degradations that result from expediencies in implementation.

6.7.1 Digital coding of the time-dependent Fourier transform

In Sections 6.1 and 6.2, it was shown that speech (in fact any signal) can be exactly represented by a set of bandpass channels of the type depicted in Fig. 6.45a, where the center frequencies, ω_k, and analysis windows, $w_k(n)$, are selected to cover the desired frequency band, and the complex constants,

$P_k = |P_k|e^{j\phi_k}$, are chosen so that the composite response of the sum of all the channels is as close as possible to the ideal of perfectly flat amplitude response and exactly linear phase. In Section 6.1.3, it was shown that since $w_k(n)$ corresponds to the impulse response of a lowpass filter, the time-dependent Fourier transform at frequency ω_k can be sampled at a lower rate than the input signal. In fact, it was shown that the total required number of samples/sec of $X_n(e^{j\omega_k})$ could be made equal to the sampling rate of the input signal. Thus to reduce the computation rate required to implement the analysis, and to reduce the bit-rate of the time-dependent Fourier transform representation, the channel signals are "sampled" at a much lower rate, and quantized and encoded for transmission or storage. This is depicted for a single channel in Fig. 6.45b. The analysis operations, which are shown to the left of the dotted line, consist of a modulator followed by a lowpass filter, followed by a decimator, and then an encoder. If $w_k(n)$ is a finite length sequence, the operation of decimation is simply incorporated into the linear filtering operation, i.e., the output is simply computed every D_k samples of the input. The encoding of the channel signals involves quantization and encoding as discussed in Chapter 5. In synthesis, the digital representation is first decoded, and then a quantized version of $X_n(e^{j\omega_k})$ is computed by interpolation. Note that it is possible to use a lower sampling rate for synthesis of the output waveform than the original input sampling rate if high-frequency channels are omitted. Thus, the interpolation factor D_k may be smaller than the decimation factor D_k. The quantized time-dependent Fourier transform channel signal, $\hat{X}_n(e^{j\omega_k})$, modulates a complex sinusoid to produce the signal $\hat{y}_k(n)$ which is then added to the other outputs to produce

$$\hat{y}(n) = \sum_{k=0}^{N-1} P_k \hat{y}_k(n) \tag{6.166}$$

To illustrate the practical considerations in coding speech in this way, we will consider an example from Ref. [6]. The computation of $X_n(e^{j\omega_k})$ was implemented using the FFT as discussed in Section 6.3. Since the FFT program required that N be a power of 2, the input was sampled at the rather

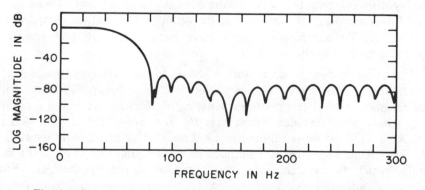

Fig. 6.46 Frequency response of the analysis window. (After Schafer and Rabiner [6].)

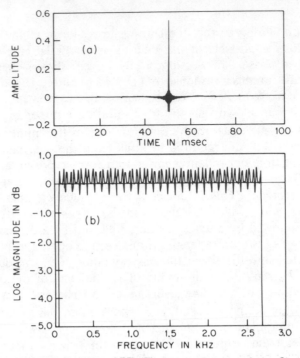

Fig. 6.47 Composite time and frequency response of the filter bank. (After Schafer and Rabiner [6].)

unusual rate of 12195 samples/sec. The value of N was 128, so that the analysis frequencies were $\omega_k = 2\pi k/128$, which correspond to analog frequencies, $F_k = (95.273k)$ Hz. The analysis window, $w(n)$ (the same for each channel), was the impulse response of a linear phase FIR filter of length 731 samples that was designed by frequency sampling methods [25]. The frequency response of this filter is shown in Fig. 6.46. Note that above 80 Hz, the attenuation of the filter is at least 60 dB. Using appropriate complex constants P_k, it was possible to obtain a composite response as shown in Fig. 6.47. Note that $P_0 = 0$, and $P_k = 0$ for $28 < k < 100$. Since the frequency band covered by the synthesizer only included frequencies up to about 2690 Hz, a sampling rate of 10004 samples/sec was used at the output. (An output sampling rate as low as about 6000 samples/sec could have been used with a correspondingly sharper analog filter at the output.)

The effect of design parameters is shown in the spectrograms of Fig. 6.48. An input speech utterance is shown in Fig. 6.48a and Fig. 6.48b shows the output of an analysis-synthesis system consisting of 28 channels of the form of Fig. 6.18b, where the complex constants P_k are unity for $1 \leqslant k \leqslant 28$ and $100 \leqslant k \leqslant 127$ and zero otherwise; i.e., no special phase compensation was used. The channel signals were sampled at the Nyquist rate (i.e., 160 times per second), which assured accurate reconstruction at the synthesizer. Furthermore, no quantization was done, i.e., $X_n(e^{j\omega_k})$ was represented with full 16-bit accuracy. A comparison of the wideband spectrograms (which, of course, have good temporal resolution) shows an effect which is consistent with a composite

326

impulse response with echoes as in Fig. 6.25. The fuzziness of the spectrogram in Fig. 6.48b is due to the delayed energy of the echo and such distortion is perceived as reverberation in the speech. Careful listening revealed a distinct "hollow barrel" effect when the utterances corresponding to Figs. 6.48a and 6.48b were compared. In contrast, with the phase properly adjusted as discussed in Section 6.2.1 (corresponding to Fig. 6.47), the spectrogram of the output in Fig. 6.48c is undistinguishable from that of the input (Fig. 6.48a) and the input and output speech signals were likewise perceptually indistinguishable.

Having given evidence that it is indeed possible to reconstruct the speech signal from its time-dependent Fourier transform, let us now turn to a con-

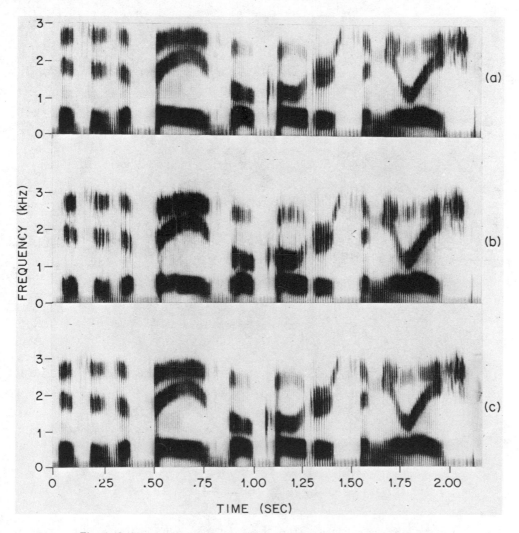

Fig. 6.48 Illustration of unquantized operation $(1/T_1 = 160 \text{ Hz})$; (a) input speech; (b) output speech with no phase adjustment; (c) output speech with best phase adjustment. (After Schafer and Rabiner [6].)

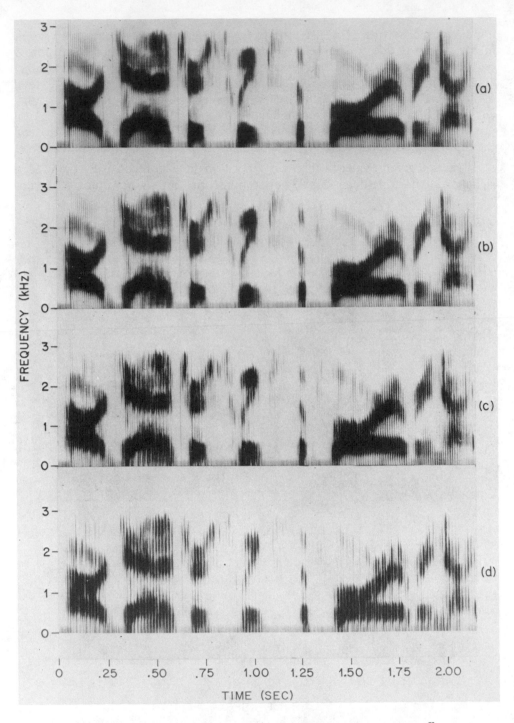

Fig. 6.49 Illustration of effect of aliasing for PCM coding; lowpass cutoff = 80 Hz; no quantization; (a) $1/T_1 = 160$ Hz; (b) $1/T_1 = 100$ Hz; (c) $1/T_1 = 80$ Hz; (d) $1/T_1 = 60$ Hz. (After Schafer and Rabiner [6].)

sideration of ways of coding the channel signals for digital transmission or storage. We recall from Chapter 5 that in coding any waveform, there are two basic concerns: the sampling rate and the number of bits required per sample. The product of these two quantities gives the bit-rate, which as a general rule will be minimized by using the minimum acceptable sampling rate and the fewest bits/sample. In this case, the total information rate is the sum of the bit rates for each complex channel signal.

Before discussing quantization, it is helpful to consider the effects of lowering the sampling rate of the channel signals. Recall that the real and imaginary parts of $X_n(e^{j\omega_k})$ are outputs of lowpass filters. Thus in the example just discussed, where the frequency response of the lowpass filter is shown in Fig. 6.46, it is certain that negligible aliasing will occur if the channel signals are sampled at a rate of 160 times per second or higher, since the filter response has at least 60 dB attenuation above 80 Hz. If a lower sampling rate is used without a corresponding reduction in filter bandwidth, aliasing occurs in the time dimension. If the bandwidth is reduced without reducing the spacing of the channels, then the synthesized speech is bound to be more reverberant since eventually the individual channel responses will not overlap, leaving "holes" in the spectrum of the synthetic speech. We cannot decrease the channel spacing without increasing the number of channels so as to cover the same band of frequencies. Therefore, if we attempt to lower the information rate by lowering the sampling rate of the channel signals, we must be prepared to tolerate either aliasing distortion in the time dimension that comes with under-sampling the channel signals, or increased reverberation distortion that comes with narrowing the filter bandwidths. These two effects are shown through the use of spectrograms in Figs. 6.49 and 6.50. Figure 6.49 illustrates the effect of aliasing due to sampling the short-time spectrum at too low a rate. The sampling rates of the channel signals in this figure are (a) 160 Hz, (b) 100 Hz, (c) 80 Hz, and (d) 60 Hz. Clearly, there is considerable distortion in cases (c) and (d), while the effect is much less evident in case (b). By comparing Fig. 6.49d to the spectrogram of the original speech utterance in Fig. 6.48a, it is apparent that in cases of severe time-dimension aliasing distortion, the pitch of the speech utterance becomes severely distorted, while the formant frequencies remain relatively intact. Figure 6.50 illustrates the effect of narrowing the bandwidths of the analysis filters, while leaving the spacing of the frequency channels the same. In all three cases, the sampling rate of the channel signals was 160 Hz. In Fig. 6.50a, the filter was as shown in Fig. 6.46; that is, above 80 Hz the attenuation of the filter was at least 60 dB. In Fig. 6.50b, the corresponding cut-off frequency was 53 Hz and in Fig. 6.50c, the corresponding cut-off frequency was 36 Hz. As expected, the spectrograms in Figs. 6.50b and 6.50c show significant degradation, which can be attributed to the increased reverberation caused by lengthening the effective duration of the window. It can also be observed that although the pitch of the signal appears to remain intact, the formant trajectories are severely damaged by the reverberation that is introduced by the narrowband filters. It would appear that from the point of

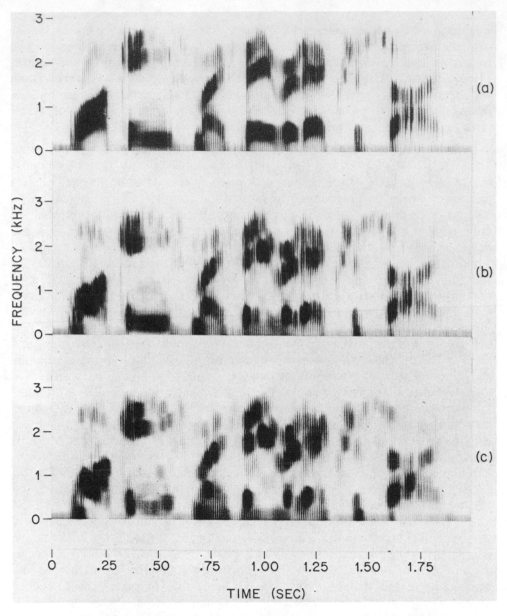

Fig. 6.50 Illustration of effect of narrowband analysis filters for PCM coding; no quantization; $1/T_1 = 160$ Hz; lowpass cutoff set to (a) 80 Hz; (b) 53 Hz; (c) 11 Hz. (After Schafer and Rabiner [6].)

view of intelligibility, then, that time dimension aliasing distortion should be preferable to the reverberation introduced by narrowing the filter bands.

In order to determine the information rate required for representing speech using the time-dependent Fourier transform, it is necessary to choose a quantization scheme. Most of the schemes discussed in Chapter 5 could be used to quantize the real and imaginary parts of the complex channel signals.

Two examples discussed in Reference [6] are the use of adaptive delta modulation, and PCM coding. Using an adaptive delta modulation system described by Jayant [26], the 28 channels were represented by 1 bit/sample. Thus, the overall bit rate for the system is 56 times the sampling rate of the channel signals, since both the real and the imaginary parts of the channel signals must be encoded. Since the adaptive delta modulation system requires a sampling rate

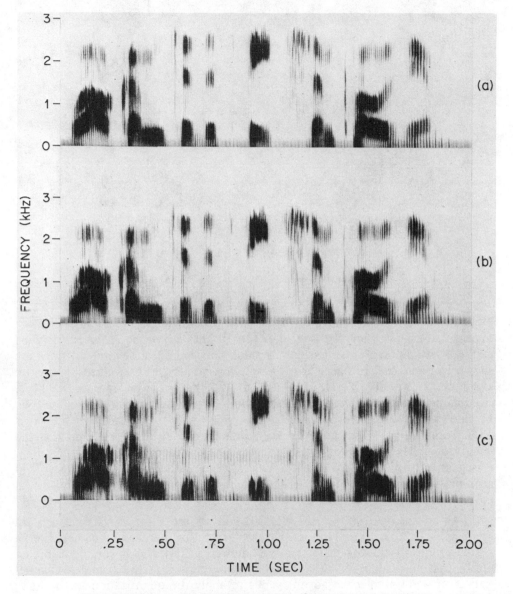

Fig. 6.51 Adaptive delta modulation coding of the spectrum parameters; (a) 28 kb/s ($1/T_1$=500 Hz); (b) 21 kb/s ($1/T_1$=375 Hz); (c) 14 kb/s ($1/T_1$=250 Hz). (After Schafer and Rabiner [6].)

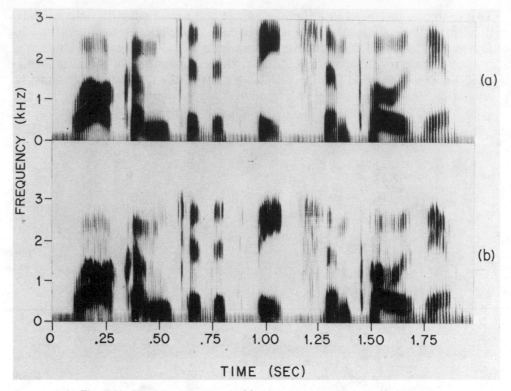

Fig. 6.52 Quantized operation; (a) input speech; (b) $1/T_1 = 100$ Hz.
Total bit rate $= 16$ kb/s. (After Schafer and Rabiner [6].)

on the order of five to ten times the Nyquist rate for good performance, we can expect that bit rates on the order of 20 to 30 kb/s would be required for good results. Examples of adaptive delta modulation coding at several bit rates are shown in Fig. 6.51. In Fig. 6.51a, the total bit rate is 28 kb/s, corresponding to a sampling rate of 500 Hz. In Fig. 6.51b, the total bit rate is 21 kb/s, corresponding to a sampling rate of 375 samples/sec, and in Fig. 6.51c, the total bit rate is 14 kb/s, corresponding to a sampling rate of 250 samples/sec. It is clear from Fig. 6.51 that good quality is maintained at 28 kb/sec, but that the quality degrades rapidly for lower bit rates. As an alternative to ADM coding, it is possible to use APCM coding. This has been utilized by Crochiere [27] in a nonuniform analysis with from 4 to 5 channels in achieving good quality at information rates on the order of 16 kb/s.

As an example of the use of PCM coding, the same 28 channel system was used with a sampling rate of 100 samples/sec for the channel signals (i.e., a small amount of aliasing distortion was allowed in order to lower the sampling rate). Instead of coding the real and imaginary parts of the complex channel signals, it was found to be advantageous to apply the quantization to the logarithm of the magnitude of the time-dependent Fourier transform and to the phase. Taking advantage of the insensitivity of the auditory system at higher frequencies, the low frequency channels were represented more accurately than the high frequency channels. Using 3 bits for the log magnitude and 4 bits for the phase of channels 1 through 10 and 2 bits on the log magnitude and 3 bits for the phase on channels 11 through 28, it was possible to obtain a representation at a bit rate of 16 kb/s without appreciable distortion. This is depicted in Fig. 6.52, which shows a wideband spectrogram of the input speech signal in Fig. 6.52a and 16 kb/s coding in Fig. 6.52b.

The information rates achieved using digital coding of the time-dependent Fourier transform are moderately high, and are comparable to bit rates that can be achieved by direct coding of the speech wave using adaptive quantization techniques. The complexity of the time-dependent Fourier representation is, of course, much greater than most waveform coding systems. The main advantage of the time-dependent Fourier transform representation is that it affords additional flexibility in manipulating the parameters of the speech signal. This will become evident in subsequent discussion.

Let us summarize what we have learned so far about speech coding schemes based upon the time-dependent Fourier transform. First, we have seen that if we sample the channel signals at a high enough rate and do not quantize the samples (12 bits/sample is adequate) then perceptually perfect reproduction of the input speech is possible. The bit rate required for such a representation is rather high, however. (For the example discussed, high quality reproduction of a 3 kHz band required an information rate of approximately 100 kb/s.) To reduce the bit-rate there are two approaches. The straightforward approach discussed in this section so far is to simply quantize the channel signals more coarsely and reduce the sampling rate. Using this approach, it is possible to reproduce the input with slight degradation at bit rates on the order of 16 kb/s. A second approach is to incorporate some of the properties of the

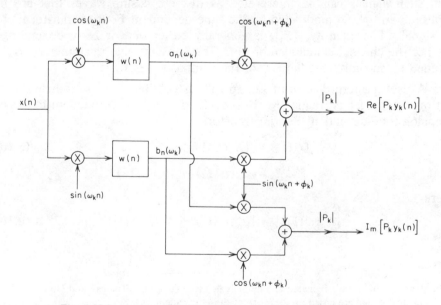

Fig. 6.53 Implementation of a single channel of the phase vocoder.

speech model as fixed components of the analysis or synthesis system, thus effectively removing some of redundancy of the speech signal. Other degradations in speech quality result from approximations introduced to simplify the implementation of the analysis-synthesis system. All of these degradations tend to be perceived as modifications of speech quality and intelligibility that are distinct from the type of distortions in waveform coding systems which are generally modelled as additive (possibly signal correlated) noise. Thus, signal-to-noise measurements as used in Chapter 5 are almost totally meaningless for vocoder systems. For this reason it has been necessary to describe vocoder performance by spectrogram comparisons and informal subjective evaluations of the type of distortion which are perceived by listeners. This will be our approach throughout the remaining chapters where vocoder systems are discussed.

6.7.2 The phase vocoder[12]

An interesting and novel approach to an analysis-system based on the short-time spectrum is the phase vocoder (Flanagan and Golden [28]). To understand how the system operates consider the response of a single channel. For this purpose it is convenient to depict the system of Fig. 6.45a entirely in terms of real operations as in Fig. 6.53. Recalling that we normally choose $\omega_{N-k} = 2\pi - \omega_k$ and $P_k = P_{N-k}^*$, then we see that the imaginary parts cancel out leaving only the real parts which can easily be shown to be equal to

$$R_e[P_k y_k(n)] = |P_k||X_n(e^{j\omega_k})|\cos[\omega_k n + \theta_n(\omega_k) + \phi_k] \qquad (6.167)$$

Thus, it is signals such as these that are summed to produce the composite output. Such signals can be interpreted as discrete cosine waves that are both amplitude and phase modulated by the time-dependent Fourier transform channel signal. The quantity, $|P_k|$, is generally either one or zero, depending on whether the channel is included or not. The constant phase parameters, ϕ_k, are included to maximize the flatness of the composite response.

A useful interpretation of Eq. (6.167) is possible if we introduce the concept of instantaneous frequency. In order to do this it is convenient to consider an analog time-dependent Fourier transform

$$X_a(t, \Omega_k) = |X_a(t, \Omega_k)|e^{j\theta_a(t, \Omega_k)} \qquad (6.168a)$$

$$= a_a(t, \Omega_k) - jb_a(t, \Omega_k) \qquad (6.168b)$$

where

$$|X_a(t, \Omega_k)| = [a_a^2(t, \Omega_k) + b_a^2(t, \Omega_k)]^{1/2} \qquad (6.169a)$$

[12]The phase vocoder was originated and intensively investigated by Flanagan and Golden [28]. The results in this section are based on the work of these investigators.

334

and

$$\theta_a(t, \Omega_k) = -\tan^{-1}\left[\frac{b_a(t, \Omega_k)}{a_a(t, \Omega_k)}\right] \qquad (6.169b)$$

This time-dependent, continuous-time, Fourier transform could be defined as

$$X_a(t, \Omega_k) = \int_{-\infty}^{\infty} x_a(\tau) w_a(t-\tau) e^{-j\Omega_k \tau} d\tau \qquad (6.170)$$

where $x_a(\tau)$ is the continuous-time waveform of the speech signal, and $w_a(\tau)$ is a continuous-time analysis window or, equivalently, the impulse response of an analog lowpass filter. The quantity

$$\dot{\theta}_a(t, \Omega_k) = \frac{d\theta_a(t, \Omega_k)}{dt} \qquad (6.171)$$

called the phase derivative, is the instantaneous frequency deviation of the kth channel from its center frequency, Ω_k. The phase derivative can be expressed in terms of $a_a(t, \Omega_k)$ and $b_a(t, \Omega_k)$ as

$$\dot{\theta}_a(t, \Omega_k) = \frac{b_a(t, \Omega_k) \dot{a}_a(t, \Omega_k) - a_a(t, \Omega_k) \dot{b}_a(t, \Omega_k)}{a_a^2(t, \Omega_k) + b_a^2(t, \Omega_k)} \qquad (6.172)$$

where the raised dot signifies differentiation. When dealing with discrete-time signal processing, we assume that $x_a(t)$ and $X_a(t, \Omega_k)$ are bandlimited, and that $X_n(e^{j\omega_k})$ is a sampled version of an analog time-dependent Fourier transform; i.e.,

$$X_n(e^{j\omega_k}) = X_a(nT, \omega_k/T) \qquad (6.173)$$

Likewise, the "phase derivative" of $X_n(e^{j\omega_k})$ is defined as a sampled version of $\dot{\theta}_a(t, \Omega_k)$; i.e.,

$$\dot{\theta}_n(\omega_k) = \frac{b_n(\omega_k) \dot{a}_n(\omega_k) - a_n(\omega_k) \dot{b}_n(\omega_k)}{a_n^2(\omega_k) + b_n^2(\omega_k)} \qquad (6.174)$$

where in this case, $\dot{a}_n(\omega_k)$ and $\dot{b}_n(\omega_k)$ are assumed to be sequences derived by sampling corresponding bandlimited analog derivative signals. These derivative signals can be obtained by digital filtering of the sequences $a_n(\omega_k)$ and $b_n(\omega_k)$. (See Problem 6.16.)

To see why phase derivative signals are of interest, consider the situation where the channel center frequencies are closely spaced. Particularly, consider the case when the pitch is constant and only a single harmonic of the fundamental is in the passband of the kth channel. In this case, we would find that $|X_n(e^{j\omega_k})|$ would reflect the slowly varying amplitude response of the vocal tract at a frequency of approximately ω_k. The phase derivative would be a constant, which would be equal to the deviation of the harmonic component from the center frequency. Now if the vocal tract response and pitch vary slowly, as in the normal speech, it is reasonable to argue that the magnitude and phase derivative will both be slowly varying. Indeed, it is reasonable to argue that the effects of aliasing in sampling the magnitude and phase derivative signals should be less severe, perceptually, than the aliasing effects when the real and imaginary parts of the time-dependent Fourier transform are sampled [28].

It should be noted that for synthesis $\theta_a(t, \Omega_k)$ is obtained from $\dot{\theta}_a(t, \Omega_k)$ by integration; i.e.,

$$\theta_a(t, \Omega_k) = \int_{t_0}^{t} \dot{\theta}_a(\tau, \Omega_k)\,d\tau + \theta_a(t_o, \Omega_k) \tag{6.175}$$

This equation suggests that $\theta_n(\omega_k)$, which is a sampled version of $\theta_a(t, \Omega_k)$, should be even smoother and more lowpass than $\dot{\theta}_n(\omega_k)$. Thus, it might be supposed that $\theta_n(\omega_k)$ could be sampled at an even lower rate than $\dot{\theta}_n(\omega_k)$. However, this neglects the fact that $\theta_n(\omega_k)$ is unbounded, and thus unsuitable for quantization. (This can easily be seen by considering the case of constant pitch.) A bounded phase can be obtained by computing the principal value; i.e., restricting values of $\theta_n(\omega_k)$ to be in a range 0 to 2π or $-\pi$ to π. Unfortunately, the principal value phase will be "discontinuous" (i.e., the principal value of $\theta_a(t, \Omega_k)$ will be a discontinuous function of t) and thus it will not be a lowpass signal. The fact that the principal value of the phase is discontinuous does not mean that phase cannot be quantized, since all that is required is that it be possible to reconstruct the corresponding real and imaginary parts of $X_n(e^{j\omega_k})$ at an appropriate sampling rate. Thus, the sampling rate of $\theta_n(\omega_k)$ must be as high as the rate required for $a_n(\omega_k)$ and $b_n(\omega_k)$. In fact, in the results described in Section 6.7.1, it was the principal value of phase that was quantized.

The phase derivative, while appearing to have the advantage of smoothness, is not without disadvantages in an analysis/synthesis system. This can be seen from Eq. (6.175) which shows that in reconstructing $\theta_n(\omega_k)$ from $\dot{\theta}_n(\omega_k)$ we must have an "initial condition". Normally, such initial conditions will not be known and arbitrarily assuming zero initial phase results effectively in an error in the fixed phase angles ϕ_k. This can cause the composite response of the complete analysis-synthesis system to deviate appreciably from the ideal flat

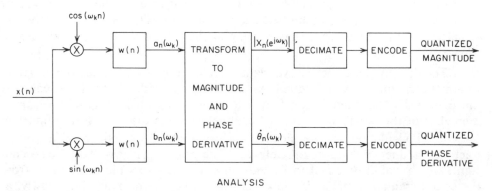

Fig. 6.54 Complete single channel of a phase vocoder analyzer.

336

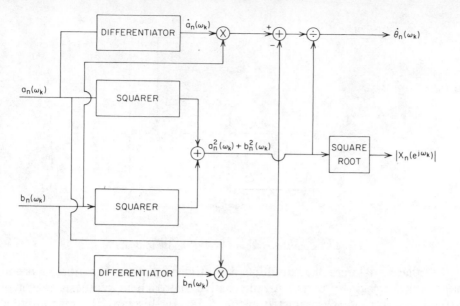

Fig. 6.55 Conversion from a and b to $\dot{\theta}$ and $|X(e^{j\omega})|$.

magnitude and linear phase, resulting in synthetic speech that may sound quite reverberant.

A vocoder analyzer based upon the magnitude and phase derivative is depicted in Fig. 6.54. Figure 6.54 shows a single channel of the analysis section for a frequency $0 < \omega_k < \pi$. All other channels have exactly the same form, although they may differ in the details of the decimation and quantization. The operations required to transform $a_n(\omega_k)$ and $b_n(\omega_k)$ into $|X_n(e^{j\omega_k})|$ and $\dot{\theta}_n(\omega_k)$ are depicted in Fig. 6.55. One approach to synthesis from magnitude and phase derivative signals is shown in Fig. 6.56. This approach involves conversion to real and imaginary parts followed by synthesis as in Fig. 6.53. The operations required to convert from magnitude and phase derivative signals to real and imaginary parts are depicted in Fig. 6.57. It can be seen that the phase derivative signals must be integrated to produce a phase signal. The cosine and sine

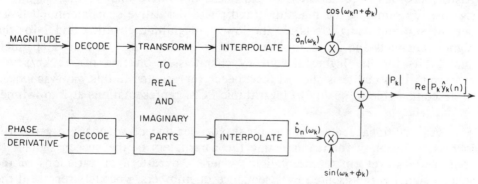

SYNTHESIS

Fig. 6.56 Implementation of the synthesizer for a single channel of the phase vocoder.

337

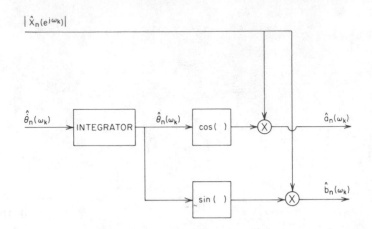

Fig. 6.57 Conversion from $|X(e^{j\omega})|$ and $\dot{\theta}$ to a and b.

of the phase angle are then multiplied by the magnitude function to produce the real and imaginary parts. An alternative approach to synthesis that avoids the conversion process is depicted in Fig. 6.58. In this case, the magnitude and phase derivative signals are interpolated with the resulting magnitude and phase sequences being used to amplitude and phase modulate the sinusoid. Thus, the conversion from magnitude and phase derivative to real and imaginary parts is replaced by the need for a phase modulator. Assuming that the implementation of such a digital phase modulator is not extremely complicated, it is clear that the synthesis scheme of Fig. 6.58 is significantly simpler than the scheme depicted in Figs. 6.56 and 6.57.

A detailed study of techniques for sampling and quantizing the magnitude and phase derivative signals in a phase vocoder has been carried out by Carlson [29]. In that study, a 28 channel phase vocoder was implemented with a channel spacing of 100 Hz. Linear quantizers were used for the phase derivative parameters and logarithmic quantizers were used on the magnitude parameters. Bits were distributed nonuniformly among the channels, with more bits being allocated to represent the lower channels and fewer bits for the upper channels. Also, more bits were allocated to the phase derivative than to the magnitude signals. By sampling the magnitude and phase derivative signals only 60 times per second, and using 2 bits for the lowest magnitude channels and 1 for the highest channels and 3 bits for the lowest frequency phase derivative channels and 2 bits for the highest frequency channels, a bit rate of 7.2 kb/s was achieved. Informal tests showed that speech represented in this way was judged to be comparable in quality to logarithmic PCM representations at 2 to 3 times the bit rate.

An additional feature of the phase vocoder and of vocoders in general is increased flexibility for manipulating the parameters of the speech signal. In contrast to waveform representations, where the pattern of variations of the speech signal is represented by a sequence of numbers, vocoders represent the speech signal in terms of parameters more closely related to the fundamental parameters of speech production. For example, as we have already argued, in the case of a phase vocoder with closely spaced channels, it is reasonable to suppose that the magnitude of the complex channel signals represents mainly information about the vocal tract transfer function, while the phase derivative signals give information about the excitation. A simple example of the manner in which basic speech parameters can be modified using a phase vocoder is suggested by Fig. 6.58. Suppose that the phase derivative signal is arbitrarily set to 0, so that the output signal is formed by the product of the magnitude of the short-time Fourier transform, and a cosine of fixed frequency, ω_k. If we assume equally spaced channels, then the composite output will appear as a periodic signal with fundamental frequency equal to the spacing of the channels. Since the magnitude function varies as a function of time, the output will not be periodic, but will be slowly varying. This type of synthesis gives a distinctly monotone output, as would be expected. Alternatively, if the phase derivative signals were allowed to vary randomly, we would expect resulting speech to sound like whispered speech

Another more useful application of the flexibility inherent in a phase vocoder system involves the alteration of the time and frequency dimensions of the speech signal as described by Flanagan and Golden [28]. Referring again to Fig. 6.58, we recall that the instantaneous frequency of the cosine is $[\omega_k + \dot{\theta}_n(\omega_k)]$. Thus, a frequency divided signal can be obtained by simply dividing ω_k and $\dot{\theta}_n(\omega_k)$ by a constant q. If each channel is synthesized in this fashion, the result is a frequency-compressed signal, where the frequency scale is compressed by the factor q. The frequency scale of the resulting signal can be restored by recording the signal at one speed and playing it back at q times the speed. Alternatively, one can use a digital-to-analog converter operating at q times the clock frequency of the sampling frequency of the input. In either case, the compression of the time scale counteracts the compression of the frequency scale introduced in the synthesis. The result is a signal with the normal frequency dimension but with a compressed time scale. Similar operations can be applied to expand the time scale. In this case, the center frequency ω_k and the phase angle are multipled by a factor q and the resulting expanded frequency scale is restored by playing back the output signal at a slower rate. The result in this case is a time-expanded signal with the normal frequency dimensions. Figure 6.59 (due to Flanagan and Golden [28]) shows an example of both time-expanded and time-compressed speech produced using the process just described for a factor $q = 2$.

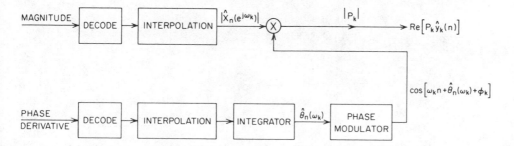

Fig. 6.58 Alternate form for synthesis.

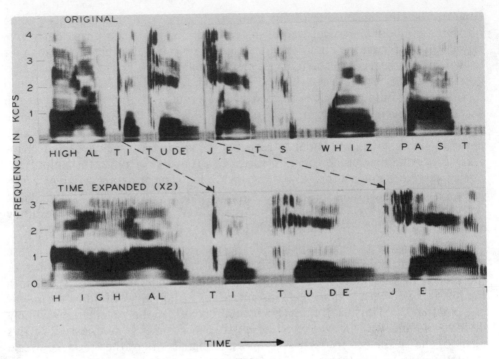

(a)

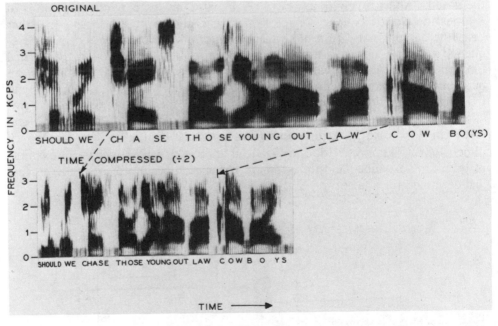

(b)

Fig. 6.59 An example of (a) time expansion; and (b) time compression using the phase vocoder. (After Flanagan and Golden [28].)

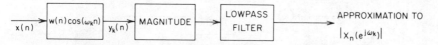

Fig. 6.60 Method for approximating the short-time spectrum.

6.7.3 The channel vocoder

The oldest form of speech coding device is the channel vocoder which was invented by Dudley [30]. The channel vocoder is similar in many respects to the systems that we have discussed in this section so far. It differs primarily in the fact that the channel vocoder incorporates more of the speech model into the analysis and synthesis configuration and, for simplicity, a number of approximations are introduced into the time-dependent Fourier analysis and synthesis. In order to see how the channel vocoder is related to the time-dependent Fourier transform representations that we have discussed so far, let us return to Eq. (6.167). We recall that this expression represents the contribution to the composite output of the kth channel. We have interpreted this expression as representing a cosine of nominal center frequency ω_k which is phase modulated and amplitude modulated, with the amplitude modulation corresponding to the magnitude of the time-dependent Fourier transform and the phase modulation corresponding to the phase angle of the time-dependent Fourier transform. We have also seen that each channel of analysis can be thought of as a bandpass filter with center frequency ω_k. This suggests that the magnitude of the time-dependent Fourier transform can be approximated by envelope detection on the output of a bandpass filter with center frequency ω_k. This is depicted in Fig. 6.60 which shows a bandpass filter with impulse response $w(n)\cos(\omega_k n)$, followed by a full-wave rectifier (the magnitude block) which is in turn followed by a lowpass filter. The full-wave rectifier and

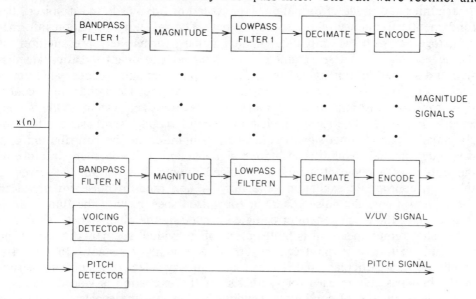

Fig. 6.61 Block diagram of channel vocoder analyzer.

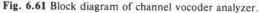

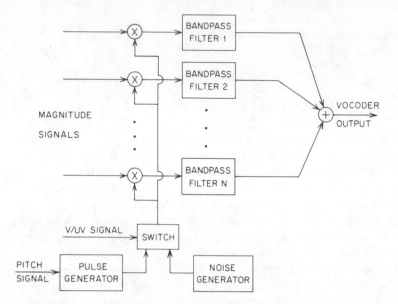

Fig. 6.62 Block diagram of channel vocoder synthesizer.

lowpass filter serve as an approximate envelope detector. Such a system is a basic component in the channel vocoder. The analyzer consists of a bank of such channels with analysis frequencies distributed across the speech band of interest. We have seen, however, that the speech signal cannot be represented by the amplitude spectrum alone but that the phase derivative signals contain information about the excitation. We have already argued that if the phase derivative signals are arbitrarily set to 0, the resulting speech will be entirely voiced and monotone. In order to represent the proper excitation for the speech signal, a channel vocoder has an additional analysis component for determining the mode of excitation, i.e., voiced or unvoiced, and if voiced, the fundamental frequency of the speech signal. The resulting excitation information, together with the amplitude channel signals, forms the representation for the speech signal. These parameters are sampled and quantized for transmission and storage in a digital system. The complete channel vocoder analyzer is depicted in Fig. 6.61. In order to synthesize the output for a channel vocoder, significant modifications must be made in the synthesizer system. This is depicted in Fig. 6.62. The basic principle of channel vocoder synthesis can be simply stated. The channel signals control the amplitude of the contribution of a particular channel, while the excitation signals control the detailed structure of the output of a given channel. The voiced/unvoiced signal simply serves to select an appropriate excitation generator, — i.e., random noise for unvoiced speech and a periodic pulse generator for voiced speech, with the fundamental frequency of the pulse generator being controlled by the pitch signal. Thus, the composite output spectrum is built up out of individual segments in which the amplitude within a given frequency band is roughly constant. In fact, the amplitude in a particular band retains the frequency selective shaping properties of the bandpass filters used for synthesis. If the excitation is voiced, then the output is composed of contiguous bands in which the fine spectral structure is

342

characteristic of periodicity while if the excitation is unvoiced, the spectrum varies continuously across each band. The resulting speech tends to be highly reverberant in nature because of complete lack of control over the merging together of adjacent bands. This can be seen from Fig. 6.63 (due to Flanagan [31]) which shows a comparison of a spectrogram of a speech signal at the input to a channel vocoder with a spectrogram of the corresponding output of a 15 channel vocoder. It can be seen that because of the coarse spacing of channels, the formant information appears highly quantized with formant frequency

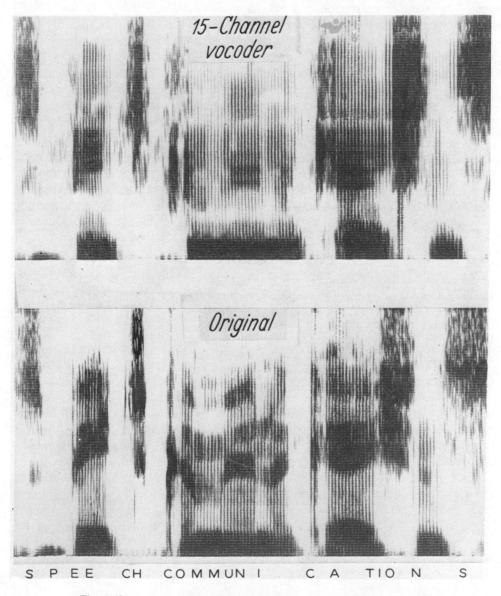

Fig. 6.63 An example of a 15 channel vocoder. (After Flanagan [31]).

variations being drastically altered in some cases. The result of making the approximations depicted in Figs. 6.61 and 6.62 is a rather drastic reduction in information rate with a concomitant increase in distortion. Channel vocoders typically operate in the range 1200 bits/sec to 9600 bits/sec with roughly 600 bits/sec devoted to the pitch and voicing information and the remaining information devoted to the channel signals. The channel vocoder, even more so than the phase vocoder, permits modification to the speech signal because the excitation and vocal tract information are represented separately. It is easy to see, for example, how pitch can be changed independently of the vocal tract information. For example, if the pulse generator always produces the same fundamental frequency, that is the pitch information is not utilized, then monotone speech will be produced. If no pulse generator excitation is used, but rather the excitation is always random noise, then whispered speech may be produced. Independent variations of the time and frequency scale can also be achieved using a channel vocoder simply by appropriate scaling of the center frequencies of the bandpass filters and of the pitch period.

A major contribution to the reduction in bit rate that is achieved with a channel vocoder is the direct representation of the pitch and voicing information. This, however, is one of the weaknesses of the channel vocoder system, since the detection of pitch and voicing is often a difficult task. Thus, the phase vocoder or representations more closely following the basic theory laid down in the earlier sections of this chapter have the advantage that pitch tracking is not required.

Both for historical reasons and because of the large number of parameters (e.g., number of filters, filter spacings, types of filters, etc.), the channel vocoder has been the subject of a number of intensive investigations. As such, a fair degree of sophistication in the implementation of channel vocoders has been obtained. The interested reader is referred to discussions of these results by Flanagan [31], Schroeder [32], and Gold and Rader [33,34].

6.8 Summary

In this chapter we have presented an intensive analysis of the short-time Fourier transform as applied to speech signals. We have shown how this representation of speech can be used effectively to estimate basic speech parameters such as pitch period and formant frequencies. Also considered were applications of the short-time Fourier transform in the design of vocoders such as the phase vocoder and the channel vocoder.

REFERENCES

1. R. W. Schafer and L. R. Rabiner, "Design of Digital Filter Banks for Speech Analysis," *Bell Syst. Tech. J.,* Vol. 50, No. 10, pp. 3097-3115, December 1971.

2. R. W. Schafer, L. R. Rabiner, and O. Herrmann, "FIR Digital Filter Banks

for Speech Analysis," *Bell Syst. Tech. J.,* Vol. 54, No. 3, pp. 531-544, March 1975.

3. J. B. Allen, "Short-Term Spectral Analysis and Synthesis and Modification by Discrete Fourier Transform," *IEEE Trans. Acoustics, Speech, and Signal Proc.,* Vol. ASSP-25, No. 3, pp. 235-238, June 1977.

4. J. B. Allen and L. R. Rabiner, "A Unified Theory of Short-Time Spectrum Analysis and Synthesis," *Proc. IEEE,* Vol. 65, No. 11, pp. 1558-1564, November 1977.

5. J. F. Kaiser, "Nonrecursive Digital Filter Design Using the I_0-SINH Window Function," *Proc. 1974 IEEE Int. Symp. on Circuits and Syst.,* pp. 20-23, April 1974. (Also in *Digital Signal Processing, II,* IEEE Press, 1976.)

6. R. W. Schafer and L. R. Rabiner, "Design and Simulation of a Speech Analysis-Synthesis System Based on Short-Time Fourier Analysis," *IEEE Trans. Audio and Electroacoustics,* Vol. AU-21, No. 3, pp. 165-174, June 1973.

7. M. R. Portnoff, "Implementation of the Digital Phase Vocoder Using the Fast Fourier Transform," *IEEE Trans. Acoustics, Speech, and Signal Proc.,* Vol. ASSP-24, No.3, pp. 243-248, June 1976.

8. R. K. Potter, G. A. Kopp, and H. G. Kopp, *Visible Speech,* Dover Publications, New York, 1966.

9. R. H. Bolt et al., "Speaker Identification by Speech Spectrograms," *Science,* 166, pp. 338-343, 1969.

10. F. Poza, "Voiceprint Identification: Its Forensic Application," *Proc. 1974 Carnahan Crime Countermeasures Conference,* April 1974.

11. R. W. Schafer and L. R. Rabiner, "System for Automatic Formant Analysis of Voiced Speech," *J. Acoust. Soc. Am.,* Vol. 47, No. 2, pp. 634-648, February 1970.

12. D. W. Tufts, S. E. Levinson, and R. Rao, "Measuring Pitch and Formant Frequencies for a Speech Understanding System," *Proc. 1976 IEEE Int. Conf. on Acoustics, Speech, and Signal Proc.,* pp. 314-317, April 1976.

13. R. Koenig, H. K. Dunn, and L. Y. Lacey, "The Sound Spectrograph," *J. Acoust. Soc. Am.,* Vol. 18, pp. 19-49, 1946.

14. A. V. Oppenheim, "Speech Spectrograms Using the Fast Fourier Transform," *IEEE Spectrum,* Vol. 7, pp. 57-62, August 1970.

15. H. F. Silverman and N. R. Dixon, "A Parametrically Controlled Spectral Analysis System for Speech," *IEEE Trans. Acoustics, Speech, and Signal Proc.,* Vol. ASSP-22, No. 5, pp. 362-381, October 1974.

16. M. R. Schroeder, "Period Histogram and Product Spectrum: New Methods for Fundamental Frequency Measurement," *J. Acoust. Soc. Am.,* Vol. 43, No. 4, pp. 829-834, April 1968.

17. A. M. Noll, "Pitch Determination of Human Speech by the Harmonic Product Spectrum, the Harmonic Sum Spectrum, and a Maximum Likelihood Estimate," *Proc. Symp. Computer Proc. in Comm.*, pp. 779-798, April 1969.

18. C. G. Bell, H. Fujisaki, J. M. Heinz, K. N. Stevens, and A. S. House, "Reduction of Speech Spectra by Analysis-by-Synthesis Techniques," *J. Acoust. Soc. Am.*, Vol. 33, pp. 1725-1736, December 1961.

19. E. N. Pinson, "Pitch Synchronous Time Domain Estimation of Formant Frequencies and Bandwidths," *J. Acoust. Soc. Am.*, Vol. 35, No. 8, pp. 1264-1273, August 1963.

20. A. V. Oppenheim, "A Speech Analysis-Synthesis System Based on Homomorphic Filtering," *J. Acoust. Soc. Am.*, Vol. 45, pp. 458-465, February 1969.

21. J. Olive, "Automatic Formant Tracking in a Newton-Raphson Technique," *J. Acoust. Soc. Am.*, Vol. 50, pp. 661-670, August 1971.

22. M. Halle and K. N. Stevens, "Analysis by Synthesis," *Proc. Sem. Speech Compression*, Vol. II, Paper D7, December 1959.

23. M. V. Mathews, J. E. Miller and E. E. David, Jr., "Pitch Synchronous Analysis of Voiced Sounds," *J. Acoust. Soc. Am.*, Vol. 33, pp. 179-186, 1961.

24. A. E. Rosenberg, "Effect of Glottal Pulse Shape on the Quality of Natural Vowels," *J. Acoust. Soc. Am.*, Vol. 49, pp. 583-590, 1971.

25. L. R. Rabiner and B. Gold, *Theory and Application of Digital Signal Processing,* Chapter 3, pp. 105-123, Prentice-Hall, Englewood Cliffs, N.J., 1975.

26. N. S. Jayant, "Adaptive Delta Modulation With a One-Bit Memory," *Bell Syst. Tech. J.,* Vol. 49, pp. 321-342, 1970.

27. R. E. Crochiere, "On the Design of Sub-Band Coders for Low Bit Rate Speech Communication," *Bell Syst. Tech. J.,* Vol. 65, No. 5, pp. 747-770, May-June 1977.

28. J. L. Flanagan and R. M. Golden, "Phase Vocoder," *Bell Syst. Tech. J.,* Vol. 45, pp. 1493-1509, 1966.

29. J. P. Carlson, "Digitalized Phase Vocoder," *Proc. Conf. on Speech Comm. and Proc.,* Boston, Mass., November 1967.

30. H. Dudley, "The Vocoder," *Bell Labs Record,* Vol. 17, pp. 122-126, 1939.

31. J. L. Flanagan, *Speech Analysis, Synthesis and Perception,* Second Edition, Chapter 8, pp. 321-385, Springer-Verlag, New York, 1972.

32. M. R. Schroeder, "Vocoders: Analysis and Synthesis of Speech," *Proc. IEEE,* Vol. 54, pp. 720-734, May 1966.

33. B. Gold and C. M. Rader, "Systems for Compressing the Bandwidth of Speech," *IEEE Trans. Audio and Electroacoustics,* Vol. AU-15, No. 3, pp. 131-135, September 1967.

34. B. Gold and C. M. Rader, "The Channel Vocoder," *IEEE Trans. Audio and Electroacoustics,* Vol. AU-15, No. 4, pp. 148-160, December 1967.

PROBLEMS

6.1 If the short-time Fourier transform is expressed as

$$X_n(e^{j\omega}) = a_n(\omega) - jb_n(\omega) = |X_n(e^{j\omega})|e^{j\theta_n(\omega)}$$

then prove that if $x(n)$ is real, then
(a) $a_n(\omega) = a_n(2\pi-\omega) = a_n(-\omega)$
(b) $b_n(\omega) = -b_n(2\pi-\omega) = -b_n(-\omega)$
(c) $|X_n(e^{j\omega})| = |X_n(e^{j(2\pi-\omega)})| = |X(e^{-j\omega})|$
(d) $\theta_n(\omega) = -\theta_n(2\pi-\omega) = -\theta_n(-\omega)$

6.2 If we define the short-time Fourier transform of the signal $x(n)$ as

$$X_n(e^{j\omega}) = \sum_{m=-\infty}^{\infty} x(m)w(n-m)e^{-j\omega m}$$

Show the following properties hold
(a) Linearity -
   if $v(n) = x(n) + y(n)$, then $V_n(e^{j\omega}) = X_n(e^{j\omega}) + Y_n(e^{j\omega})$
(b) Shifting property -
   if $v(n) = x(n-n_0)$, then $V_n(e^{j\omega}) = X_{n-n_0}(e^{j\omega})e^{-j\omega n_0}$
(c) Scaling property -
   if $v(n) = \alpha x(n)$, then $V_n(e^{j\omega}) = \alpha X_n(e^{j\omega})$
(d) Exponential weighting -
   if $v(n) = a^n x(n)$, then $V_n(e^{j\omega}) = X_n(a^{-1}e^{j\omega})$
(e) Conjugate symmetry -
   if $x(n)$ is real, then $X_n(e^{j\omega}) = X_n^*(e^{-j\omega})$

6.3 By definition

$$X_n(e^{j\omega}) = a_n(\omega) - jb_n(\omega) = |X_n(e^{j\omega})|e^{j\theta_n(\omega)}$$

(a) Obtain expressions for $|X_n(e^{j\omega})|$ and $\theta_n(\omega)$ in terms of $a_n(\omega)$ and $b_n(\omega)$.
(b) Obtain expressions for $a_n(\omega)$ and $b_n(\omega)$ in terms of $|X_n(e^{j\omega})|$ and $\theta_n(\omega)$.

6.4 If the sequences $x(n)$ and $w(n)$ have normal Fourier transforms $X(e^{j\omega})$ and $W(e^{j\omega})$, then prove that the short-time Fourier transform

$$X_n(e^{j\omega}) = \sum_{m=-\infty}^{\infty} x(m)w(n-m)e^{-j\omega m}$$

can be put in the form

$$X_n(e^{j\omega}) = \frac{1}{2\pi} \int_{-\pi}^{\pi} W(e^{j\theta}) e^{j\theta n} X(e^{j(\omega+\theta)}) d\theta$$

i.e., $X_n(e^{j\omega})$ is a smoothed spectral estimate of $X(e^{j\omega})$ at frequency ω.

6.5 If we define a short-time power density spectrum of a signal in terms of its short-time Fourier transform as

$$S_n(e^{j\omega}) = |X_n(e^{j\omega})|^2$$

and we define the short-time autocorrelation of the signal as

$$R_n(k) = \sum_{m=-\infty}^{\infty} w(n-m)x(m)w(n-k-m)x(m+k)$$

then show that if

$$X_n(e^{j\omega}) = \sum_{m=-\infty}^{\infty} x(m)w(n-m)e^{-j\omega m}$$

$R_n(k)$ and $S_n(e^{j\omega})$ are related as a Fourier transform pair — i.e., show that $S_n(e^{j\omega})$ is the Fourier transform of $R_n(k)$ and vice versa.

6.6 Suppose that the window sequence, $w(n)$ used in short-time Fourier analysis is causal and has a rational z-transform of the form

$$W(z) = \frac{\displaystyle\sum_{r=0}^{N_z} b_r z^{-r}}{1 - \displaystyle\sum_{k=1}^{N_p} a_k z^{-k}}$$

(a) What properties should $W(z)$ (or equivalently, $W(e^{j\omega})$) have in order that it be suitable for this application?

(b) Obtain a recurrence formula for $X_n(e^{j\omega})$ in terms of the signal $x(n)$ and previous values of $X_n(e^{j\omega})$.

(c) Consider the case

$$W(z) = \frac{1}{1 - az^{-1}}$$

How should a be chosen to obtain a frequency resolution of approximately 100 Hz at a sampling rate of 10 kHz?

(d) The value of a required in (c) suggests that problems may arise in implementing very narrowband time-dependent Fourier analysis recursively. Briefly discuss the nature of these problems.

6.7 Prove that

$$\sum_{k=0}^{N-1} e^{j\frac{2\pi}{N}kn} = N \sum_{r=-\infty}^{\infty} \delta(n-rN) .$$

$$= N \qquad n = rN \qquad r = 0, \pm 1,...$$

$$= 0 \qquad \textit{otherwise}$$

348

In proving this result, make use of the identity

$$\sum_{k=0}^{N-1} \alpha^k = \frac{1 - \alpha^N}{1 - \alpha}.$$

6.8 In implementing time-dependent Fourier representations, we employ sampling in both the time and frequency dimensions. In this problem we investigate the effects of both types of sampling.

Consider a sequence $x(n)$ with conventional Fourier transform

$$X(e^{j\omega}) = \sum_{m=-\infty}^{\infty} x(m)e^{-j\omega m}$$

(a) If the periodic function $X(e^{j\omega})$ is sampled at frequencies $\omega_k = 2\pi k/N, \; k = 0, 1, ..., N - 1$, we obtain

$$\tilde{X}(k) = \sum_{m=-\infty}^{\infty} x(m)e^{-j\frac{2\pi}{N}km}$$

These samples can be thought of as the discrete Fourier transform of the sequence $\tilde{x}(n)$ given by

$$\tilde{x}(n) = \frac{1}{N} \sum_{k=0}^{N-1} \tilde{X}(k)e^{j\frac{2\pi}{N}kn}$$

Show that

$$\tilde{x}(n) = \sum_{r=-\infty}^{\infty} x(n+rN)$$

(b) What are the conditions on $x(n)$ so that no aliasing distortion occurs in the time domain when $X(e^{j\omega})$ is sampled?

(c) Now consider "sampling" the sequence $x(n)$; i.e. let us form the new sequence

$$y(n) = x(nM)$$

consisting of every M^{th} sample of $x(n)$. Show that the Fourier transform of $y(n)$ is

$$Y(e^{j\omega}) = \frac{1}{M} \sum_{k=0}^{M-1} X(e^{j(\omega-2\pi k)/M})$$

In proving this result you may wish to begin by considering the sequence

$$v(n) = x(n)p(n)$$

where

$$p(n) = \sum_{r=-\infty}^{\infty} \delta(n+rM)$$

349

Then note that $y(n) = v(nM) = x(nM)$.

(d) What are the conditions on $X(e^{j\omega})$ so that no aliasing distortion in the frequency domain occurs when $x(n)$ is sampled?

6.9 Consider a window $w(n)$ with Fourier transform $W(e^{j\Omega T})$ which is bandlimited to the range $0 \leqslant \Omega \leqslant \Omega_c$. We wish to show that

$$\sum_{r=-\infty}^{\infty} w(rR-n) = W(e^{j0})/R$$

independent of n if R is a sufficiently small (nonzero) integer.

(a) Let $\hat{w}(r) = w(rR-n)$. Obtain an expression for $\hat{W}(e^{j\Omega T'})$ in terms of R and $W(e^{j\Omega T})$ where T is the sampling rate of $w(n)$, and $T' = RT$ is the sampling rate of $\hat{w}(r)$. (Hint: Recall the problem of decimating a signal by an R to 1 factor or see Problem 6.8c.)

(b) Assuming $W(e^{j\Omega T}) = 0$ for $|\Omega| > \Omega_c$, derive an expression for the maximum value of R (as a function of Ω_c) such that

$$\hat{W}(e^{j0}) = W(e^{j0})/R$$

(c) Recalling that $\sum_{r=-\infty}^{\infty} \hat{w}(r)e^{-j\Omega T'r} = \hat{W}(e^{j\Omega T'})$ show that if the conditions of part (b) are met, then the relation given at the beginning of this problem is valid.

6.10 (a) Show that the impulse response of the system of Fig. P6.10 is

$$h_k(n) = h(n)\cos(\omega_k n)$$

(b) Find an expression for the frequency response of the system of Fig. P6.10

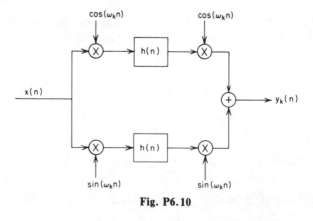

cos(ω_kn)   cos(ω_kn)

x(n)

y_k(n)

sin(ω_kn)   sin(ω_kn)

Fig. P6.10

6.11 Emphasis of the high frequency region of the spectrum is often accomplished using a first difference. In this problem we examine the effect of such operations on the short-time Fourier transform.

(a) Let $y(n) = x(n) - x(n-1)$. Show that

$$Y_n(e^{j\omega}) = X_n(e^{j\omega}) - e^{-j\omega}X_{n-1}(e^{j\omega})$$

350

(b)   Under what conditions can we make the approximation

$$Y_n(e^{j\omega}) \approx (1-e^{-j\omega})X_n(e^{j\omega})$$

In general, $x(n)$ may be linearly filtered as in

$$y(n) = \sum_{k=0}^{N-1} h(k)x(n-k)$$

(c)   Show that $Y_n(e^{j\omega})$ is related to $X_n(e^{j\omega})$ by an expression of the form

$$Y_n(e^{j\omega}) = X_n(e^{j\omega}) * h_\omega(n)$$

Find $h_\omega(n)$ in terms of $h(n)$.

(d)   Is it reasonable to expect that

$$Y_n(e^{j\omega}) = H(e^{j\omega})X_n(e^{j\omega})$$

6.12 A filter bank design with N filters has the following specifications

1.   The center frequencies of the bands are ω_k.
2.   The bands are symmetric around $\omega = \pi$, i.e. $\omega_k = 2\pi - \omega_{N-k}$, $P_k = P_{N-k}^*$, $w_k(n) = w_{N-k}(n)$.
3.   A channel exists for $\omega_k = 0$.

For both N even and N odd:

(a)   Sketch the locations of the N filter bands.
(b)   Derive an expression for the composite impulse response of the filter bank in terms of $w_k(n)$, ω_k, P_k and N.

6.13 To illustrate the reverberation obtained in filter banks using IIR filters, consider the composite impulse response

$$h(n) = \alpha_1\delta(n) + \alpha_2\delta(n-N) + \alpha_3\delta(n-2N)$$

which represents echos spaced N samples apart.

(a)   Determine the system function $H(e^{j\omega})$ for this example, and show that the squared magnitude response can be written as

$$|H(e^{j\omega})|^2 = (\alpha_2+(\alpha_1+\alpha_3)\cos(\omega N))^2 + (\alpha_1-\alpha_3)^2\sin^2(\omega N)$$

(b)   Show that the phase response of the system can be written as

$$\theta(\omega) = -\omega N + \tan^{-1}\left[\frac{(\alpha_1-\alpha_3)\sin(\omega N)}{\alpha_2 + (\alpha_1+\alpha_3)\cos(\omega N)}\right]$$

(c)   To determine locations of amplitude maxima and minima, $|H(e^{j\omega})|^2$ is differentiated with respect to ω and the result is set to zero. Show that for $|\alpha_1+\alpha_3| \ll |\alpha_2|$ the locations of the maxima and minima are

$$\omega = \pm \frac{k\pi}{N} \qquad k = 0, 1, 2, \ldots$$

(d) Using the results of part (c), show that the peak-to-peak amplitude ripple (in dB) can be expressed as

$$R_A = 20 \log_{10} \left[\frac{|\alpha_2 + \alpha_1 + \alpha_3|}{|\alpha_2 - \alpha_1 - \alpha_3|} \right]$$

(e) Solve for R_A for the cases

(i) $\alpha_1 = 0.1,$ $\alpha_2 = 1.0,$ $\alpha_3 = 0.2$
(ii) $\alpha_1 = 0.15,$ $\alpha_2 = 1.0,$ $\alpha_3 = 0.15$
(iii) $\alpha_1 = 0.1,$ $\alpha_2 = 1.0,$ $\alpha_3 = 0.1$

(f) By differentiating $\theta(\omega)$ with respect to ω, it can be shown that maxima and minima of θ occur at values of ω satisfying

$$\cos(\omega N) = - \left| \frac{\alpha_1 + \alpha_3}{\alpha_2} \right|$$

Show that the peak-to-peak phase ripple is given by

$$R_p = 2 \tan^{-1} \left[\frac{\alpha_1 - \alpha_3}{(\alpha_2^2 - (\alpha_1 + \alpha_3)^2)^{1/2}} \right]$$

(g) Solve for R_p for the cases of part (e). Discuss the effects of varying α_1 and α_3 on R_A and R_p.

6.14 A proposed digital filter bank pitch detector consists of a bank of digital bandpass filters with lower cutoff frequencies given as

$$F_k = 2^{k-1} F_1 \quad k = 1, 2, \ldots, M$$

and upper cutoff frequencies given as

$$F_{k+1} = 2^k F_1 \quad k = 1, 2, \ldots, M$$

This choice of cutoff frequencies gives the filter bank the property that if the input is periodic with fundamental frequency F_0, such that

$$F_k < F_0 < F_{k+1}$$

then the filter outputs of bands 1 to $k - 1$ will have little energy, the output of band k will contain the fundamental frequency, and bands $k + 1$ to M will contain 1 or more harmonics. Thus, by following each filter output by a detector which can detect pure tones a good indication of pitch can be obtained.

(a) Determine F_1 and M such that this method would work for pitch frequency from 50 Hz to 800 Hz.
(b) Sketch the required frequency response of each of the M bandpass filters.
(c) Can you suggest simple ways to implement the tone detector required at the output of each filter?
(d) What types of problems would you anticipate in implementing this method using nonideal bandpass filters?

(e) What would happen if the input speech were bandlimited from 300 Hz to 3000 Hz, e.g., telephone line input? Can you suggest improvements in these cases?

6.15 Consider a periodic sequence

$$\tilde{x}(n) = \sum_{r=-\infty}^{\infty} h_v(n+rN_p)$$

representing a voiced speech sound.

(a) Show that the Fourier series for $\tilde{x}(n)$ can be represented as the Fourier series

$$\tilde{x}(n) = \frac{1}{N_p} \sum_{k=0}^{N_p-1} \tilde{X}(k) e^{j\frac{2\pi}{N_p}kn}$$

where the Fourier coefficients $\tilde{X}(k)$ are samples of the Fourier transform of the voiced speech impulse response; i.e.,

$$\tilde{X}(k) = H_v(e^{j\frac{2\pi}{N_p}k})$$

(See Prob. 6.8.)

(b) Show that the short-time Fourier transform of $\tilde{x}(n)$ can be expressed as

$$\tilde{X}(e^{j\omega}) = \frac{1}{N_p} \sum_{k=0}^{N_p-1} H_v(e^{j\frac{2\pi}{N_p}k}) W_n(e^{j(\omega-2\pi k/N_p)})$$

where $W_n(e^{j\omega})$ is the Fourier transform of $w(n-m)$.

(c) How many *different* values can $\tilde{X}_n(e^{j\omega})$ take on for a given frequency, ω.

(d) For the rectangular window

$$w(n) = 1 \quad 0 \leqslant n \leqslant N_p - 1$$
$$= 0 \quad otherwise$$

find the function $W_n(e^{j\omega})$.

(e) For the rectangular window of length N_p, for what values of n will it be true that

$$\tilde{X}_n(e^{j\frac{2\pi}{N_p}k}) = H_v(e^{j\frac{2\pi}{N_p}k})$$

6.16 Consider the analysis and synthesis of the signal $x(n) = cos(\omega_0 n)$. The analysis network is shown in Fig. P6.16a for the kth channel.

(a) Determine $a_n(\omega_k)$ and $b_n(\omega_k)$ for the given input signal.

(b) Assuming $h(n)$ is a narrowband lowpass filter, simplify your expressions for $a_n(\omega_k)$ and $b_n(\omega_k)$ assuming that $(\omega_0-\omega_k)$ falls within the band of the filter, and that $H(e^{j\omega}) \approx 1$ for such frequencies.

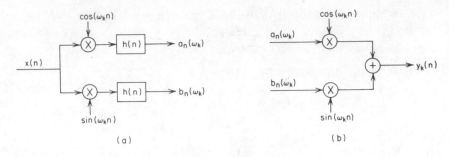

Fig. P6.16

(c) The signals $a_n(\omega_k)$ and $b_n(\omega_k)$ are combined to give magnitude, $M_n(\omega_k)$, and phase derivative, $\dot{\phi}_n(\omega_k)$. Determine $M_n(\omega_k)$ and $\dot{\phi}_n(\omega_k)$ for this example.

(d) Show that using the synthesis network of Fig. P6.16b, the output signal is essentially identical to the input signal.

(e) The phase derivative $\dot{\phi}_n(\omega_k)$ is computed using the relation

$$\dot{\phi}_n(\omega_k) = \frac{b_n(\omega_k)\dot{a}_n(\omega_k) - a_n(\omega_k)\dot{b}_n(\omega_k)}{[a_n(\omega_k)]^2 + [b_n(\omega_k)]^2}$$

Solve for $\dot{\phi}_n(\omega_k)$ for this example, and compare your results with those of part (c).

(f) Now assume that the derivatives of part (e) are computed using a simple first difference, i.e.,

$$\dot{a}_n(\omega_k) \approx \frac{1}{T}\left(a_n(\omega_k) - a_{n-1}(\omega_k)\right)$$

where T is the sampling period in the time dimension. Now solve for $\dot{\phi}_n(\omega_k)$ and compare your results with part (c). Under what conditions are they approximately the same?

7

Homomorphic
Speech Processing

7.0 Introduction

One of the fundamental assumptions that we have used throughout this book is that speech can be represented as the output of a linear time-varying system whose properties vary slowly with time. This has led us to the basic principle of speech analysis which says that if we consider short segments of the speech signal, then each segment can effectively be modelled as having been generated by exciting a linear time-invariant system either by a quasi-periodic impulse train or a random noise signal. As we have seen, the problem of speech analysis is to estimate the parameters of the speech model and to measure their variations with time. Since the excitation and impulse response of a linear time-invariant system are combined in a convolutional manner, the problem of speech analysis can also be viewed as a problem in separating the components of a convolution. This problem is often called "deconvolution." In Chapter 6, we studied techniques for performing such a deconvolution based upon the time-dependent Fourier representation of speech. In this chapter, we use the theory developed in Chapter 6 and extend the techniques by the introduction of the concept of homomorphic filtering. After a brief introduction to the general theory of homomorphic systems for convolution, we shall then discuss a variety of applications of homomorphic deconvolution in the analysis of speech signals.

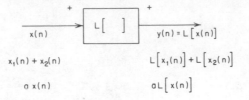

Fig. 7.1 Representation of a system obeying the superposition principle.

7.1 Homomorphic Systems for Convolution

Homomorphic systems for convolution obey a generalized principle of superposition. The principle of superposition as it is normally expressed for conventional linear systems is given by

$$L[x(n)] = L[x_1(n) + x_2(n)]$$
$$= L[x_1(n)] + L[x_2(n)]$$
$$= y_1(n) + y_2(n) = y(n) \qquad (7.1a)$$

and

$$L[ax(n)] = aL[x(n)] = ay(n) \qquad (7.1b)$$

where L represents the linear operator. The principle of superposition simply states that if an input signal is composed of a linear combination of elementary signals, then the output is a linear combination of corresponding outputs. This is depicted in Fig. 7.1, where the $+$ symbol at the input and output implies that an additive combination at the input produces an additive combination at the output.

As shown in Chapter 2, a direct result of the principle of superposition is the fact that the output of a linear time-invariant system can be expressed as the convolution sum

$$y(n) = \sum_{k=-\infty}^{\infty} h(n-k)x(k) = h(n) * x(n) \qquad (7.2)$$

The $*$ symbol will henceforth denote the operation of discrete-time convolution. By analogy with the principle of superposition for conventional linear systems, we can define a class of systems which obey a generalized principle of superposition where addition is replaced by convolution. (It can easily be shown that convolution has the same algebraic properties as addition [1].) That is,

$$H[x(n)] = H[x_1(n) * x_2(n)]$$
$$= H[x_1(n)] * H[x_2(n)] \qquad (7.3)$$
$$= y_1(n) * y_2(n) = y(n)$$

An equation similar to Eq. (7.1b), which expresses scalar multiplication in the generalized sense, can also be given [2]; however, the notion of generalized scalar multiplication is not needed for the applications that we shall discuss.

356

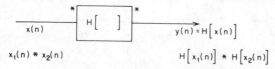

$y(n) = H\big[x(n)\big]$

$H\big[x_1(n)\big] * H\big[x_2(n)\big]$

Fig. 7.2 Representation of a homomorphic system for convolution.

Systems having the property expressed by Eq. (7.3) are termed "homomorphic systems for convolution." This terminology stems from the fact that such transformations can be shown to be homomorphic transformations in the sense of linear vector spaces [3]. Such systems are depicted as shown in Fig. 7.2, where the operation of convolution is noted explicitly at the input and output of the system. A homomorphic filter is simply a homomorphic system having the property that one component (the desired component) passes through the system essentially unaltered, while the undesired component is removed. In Eq. (7.3), for example, if $x_1(n)$ were the undesirable component, we would require that the output corresponding to $x_1(n)$ be a unit sample, while the output corresponding to $x_2(n)$ would closely approximate $x_2(n)$. This is entirely analogous to the situation with conventional linear systems where we are faced with the problem of separating a desired signal from an additive combination of signal and noise.

An important aspect of the theory of homomorphic systems is that any homomorphic system can be represented as a cascade of three homomorphic systems, as depicted in Fig. 7.3 for the case of homomorphic systems for convolution [3]. The first system takes inputs combined by convolution and transforms them into an additive combination of corresponding outputs. The second system is a conventional linear system obeying the principle of superposition as given in Eq. (7.1). The third system is the inverse of the first system; i.e., it transforms signals combined by addition back into signals combined by convolution. The importance of the existence of such a canonic form for homomorphic systems lies in the fact that the design of such systems reduces to the problem of the design of a linear system. The system $D_*[\ ]$ is called the characteristic system for homomorphic deconvolution and it is fixed in the canonic form of Fig. 7.3. Likewise, its inverse is also a fixed system. The characteristic system for homomorphic deconvolution obeys a generalized principle of superposition where the input operation is convolution and the output operation is ordinary addition. The properties of the characteristic system are defined as

$$D_*[x(n)] = D_*[x_1(n)*x_2(n)]$$

$$= D_*[x_1(n)] + D_*[x_2(n)] \tag{7.4}$$

$$= \hat{x}_1(n) + \hat{x}_2(n) = \hat{x}(n)$$

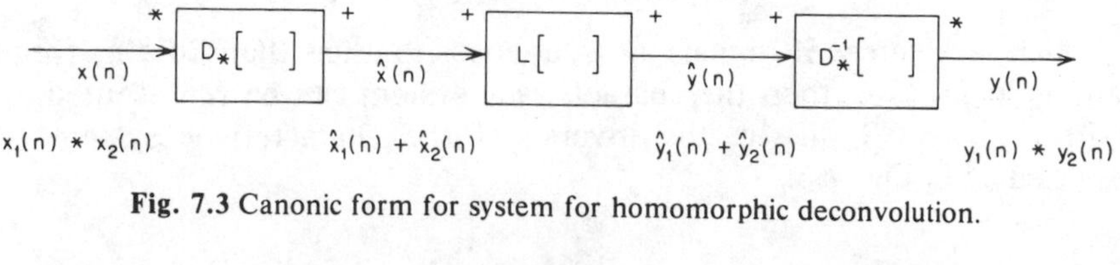

Fig. 7.3 Canonic form for system for homomorphic deconvolution.

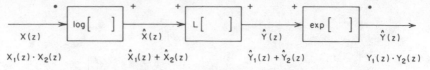

$$X(z) \qquad \hat{X}(z) \qquad \hat{Y}(z) \qquad \hat{Y}(z)$$
$$X_1(z) \cdot X_2(z) \qquad \hat{X}_1(z) + \hat{X}_2(z) \qquad \hat{Y}_1(z) + \hat{Y}_2(z) \qquad Y_1(z) \cdot Y_2(z)$$

Fig. 7.4 Frequency domain representation of a homomorphic system for convolution.

Likewise, the inverse characteristic system D_*^{-1} is defined as

$$D_*^{-1}[\hat{y}(n)] = D_*^{-1}[\hat{y}_1(n) + \hat{y}_2(n)]$$

$$= D_*^{-1}[\hat{y}_1(n)] * D_*^{-1}[\hat{y}_2(n)] \qquad (7.5)$$

$$= y_1(n) * y_2(n) = y(n)$$

The mathematical representation of the characteristic system is dependent upon the fact that we require that if the input is a convolution

$$x(n) = x_1(n) * x_2(n), \qquad (7.6)$$

then the z-transform of the input is the product of the corresponding z-transforms.

$$X(z) = X_1(z) \cdot X_2(z) \qquad (7.7)$$

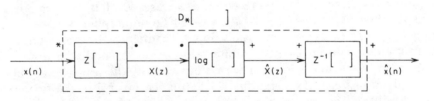

Fig. 7.5 Representation of the characteristic system for homomorphic deconvolution.

From Eq. (7.4), it is clear that the z-transform of the output of the characteristic system must be an additive combination of z-transforms. Thus, the frequency domain behavior of the characteristic system for convolution must have the property that if a signal is represented as a product of z-transforms at the input, then the output must be a sum of corresponding output z-transforms. One approach to the representation of such a system is depicted in Fig. 7.4. This approach is based upon the fact that the logarithm of a product can be defined so that it is equal to the sum of the logarithms of the individual terms. That is,

$$\hat{X}(z) = \log[X(z)] = \log[X_1(z) \cdot X_2(z)]$$

$$= \log[X_1(z)] + \log[X_2(z)] \qquad (7.8)$$

If we wish to represent signals as sequences, rather than in the frequency domain as in Fig. 7.4, then the characteristic system can be represented as depicted in Fig. 7.5. Similarly, the inverse of the characteristic system can be represented as in Fig. 7.6.

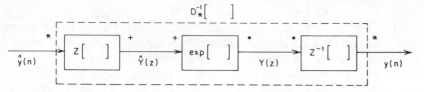

Fig. 7.6 Representation of the inverse of the characteristic system for homomorphic deconvolution.

The representation of the characteristic system and its inverse as depicted in Figs. 7.5 and 7.6, respectively, is dependent upon the validity of Eq. (7.8). That is, the logarithm must be defined so that it has the property that the logarithm of a product is equal to the sum of the logarithms. This is trivially true for real positive quantities. However, the z-transform is in general a complex quantity and there are important considerations of uniqueness when dealing with the logarithm of a complex number. For computational purposes we shall be primarily concerned with ensuring that Eq. (7.8) is valid when evaluated upon the unit circle; i.e., for $z = e^{j\omega}$. A detailed discussion of the problems of uniqueness for Eq. (7.8) is given in Ref. [2]. For our purposes here, it is sufficient to state that an appropriate definition of the complex logarithm is

$$\hat{X}(e^{j\omega}) = \log|X(e^{j\omega})| + j \arg[X(e^{j\omega}] \tag{7.9}$$

In this equation the real part (i.e., $\log|X(e^{j\omega})|$) causes no particular difficulty. However, problems of uniqueness arise in defining the imaginary part (i.e., $\arg[X(e^{j\omega})]$), which is simply the phase angle of the z-transform evaluated on the unit circle. In Ref. [2] it is shown that one approach to dealing with the problems of uniqueness of the phase angle is to require that the phase angle be a continuous odd function of ω. With this condition, Eq. (7.8) is satisfied.

Given that it is possible to compute the complex logarithm so as to satisfy Eq. (7.8), the inverse transform of the complex logarithm of the Fourier transform of the input is the output of the characteristic system for convolution, i.e.,

$$\hat{x}(n) = \frac{1}{2\pi} \int_{-\pi}^{\pi} \hat{X}(e^{j\omega}) e^{j\omega n} d\omega \tag{7.10}$$

The output of the characteristic system, $\hat{x}(n)$, is called the "complex cepstrum." (The term "cepstrum" was introduced by Bogert et al. [4] and has come to be accepted terminology for the inverse Fourier transform of the logarithm of the power spectrum of a signal. The term "complex cepstrum" implies that the complex logarithm is involved.) We shall use the term "cepstrum" for the quantity

$$c(n) = \frac{1}{2\pi} \int_{-\pi}^{\pi} \log|X(e^{j\omega})| e^{j\omega n} d\omega \tag{7.11}$$

(The sequence $c(n)$ can be shown to be equal to the even part of the complex cepstrum $\hat{x}(n)$.) (See Problem 7.1.)

In the above discussion we have defined the characteristic system for homomorphic convolution and therefore have defined the canonic form for all homomorphic systems for convolution. All systems of this class differ only in the linear part of the system. The choice of the linear system necessarily depends upon the properties of its input signals. Thus, in order to see how to design the linear system, it is necessary next to consider the nature of the output of the characteristic system; i.e., we must consider the properties of the complex cepstrum for typical input signals.

7.1.1 Properties of the complex cepstrum

In order to determine the properties of the complex cepstrum, it is sufficient to consider the case of rational z-transforms. The most general form that it is necessary to consider is

$$X(z) = \frac{Az^r \prod_{k=1}^{M_i} (1-a_k z^{-1}) \prod_{k=1}^{M_o} (1-b_k z)}{\prod_{k=1}^{N_i} (1-c_k z^{-1}) \prod_{k=1}^{N_o} (1-d_k z)} \qquad (7.12)$$

where the magnitudes of the quantities a_k, b_k, c_k, and d_k are all less than 1. Thus, the terms $(1-a_k z^{-1})$ and $(1-c_k z^{-1})$ correspond to zeros and poles inside the unit circle, and the terms $(1-b_k z)$ and $(1-d_k z)$ correspond to zeros and poles outside the unit circle. The factor z^r represents simply a shift in the time origin. Under the assumption of Eq. (7.8), the complex logarithm of $X(z)$ is

$$\hat{X}(z) = \log[A] + \log[z^r] + \sum_{k=1}^{M_i} \log(1-a_k z^{-1}) + \sum_{k=1}^{M_o} \log(1-b_k z)$$

$$- \sum_{k=1}^{N_i} \log(1-c_k z^{-1}) - \sum_{k=1}^{N_o} \log(1-d_k z) \qquad (7.13)$$

When Eq. (7.13) is evaluated on the unit circle, it can be seen that the term $\log[e^{j\omega r}]$ will contribute only to the imaginary part of the complex logarithm. Since this term carries only information about time origin, it is generally removed in the process of computing the complex cepstrum [2]. Thus, we shall neglect this term in our discussion of the properties of the complex cepstrum. Using the fact that the logarithmic terms can be written as a power series expansion, it is relatively straightforward to show that the complex cepstrum has the form

$$\hat{x}(n) = \log[A] \qquad\qquad n = 0$$

$$= \sum_{k=1}^{N_i} \frac{c_k^n}{n} - \sum_{k=1}^{M_i} \frac{a_k^n}{n} \qquad n > 0 \qquad (7.14)$$

$$= \sum_{k=1}^{M_o} \frac{b_k^{-n}}{n} - \sum_{k=1}^{N_o} \frac{d_k^{-n}}{n} \qquad n < 0$$

360

Equation (7.14) allows us to see a number of important properties of the complex cepstrum. First, we observe that, in general, the complex cepstrum is nonzero and of infinite extent for both positive and negative n, even though $x(n)$ may be causal, stable, and even of finite duration. Furthermore, it is apparent that the complex cepstrum is a decaying sequence which is bounded by

$$|\hat{x}(n)| < \beta \frac{\alpha^{|n|}}{|n|} \quad \text{for} \quad |n| \to \infty \quad (7.15)$$

where α is the maximum absolute value of the quantities a_k, b_k, c_k, and d_k, and β is a constant multiplier.

If $X(z)$ has no poles or zeros outside the unit circle (i.e., $b_k = d_k = 0$), then

$$\hat{x}(n) = 0 \quad \text{for} \quad n < 0 \quad (7.16)$$

Such signals are called "minimum phase" signals [5]. A general result for sequences of the form Eq. (7.16) is that such sequences are completely represented by the real parts of their Fourier transforms. Thus, we should be able to represent the complex cepstrum of minimum phase signals by the logarithm of the magnitude of the Fourier transform alone. This can easily be shown by remembering that the real part of the Fourier transform is the Fourier transform of the even part of the sequence; i.e., since $\log|X(e^{j\omega})|$ is the Fourier transform of the cepstrum, then

$$c(n) = \frac{\hat{x}(n) + \hat{x}(-n)}{2} \quad (7.17)$$

Using Eqs. (7.16) and (7.17), it is easily shown that

$$\hat{x}(n) = 0 \quad\quad n < 0$$
$$= c(n) \quad\quad n = 0 \quad (7.18)$$
$$= 2c(n) \quad n > 0$$

Thus, for minimum phase sequences the complex cepstrum can be obtained by computing the cepstrum and then using Eq. (7.18). Another important result for minimum phase sequences is that the complex cepstrum can be computed recursively from the input signal [1,2,5]. The recursion formula is

$$\hat{x}(n) = 0 \quad\quad\quad\quad\quad\quad\quad\quad\quad n < 0$$
$$= \log[x(0)] \quad\quad\quad\quad\quad\quad n = 0 \quad (7.19)$$
$$= \frac{x(n)}{x(0)} - \sum_{k=0}^{n-1} \left(\frac{k}{n}\right)\hat{x}(k)\,\frac{x(n-k)}{x(0)} \quad n > 0$$

Similar results can be obtained in the case when $X(z)$ has no poles or zeros inside the unit circle. Such signals are called "maximum phase." In this case, it can be seen from Eq. (7.14) that

$$\hat{x}(n) = 0, \quad n > 0 \quad (7.20)$$

361

If we again use Eq. (7.16) and Eq. (7.17) together, we see that

$$\hat{x}(n) = 0 \qquad n > 0$$
$$= c(n) \qquad n = 0 \qquad\qquad (7.21)$$
$$= 2c(n) \qquad n < 0$$

As in the case of minimum-phase sequences, we can also obtain a recursion formula for the complex cepstrum, of the form

$$\hat{x}(n) = \frac{x(n)}{x(0)} - \sum_{k=n+1}^{0} \left(\frac{k}{n}\right)\hat{x}(k)\frac{x(n-k)}{x(0)} \qquad n < 0$$

$$= \log[x(0)] \qquad\qquad n = 0 \qquad\qquad (7.22)$$

$$= 0 \qquad\qquad n > 0$$

An important special case is that of an input of the form

$$p(n) = \sum_{r=0}^{M} \alpha_r \delta(n-rN_p) \qquad\qquad (7.23)$$

i.e., a train of impulses. The z-transform of Eq. (7.23) is

$$P(z) = \sum_{r=0}^{M} \alpha_r z^{-rN_p} \qquad\qquad (7.24)$$

From Eq. (7.24) it is evident that $P(z)$ is really a polynomial in the variable z^{-N_p} rather than z^{-1}. Thus, $P(z)$ can be expressed as a product of factors of the form $(1-az^{-N_p})$ and $(1-bz^{N_p})$ and therefore it is easily seen that the complex cepstrum, $\hat{p}(n)$, will be nonzero only at integer multiples of N_p. For example, suppose $p(n)$ is

$$p(n) = \delta(n) + \alpha\delta(n-N_p) \qquad\qquad (7.25)$$

where $0 < \alpha < 1$. Then

$$P(z) = 1 + \alpha z^{-N_p} \qquad\qquad (7.26)$$

and

$$\hat{P}(z) = \log(1+\alpha z^{-N_p}) = \sum_{n=1}^{\infty} (-1)^{n+1}\frac{\alpha^n}{n} z^{-nN_p} \qquad\qquad (7.27)$$

Therefore, $\hat{p}(n)$ is an impulse train with impulses spaced by N_p

$$p(n) = \sum_{r=1}^{\infty} (-1)^{r+1}\frac{\alpha^r}{r} \delta(n-rN_p) \qquad\qquad (7.28)$$

The fact that the complex cepstrum of a train of uniformly spaced impulses is also a uniformly spaced impulse train with the same spacing is a very important result for speech analysis as we shall see in Section 7.2. However, before discussing the details of homomorphic speech processing, let us briefly consider the implementation of homomorphic filters for convolved signals.

7.1.2 Computational considerations

The mathematical representations of the characteristic system and its inverse depicted in Figs. 7.5 and 7.6, respectively, suggest a means for implementing homomorphic systems for convolution. If we restrict our attention to input sequences that are absolutely summable, then the z-transform of the input signal will have a region of convergence that includes the unit circle. That is, the sequence will have a Fourier transform. In such cases, it is appropriate to replace the z-transform operations in Figs. 7.5 and 7.6 by Fourier transform operations. In particular, for the important special case of a finite length input sequence, the mathematical representation of the characteristic system for convolution is given as

$$X(e^{j\omega}) = \sum_{n=0}^{N-1} x(n) e^{-j\omega n} \tag{7.29a}$$

$$\hat{X}(e^{j\omega}) = \log[X(e^{j\omega})] = \log|X(e^{j\omega})| + j \arg[X(e^{j\omega})] \tag{7.29b}$$

$$\hat{x}(n) = \frac{1}{2\pi} \int_{-\pi}^{\pi} \hat{X}(e^{j\omega}) e^{j\omega n} d\omega \tag{7.29c}$$

Equation (7.29a) is the Fourier transform of the input sequence, Eq. (7.29b) gives the complex logarithm of the Fourier transform of the input, and Eq. (7.29c) is the inverse Fourier transform of the complex logarithm of the Fourier transform of the input. As we have already observed, there are questions of uniqueness of this set of equations. In order to clearly define the complex cepstrum by Eq. (7.29), we must uniquely define the complex logarithm of the Fourier transform. To do this, it is helpful to impose the constraint that the complex cepstrum of a real input sequence be also a real sequence. Recall that for a real sequence the Fourier transform is an even function and the imaginary part is odd. Therefore, if the complex cepstrum is to be a real sequence, we must define the log magnitude function to be an even function of ω and the phase must be defined to be an odd function of ω. It can be shown that a further sufficient condition for the complex logarithm to be unique is that the phase be computed so that it is a continuous periodic function of ω with period of 2π [1,2]. (This continuity condition is also necessary for $\hat{X}(e^{j\omega})$ to be a valid Fourier transform.) Algorithms for the computation of an appropriate phase function have been developed and are described in Refs. [2,6].

Equation (7.29) is still not in a form that is amenable to computation, since Eq. (7.29) requires the evaluation of an integral. However, we can approximate Eq. (7.29) by using the discrete Fourier transform. The discrete Fourier transform (DFT) of a finite length sequence is identical to a sampled version of the Fourier transform of that same sequence [5]. Furthermore, the discrete Fourier transform can be efficiently computed by a fast Fourier transform algorithm [5]. Thus, the approach that is suggested for computing the complex cepstrum is to replace all of the Fourier transform operations by corresponding discrete Fourier transform operations. The resulting equations are given as

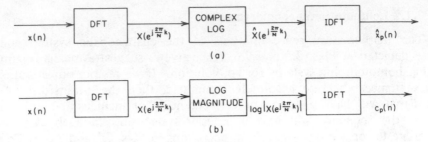

(a)

(b)

Fig. 7.7 Practical implementations of systems for obtaining (a) the complex cepstrum; and (b) the cepstrum.

$$X_p(k) = \sum_{n=0}^{N-1} x(n)e^{-j\frac{2\pi}{N}kn} \qquad N \leqslant k \leqslant N-1 \qquad (7.30a)$$

$$\hat{X}_p(k) = \log\,[X_p(k)] \qquad\qquad 0 \leqslant k \leqslant N-1 \qquad (7.30b)$$

$$\hat{x}_p(n) = \frac{1}{N}\sum_{k=0}^{N-1}\hat{X}_p(k)e^{j\frac{2\pi}{N}kn} \qquad 0 \leqslant n \leqslant N-1 \qquad (7.30c)$$

Equation (7.30c) represents the inverse discrete Fourier transform (IDFT) of the complex logarithm of the discrete Fourier transform of a finite length input sequence. The subscript p explicitly denotes the fact that the resulting sequence is not precisely equal to the complex cepstrum as defined in Eq. (7.29). This is due to the fact that the complex logarithm used in the DFT calculations is a sampled version of $\hat{X}(e^{j\omega})$ and thus, the resulting inverse transform is an aliased version of the true complex cepstrum (see Refs. [1,2,5].) That is, the complex cepstrum computed by Eqs. (7.30) is related to the true complex cepstrum by [5]

$$\hat{x}_p(n) = \sum_{r=-\infty}^{\infty} \hat{x}(n+rN) \qquad\qquad (7.31)$$

The computational operations for the implementation of the characteristic system for convolution are depicted in Fig. 7.7a.

We have observed that the complex cepstrum involves the use of the complex logarithm and that the cepstrum, as it has traditionally been defined, involves only the logarithm of the magnitude of the Fourier transform; that is, the cepstrum $c(n)$ is given by

$$c(n) = \frac{1}{2\pi} \int_{-\pi}^{\pi} \log|X(e^{j\omega})|e^{j\omega n}d\omega, \qquad -\infty < n < \infty \qquad (7.32)$$

An approximation to the cepstrum can be obtained by computing the inverse discrete Fourier transform of the logarithm of the magnitude of the discrete Fourier transform of the input sequence; i.e.,

$$c_p(n) = \frac{1}{N}\sum_{k=0}^{N-1} \log|X_p(k)|e^{j\frac{2\pi}{N}kn}, \qquad 0 \leqslant n \leqslant N-1 \qquad (7.33)$$

As before, the cepstrum computed using the discrete Fourier transform is related to the true cepstrum by

$$c_p(n) = \sum_{r=-\infty}^{\infty} c(n+rN) \qquad (7.34)$$

Figure 7.7b shows how the computations leading to Eq. (7.34) are implemented using the DFT and the inverse DFT.

Because of the aliasing inherent in the use of the discrete Fourier transform for cepstrum computations, it is often necessary to use a rather large value of N. As discussed in Refs. [1,2,5,6] a large value for N (that is, a high rate of sampling of the Fourier transform) is also required for accurate computation of the complex logarithm. However, the existence of fast Fourier transform (FFT) algorithms makes it feasible to use values of $N = 512$ or larger.

An alternative approach to the computation of the complex cepstrum of a finite duration sequence without the undesirable aliasing affects discussed above has recently been proposed [7]. The basic idea is to use Eq. (7.14) directly to define the complex cepstrum of a sequence in terms of the locations of the zeros (roots) of the z-transform polynomial. This method presupposes that one can accurately and efficiently find the roots of high order polynomials (e.g., polynomials of degree 500 are not unusual for speech applications). However, when one can perform the required root finding accurately the resulting complex cepstrum is in theory free of the aliasing which is inherent in using finite length transforms for computation. Good results have been reported on a few test cases using this method [7].

The mathematical and computational representations of homomorphic systems for deconvolution have been discussed in this section. We have not elaborated on the mathematical and computational details of such systems, since these are adequately covered in other references [1,2,5-7]. We shall now turn to a discussion of speech analysis applications of homomorphic systems for convolution.

7.2 The Complex Cepstrum of Speech

The now familiar model for the speech waveform and the time-dependent analysis principle that we have repeatedly invoked can be combined with the theory of homomorphic filtering in a very useful way. Recall that the model for speech production consists essentially of a slowly time-varying linear system excited by either a quasi-periodic impulse train or by random noise. Thus, it is appropriate to think of a short segment of voiced speech as having been generated by exciting a linear time-invariant system by a periodic impulse train. Similarly, a short segment of unvoiced speech can be thought of as resulting from the excitation of a linear time-invariant system by random noise. That is, a short segment of voiced speech can be thought of as a segment from the waveform

$$s(n) = p(n)*g(n)*v(n)*r(n) = p(n)*h_v(n)$$

$$= \sum_{r=-\infty}^{\infty} h_v(n-rN_p) \tag{7.35}$$

where $p(n)$ is a periodic impulse train of period N_p samples and $h_v(n)$ is the impulse response of a linear system that combines the effects of the glottal wave shape, $g(n)$, the vocal tract impulse response, $v(n)$, and the radiation impulse response, $r(n)$. Similarly, a short segment of unvoiced speech can be thought of as a segment from the waveform

$$s(n) = u(n)*v(n)*r(n) = u(n)*h_u(n) \tag{7.36}$$

where $u(n)$ is a random noise excitation and $h_u(n)$ is the impulse response of a system that represents the combined effects of the vocal tract and the radiation. For the case of voiced speech, the transfer function of the linear system is of the form

$$H_v(z) = G(z)V(z)R(z) \tag{7.37}$$

and for unvoiced speech, it is

$$H_u(z) = V(z)R(z). \tag{7.38}$$

Let us briefly review the nature of the components of Eqs. (7.37) and (7.38). From Chapter 3, we recall that a general model for the vocal tract transfer function is of the form

$$V(z) = \frac{Az^{-M} \sum_{k=1}^{M_i} (1-a_k z^{-1}) \sum_{k=1}^{M_o} (1-b_k z)}{\sum_{k=1}^{N_i} (1-c_k z^{-1})} \tag{7.39}$$

For voiced speech, except nasals, an adequate model includes only poles, i.e., $a_k = 0$, $b_k = 0$, for all k. For nasals and for unvoiced speech, it is necessary to include both poles and zeros. Some of the zeros may lie outside the unit circle, but, of course, for stability all the poles, c_k, must lie inside the unit circle. Also, since $v(n)$ is real, all the complex poles and zeros must occur in complex conjugate pairs. The radiation effects were seen in Chapter 3 to result in a high frequency emphasis which can be roughly modelled by

$$R(z) \approx 1 - z^{-1} \tag{7.40}$$

Finally, for voiced speech, the glottal pulse shape is of finite duration. Thus, $G(z)$ will have the form

$$G(z) = B \sum_{k=1}^{L_i} (1-\alpha_k z^{-1}) \sum_{k=1}^{L_o} (1-\beta_k z) \tag{7.41}$$

where the zeros, α_k and β_k, can be both inside and outside the unit circle.

Using the above model and the results of Section 7.1.2, we can now begin to see the general form of the complex cepstrum of a short segment of speech.

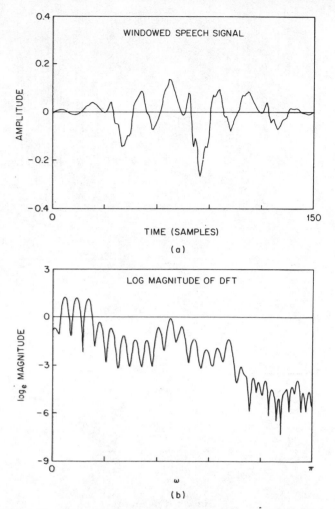

Fig. 7.8 Homomorphic analysis of voiced speech; (a) windowed time waveform; (b) log of magnitude of short-time Fourier transform; (c) principal value of phase; (d) "unwrapped" phase; (e) complex cepstrum; (f) cepstrum.

(See Ref. [8] for a detailed development.) For voiced speech, we note that the combined contributions of the vocal tract, glottal pulse, and radiation will in general be nonminimum phase, and thus the complex cepstrum will be nonzero for both positive and negative time. Note from Eq. (7.14) that the complex cepstrum will decay rapidly for large n. Also, note that the contribution to the complex cepstrum due to the periodic excitation will occur at integer multiples of the spacing between impulses; i.e., we should expect to see impulses in the complex cepstrum at multiples of the fundamental period. The example depicted in Fig. 7.8 illustrates the important features for voiced speech. Figure 7.8a shows a segment of voiced speech multiplied by a Hamming window. Figure 7.8b shows the log magnitude of the discrete Fourier transform. The periodic component in this function is, of course, due to the periodic nature of

367

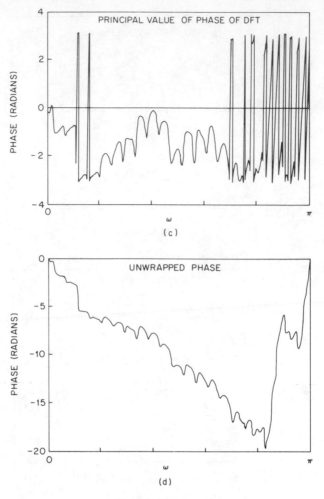

Fig. 7.8 (Continued)

the input. Figure 7.8c shows the discontinuous nature of the principal value of the phase, while Fig. 7.8d shows the phase curve without discontinuities. Figures 7.8b and 7.8d together are the Fourier transform of the complex cepstrum shown in Fig. 7.8e. Notice the peaks at both positive and negative times equal to the pitch period, and notice the rapidly decaying low-time components representing the combined effects of the vocal tract, glottal pulse and radiation. The cepstrum, which is simply the inverse transform of only the log magnitude (i.e., the phase is effectively set to zero), is shown in Fig. 7.8f. Note that the cepstrum also displays the same general properties as the complex cepstrum as indeed it should, since the cepstrum is the even part of the complex cepstrum.

The sequence of graphs in Fig. 7.8 suggests how homomorphic filtering can be applied to speech analysis. We note first of all that the impulses due to the periodic excitation tend to be separated from the remaining components of the complex cepstrum. This suggests that the appropriate system for homomorphic filtering of speech is as depicted in Fig. 7.9. That is, a segment

368

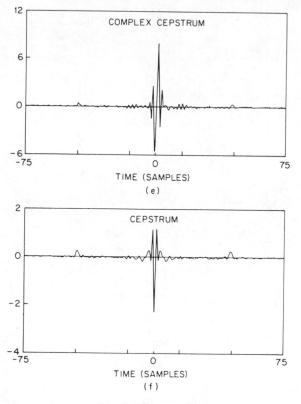

TIME (SAMPLES)

(e)

TIME (SAMPLES)

(f)

Fig. 7.8 (Continued)

of speech is selected by the window, $w(n)$; the cepstrum is computed as discussed in Section 7.1.3; and the desired component is selected by a "cepstrum window," $l(n)$. This type of filtering is appropriately termed "frequency invariant linear filtering." The resulting windowed complex cepstrum is processed by the inverse characteristic system to recover the desired component. This is illustrated in Fig. 7.10. Figures 7.10a and 7.10b show the log magnitude and phase obtained in the process of implementing the inverse characteristic system for the case when $l(n)$ is of the form

$$l(n) = 1, \quad |n| < n_0$$
$$= 0, \quad |n| \geqslant n_0 \quad (7.42)$$

where n_0 is chosen to be less than the pitch period, N_p. The corresponding output waveform is shown in Fig. 7.10c. (Note that a constant phase shift of π radians has been discarded in the implementation of the cepstrum computa-

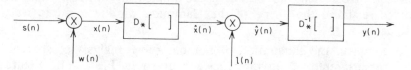

Fig. 7.9 Implementation of a system for homomorphic filtering of speech.

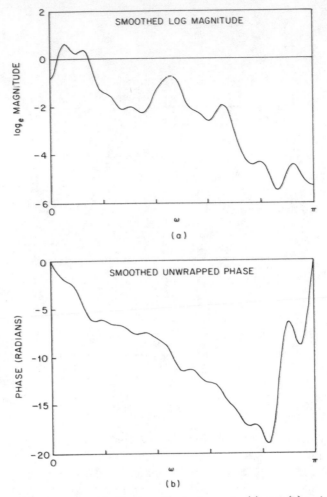

Fig. 7.10 Homomorphic filtering of voiced speech; (a) and (b) estimate of magnitude and phase of $H_v(e^{j\omega})$; (c) estimate of $h_v(n)$; (d) and (e) estimate of magnitude and phase of $P(e^{j\omega})$; (f) estimate of $p(n)$.

tion.) This waveform approximates the impulse response $h_v(n)$ defined in Eq. (7.35). If $l(n)$ is chosen so as to retain the excitation components; i.e.,

$$l(n) = 0, \quad |n| < n_0$$
$$= 1, \quad |n| \geqslant n_0 \tag{7.43}$$

then Figs. 7.10d, e, and f are obtained for the log magnitude, phase, and output respectively. Note that the output approximates an impulse train with the amplitudes retaining the shape of the Hamming window used to weight the input signal.

To complete the illustration of homomorphic analysis of speech, let us consider the example of unvoiced speech given in Fig. 7.11. Figure 7.11a shows a segment of unvoiced speech multiplied by a Hamming window. Figure 7.11b shows the corresponding log magnitude function and Fig. 7.11c shows the

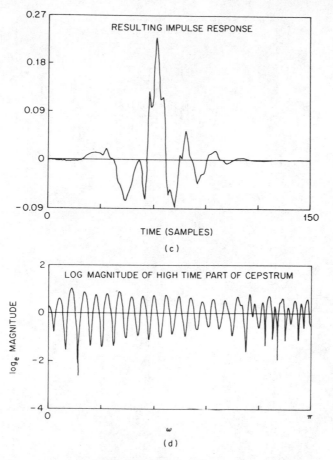

RESULTING IMPULSE RESPONSE

TIME (SAMPLES)

(c)

LOG MAGNITUDE OF HIGH TIME PART OF CEPSTRUM

$\log_e$ MAGNITUDE

ω

(d)

Fig. 7.10 (Continued)

corresponding cepstrum. Note the erratic variation of the log magnitude function. This is due to the fact that the excitation is random and thus the Fourier transform of a short segment contains a random component. In this case it makes little sense to compute the phase. It is clear from Fig. 7.11c that the cepstrum does not display any sharp peaks as was the case for voiced speech; however, the low-time portion of the cepstrum does contain information about $H_u(e^{j\omega})$. This is illustrated by Fig. 7.11d which shows the log magnitude function obtained by applying the cepstrum window of Eq. (7.42) to the cepstrum of Fig. 7.11c.

The previous discussion and examples show that it is indeed possible to obtain approximations to some of the basic components of the speech waveform by homomorphic filtering. This approach has not been carried very far, primarily because in most speech analysis applications, the complete deconvolution of the speech waveform is not necessary. Rather, we are generally content with estimates of basic parameters such as pitch period and formant frequencies. For this purpose, the cepstrum is entirely sufficient. Thus in most speech analysis applications, we are freed from the burdensome phase computation. Notice, for example, by comparing Figs. 7.8f and 7.11c that the

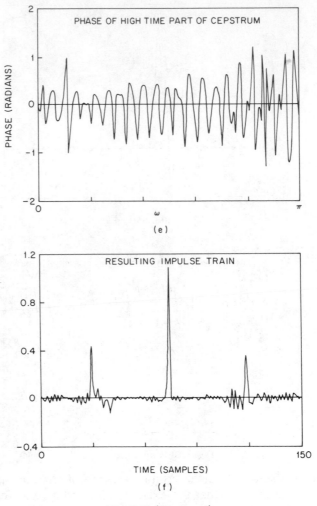

Fig. 7.10 (Continued)

cepstrum provides a way of distinguishing between voiced and unvoiced speech, and furthermore, the pitch period of voiced speech is placed clearly in evidence in the cepstrum. Also note that the formant frequencies show up clearly in the log magnitude of the vocal tract transfer function which can be obtained by applying the window of Eq. (7.42) to the cepstrum. In the remaining sections of this chapter we shall explore the use of the cepstrum in estimating pitch and formant frequencies and as a basis for a complete vocoder system.

7.3 Pitch Detection

Figures 7.8f and 7.11c suggest a powerful means for pitch estimation based on homomorphic processing. We observe that for the voiced speech example, there is a peak in the cepstrum at the fundamental period of the input speech segment. No such peak appears in the cepstrum for the unvoiced speech seg-

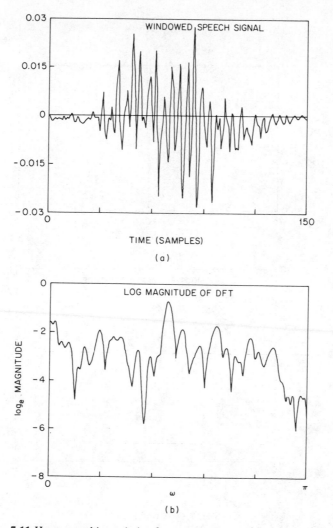

Fig. 7.11 Homomorphic analysis of unvoiced speech; (a) windowed time waveform; (b) log magnitude of short-time Fourier transform; (c) cepstrum; (d) estimate of $H_u(e^{j\omega})$.

ment. These properties of the cepstrum can be used as a basis for determining whether a speech segment is voiced or unvoiced and for estimating the fundamental period of voiced speech.

The outline of the pitch estimation procedure based on the cepstrum is rather simple. The cepstrum, computed as discussed in Section 7.1.3, is searched for a peak in the vicinity of the expected pitch period. If the cepstrum peak is above a pre-set threshold, the input speech segment is likely to be voiced, and the position of the peak is a good estimate of the pitch period. If the peak does not exceed the threshold, it is likely that the input speech segment is unvoiced. The time variation of the mode of excitation and the pitch period can be estimated by computing a time-dependent cepstrum based upon a time dependent Fourier transform. Typically, the cepstrum is computed once

373

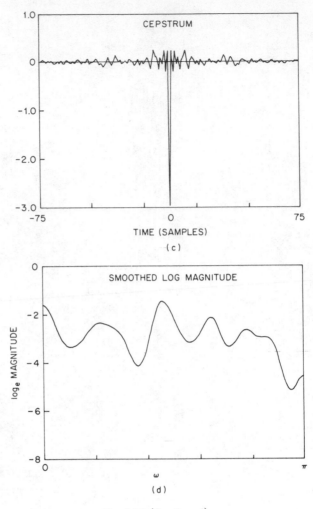

Fig. 7.11 (Continued)

every 10-20 msec since the excitation parameters do not change rapidly in normal speech.

Figures 7.12 and 7.13 show examples due to A. M. Noll [9], who first described a procedure for estimating pitch using the cepstrum. Figure 7.12 shows a series of log spectra and corresponding cepstra for a male speaker. The cepstra plotted in this example are the square of $c(n)$ as we have defined it. In this example, the sampling rate of the input was 10 kHz. A 40 msec (400 samples) Hamming window was moved in jumps of 10 msec; i.e., log spectra on the left and corresponding cepstra on the right are computed at 10 msec intervals. It can be seen from Fig. 7.12 that the first seven 40 msec intervals correspond to unvoiced speech, while the remaining cepstra indicate that the pitch period increases with time (i.e., fundamental frequency decreases). Figure 7.13 shows an example for a female talker. In this case, the speech waveform corresponding to this sequence of spectra and cepstra is voiced at the

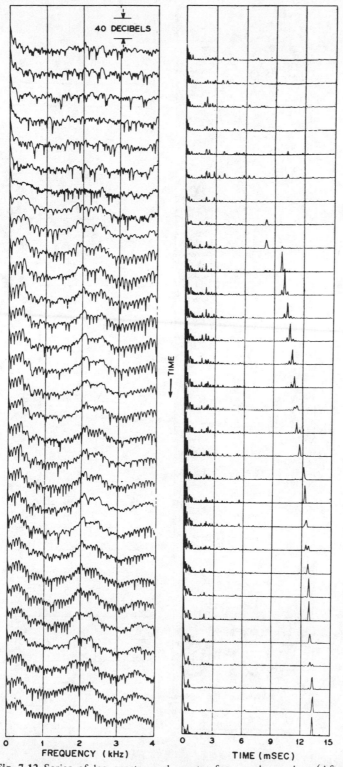

40 DECIBELS

← TIME

0   1   2   3   4
FREQUENCY (kHz)

0   3   6   9   12   15
TIME (mSEC)

Fig. 7.12 Series of log spectra and cepstra for a male speaker. (After Noll [9].)

375

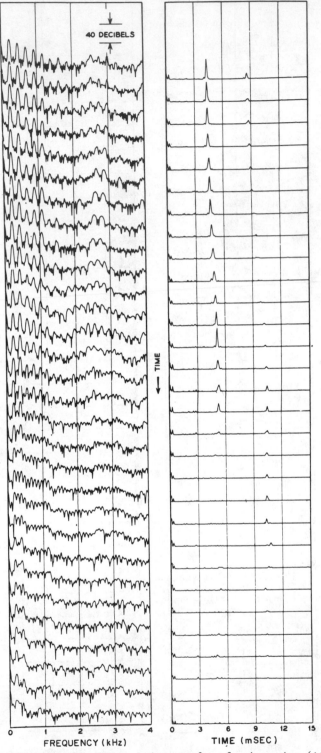

Fig. 7.13 Series of log spectra and cepstra for a female speaker. (After Noll [9].)

beginning and then becomes unvoiced at the end. It can be seen that at the end of the voiced interval, the pitch period doubles as sometimes occurs at the end of voicing. It can be seen by comparing Figs. 7.12 and 7.13 that the pitch frequency is much higher for the female talker than for the male talker.

These two examples, although impressive in the prominence with which the pitch information is displayed, may lead us to assume that an overly simplistic algorithm will produce high quality estimates of pitch and voicing. Unfortunately, as is usually the case in speech analysis, there are numerous special cases and trade-offs that must be considered in designing a cepstrum pitch detection algorithm. Noll [9] has given a flow chart of one such algorithm; however, a variety of schemes based upon the cepstrum have been used with success. Rather than give the details of any one procedure here, we feel that it is more useful to call attention to some of the essential difficulties in using the cepstrum for pitch detection.

First, the presence of a strong peak in the cepstrum in the range 3-20 msec is a very strong indication that the input speech segment is voiced. However, the absence of a peak or the existence of a low level peak is not necessarily a strong indication that the input speech segment is unvoiced. That is, the strength of or even the existence of a cepstrum peak for voiced speech is dependent on a variety of factors, including the length of the window applied to the input signal and the formant structure of the input signal. It is easily shown (see Problem 7.10) that the maximum height of the "pitch peak" is unity. This can be achieved only in the case of absolutely identical pitch periods. This is, of course, highly unlikely in natural speech, even in the case of a rectangular window which encloses exactly an integer number of periods. Rectangular windows are rarely used due to the inferior spectrum estimates that result, and in the case of, for example, a Hamming window, it is clear that both window length and the relative positions of the window and the speech signal will have considerable effect upon the height of the cepstrum peak. As an extreme example, suppose that the window is less than two pitch periods long. Clearly it is not reasonable to expect any strong indication of periodicity in the spectrum or the cepstrum in this case. Thus, the window duration is usually set so that, taking account of the tapering of the data window, at least two clearly defined periods remain in the windowed speech segment. For low pitched male speech, this requires a window on the order of 40 msec in duration. For higher pitched voices, proportionately shorter windows can be used. It is, of course, desirable to maintain the window as short as possible so as to minimize the variation of speech parameters across the analysis interval. The longer the window, the greater the variation from beginning to end and the greater will be the deviation from the model upon which the analysis is based. One approach to maintaining a window that is neither too short or too long is to adapt the window length based upon the previous (or possibly average) pitch estimates [10,11].

Another way in which the signal can deviate from the model is if it is extremely bandlimited. An extreme example is the case of a pure sinusoid. In

377

this case there is only one peak in the log spectrum. If there is no periodic oscillation in the log spectrum, there will be no peak in the cepstrum. In speech, voiced stops are generally extremely bandlimited, with no clearly defined harmonic structure at frequencies above a few hundred Hertz. In such cases there is essentially no peak in the cepstrum. Fortunately, for all but the shortest pitch periods, the pitch peak occurs in a region where the other cepstrum components have died out appreciably. Therefore, a rather low threshold can be used in searching for the pitch peak (e.g., on the order of 0.1).

With appropriate window length at input, the location and amplitude of the cepstrum peak provide a reliable pitch and voicing estimate most of the time. In the cases where the cepstrum fails to clearly display the pitch and voicing, the reliability can be improved by the addition of other information such as zero-crossing rate and energy, and by forcing the pitch and voicing estimates to vary smoothly [11]. The extra logic required to take care of special cases often requires considerable code in software implementations, but this part of a cepstrum pitch detection scheme is a small portion of the total computational effort and is well worthwhile.

7.4 Formant Estimation

From the examples of Section 7.2 we have seen that it is reasonable to assume that the low-time part of the cepstrum corresponds primarily to the vocal tract, glottal pulse and radiation information, while the high-time part is due primarily to the excitation. This is exploited in pitch and voicing estimation by searching only the high-time portion for peaks. The examples of Section 7.2 also suggest ways of using the cepstrum to estimate vocal tract response parameters. Specifically, recall that the "smoothed" log magnitude functions of Figs. 7.10a and 7.11d can be obtained by windowing the cepstrum. These smoothed log spectra display the resonant structure of the particular input speech segment; i.e., the peaks in the spectrum correspond essentially to the formant frequencies. This suggests that formants can be estimated by locating the peaks in the "cepstrally smoothed" log spectra.

Consider a model for speech production as given in Fig. 7.14. This very parsimonious model represents voiced speech by pitch period, amplitude, and the lowest three formant frequencies and unvoiced speech by simply amplitude and a single zero and pole. Additional fixed compensation accounts for the high frequency properties of the speech signal. All of the indicated parameters, of course, vary with time. A scheme for estimating these parameters is based on the computation of a cepstrally smoothed log magnitude function once every 10-20 msec [11,12]. The peaks of the log spectrum are located and a voicing decision is made from the cepstrum. If the speech segment is voiced, the pitch period is estimated from the cepstrum and the first three formant frequencies are estimated from the set of peaks in the log spectrum by logic based upon the model for speech production [11,12]. In the case of unvoiced speech, the pole is set at the location of the highest peak in the log spectrum and the zero

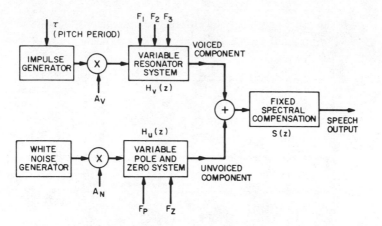

Fig. 7.14 Digital model for speech production.

located so that the relative amplitude between low and high frequencies is preserved [12].

An illustration of estimation of pitch and formant frequencies for voiced speech is given in Fig. 7.15. The left-hand half of the figure shows a sequence of cepstra computed at 20 msec intervals. On the right, the log magnitude spectrum is plotted with the corresponding cepstrally smoothed log spectrum superimposed. The lines connect the peaks that were selected by the algorithm described in Ref. [11] as the first three formant frequencies. It can be seen in Fig. 7.15 that two formant frequencies occasionally come so close together that there are no longer two distinct peaks. These situations can be detected and the resolution can be improved by evaluating the z-transform of $H_v(z)$ on a contour that passes closer to the poles. This evaluation is faciliated by a spectrum analysis algorithm called the chirp z-transform (CZT) [13]. An example of the improved resolution is shown in Figure 7.16.

Another approach to formant estimation from cepstrally smoothed log spectra was explored by Olive [14], who used an iterative procedure reminiscent of the analysis-by-synthesis method discussed in Chapter 6 to find a set of poles for a transfer function to match the smoothed log spectrum with minimum squared error.

Speech can be synthesized from the formant and pitch data estimated as described above by simply controlling the model of Fig. 7.14 with the estimated parameters. In this case, for voiced speech, the steady state vocal tract transfer function is modelled as

$$V(z) = \prod_{k=1}^{4} \frac{1 - 2e^{-\alpha_k T}\cos(2\pi F_k T) + e^{-2\alpha_k T}}{1 - 2e^{-\alpha_k T}\cos(2\pi F_k T)z^{-1} + e^{-2\alpha_k T}z^{-2}} \tag{7.44}$$

This equation describes a cascade of digital resonators which has unity gain at zero frequency so that the speech amplitude depends only on the amplitude control, A_v. The first three formant frequencies, F_1, F_2, and F_3, vary with time while F_4 is fixed at about 4000 Hz and $T = 0.0001$ sec (i.e., 10 kHz sampling frequency). The formant bandwidths α_k are also fixed at average values for

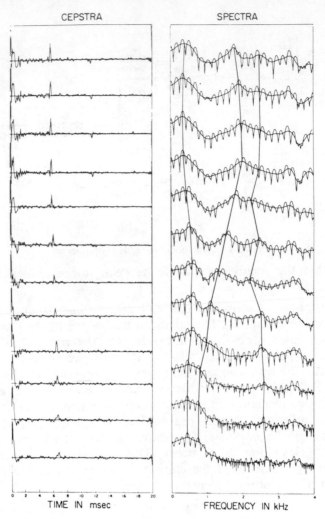

CEPSTRA   SPECTRA

TIME IN msec   FREQUENCY IN kHz

Fig. 7.15 Automatic formant estimation from cepstrally smoothed log spectra. (After Schafer and Rabiner [11].)

speech. The fixed spectral compensation, which approximates the glottal pulse and radiation contributions, is

$$S(z) = \frac{(1 - e^{-aT})(1 + e^{-bT})}{(1 - e^{-aT}z^{-1})(1 + e^{-bT}z^{-1})} \qquad (7.45)$$

where a and b are chosen to provide a good spectral match. Representative values of a and b are 400π and 5000π, respectively. More accurate values for a given speaker may be determined from a long-term average spectrum for that speaker.

For unvoiced speech, the vocal tract contributions are simulated by a system having the steady state transfer function

$$V(z) = \frac{(1 - 2e^{-\beta T}\cos(2\pi F_p T) + e^{-2\beta T})(1 - 2e^{-\beta T}\cos(2\pi F_z T)z^{-1} + e^{-2\beta T}z^{-2})}{(1 - 2e^{-\beta T}\cos(2\pi F_p T)z^{-1} + e^{-2\beta T}z^{-2})(1 - 2e^{-\beta T}\cos(2\pi F_z T) + e^{-2\beta T})}$$

where F_p has been chosen as the greatest peak of the smoothed log spectrum above 1000 Hz and F_z satisfies the empirical formula

$$F_z = (0.0065F_p + 4.5 - \Delta)(0.014F_p + 28) \qquad (7.46)$$

where

$$\Delta = 20 \log_{10}\left|H\left(e^{j2\pi F_p T}\right)\right| - 20 \log_{10}\left|H(e^{j0})\right| \qquad (7.47)$$

which ensures that the approximate relative amplitude relationship is preserved [12]. That this rather simple model can preserve the essential spectral features is shown in Figs. 7.17 and 7.18 which show comparisons between the smoothed log spectrum and the model defined by Fig. 7.14 and Eqs. (7.44)-(7.47) for both voiced and unvoiced speech, respectively.

An example of speech synthesized using this model is shown in Fig. 7.19. The upper part of the figure shows the parameters estimated from the utterance whose spectrogram is given in Fig. 7.19b. Figure 7.19c shows a spectrogram of synthetic speech created by controlling the model of Fig. 7.14 with the parameters of Fig. 7.19a. It is clear that the essential features of the signal are well preserved in the synthetic speech. Indeed, even though the representation in terms of the model is very crude, the synthetic speech is very intelligible and retains many of the identifying features of the original speaker. In fact, the pitch and formant frequencies estimated by this procedure formed the basis for extensive experiments in speaker verification. (See Chapter 9, Section 9.2.)

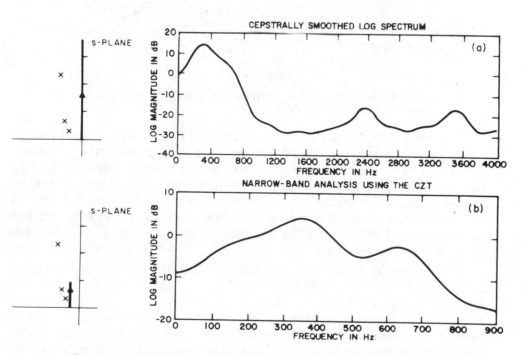

Fig. 7.16 Improved frequency resolution obtained by using the CZT algorithm. (After Schafer and Rabiner [11].)

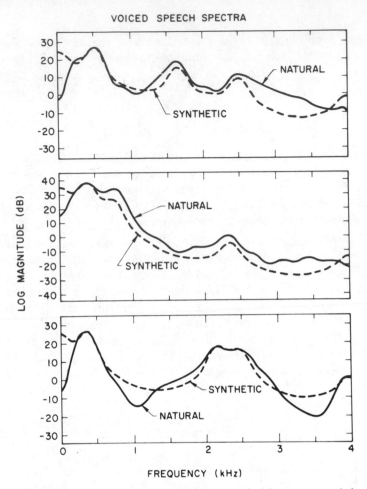

Fig. 7.17 Comparisons between cepstrally smoothed log spectra and the speech model spectra for voiced speech.

An important property of the representation that we have been discussing is that the information rate can be very low. A complete analysis/synthesis system (or formant vocoder) based upon this representation is shown in Fig. 7.20. The model parameters are estimated 100 times/sec and lowpass filtered to remove noise. The sampling rate is reduced to twice the filter cutoff frequency and the parameters are quantized. For synthesis, each parameter is interpolated back to a rate of 100 samples/sec and supplied to a synthesizer as depicted in Fig. 7.14.

A perceptual study was performed to determine appropriate system parameters [15]. The analysis and synthesis sections were first connected directly to produce references. Then the parameters were lowpass filtered to determine the lowest bandwidth for which no perceptual difference could be observed between synthesis with the filtered and unfiltered parameters. It was found that the bandwidth could be reduced to about 16 Hz with no noticeable change in quality. The filtered parameters could then be sampled at about 33 Hz (3-to-1 decimation). Then an experiment was performed to determine the

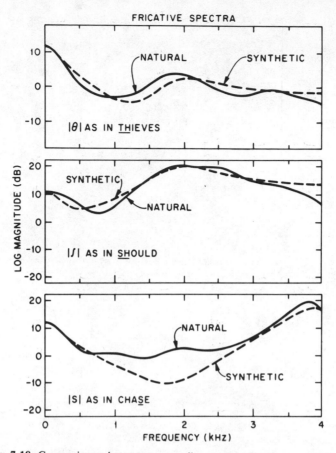

Fig. 7.18 Comparisons between cepstrally smoothed spectra and the speech model spectra for unvoiced speech.

required information rate. The formant and pitch parameters were quantized with a linear quantizer (adjusted to the range of each parameter), and the amplitude parameters were quantized with a logarithmic quantizer. A summary of the results of the perceptual test is given in Table 7.1. Using a sampling rate of 33 samples/sec and the numbers in Table 7.1, it was found that for the all-voiced utterances of the experiment no degradation in quality over the unquantized synthesis occurs for a total bit rate of about 600 bits/sec. (Note that an additional one-bit voiced/unvoiced parameter transmitted 100 times/sec was required to adequately present voiced/unvoiced transitions.)

Table 7.1 Results of Perceptual Evaluation of
a Formant Vocoder [15].

Parameter	Required Bits/Sample
τ	6
F_1	3
F_2	4
F_3	3
$\log[A_v]$	2

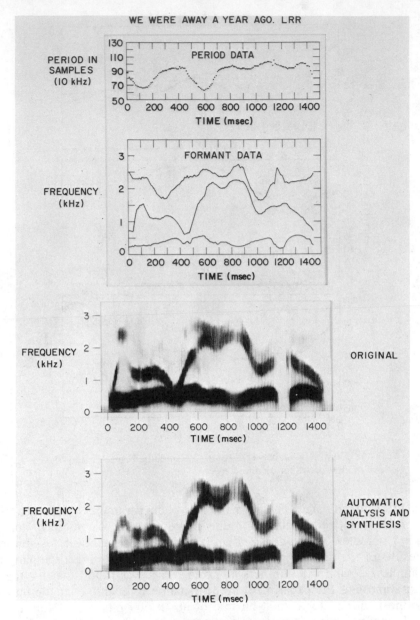

Fig. 7.19 Automatic analysis and synthesis of "We were away a year ago"; (a) pitch period and formant data as plotted by computer; (b) wideband spectrogram of original speech; (c) wideband spectrogram of synthetic speech generated from the data in (a). (After Schafer and Rabiner [11].)

Figure 7.21a shows an example of the model parameters estimated from natural speech at a rate of 100 times/sec. Figure 7.21b shows these parameters after smoothing by a 16 Hz linear phase, FIR lowpass filter. Figure 7.21c shows the parameters after decimation by a factor 3, quantization according to Table 7.1, and interpolation by a factor of 3. Although clear differences are apparent among the three representations, there is little or no perceptual difference between the synthetic speech produced from the three sets of parameters. This representation of speech at 600 bits/sec has been used in experiments on speech synthesis for computer voice response [16], as further discussed in Section 9.1.3 of Chapter 9.

7.5 The Homomorphic Vocoder

We have seen that time-dependent homomorphic processing leads to a convenient representation in which the basic speech parameters are clearly displayed and isolated from one another; i.e., excitation information in the high-time region of the cepstrum and vocal tract and glottal waveshape information in the low-time region. The time-dependent complex cepstrum, indeed, retains all the information of the time-dependent Fourier transform, which we have seen in Chapter 6 is an exact representation of the speech wave. The cepstrum, however, ignores the phase of the time-dependent Fourier representation and therefore, the time-dependent cepstrum cannot uniquely represent the speech waveform. Nevertheless, we have seen that the cepstrum is a convenient basis for estimating pitch, voicing, and formant frequencies. The cepstrum has also been used directly as a representation of speech in a system that has been called a homomorphic vocoder [17].

In the homomorphic vocoder, the cepstrum is computed once every 10-20 msec. Pitch and voicing are estimated from the cepstrum and the low-time part of each cepstrum (e.g., approximately the first 30 samples) is quantized and encoded for transmission or storage. At the synthesizer, an approximation to the impulse response $h_v(n)$ or $h_u(n)$ is computed from the quantized low-time cepstrum and explicitly convolved with an excitation function created at the

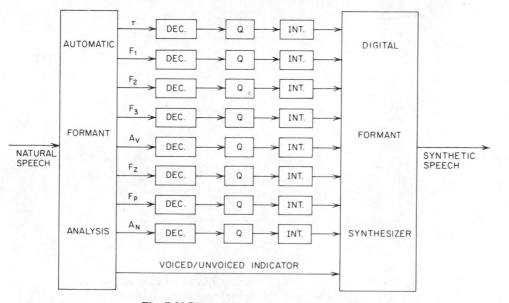

Fig. 7.20 Block diagram of a formant vocoder.

385

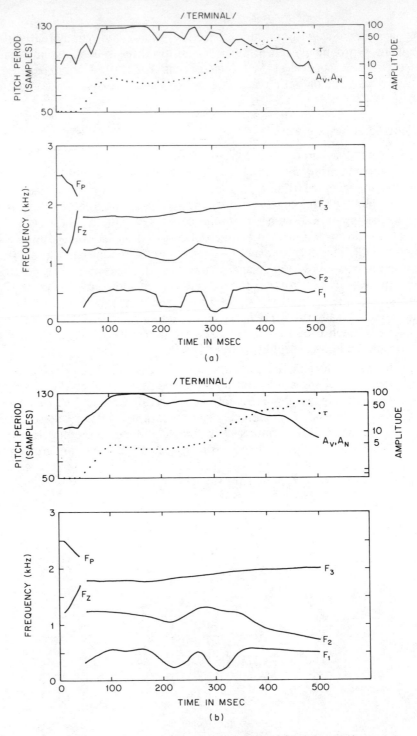

Fig. 7.21 Illustration of process of quantization of formant vocoder control signals; (a) raw data; (b) smoothed data; (c) quantized and smoothed data.

386

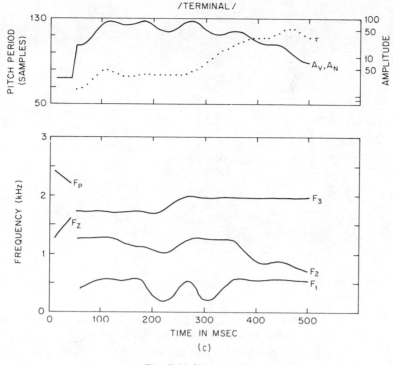

Fig. 7.21 (Continued)

synthesizer from the pitch, voicing and amplitude information. This is depicted in Fig. 7.22. Figure 7.22a shows the analysis system. The cepstrum is computed as discussed in Section 7.1.3. Then the low-time portion is selected by the cepstrum window, $l(n)$. In the simulations reported in Ref. [17], the first 26 cepstrum values were retained for quantization. The full cepstrum was also used to estimate pitch and voicing as discussed in Section 7.3. The excitation information together with the quantized cepstrum values form a digital representation with a sampling rate of 50-100 samples/sec. To synthesize an approximation to the input speech, an impulse response is computed from the cepstrum. To see how this is done, recall that the cepstrum is an even function and thus, knowledge of the positive-time part of the cepstrum permits the reconstruction of the negative-time part at the synthesizer simply by symmetry. The Fourier transform of the quantized low-time portion of the cepstrum is an approximation to the log magnitude of the combined vocal tract, glottal pulse, and radiation system function. However, the phase in this case is zero. In Fig. 7.22b, the transform is exponentiated, producing a real, even transform whose inverse transform is an "impulse response" which is an even sequence. A zero phase impulse response produced in this way from the cepstrum of Fig. 7.8f is shown in Fig. 7.23a. This impulse response can be convolved with an excitation function consisting of impulses spaced by the pitch period for voiced speech, and uniformly spaced, random polarity, impulses for unvoiced speech. (In the implementation of Ref. [17], the spacing between impulses was greater than unity for unvoiced synthesis so as to minimize computation.)

387

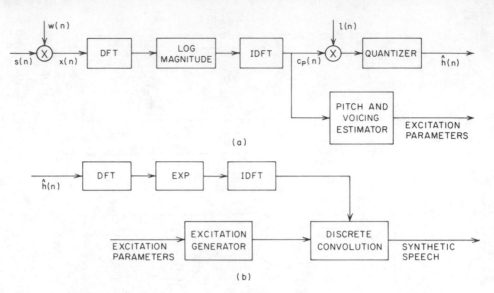

Fig. 7.22 Block diagram of a homomorphic vocoder; (a) analyzer; (b) synthesizer.

An impulse response whose log magnitude function is identical to the zero phase impulse response can be computed by arbitrarily forcing the input, $\hat{h}(n)$, to correspond to a minimum phase signal. That is, the cepstrum window is (see Eq. (7.18))

$$
\begin{aligned}
l(n) &= 1 \quad n = 0 \\
&= 2 \quad 0 < n \leqslant n_0 \\
&= 0 \quad \textit{otherwise}
\end{aligned}
\tag{7.48}
$$

The transform of the resulting $\hat{h}(n)$ in Fig. 7.22b will be minimum phase [5]. Figure 7.23b shows a minimum phase impulse response which has the same log magnitude as the zero phase impulse response of Fig. 7.23a. Oppenheim [17] also considered the case of a maximum phase reconstruction of the impulse response; i.e.,

$$
\begin{aligned}
l(n) &= 1 \quad n = 0 \\
&= 2 \quad -n_0 \leqslant n < 0 \\
&= 0 \quad \textit{otherwise}
\end{aligned}
\tag{7.49}
$$

The corresponding maximum phase impulse response is shown in Fig. 7.23c. Perceptual tests using all three phases showed the minimum phase synthesis to be preferred to the other two possibilities. This is reasonable in view of the fact that the minimum phase is closest to the phase of natural speech.

The homomorphic vocoder was reported to produce "very high quality, natural sounding speech," [17] with 26 cepstrum values quantized with 6 bits and sampled 50 times/sec. Subsequent studies have shown that the information rate can be lowered significantly by further transformations on the cep-

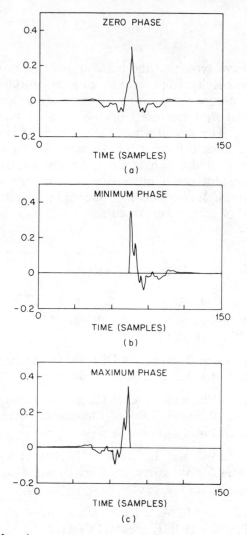

Fig. 7.23 Impulse responses computed from the cepstrum; (a) zero phase; (b) minimum phase; (c) maximum phase.

strum values before quantization [18]. Other studies have demonstrated the effectiveness of adapting the duration of the input spectrum window to the characteristics of the speech wave [19].

The homomorphic vocoder, as in the case of all vocoder systems which attempt to separate the speech parameters into excitation and vocal tract parameters, achieves low information rate and provides added flexibility in manipulating the speech signal at the expense of added complexity in the representation and degradation in quality. This particular system has the advantage that the cepstrum, which requires the greatest computational expenditure, can be used in the estimation of both the excitation parameters and the vocal tract parameters. Such a scheme is particularly attractive when a device for computing the DFT is readily available.

7.6 Summary

In this chapter we have presented the basic ideas of homomorphic signal processing as applied to speech. The main idea of homomorphic speech processing is the separation or deconvolution of a segment of speech into a component representing the vocal tract impulse response, and a component representing the excitation source. The way in which such separation is achieved is through linear filtering of the inverse Fourier transform of the log spectrum of the signal, i.e., the cepstrum of the signal. Computational considerations in implementing a homomorphic speech processing system were described. Finally, some typical methods for estimating speech parameters based on the homomorphic model of speech were discussed.

REFERENCES

1. A. V. Oppenheim, R. W. Schafer, and T. G. Stockham, Jr., "Nonlinear Filtering of Multiplied and Convolved Signals," *Proc. IEEE,* Vol. 56, No. 8, pp. 1264-1291, August 1968.

2. R. W. Schafer, "Echo Removal by Discrete Generalized Linear Filtering," Technical Report 466, Research Lab of Electronics, MIT, February 1969.

3. A. V. Oppenheim, "Superposition in a Class of Nonlinear Systems," Tech. Report No. 432, Research Lab. of Electronics, MIT, Cambridge, Massachusetts, March 1965.

4. B. Bogert, M. Healy, and J. Tukey, "The Quefrency Alanysis of Time Series for Echoes," *Proc. Symp. on Time Series Analysis,* M. Rosenblatt, Ed., Ch. 15, pp. 209-243, J. Wiley, New York, 1963.

5. A. V. Oppenheim and R. W. Schafer, *Digital Signal Processing,* Chapter 10, pp. 480-531, Prentice-Hall, Englewood Cliffs, N.J., 1975.

6. J. M. Tribolet, "A New Phase Unwrapping Algorithm," *IEEE Trans. on Acoustics, Speech, and Signal Proc.,* Vol. ASSP-25, No. 2, pp. 170-177, April 1977.

7. K. Steiglitz and B. Dickinson, "Computation of the Complex Cepstrum by Factorization of the z-Transform," *Proc. 1977 ICASSP,* pp. 723-726, May 1977.

8. A. V. Oppenheim and R. W. Schafer, "Homomorphic Analysis of Speech," *IEEE Trans. on Audio and Electroacoustics,* Vol. AU-16, No. 2, pp. 221-226, June 1968.

9. A. M. Noll, "Cepstrum Pitch Determination," *J. Acoust. Soc. Am.,* Vol. 41, pp. 293-309, February 1967.

10. L. R. Rabiner, "On the Use of Autocorrelation Analysis for Pitch Detec-

tion," *IEEE Trans on Acoustics, Speech, and Signal Proc.,* Vol. ASSP-26, No. 1, pp. 24-33, February 1977.

11. R. W. Schafer and L. R. Rabiner, "System for Automatic Formant Analysis of Voiced Speech," *J. Acoust. Soc. Am.,* Vol. 47, No. 2, pp. 634-648, February 1970.

12. J. L. Flanagan, C. H. Coker, L. R. Rabiner, R. W. Schafer, and N. Umeda, "Synthetic Voices for Computers," *IEEE Spectrum,* Vol. 7, No. 10, pp. 22-45, October 1970.

13. L. R. Rabiner, R. W. Schafer, and C. M. Rader, "The Chirp z-Transform Algorithm and Its Application," *Bell System Tech. J.,* Vol. 48, pp. 1249-1292, 1969.

14. J. Olive, "Automatic Formant Tracking in a Newton- Raphson Technique," *J. Acoust. Soc. Am.,* Vol. 50, pp. 661-670. August 1971.

15. A. E. Rosenberg, R. W. Schafer, and L. R. Rabiner, "Effects of Smoothing and Quantizing the Parameters of Formant-Coded Voiced Speech," *J. Acoust. Soc. Am.,* Vol. 50, No. 6, pp. 1532-1538, December 1971.

16. L. R. Rabiner, R. W. Schafer, and J. L. Flanagan, "Computer Synthesis of Speech by Concatentation of Formant-Coded Words," *Bell System Tech. J.,* Vol. 50, No. 5, pp. 1541-1558, May-June 1971.

17. A. V. Oppenheim, "A Speech Analysis-Synthesis System Based on Homomorphic Filtering," *J. Acoust. Soc. Am.,* Vol. 45, pp. 458-465, February 1969.

18. C. J. Weinstein and A. V. Oppenheim, "Predictive Coding in a Homomorphic Vocoder," *IEEE Trans. on Audio and Electroacoustics,* Vol. AU-19, No. 3, pp. 243-248, September 1971.

19. C. R. Patisaul and J. C. Hammett, "Time-Frequency Resolution Experiment in Speech Analysis and Synthesis," *J. Acoust. Soc. Am.,* Vol. 58, No. 6, pp. 1296-1307, December 1975.

PROBLEMS

7.1 The complex cepstrum, $\hat{x}(n)$, of a sequence $x(n)$ is the inverse Fourier transform of the complex log spectrum

$$\hat{X}(e^{j\omega}) = \log|X(e^{j\omega})| + j \arg[X(e^{j\omega})]$$

Show that the cepstrum, $c(n)$, defined as the inverse Fourier transform of the log magnitude, is the even part of $\hat{x}(n)$; i.e., show that

$$c(n) = \frac{\hat{x}(n) + \hat{x}(-n)}{2}$$

7.2 Consider an all-pole model of the vocal tract transfer function of the form

$$V(z) = \frac{1}{\displaystyle\prod_{k=1}^{q} (1-c_k z^{-1})(1-c_k^* z^{-1})}$$

where

$$c_k = r_k e^{j\theta_k} .$$

Show that the corresponding cepstrum is

$$\hat{v}(n) = 2 \sum_{k=1}^{q} \frac{(r_k)^n}{n} \cos(\theta_k n)$$

7.3 Consider an all-pole model for the combined vocal tract, glottal pulse, and radiation system of the form

$$H(z) = \frac{G}{1 - \displaystyle\sum_{k=1}^{p} \alpha_k z^{-k}}$$

Assume that all the poles of $H(z)$ are inside the unit circle. Use Eq. (7.22) to obtain a recursion relation between the complex cepstrum, $\hat{h}(n)$, and the coefficients $\{\alpha_k\}$. (Hint: How is the complex cepstrum of $1/H(z)$ related to $\hat{h}(n)$?)

7.4 Consider a *finite length* minimum phase sequence $x(n)$ with complex cepstrum $\hat{x}(n)$, and a sequence

$$y(n) = \alpha^n x(n)$$

with complex cepstrum $\hat{y}(n)$.
(a)   If $0 < \alpha < 1$, how will $\hat{y}(n)$ be related to $\hat{x}(n)$?
(b)   How should α be chosen so that $y(n)$ would no longer be minimum phase?
(c)   How should α be chosen so that $y(n)$ is maximum phase?

7.5 Show that if $x(n)$ is minimum phase, then $x(-n)$ is maximum phase.

7.6 Consider a sequence, $x(n)$, with complex cepstrum $\hat{x}(n)$. The z-transform of $\hat{x}(n)$ is

$$\hat{X}(z) = \log[X(z)] = \sum_{m=-\infty}^{\infty} \hat{x}(m) z^{-m}$$

where $X(z)$ is the z-transform of $x(n)$. The z-transform $\hat{X}(z)$ is sampled at N equally spaced points on the unit circle, to obtain

$$\hat{X}_p(k) = \hat{X}(e^{j\frac{2\pi}{N}k}) \qquad 0 \leqslant k \leqslant N - 1$$

Using the inverse DFT, we compute

$$\hat{x}_p(n) = \frac{1}{N} \sum_{k=0}^{N-1} \hat{X}_p(k) e^{j\frac{2\pi}{N}kn} \qquad 0 \leqslant n \leqslant N - 1$$

which serves as an approximation to the complex cepstrum.

(a) Express $\hat{X}_p(k)$ in terms of the true complex cepstrum, $\hat{x}(m)$.

(b) Substitute the expression obtained in (a) into the inverse DFT expression for $\hat{x}_p(n)$ and show that

$$\hat{x}_p(n) = \sum_{r=-\infty}^{\infty} \hat{x}(n+rN)\cdot$$

7.7 Consider the sequence

$$x(n) = \delta(n) + \alpha\delta(n-N_p)$$

(a) Find the complex cepstrum of $x(n)$. Sketch your result.

(b) Sketch the cepstrum, $c(n)$, for $x(n)$.

(c) Now suppose that the approximation $\hat{x}_p(n)$ is computed using Eq. (7.30). Sketch $\hat{x}_p(n)$ for $0 \leqslant n \leqslant N-1$, for the case $N_p = N/6$. What if N is not divisible by N_p?

(d) Repeat (c) for the cepstrum approximation $c_p(n)$ for $0 \leqslant n \leqslant N-1$, as computed using Eq. (7.33).

(e) If the largest impulse in the cepstrum approximation, $c_p(n)$, is used to detect N_p, how large must N be in order to avoid confusion?

7.8 In order to smooth the log magnitude spectrum of a signal, its cepstrum is often windowed and Fourier transformed as shown in Fig. P7.8.

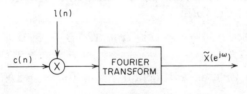

Fig. P7.8

(a) Write an expression relating $\tilde{X}(e^{j\omega})$ to $\log|X(e^{j\omega})|$ and $L(e^{j\omega})$ where $L(e^{j\omega})$ is the Fourier transform of $l(n)$.

(b) To smooth $\log|X(e^{j\omega})|$ what type of cepstral window, $l(n)$, should be used?

(c) Compare the use of a rectangular cepstral window, and a Hamming cepstral window.

(d) How long should the cepstral window be? Why?

7.9 Consider a segment of voiced speech to be represented as

$$s(m) = p(m)*h_v(m)$$

where

$$p(m) = \sum_{r=-\infty}^{\infty} \delta(m-rN_p)$$

In computing the complex cepstrum (or cepstrum) the first step is to multiply $s(m)$ by a window, $w(m)$, thereby obtaining the segment.

$$x_n(m) = s(m)w(n-m)$$

as the input to the homomorphic processing system.

(a) State the condition under which we can approximate $x_n(m)$ as

$$x(m) = p_n(m)*h_v(m)$$

where

$$p_n(m) = p(m)w(n-m)$$

(b) For the special case $n = 0$, find the z-transform of $p_0(m)$ in terms of the z-transform of $w(m)$.

(c) Express the complex cepstrum, $\hat{p}_0(m)$, in terms of $\hat{w}(m)$.

7.10 In Problem 7.9 it is shown that the periodicity of a windowed segment of voiced speech can be approximately represented by

$$p_n(m) = p(m)w(n-m)$$

where

$$p(m) = \sum_{r=-\infty}^{\infty} \delta(m-rN_p)$$

In this problem we investigate the effect of the window position on the resulting complex cepstrum $\hat{p}_n(m)$. Assume that the window is a Hamming window of the form

$$w(m) = .54 - .46\cos(2\pi m/(2N_p)) \qquad 0 \leqslant n \leqslant 2N_p$$
$$= 0 \qquad\qquad\qquad\qquad\qquad otherwise$$

(a) Sketch $p_n(m)$ as a function of m for $n = 3N_p/4,\ 9N_p/8,\ 5N_p/4,\ 3N_p/2$.

(b) In each of the above cases, give an expression for $p_n(m)$ and show that the corresponding z-transforms are of the form

$$P_n(z) = \alpha_1 z^{N_p} + \alpha_2 + \alpha_3 z^{-N_p}$$

(c) In each of the above cases, find and sketch the complex cepstrum $\hat{p}_n(m)$. (Hint: Use the power series expansion for $\log[P_n(z)]$.) Ignore terms of the form $\log[z^{\pm N_p}]$.

(d) For what position of the window is
  (i)   the sequence $p_n(m)$ minimum phase?
  (ii) the sequence $p_n(m)$ maximum phase?
  (iii) the first cepstrum peak largest?
  (iv) the first cepstrum peak smallest?

(e) How would your answers to the above questions change if the window is lengthened? Shortened?

7.11 The z-transform of a signal $x(n)$ is defined as

$$X(z) = \sum_{n=0}^{N-1} x(n)z^{-n}$$

We evaluate $X(z)$ at a set of points

$$z_k = AW^{-k} \qquad k = 0, 1, ..., M - 1$$

where A and W are arbitrary complex numbers. If we make the simple substitution

$$nk = \frac{[n^2 + k^2 - (k-n)^2]}{2}$$

then $X(z_k)$ can be written in the form

$$X(z_k) = P(k) \sum_{n=0}^{N-1} y(n) g(k-n)$$

i.e., $X(z_k)$ is a convolution of $y(n)$ and $g(n)$.
(a) Determine $P(k)$, $y(n)$ and $g(n)$ in terms of $x(n)$, A, and W.
(b) Sketch the points z_k in the z-plane.
(c) Can you suggest how the FFT can be used to evaluate the above expression for $X(z_k)$?

8

Linear Predictive Coding
of Speech

8.0 Introduction

One of the most powerful speech analysis techniques is the method of linear predictive analysis. This method has become the predominant technique for estimating the basic speech parameters, e.g., pitch, formants, spectra, vocal tract area functions, and for representing speech for low bit rate transmission or storage. The importance of this method lies both in its ability to provide extremely accurate estimates of the speech parameters, and in its relative speed of computation. In this chapter, we present a formulation of the ideas behind linear prediction, and discuss some of the issues which are involved in using it in practical speech applications.

The basic idea behind linear predictive analysis is that a speech sample can be approximated as a linear combination of past speech samples. By minimizing the sum of the squared differences (over a finite interval) between the actual speech samples and the linearly predicted ones, a unique set of predictor coefficients can be determined. (The predictor coefficients are the weighting coefficients used in the linear combination.)

The philosophy of linear prediction is intimately related to the basic speech synthesis model discussed in Chapter 3 in which it was shown that speech can be modelled as the output of a linear, time-varying system excited by either quasi-periodic pulses (during voiced speech), or random noise (during unvoiced speech). The linear prediction method provides a robust, reliable, and accurate method for estimating the parameters that characterize the linear, time-varying system.

Linear predictive techniques have already been discussed in the context of the waveform quantization methods of Chapter 5. There it was suggested that a linear predictor could be applied in a differential quantization scheme to reduce the bit rate of the digital representation of the speech waveform. In fact, the mathematical basis for an adaptive high order predictor used for DPCM waveform coding is identical to the analysis that we shall present in this chapter. In adaptive DPCM coding the emphasis is on finding a predictor that will reduce the variance of the difference signal so that quantization error can also be reduced. In this chapter we take a more general viewpoint and show how the basic linear prediction idea leads to a set of analysis techniques that can be used to estimate parameters of a speech model. This general set of linear predictive analysis techniques is often referred to as linear predictive coding or LPC.

The techniques and methods of linear prediction have been available in the engineering literature for a long time. The ideas of linear prediction have been in use in the areas of control, and information theory under the names of system estimation and system identification. The term system identification is particularly descriptive of LPC methods in that once the predictor coefficients have been obtained, the system has been uniquely identified to the extent that it can be modelled as an all-pole linear system.

As applied to speech processing, the term linear prediction refers to a variety of essentially equivalent formulations of the problem of modelling the speech waveform [1-18]. The differences among these formulations are often those of philosophy or way of viewing the problem. In other cases the differences concern the details of the computations used to obtain the predictor coefficients. Thus as applied to speech, the various (often equivalent) formulations of linear prediction analysis have been:

1. the covariance method [3]
2. the autocorrelation formulation [1,2,9]
3. the lattice method [11,12]
4. the inverse filter formulation [1]
5. the spectral estimation formulation [12]
6. the maximum likelihood formulation [4,6]
7. the inner product formulation [1]

In this chapter we will examine in detail the similarities and differences among only the first three basic methods of analysis listed above, since all the other formulations are equivalent to one of these three.

The importance of linear prediction lies in the accuracy with which the basic model applies to speech. Thus a major part of this chapter is devoted to a discussion of how a variety of speech parameters can be reliably estimated using linear prediction methods. Furthermore some typical examples of speech applications which rely primarily on linear predictive analysis are discussed here, and in Chapter 9, to show the wide range of problems to which LPC has been successfully applied.

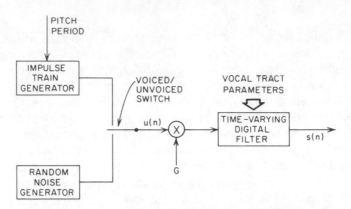

Fig. 8.1 Block diagram of simplified model for speech production.

8.1 Basic Principles of Linear Predictive Analysis

Throughout this book we have repeatedly referred to the basic discrete-time model for speech production that was developed in Chapter 3. The particular form of this model that is appropriate for the discussion of linear predictive analysis is depicted in Fig. 8.1. In this case, the composite spectrum effects of radiation, vocal tract, and glottal excitation are represented by a time-varying digital filter whose steady-state system function is of the form

$$H(z) = \frac{S(z)}{U(z)} = \frac{G}{1 - \sum_{k=1}^{p} a_k z^{-k}} \tag{8.1}$$

This system is excited by an impulse train for voiced speech or a random noise sequence for unvoiced speech. Thus, the parameters of this model are: voiced/unvoiced classification, pitch period for voiced speech, gain parameter G, and the coefficients $\{a_k\}$ of the digital filter. These parameters, of course, all vary slowly with time.

The pitch period and voiced/unvoiced classification can be estimated using one of the many methods already discussed in this book or by methods based on linear predictive analysis to be discussed later in this chapter. As discussed in Chapter 3, this simplified all-pole model is a natural representation of non-nasal voiced sounds, but for nasals and fricative sounds, the detailed acoustic theory calls for both poles and zeros in the vocal tract transfer function. We shall see, however, that if the order p is high enough, the all-pole model provides a good representation for almost all the sounds of speech. The major advantage of this model is that the gain parameter, G, and the filter coefficients $\{a_k\}$ can be estimated in a very straightforward and computationally efficient manner by the method of linear predictive analysis.

For the system of Fig. 8.1, the speech samples $s(n)$ are related to the excitation $u(n)$ by the simple difference equation

$$s(n) = \sum_{k=1}^{p} a_k s(n-k) + G u(n) \tag{8.2}$$

A linear predictor with prediction coefficients, α_k is defined as a system whose output is

$$\tilde{s}(n) = \sum_{k=1}^{p} \alpha_k s(n-k) \tag{8.3}$$

Such systems were used in Chapter 5 to reduce the variance of the difference signal in differential quantization schemes. The system function of a pth order linear predictor is the polynomial

$$P(z) = \sum_{k=1}^{p} \alpha_k z^{-k} \tag{8.4}$$

The prediction error, $e(n)$, is defined as

$$e(n) = s(n) - \tilde{s}(n) = s(n) - \sum_{k=1}^{p} \alpha_k s(n-k) \tag{8.5}$$

From Eq. (8.5) it can be seen that the prediction error sequence is the output of a system whose transfer function is

$$A(z) = 1 - \sum_{k=1}^{p} \alpha_k z^{-k} \tag{8.6}$$

It can be seen by comparing Eqs. (8.2) and (8.5) that if the speech signal obeys the model of Eq. (8.2) exactly, and if $\alpha_k = a_k$, then $e(n) = Gu(n)$. Thus, the *prediction error filter*, $A(z)$, will be an *inverse filter* for the system, $H(z)$, of Eq. (8.1), i.e.,

$$H(z) = \frac{G}{A(z)} \tag{8.7}$$

The basic problem of linear prediction analysis is to determine a set of predictor coefficients $\{\alpha_k\}$ directly from the speech signal in such a manner as to obtain a good estimate of the spectral properties of the speech signal through the use of Eq. (8.7). Because of the time-varying nature of the speech signal the predictor coefficients must be estimated from short segments of the speech signal. The basic approach is to find a set of predictor coefficients that will minimize the mean-squared prediction error over a short segment of the speech waveform. The resulting parameters are then *assumed* to be the parameters of the system function, $H(z)$, in the model for speech production.

That this approach will lead to useful results may not be immediately obvious, but it can be justified in several ways. First, recall that if $\alpha_k = a_k$, then $e(n) = Gu(n)$. For voiced speech this means that $e(n)$ would consist of a train of impulses; i.e., $e(n)$ would be small most of the time. Thus, finding α_k's that minimize prediction error seems consistent with this observation. A second motivation for this approach follows from the fact that if a signal is generated by Eq. (8.2) with non-time-varying coefficients and excited either by a single impulse or by a stationary white noise input, then it can be shown that the predictor coefficients that result from minimizing the mean squared prediction error (over all time) are identical to the coefficients of Eq. (8.2). A third very pragmatic justification for using the minimum mean-squared prediction error as a basis for estimating the model parameters is that this approach leads to a set of linear equations that can be efficiently solved to obtain the predictor parameters. More importantly the resulting parameters comprise a very useful and accurate representation of the speech signal as we shall see in this chapter.

The short-time average prediction error is defined as

$$E_n = \sum_m e_n^2(m) \tag{8.8}$$

$$= \sum_m (s_n(m) - \tilde{s}_n(m))^2 \tag{8.9}$$

$$= \sum_m \left[s_n(m) - \sum_{k=1}^p \alpha_k s_n(m-k) \right]^2 \tag{8.10}$$

where $s_n(m)$ is a segment of speech that has been selected in the vicinity of sample n, i.e.,

$$s_n(m) = s(m+n) \tag{8.11}$$

The range of summation in Eqs. (8.8)-(8.10) is temporarily left unspecified, but since we wish to develop a short-time analysis technique, the sum will always be over a finite interval. Also note that to obtain an average we should divide by the length of the speech segment. However, this constant is irrelevant to the set of linear equations that we will obtain and therefore is omitted. We can find the values of α_k that minimize E_n in Eq. (8.10) by setting $\partial E_n/\partial \alpha_i = 0$, $i=1,2,\ldots,p$, thereby obtaining the equations

$$\sum_m s_n(m-i)s_n(m) = \sum_{k=1}^p \hat{\alpha}_k \sum_m s_n(m-i)s_n(m-k) \qquad 1 \leqslant i \leqslant p \tag{8.12}$$

where $\hat{\alpha}_k$ are the values of α_k that minimize E_n. (Since $\hat{\alpha}_k$ is unique, we will drop the caret and use the notation α_k to denote the values that minimize E_n.) If we define

$$\phi_n(i,k) = \sum_m s_n(m-i)s_n(m-k) \tag{8.13}$$

then Eq. (8.12) can be written more compactly as

$$\sum_{k=1}^p \alpha_k \phi_n(i,k) = \phi_n(i,0) \qquad i=1,2,\ldots,p \tag{8.14}$$

This set of p equations in p unknowns can be solved in an efficient manner for the unknown predictor coefficients $\{\alpha_k\}$ that minimize the average squared prediction error for the segment $s_n(m)$.[1] Using Eqs. (8.10) and (8.12), the minimum mean-squared prediction error can be shown to be

---

[1]It is clear that the α_k's are functions of n (the time index at which they are estimated) although this dependence will not be explicitly shown. We shall also find it advantageous to drop the subscripts n on E_n, $s_n(m)$, and $\phi_n(i,k)$ when no confusion will result.

$$E_n = \sum_m s_n^2(m) - \sum_{k=1}^{p} \alpha_k \sum_m s_n(m) s_n(m-k) \qquad (8.15)$$

and using Eq. (8.14) we can express E_n as

$$E_n = \phi_n(0,0) - \sum_{k=1}^{p} \alpha_k \phi_n(0,k) \qquad (8.16)$$

Thus the total minimum error consists of a fixed component, and a component which depends on the predictor coefficients.

To solve for the optimum predictor coefficients, we must first compute the quantities $\phi_n(i,k)$ for $1 \leqslant i \leqslant p$ and $0 \leqslant k \leqslant p$. Once this is done we only have to solve Eq. (8.14) to obtain the α_k's. Thus, in principle, linear prediction analysis is very straightforward. However, the details of the computation of $\phi_n(i,k)$ and the subsequent solution of the equations are somewhat intricate and further discussion is required.

So far we have not explicitly indicated the limits on the sums in Eqs. (8.8)-(8.10) and in Eq. (8.12); however it should be emphasized that the limits on the sum in Eq. (8.12) are identical to the limits assumed for the mean squared prediction error in Eqs. (8.8)-(8.10). As we have stated, if we wish to develop a short-time analysis procedure, the limits must be over a finite interval. There are two basic approaches to this question, and we shall see below that two methods for linear predictive analysis emerge out of a consideration of the limits of summation and the definition of the waveform segment $s_n(m)$.

8.1.1 The autocorrelation method [1,2,5]

One approach to determining the limits on the sums in Eqs. (8.8)-(8.10) and Eq. (8.12) is to assume that the waveform segment, $s_n(m)$, is identically zero outside the interval $0 \leqslant m \leqslant N - 1$. This can be conveniently expressed as

$$s_n(m) = s(m+n) w(m) \qquad (8.17)$$

where $w(m)$ is a finite length window (e.g. a Hamming window) that is identically zero outside the interval $0 \leqslant m \leqslant N - 1$.

The effect of this assumption on the question of limits of summation for the expressions for E_n can be seen by considering Eq. (8.5). Clearly, if $s_n(m)$ is nonzero only for $0 \leqslant m \leqslant N - 1$, then the corresponding prediction error, $e_n(m)$, for a pth order predictor will be nonzero over the interval $0 \leqslant m \leqslant N - 1 + p$. Thus, for this case E_n is properly expressed as

$$E_n = \sum_{m=0}^{N+p-1} e_n^2(m) \qquad (8.18)$$

Alternatively, we could have simply indicated that the sum should be over all nonzero values by summing from $-\infty$ to $+\infty$ [2].

Returning to Eq. (8.5), it can be seen that the prediction error is likely to be large at the beginning of the interval (specifically $0 \leqslant m \leqslant p-1$) because we

401

are trying to predict the signal from samples that have arbitrarily been set to zero. Likewise the error can be large at the end of the interval (specifically $N \leqslant m \leqslant N+p-1$) because we are trying to predict zero from samples that are nonzero. For this reason, a window which tapers the segment, $s_n(m)$, to zero is generally used for $w(m)$ in Eq. (8.17).

The limits on the expression for $\phi_n(i,k)$ in Eq. (8.13) are identical to those of Eq. (8.18). However, because $s_n(m)$ is identically zero outside the interval $0 \leqslant m \leqslant N-1$, it is simple to show that

$$\phi_n(i,k) = \sum_{m=0}^{N+p-1} s_n(m-i)s_n(m-k) \quad \begin{matrix} 1 \leqslant i \leqslant p \\ 0 \leqslant k \leqslant p \end{matrix} \tag{8.19a}$$

can be expressed as

$$\phi_n(i,k) = \sum_{m=0}^{N-1-(i-k)} s_n(m)s_n(m+i-k) \quad \begin{matrix} 1 \leqslant i \leqslant p \\ 0 \leqslant k \leqslant p \end{matrix} \tag{8.19b}$$

Furthermore it can be seen that in this case $\phi_n(i,k)$ is identical to the short-time autocorrelation function of Eq. (4.30) evaluated for $(i-k)$. That is

$$\phi_n(i,k) = R_n(i-k) \tag{8.20}$$

where

$$R_n(k) = \sum_{m=0}^{N-1-k} s_n(m)s_n(m+k) \tag{8.21}$$

The computation of $R_n(k)$ is covered in detail in Section 4.6 and thus we shall not consider such details here. Since $R_n(k)$ is an even function, it follows that

$$\phi_n(i,k) = R_n(|i-k|) \quad \begin{matrix} i = 1, 2, \ldots, p \\ k = 0, 1, \ldots, p \end{matrix} \tag{8.22}$$

Therefore Eq. (8.14) can be expressed as

$$\sum_{k=1}^{p} \alpha_k R_n(|i-k|) = R_n(i) \quad 1 \leqslant i \leqslant p \tag{8.23}$$

Similarly, the minimum mean squared prediction error of Eq. (8.16) takes the form

$$E_n = R_n(0) - \sum_{k=1}^{p} \alpha_k R_n(k) \tag{8.24}$$

The set of equations given by Eqs. (8.23) can be expressed in matrix form as

$$\begin{bmatrix} R_n(0) & R_n(1) & R_n(2) & \cdots & R_n(p-1) \\ R_n(1) & R_n(0) & R_n(1) & \cdots & R_n(p-2) \\ R_n(2) & R_n(1) & R_n(0) & \cdots & R_n(p-3) \\ \cdots & \cdots & \cdots & \cdots & \cdots \\ \cdots & \cdots & \cdots & \cdots & \cdots \\ R_n(p-1) & R_n(p-2) & R_n(p-3) & \cdots & R_n(0) \end{bmatrix} \begin{bmatrix} \alpha_1 \\ \alpha_2 \\ \alpha_3 \\ \cdots \\ \cdots \\ \alpha_p \end{bmatrix} = \begin{bmatrix} R_n(1) \\ R_n(2) \\ R_n(3) \\ \cdots \\ \cdots \\ R_n(p) \end{bmatrix} \tag{8.25}$$

The $p \times p$ matrix of autocorrelation values is a Toeplitz matrix; i.e., it is symmetric and all the elements along a given diagonal are equal. This special property will be exploited in Section 8.3 to obtain an efficient algorithm for the solution of Eq. (8.23).

8.1.2 The covariance method [3]

The second basic approach to defining the speech segment $s_n(m)$ and the limits on the sums is to fix the interval over which the mean-squared error is computed and then consider the effect on the computation of $\phi_n(i,k)$. That is, if we define

$$E_n = \sum_{m=0}^{N-1} e_n^2(m) \qquad (8.26)$$

then $\phi_n(i,k)$ becomes

$$\phi_n(i,k) = \sum_{m=0}^{N-1} s_n(m-i) s_n(m-k) \qquad \begin{array}{c} 1 \leqslant i \leqslant p \\ 0 \leqslant k \leqslant p \end{array} \qquad (8.27)$$

In this case, if we change the index of summation we can express $\phi_n(i,k)$ as either

$$\phi_n(i,k) = \sum_{m=-i}^{N-i-1} s_n(m) s_n(m+i-k) \qquad \begin{array}{c} 1 \leqslant i \leqslant p \\ 0 \leqslant k \leqslant p \end{array} \qquad (8.28a)$$

or

$$\phi_n(i,k) = \sum_{m=-k}^{N-k-1} s_n(m) s_n(m+k-i) \qquad \begin{array}{c} 1 \leqslant i \leqslant p \\ 0 \leqslant k \leqslant p \end{array} \qquad (8.28b)$$

Although the equations look very similar to Eq. (8.19b), we see that the limits of summation are not the same. Equations (8.28) call for values of $s_n(m)$ outside the interval $0 \leqslant m \leqslant N - 1$. Indeed, to evaluate $\phi_n(i,k)$ for all of the required values of i and k requires that we use values of $s_n(m)$ in the interval $-p \leqslant m \leqslant N - 1$. If we are to be consistent with the limits on E_n in Eq. (8.26) then we have no choice but to supply the required values. In this case it does not make sense to taper the segment of speech to zero at the ends as in the autocorrelation method since the necessary values are made available from outside the interval $0 \leqslant m \leqslant N - 1$. Clearly, this approach is very similar to what was called the modified autocorrelation function in Chapter 4. As pointed out in Section 4.6, this approach leads to a function which is not a true autocorrelation function, but rather, the cross-correlation between two very similar, but not identical, finite length segments of the speech wave. Although the differences between Eq. (8.28) and Eq. (8.19b) appear to be minor computational details, the set of equations

$$\sum_{k=1}^{p} \alpha_k \phi_n(i,k) = \phi_n(i,0) \qquad i = 1, 2, \ldots, p \qquad (8.29a)$$

has significantly different properties that strongly affect the method of solution and the properties of the resulting optimum predictor. In matrix form these equations become

$$
\begin{bmatrix}
\phi_n(1,1) & \phi_n(1,2) & \phi_n(1,3) & \cdots & \phi_n(1,p) \\
\phi_n(2,1) & \phi_n(2,2) & \phi_n(2,3) & \cdots & \phi_n(2,p) \\
\phi_n(3,1) & \phi_n(3,2) & \phi_n(3,3) & \cdots & \phi_n(3,p) \\
\cdots & \cdots & \cdots & \cdots & \cdots \\
\cdots & \cdots & \cdots & \cdots & \cdots \\
\phi_n(p,1) & \phi_n(p,2) & \phi_n(p,3) & \cdots & \phi_n(p,p)
\end{bmatrix}
\begin{bmatrix}
\alpha_1 \\ \alpha_2 \\ \alpha_3 \\ \cdots \\ \cdots \\ \alpha_p
\end{bmatrix}
=
\begin{bmatrix}
\phi_n(1,0) \\ \phi_n(2,0) \\ \phi_n(3,0) \\ \cdots \\ \cdots \\ \phi_n(p,0)
\end{bmatrix}
\qquad (8.29b)
$$

In this case, since $\phi_n(i,k) = \phi_n(k,i)$ (see Eq. (8.28)), the $p \times p$ matrix of correlation-like values is symmetric but *not* Toeplitz. Indeed, it can be seen that the diagonal elements are related by the equation

$$
\phi_n(i+1,k+1) = \phi_n(i,k) + s_n(-i-1)s_n(-k-1)
$$
$$
- s_n(N-1-i)s_n(N-1-k) \qquad (8.30)
$$

The method of analysis based upon this method of computation of $\phi_n(i,k)$ has come to be known as the *covariance method* because the matrix of values $\{\phi_n(i,k)\}$ has the properties of a covariance matrix [5].[2]

8.1.3 Summary

It has been shown that by using different definitions of the segments of the signal to be analyzed, two distinct sets of analysis equations can be obtained. For the autocorrelation method, the signal is windowed by an N-point window, and the quantities $\phi_n(i,k)$ are obtained using a short-time autocorrelation function. The resulting matrix of correlations is Toeplitz leading to one type of solution for the predictor coefficients. For the covariance method, the signal is assumed to be known for the set of values $-p \leqslant n \leqslant N-1$. Outside this interval no assumptions need be made about the signal, since these are the only values needed in the computation. The resulting matrix of correlations in this case is symmetric but not Toeplitz. The result is that the two methods of computing the correlations lead to different methods of solution of the analysis equations and to sets of predictor coefficients with somewhat different properties.

In later sections we will compare and contrast computational details and results for both these techniques as well as for another method yet to be discussed. First, however, we will show how the gain, G, in Fig. 8.1, can be determined from the prediction error expression.

8.2 Computation of the Gain for the Model [2]

It is reasonable to expect that the gain, G, could be determined by matching the

---

[2]This terminology, which is firmly entrenched, is somewhat confusing since the term covariance usually refers to the correlation of a signal with its mean removed.

energy in the signal with the energy of the linearly predicted samples. This indeed is true when appropriate assumptions are made about the excitation signal to the LPC system.

It is possible to relate the gain constant G to the excitation signal and the error in prediction by referring back to Eqs. (8.2) and (8.5).[3] The excitation signal, $Gu(n)$, can be expressed as

$$Gu(n) = s(n) - \sum_{k=1}^{p} a_k s(n-k) \tag{8.31a}$$

whereas the prediction error signal $e(n)$ is expressed as

$$e(n) = s(n) - \sum_{k=1}^{p} \alpha_k s(n-k) \tag{8.31b}$$

In the case where $a_k = \alpha_k$, i.e., the actual predictor coefficients, and those of the model are identical, then

$$e(n) = Gu(n) \tag{8.32}$$

i.e., the input signal is proportional to the error signal with the constant of proportionality being the gain constant, G. A detailed discussion of the properties of the prediction error signal is given in Section 8.5.

Since Eq. (8.32) is only approximate (i.e., it is valid to the extent that the ideal and the actual linear prediction parameters are identical) it is generally not possible to solve for G in a reliable way directly from the error signal itself. Instead the more reasonable assumption is made that the energy in the error signal is equal to the energy in the excitation input, i.e.,

$$G^2 \sum_{m=0}^{N-1} u^2(m) = \sum_{m=0}^{N-1} e^2(m) = E_n \tag{8.33}$$

At this point we must make some assumptions about $u(n)$ so as to be able to relate G to the known quantities, e.g., the α_k's and the correlation coefficients. There are two cases of interest for the excitation. For voiced speech it is reasonable to assume $u(n) = \delta(n)$, i.e., the excitation is a unit sample at $n = 0$.[4] For this assumption to be valid requires that the effects of the glottal pulse shape used in the actual excitation for voiced speech be lumped together with the vocal tract transfer function, and therefore both of these effects are essentially modelled by the time-varying linear predictor. This requires that the predictor order, p, be large enough to account for both the vocal tract and glottal pulse effects. We will discuss the choice of predictor order in a later section. For unvoiced speech it is most reasonable to assume that $u(n)$ is a zero mean, unity variance, stationary, white noise process.

Based on these assumptions we can now determine the gain constant G by utilizing Eq. (8.33). For voiced speech, we have as input $G\delta(n)$. If we call the

---

[3]Note that the gain is also a function of time.

[4]Note that for this assumption to be valid requires that the analysis interval be about the same length as a pitch period.

resulting output for this particular input $h(n)$ (since it is actually the impulse response of the system with transfer function $H(z)$ as in Eq. (8.1)) we get the relation

$$h(n) = \sum_{k=1}^{p} \alpha_k h(n-k) + G\delta(n) \qquad (8.34)$$

It is readily shown [Problem 8.1] that the autocorrelation function of $h(n)$, defined as

$$\tilde{R}(m) = \sum_{n=0}^{\infty} h(n)h(m+n) \qquad (8.35)$$

satisfies the relations

$$\tilde{R}(m) = \sum_{k=1}^{p} \alpha_k \tilde{R}(|m-k|) \qquad m=1,2,\ldots,p \qquad (8.36a)$$

and

$$\tilde{R}(0) = \sum_{k=1}^{p} \alpha_k \tilde{R}(k) + G^2 \qquad (8.36b)$$

Since Eqs. (8.36) are identical to Eqs. (8.23) it follows that

$$\tilde{R}(m) = R_n(m) \qquad 1 \leqslant m \leqslant p \qquad (8.37)$$

Since the total energies in the signal $(R(0))$ and the impulse response $(\tilde{R}(0))$ must be equal we can use Eqs. (8.24), (8.33) and (8.36b) to obtain

$$G^2 = R_n(0) - \sum_{k=1}^{p} \alpha_k R_n(k) = E_n \qquad (8.38)$$

It is interesting to note that Eq. (8.37) and the requirement that the energy of the impulse response be equal to the energy of the signal together require that the first $p+1$ coefficients of the autocorrelation function of the impulse response of the model are identical to the first $p+1$ coefficients of the auto-correlation function of the speech signal.

For the case of unvoiced speech, the correlations are defined as statistical averages. It is assumed that the input is white noise with zero mean and unity variance; i.e.,

$$E[u(n)u(n-m)] = \delta(m) \qquad (8.39)$$

If we excite the system with the random input $Gu(n)$ and call the output $g(n)$ then

$$g(n) = \sum_{k=1}^{p} \alpha_k g(n-k) + Gu(n) \qquad (8.40)$$

If we now let $\tilde{R}(m)$ denote the autocorrelation function of $g(n)$, then

$$\tilde{R}(m) = E[g(n)g(n-m)] = \sum_{k=1}^{p} \alpha_k E[g(n-k)g(n-m)] + E[Gu(n)g(n-m)]$$

$$= \sum_{k=1}^{p} \alpha_k \tilde{R}(m-k) \qquad m \neq 0 \qquad (8.41)$$

since $E[u(n)g(n-m)]=0$ for $m > 0$ because $u(n)$ is uncorrelated with any signal prior to $u(n)$. For $m = 0$ we get

$$
\tilde{R}(0) = \sum_{k=1}^{p} \alpha_k \tilde{R}(k) + GE[u(n)g(n)]
$$

$$
= \sum_{k=1}^{p} \alpha_k \tilde{R}(k) + G^2 \tag{8.42}
$$

since $E[u(n)g(n)] = E[u(n)(Gu(n) + \textit{terms prior to } n)] = G$. Since the energy in the response to $Gu(n)$ must equal the energy in the signal, we get

$$
\tilde{R}(m) = R_n(m) \quad 0 \leqslant m \leqslant p \tag{8.43}
$$

or

$$
G^2 = R_n(0) - \sum_{k=1}^{p} \alpha_k R_n(k) \tag{8.44}
$$

as was the case for the impulse excitation for voiced speech.

8.3 Solution of the LPC Equations

In order to effectively implement a linear predictive analysis system, it is necessary to solve the linear equations in an efficient manner. Although a variety of techniques can be applied to solve a system of p linear equations in p unknowns, these techniques are not equally efficient. Because of the special properties of the coefficient matrices it is possible to solve the equations much more efficiently than is possible in general. In this section we will discuss in detail two methods for obtaining the predictor coefficients, and then we will compare and contrast several properties of these solutions.

8.3.1 Cholesky decomposition solution for the covariance method [3]

For the covariance method, the set of equations which must be solved is of the form:

$$
\sum_{k=1}^{p} \alpha_k \phi_n(i,k) = \phi_n(i,0) \quad i=1,2, \ldots, p \tag{8.45}
$$

or in matrix notation

$$
\Phi \alpha = \psi \tag{8.46}
$$

where Φ is a positive definite symmetric matrix with $(i,j)^{th}$ element $\phi_n(i,j)$, and α and ψ are column vectors with elements α_i, and $\phi_n(i,0)$ respectively. The system of equations given by Eq. (8.45) can be solved in an efficient manner since the matrix Φ is a symmetric, positive definite matrix. The resulting method of solution is called the Cholesky decomposition (or sometimes it is called the square root method) [3]. For this method the matrix $\mathbf{\Phi}$ is expressed in the form

$$\mathbf{\Phi} = \mathbf{VDV}^t \tag{8.47}$$

where $\mathbf{V}$ is a lower triangular matrix (whose main diagonal elements are all 1's), and $\mathbf{D}$ is a diagonal matrix. The superscript t denotes matrix transpose. The elements of the matrices $\mathbf{V}$ and $\mathbf{D}$ are readily determined from Eq. (8.47) by solving for the $(i,j)^{th}$ element of both sides of Eq. (8.47) giving

$$\phi_n(i,j) = \sum_{k=1}^{j} V_{ik} d_k V_{jk} \qquad 1 \leqslant j \leqslant i-1 \tag{8.48}$$

or

$$V_{ij} d_j = \phi_n(i,j) - \sum_{k=1}^{j-1} V_{ik} d_k V_{jk} \qquad 1 \leqslant j \leqslant i-1 \tag{8.49}$$

and, for the diagonal elements

$$\phi_n(i,i) = \sum_{k=1}^{i} V_{ik} d_k V_{ik} \tag{8.50}$$

or

$$d_i = \phi_n(i,i) - \sum_{k=1}^{i-1} V_{ik}^2 d_k \qquad i \geqslant 2 \tag{8.51}$$

with

$$d_1 = \phi_n(1,1) \tag{8.52}$$

To illustrate the use of Eqs. (8.47)-(8.52) consider an example with $p = 4$, and matrix elements $\phi_n(i,j) = \phi_{ij}$. Equation (8.47) is thus of the form

$$
\begin{bmatrix}
\phi_{11} & \phi_{21} & \phi_{31} & \phi_{41} \\
\phi_{21} & \phi_{22} & \phi_{32} & \phi_{42} \\
\phi_{31} & \phi_{32} & \phi_{33} & \phi_{43} \\
\phi_{41} & \phi_{42} & \phi_{43} & \phi_{44}
\end{bmatrix} =
$$

$$
\begin{bmatrix}
1 & 0 & 0 & 0 \\
V_{21} & 1 & 0 & 0 \\
V_{31} & V_{32} & 1 & 0 \\
V_{41} & V_{42} & V_{43} & 1
\end{bmatrix}
\begin{bmatrix}
d_1 & 0 & 0 & 0 \\
0 & d_2 & 0 & 0 \\
0 & 0 & d_3 & 0 \\
0 & 0 & 0 & d_4
\end{bmatrix}
\begin{bmatrix}
1 & V_{21} & V_{31} & V_{41} \\
0 & 1 & V_{32} & V_{42} \\
0 & 0 & 1 & V_{43} \\
0 & 0 & 0 & 1
\end{bmatrix}
$$

To solve for d_1 to d_4, and the V_{ij}'s we begin with Eq. (8.52) for $i = 1$ giving

$$d_1 = \phi_{11}$$

Using Eq. (8.49) for $i = 2, 3, 4$ we solve for V_{21}, V_{31}, and V_{41} as

$$V_{21} d_1 = \phi_{21} \quad , \quad V_{31} d_1 = \phi_{31} \quad , \quad V_{41} d_1 = \phi_{41}$$

$$V_{21} = \phi_{21}/d_1 , \quad V_{31} = \phi_{31}/d_1 , \quad V_{41} = \phi_{41}/d_1$$

Using Eq. (8.51) for $i = 2$ gives

$$d_2 = \phi_{22} - V_{21}^2 d_1$$

Using Eq. (8.49) for $i = 3$ and 4 gives

$$V_{32}d_2 = \phi_{32} - V_{31}d_1V_{21}$$
$$V_{42}d_2 = \phi_{42} - V_{41}d_1V_{21}$$

or

$$V_{32} = (\phi_{32} - V_{31}d_1V_{21})/d_2$$
$$V_{42} = (\phi_{42} - V_{41}d_1V_{21})/d_2$$

Equation (8.51) is now used for $i = 3$ to solve for d_3, then Eq. (8.49) is used for $i = 4$ to solve for V_{43}, and finally Eq. (8.51) is used for $i = 4$ to solve for d_4.

Once the matrices $\mathbf{V}$ and $\mathbf{D}$ have been determined, it is relatively simple to solve for the column vector $\boldsymbol{\alpha}$ in a two-step procedure. From Eqs. (8.46) and (8.47) we get

$$\mathbf{VDV^t\alpha} = \boldsymbol{\psi} \tag{8.53}$$

which can be written as

$$\mathbf{VY} = \boldsymbol{\psi} \tag{8.54}$$

and

$$\mathbf{DV^t\alpha} = \mathbf{Y} \tag{8.55}$$

or

$$\mathbf{V^t\alpha} = \mathbf{D^{-1}Y} \tag{8.56}$$

Thus from the matrix $\mathbf{V}$, Eq. (8.54) can be solved for the column vector $\mathbf{Y}$ using a simple recursion of the form

$$Y_i = \psi_i - \sum_{j=1}^{i-1} V_{ij}Y_j, \qquad p \geqslant i \geqslant 2 \tag{8.57}$$

with initial condition

$$Y_1 = \psi_1 \tag{8.58}$$

Similarly having solved for $\mathbf{Y}$, Eq. (8.56) can be solved recursively for $\boldsymbol{\alpha}$ using the relation

$$\alpha_i = Y_i/d_i - \sum_{j=i+1}^{p} V_{ji}\alpha_j \quad 1 \leqslant i \leqslant p-1 \tag{8.59}$$

with initial condition

$$\alpha_p = Y_p/d_p \tag{8.60}$$

It should be noted that the index i in Eq. (8.59) proceeds backwards from $i = p - 1$ down to $i = 1$.

To illustrate the use of Eqs. (8.57)-(8.60) we continue our previous example and first solve for the $Y_i's$ assuming $\mathbf{V}$ and $\mathbf{D}$ are now known. In matrix form we have the equation

$$\begin{bmatrix} 1 & 0 & 0 & 0 \\ V_{21} & 1 & 0 & 0 \\ V_{31} & V_{32} & 1 & 0 \\ V_{41} & V_{42} & V_{43} & 1 \end{bmatrix} \begin{bmatrix} Y_1 \\ Y_2 \\ Y_3 \\ Y_4 \end{bmatrix} = \begin{bmatrix} \psi_1 \\ \psi_2 \\ \psi_3 \\ \psi_4 \end{bmatrix}$$

From Eqs. (8.57) and (8.58) we get

$$Y_1 = \psi_1$$

$$Y_2 = \psi_2 - V_{21}Y_1$$

$$Y_3 = \psi_3 - V_{31}Y_1 - V_{32}Y_2$$

$$Y_4 = \psi_4 - V_{41}Y_1 - V_{42}Y_2 - V_{43}Y_3$$

From the $Y_i's$ we solve Eq. (8.56) which is of the form

$$\begin{bmatrix} 1 & V_{21} & V_{31} & V_{41} \\ 0 & 1 & V_{32} & V_{42} \\ 0 & 0 & 1 & V_{43} \\ 0 & 0 & 0 & 1 \end{bmatrix} \begin{bmatrix} \alpha_1 \\ \alpha_2 \\ \alpha_3 \\ \alpha_4 \end{bmatrix} =$$

$$\begin{bmatrix} 1/d_1 & 0 & 0 & 0 \\ 0 & 1/d_2 & 0 & 0 \\ 0 & 0 & 1/d_3 & 0 \\ 0 & 0 & 0 & 1/d_4 \end{bmatrix} \begin{bmatrix} Y_1 \\ Y_2 \\ Y_3 \\ Y_4 \end{bmatrix} = \begin{bmatrix} Y_1/d_1 \\ Y_2/d_2 \\ Y_3/d_3 \\ Y_4/d_4 \end{bmatrix}$$

From Eqs. (8.59) and (8.60) we get

$$\alpha_4 = Y_4/d_4$$

$$\alpha_3 = Y_3/d_3 - V_{43}\alpha_4$$

$$\alpha_2 = Y_2/d_2 - V_{32}\alpha_3 - V_{42}\alpha_4$$

$$\alpha_1 = Y_1/d_1 - V_{21}\alpha_2 - V_{31}\alpha_3 - V_{41}\alpha_4$$

thus completing the solution to the covariance equations.

The use of the Cholesky decomposition procedure leads to a very simple expression for the minimum error of the covariance method in terms of the column vector $\mathbf{Y}$ and the matrix $\mathbf{D}$. We recall that for the covariance method, the prediction error E_n was of the form

$$E_n = \phi_n(0,0) - \sum_{k=1}^{p} \alpha_k \phi_n(0,k) \tag{8.61}$$

or in matrix notation

$$E_n = \phi_n(0,0) - \alpha'\psi \tag{8.62}$$

410

From Eq. (8.56) we can substitute for α' the expression $\mathbf{Y}'\mathbf{D}^{-1}\mathbf{V}^{-1}$ giving

$$E_n = \phi_n(0,0) - \mathbf{Y}'\mathbf{D}^{-1}\mathbf{V}^{-1}\psi \tag{8.63}$$

Using Eq. (8.54) we get

$$E_n = \phi_n(0,0) - \mathbf{Y}'\mathbf{D}^{-1}\mathbf{Y} \tag{8.64}$$

or

$$E_n = \phi_n(0,0) - \sum_{k=1}^{p} Y_k^2/d_k \tag{8.65}$$

Thus the mean-squared prediction error E_n can be determined directly from the column vector $\mathbf{Y}$ and the matrix $\mathbf{D}$. Furthermore Eq. (8.65) can be used to give the value of E_n for any value of p up to the value of p used in solving the matrix equations. Thus one can get an idea as to how the mean-squared prediction error varies with the number of predictor coefficients used in the solution.

8.3.2 Durbin's recursive solution for the autocorrelation equations [2]

For the autocorrelation method the matrix equation for solving for the predictor coefficients is of the form

$$\sum_{k=1}^{p} \alpha_k R_n(|i-k|) = R_n(i) \qquad 1 \le i \le p \tag{8.66}$$

By exploiting the Toeplitz nature of the matrix of coefficients, several efficient recursive procedures have been devised for solving this system of equations. Although the most popular and well known of these methods are the Levinson and Robinson algorithms [1], the most efficient method known for solving this particular system of equations is Durbin's recursive procedure [2] which can be stated as follows (for convenience of notation we shall omit the subscript on the autocorrelation function):

$$E^{(0)} = R(0) \tag{8.67}$$

$$k_i = \left[R(i) - \sum_{j=1}^{i-1} \alpha_j^{(i-1)} R(i-j) \right] / E^{(i-1)} \qquad 1 \le i \le p \tag{8.68}$$

$$\alpha_i^{(i)} = k_i \tag{8.69}$$

$$\alpha_j^{(i)} = \alpha_j^{(i-1)} - k_i \alpha_{i-j}^{(i-1)} \qquad 1 \le j \le i-1 \tag{8.70}$$

$$E^{(i)} = (1-k_i^2) E^{(i-1)} \tag{8.71}$$

Equations (8.68)-(8.71) are solved recursively for $i = 1, 2, \ldots, p$ and the final solution is given as

$$\alpha_j = \alpha_j^{(p)} \qquad 1 \le j \le p \tag{8.72}$$

Note that in the process of solving for the predictor coefficients for a predictor of order p, the solutions for the predictor coefficients of all orders less than p

411

have also been obtained — i.e., $\alpha_j^{(i)}$ is the jth predictor coefficient for a predictor of order i.

To illustrate the above procedure, consider an example of obtaining the predictor coefficients for a predictor of order 2. The original matrix equation is of the form

$$\begin{bmatrix} R(0) & R(1) \\ R(1) & R(0) \end{bmatrix} \begin{bmatrix} \alpha_1 \\ \alpha_2 \end{bmatrix} = \begin{bmatrix} R(1) \\ R(2) \end{bmatrix}$$

Using Eqs. (8.67)-(8.72), we get

$$E^{(0)} = R(0)$$

$$k_1 = R(1)/R(0)$$

$$\alpha_1^{(1)} = R(1)/R(0)$$

$$E^{(1)} = \frac{R^2(0) - R^2(1)}{R(0)}$$

$$k_2 = \frac{R(2)R(0) - R^2(1)}{R^2(0) - R^2(1)}$$

$$\alpha_2^{(2)} = \frac{R(2)R(0) - R^2(1)}{R^2(0) - R^2(1)}$$

$$\alpha_1^{(2)} = \frac{R(1)R(0) - R(1)R(2)}{R^2(0) - R^2(1)}$$

$$\alpha_1 = \alpha_1^{(2)}$$

$$\alpha_2 = \alpha_2^{(2)}$$

It should be noted that the quantity $E^{(i)}$ in Eq. (8.71) is the prediction error for a predictor of order i. Thus at each stage of the computation the prediction error for a predictor of order i can be monitored. Also, if the autocorrelation coefficients $R(i)$ are replaced by a set of normalized autocorrelation coefficients, i.e., $r(k) = R(k)/R(0)$, then the solution to the matrix equation remains unchanged. However, the error $E^{(i)}$ is now interpreted as a normalized error. If we call this normalized error $V^{(i)}$, then

$$V^{(i)} = \frac{E^{(i)}}{R(0)} = 1 - \sum_{k=1}^{i} \alpha_k r(k) \tag{8.73}$$

with

$$0 < V^{(i)} \leqslant 1 \quad i \geqslant 0 \tag{8.74}$$

It can be shown that the normalized error for $i = p$ (i.e., $V^{(p)}$) can be written in the form

$$V^{(p)} = \prod_{i=1}^{p} (1 - k_i^2) \tag{8.75}$$

where the quantities k_i are in the range

$$-1 \leqslant k_i \leqslant 1 \tag{8.76}$$

This condition on the parameters k_i is important since it can be shown [1,18] that it is a necessary and sufficient condition for all of the roots of the polynomial $A(z)$ to be inside the unit circle, thereby guaranteeing the stability of the system $H(z)$. Unfortunately a proof of this result would take us too far afield; however, the fact that we do not give a proof does not diminish the importance of this result. Furthermore, it is possible to show that no such guarantee of stability is available in the covariance method.

8.3.3 Lattice formulations and solutions [11]

As we have seen, both the covariance and the autocorrelation methods consist of two steps:

1. Computation of a matrix of correlation values.
2. Solution of a set of linear equations.

These methods have been widely used with great success in speech processing applications. However, another class of methods, called *lattice methods,* has evolved in which the above two steps have in a sense been combined into a recursive algorithm for determining the linear predictor parameters. To see how these methods are related, it is helpful to begin with the Durbin algorithm. First, let us recall that at the ith stage of this procedure, the set of coefficients $\{\alpha_j^{(i)} \ j=1,2, \ldots, i\}$ are the coefficients of the ith order optimum linear predictor. Using these coefficients we can define

$$A^{(i)}(z) = 1 - \sum_{k=1}^{i} \alpha_k^{(i)} z^{-k} \tag{8.77}$$

to be the system function of the ith-order inverse filter (or prediction error filter). If the input to this filter is the segment of the signal, $s_n(m) = s(n+m)w(m)$, then the output would be the prediction error, $e_n^{(i)}(m) = e^{(i)}(n+m)$, where

$$e^{(i)}(m) = s(m) - \sum_{k=1}^{i} \alpha_k^{(i)} s(m-k) \tag{8.78}$$

Note that for the sake of simplicity we shall henceforth drop the subscript n which denotes the fact that we are considering a segment of the signal located at sample n. In terms of z-transforms Eq. (8.78) is

$$E^{(i)}(z) = A^{(i)}(z)S(z) \tag{8.79}$$

By substituting Eq. (8.70) into Eq. (8.77) we obtain a recurrence formula for $A^{(i)}(z)$ in terms of $A^{(i-1)}(z)$; i.e.,

$$A^{(i)}(z) = A^{(i-1)}(z) - k_i z^{-i} A^{(i-1)}(z^{-1}) \tag{8.80}$$

(See Problem 8.5.) Substituting Eq. (8.80) into Eq. (8.79) we obtain

$$E^{(i)}(z) = A^{(i-1)}(z)S(z) - k_i z^{-i} A^{(i-1)}(z^{-1})S(z) \tag{8.81}$$

413

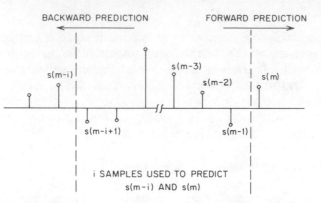

Fig. 8.2 Illustration of forward and backward prediction using an ith order predictor.

The first term in Eq. (8.81) is obviously the z-transform of the prediction error for an $(i-1)^{th}$ order predictor. The second term can be given a similar interpretation if we define

$$B^{(i)}(z) = z^{-i}A^{(i)}(z^{-1})S(z) \tag{8.82}$$

It is easily shown that the inverse transform of $B^{(i)}(z)$ is

$$b^{(i)}(m) = s(m-i) - \sum_{k=1}^{i} \alpha_k^{(i)} s(m+k-i) \tag{8.83}$$

This equation suggests that we are attempting to predict $s(m-i)$ from the i samples of the input $\{s(m-i+k),\ k=1,2,...,i\}$ that follow $s(m-i)$. Thus $b^{(i)}(m)$ is called the backward prediction error sequence. In Fig. 8.2 it is shown that the i samples involved in the prediction are the same ones used to predict $s(m)$ in terms of i *past* samples in Eq. (8.78). Now returning to Eq. (8.81) we see that the prediction error sequence $e^{(i)}(m)$ can be expressed as

$$e^{(i)}(m) = e^{(i-1)}(m) - k_i b^{(i-1)}(m-1) \tag{8.84}$$

By substituting Eq. (8.80) into Eq. (8.82) we obtain

$$B^{(i)}(z) = z^{-i}A^{(i-1)}(z^{-1})S(z) - k_i A^{(i-1)}(z)S(z) \tag{8.85}$$

or

$$B^{(i)}(z) = z^{-1}B^{(i-1)}(z) - k_i E^{(i-1)}(z) \tag{8.86}$$

Thus the ith stage backward prediction error is

$$b^{(i)}(m) = b^{(i-1)}(m-1) - k_i e^{(i-1)}(m) \tag{8.87}$$

Now Eqs. (8.84) and (8.87) define the forward and backward prediction error sequences for an ith order predictor in terms of the corresponding prediction errors of an $(i-1)^{th}$ order predictor. Using a zeroth order predictor is equivalent to using no predictor at all so that

$$e^{(0)}(m) = b^{(0)}(m) = s(m) \tag{8.88}$$

Thus we can depict Eqs. (8.84) and (8.87) by the flow graph of Fig. 8.3. Such

414

a structure is called a lattice network. It is clear that if we extend the lattice to p sections, the output of the last upper branch will be the forward prediction error as shown in Fig. 8.3. Thus, Fig. 8.3 is a digital network implementation of the prediction error filter with transfer function $A(z)$.

At this point we should emphasize that this structure is a direct consequence of the Durbin algorithm, and the parameters k_i can be obtained as in Eqs. (8.67)-(8.72). Note also that the predictor coefficients do not appear explicitly in Fig. 8.3. Itakura [4,6] has shown that the k_i parameters can be directly related to the forward and backward prediction errors and because of the nature of the lattice structure the entire set of coefficients k_i, $i=1,2,\ldots,p$} can be computed without computing the predictor coefficients. The relationship is [11]

$$k_i = \frac{\sum_{m=0}^{N-1} e^{(i-1)}(m)\, b^{(i-1)}(m-1)}{\left\{\sum_{m=0}^{N-1} (e^{(i-1)}(m))^2 \sum_{m=0}^{N-1} (b^{(i-1)}(m-1))^2\right\}^{1/2}} \qquad (8.89)$$

This expression is in the form of a normalized cross-correlation function; i.e., it is indicative of the degree of correlation between the forward and backward prediction error. For this reason the parameters k_i are called the partial correlation coefficients or PARCOR coefficients [4,6]. It is relatively straightforward to verify that Eq. (8.89) is identical to Eq. (8.68) by substituting Eqs. (8.78) and (8.83) into Eq. (8.89).

It can be seen that if Eq. (8.89) replaces Eq. (8.68) in the Durbin algorithm, the predictor coefficients can be computed recursively as before. Thus the PARCOR analysis leads to an alternative to the inversion of a matrix and

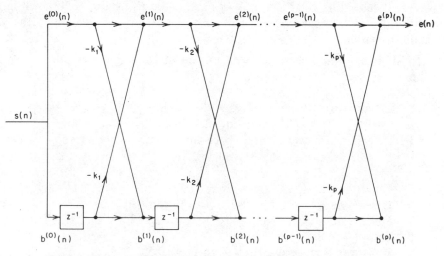

Fig. 8.3 Block diagram of a realizable implementation of the lattice method.

415

gives results identical to the autocorrelation method; i.e., the set of PARCOR coefficients is equivalent to a set of predictor coefficients that minimize the mean-squared forward prediction error. More importantly, this approach opens up a whole new class of procedures based upon the lattice configuration of Fig. 8.3 [11].

In particular, Burg [12] has developed a procedure based upon minimizing the *sum* of the mean-squared forward and backward prediction errors in Fig. 8.3; i.e.,

$$\tilde{E}^{(i)} = \sum_{m=0}^{N-1} \left[(e^{(i)}(m))^2 + (b^{(i)}(m))^2 \right] \tag{8.90}$$

Substituting Eqs. (8.84) and (8.87) into Eq. (8.90) and differentiating $\tilde{E}^{(i)}$ with respect to k_i, we obtain

$$\frac{\partial \tilde{E}^{(i)}}{\partial k_i} = -2 \sum_{m=0}^{N-1} \left[e^{(i-1)}(m) - k_i b^{(i-1)}(m-1) \right] b^{(i-1)}(m-1)$$

$$-2 \sum_{m=0}^{N-1} \left[b^{(i-1)}(m-1) - k_i e^{(i-1)}(m) \right] e^{(i-1)}(m) \tag{8.91}$$

Setting the derivative equal to zero and solving for k_i gives

$$k_i = \frac{2 \sum\limits_{m=0}^{N-1} \left[e^{(i-1)}(m) \, b^{(i-1)}(m-1) \right]}{\sum\limits_{m=0}^{N-1} \left[e^{(i-1)}(m) \right]^2 + \sum\limits_{m=0}^{N-1} \left[b^{(i-1)}(m-1) \right]^2} \tag{8.92}$$

It can be shown [1] that if k_i is estimated using Eq. (8.92) then

$$-1 \leqslant k_i \leqslant 1 \tag{8.93}$$

However, it should be clear that the k_i's estimated using Eq. (8.92) will in general differ from those estimated using Eq. (8.89), or equivalently, the autocorrelation method.

In summary, the steps involved in determining the predictor coefficients and the k parameters are as follows:

1. Initially set $e^{(0)}(m) = s(m) = b^{(0)}(m)$.
2. Compute $k_1 = \alpha_1^{(1)}$ from Eq. (8.92).
3. Determine forward and backward prediction errors $e^{(1)}(m)$ and $b^{(1)}(m)$ from Eqs. (8.84) and (8.87).
4. Set $i = 2$.
5. Determine $k_i = \alpha_i^{(i)}$ from Eq. (8.92).
6. Determine $\alpha_j^{(i)}$ for $j = 1, 2, ..., i - 1$ from Eq. (8.70).
7. Determine $e^{(i)}(m)$ and $b^{(i)}(m)$ from Eq. (8.84) and (8.87).
8. Set $i = i + 1$.
9. If i is less than or equal to p, go to step 5.
10. Procedure is terminated.

There are clearly several differences in implementation between the lattice method and the covariance and autocorrelation implementations discussed earlier. One major difference is that in the lattice method the predictor coefficients are obtained directly from the speech samples without an intermediate calculation of an autocorrelation function. At the same time the method is guaranteed to yield a stable filter without requiring the use of a window. For these reasons the lattice formulation has become an important and viable approach to the implementation of linear predictive analysis.

8.4 Comparisons Between the Methods of Solution of the LPC Analysis Equations

We have already discussed the differences in the theoretical formulations of the covariance, autocorrelation, and lattice formulations of the linear predictive analysis equations. In this section we discuss the issues involved in practical implementations of the analysis equations. Included among these issues are computational considerations, numerical and physical stability of the solutions, and the question of how to choose the number of poles and section length used in the analysis. We begin first with the computational considerations involved in obtaining the predictor coefficients from the speech waveform.

The two major issues in the computation of the predictor coefficients are the amount of storage, and the number of multiplications. Table 8.1 (due to Portnoff et al. [13] and Makhoul [11]) shows the required computation for the covariance, the autocorrelation and the lattice methods. In terms of storage, for the covariance method, the requirements are essentially N_1 locations for the data, and on the order of $p^2/2$ locations for the correlation matrix, where N_1 is the number of points in the analysis. For the autocorrelation method the requirements are N_2 locations for both the data and the window, and a number of locations proportional to p for the autocorrelation matrix. For the lattice method the requirements are $3N_3$ locations for the data and the forward and backward prediction errors. For emphasis we have assumed that the N_1 for the covariance method, the N_2 for the autocorrelation method, and the N_3 for the lattice method need not be the same. We will discuss this question later in this section. Thus in terms of storage (assuming N_1, N_2, and N_3 are comparable) the covariance and autocorrelation methods require somewhat less storage than the lattice method.

The computational requirements for the three methods, in terms of multiplications, are shown at the bottom of Table 8.1. For the covariance method, the computation of the correlation matrix requires about $N_1 p$ multiplications, whereas the solution to the matrix equation (using the Cholesky decomposition procedure) requires a number of multiplications proportional to p^3. (Portnoff et al. give an exact figure of $(p^3 + 9p^2 + 2p)/6$ multiplications, p divides, and p square roots.) For the autocorrelation method, the computation of the auto-correlation matrix requires about $N_2 p$ multiplications, whereas the solution to the matrix equations requires about p^2 multiplications. Thus if N_1 and N_2 are

Table 8.1 Computational Considerations in the LPC Solutions

	Covariance Method	Autocorrelation Method	Lattice Method
	(Cholesky Decomposition)	(Durbin Method)	(Burg Method)
Storage			
Data	N_1	N_2	$3N_3$
Matrix	proportional to $p^2/2$	proportional to p	—
Window	0	N_2	—
Computation (Multiplications)			
Windowing	0	N_2	—
Correlation	proportional to N_1p	proportional to N_2p	—
Matrix Solution	proportional to p^3	proportional to p^2	$5N_3p$

approximately equal, and with $N_1 >> p$, $N_2 >> p$, then the autocorrelation method will require somewhat less computation than the covariance method. However, since in most speech problems the number of multiplications required to compute the correlation function far exceeds the number of multiplications to solve the matrix equations, the computation times for both these formulations are quite comparable. For the lattice method a total of $5N_3p$ multiplications are needed to compute the set of partial correlation coefficients.[5] Thus the lattice method is the least computationally efficient method for solving the LPC equations. However, the other advantages of the lattice method must be kept in mind when considering the use of this method.

Another consideration in comparing these three formulations is the stability of the resulting system

$$H(z) = \frac{G}{A(z)} \qquad (8.94)$$

This system is stable if all its poles lie strictly inside the unit circle in the z-plane. The poles of the system, $H(z)$, are the zeros of denominator polynomial $A(z)$, where

$$A(z) = 1 - \sum_{k=1}^{p} \alpha_k z^{-k} \qquad (8.95)$$

As we have asserted, for the autocorrelation method all the roots of $A(z)$ lie inside the unit circle — i.e., $H(z)$ is guaranteed to be stable. It should be noted that this theoretical guarantee of stability for the autocorrelation method may not hold in practice if the autocorrelation function is computed without sufficient accuracy. In such cases the roundoff encountered in computing the autocorrelation can cause the matrix to become ill conditioned. Markel and Gray have shown that these undesirable effects can be minimized by pre-emphasizing the speech to make its spectrum as flat as possible [1]. With the use of a pre-emphasizing filter, smaller wordlengths can be used in practice and

---

[5]Makhoul has discussed a modified lattice method for obtaining the partial correlation coefficients with the same efficiency as the normal covariance method [11].

the resulting predictor polynomials will generally remain stable. The Durbin algorithm provides a convenient test for stability since it is necessary and sufficient that the parameters k_i (PARCOR's) must satisfy the condition

$$-1 \leqslant k_i \leqslant 1 \qquad (8.96)$$

Thus if, in the process of determining the predictor coefficients $\{\alpha_i\}$, any of the quantities k_i violate Eq. (8.96) then it is known that there are roots of $A(z)$ outside the unit circle.

For the covariance method, the stability of the predictor polynomial cannot be guaranteed. However, in practice, if the number of samples in the frame is sufficiently large, then the resulting predictor polynomials will almost always be stable. This is due to the fact that for a large number of samples in the analysis frame, the covariance and autocorrelation methods yield almost identical results.

For the lattice method the predictor polynomial is guaranteed to be stable since the predictor coefficients are obtained from the partial correlation coefficients which, by definition, satisfy Eq. (8.96). In addition, the stability is preserved even when the computation is performed using finite word length computations [1].

In the case when the predictor polynomial stability is uncertain, it is generally required that the roots of the predictor polynomial be determined and tested for stability. If a root is found to be outside the unit circle, a simple correction procedure is to reflect the root inside the unit circle, thereby ensuring a stable predictor polynomial with the same frequency response as the unstable polynomial.

Two other considerations in comparing and contrasting the three formulations of the LPC equations are the choice of number of predictor parameters, p, and the choice of the frame length N. The choice of p depends primarily on the sampling rate and is essentially independent of the LPC method being used. Since the speech spectrum being analyzed can generally be represented as having an average density of 2 poles (i.e., one complex pole) per kiloHertz due to the vocal tract contribution, then a total of F_s poles are required to represent this contribution to the speech spectrum, where F_s is the sampling rate in kiloHertz. Thus for a 10 kHz sampling rate, a total of 10 poles is required to represent the vocal tract. In addition, a total of 3-4 poles is required to adequately represent the source excitation spectrum and the radiation load. Thus for a 10 kHz simulation, a value of p of about 13 or 14 is required. To verify this conclusion, Figure 8.4 shows a plot (due to Atal and Hanauer [3]) of the normalized rms prediction error versus the predictor order p for sections of voiced and unvoiced speech for a 10 kHz simulation. Although the prediction error steadily decreases as p increases, for p on the order of 13-14 the error has essentially flattened off showing only small decreases as p is increased further. It is interesting to note from this figure that the normalized rms prediction error for unvoiced speech is significantly higher than for voiced speech. This is of course as expected since the model for unvoiced speech is nowhere near as

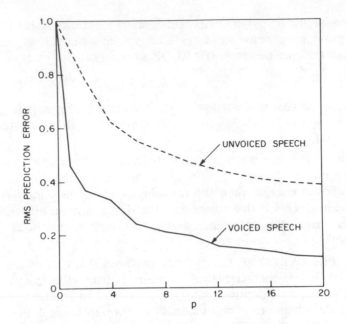

Fig. 8.4 Variation of the RMS prediction error with the number of predictor coefficients, p. (After Atal and Hanauer [3].)

accurate as it is for voiced speech. Additional experimental evidence of the behavior of the prediction error as a function of p is given in the next section.

The choice of section length N is a very important consideration in implementation of most LPC analysis systems. Clearly, it is advantageous to keep N as small as possible since the total computation load, for all three methods, is essentially proportional to N. For the autocorrelation method it has been shown that N must be on the order of several pitch periods to ensure reliable results [1,2]. Since a window is used to weight the speech in the autocorrelation method, the section duration must be sufficiently long so that the tapering effects of the window do not seriously affect the results. Thus analysis durations from $N = 100$ to $N = 400$ samples (at a 10 kHz rate) have been used in LPC implementations of the autocorrelation method, with most systems leaning toward the larger values of N. For both the covariance and lattice methods, the choice of section length is governed by several considerations. Since no windowing is required, there are no real limitations on how small the section size can be. If the analysis can be restricted to regions within each pitch period (i.e., a pitch synchronous analysis is performed) then values of N on the order of $2p$ have been used successfully. However if such small values of N are used and if a pitch pulse occurs within the analysis interval, unsatisfactory results are obtained. Thus in most practical systems in which it is not possible to perform a pitch synchronous analysis, values of N for the covariance and lattice methods are comparable to those for the autocorrelation method. In the next few sections we show results from experimental evaluations of the effects of section length, and section position on the prediction error for the covariance and autocorrelation methods.[6] We first digress into a brief discussion of the LPC error signal and the normalized error derived from it.

8.5 The Prediction Error Signal

A by-product of the LPC analysis is the generation of the error signal, $e(n)$, defined as

$$e(n) = s(n) - \sum_{k=1}^{p} \alpha_k s(n-k) = Gu(n) \qquad (8.97)$$

To the extent that the actual speech signal is generated by a system that is well modelled by a time-varying linear predictor of order p, then $e(n)$ is equally a good approximation to the excitation source. Based on this reasoning, it is expected that the prediction error will be large (for voiced speech) at the beginning of each pitch period. Thus the pitch period can be determined by detecting the positions of the samples of $e(n)$ which are large, and defining the period as the difference between pairs of samples of $e(n)$ which exceed a reasonable threshold. Alternatively the pitch period can be estimated by performing an autocorrelation analysis on $e(n)$ and detecting the largest peak in the

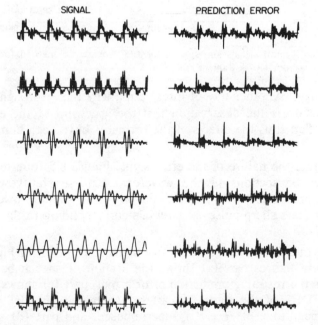

Fig. 8.5 Examples of signal (differentiated) and prediction error for vowels (i, e, a, o, u, y). (After Strube [14].)

---

[6]Investigations by Rabiner et al. [16] have found that a good choice of parameters for the lattice method are essentially those used for the covariance method. Thus we do not differentiate between these methods in the following sections.

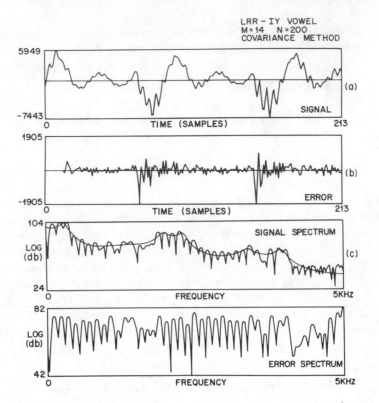

Fig. 8.6 Typical signals and spectra for LPC covariance method for a male speaker. (After Rabiner et al. [16].)

appropriate range. Another way of interpreting why the error signal is valuable for pitch detection is the observation that the spectrum of the error signal is approximately flat; thus the effects of the formants have been eliminated in the error signal.

To illustrate the nature of the error signal Figure 8.5 (due to Strube [14]) shows a series of sections of the waveforms for several vowels, and the corresponding prediction error signals. For all these simple vowel sounds the error signal exhibits sharp pulses at intervals corresponding to the pitch periods of these vowels.

Some further examples of LPC error signals are given in Figures 8.6-8.9. In each of these figures part (a) shows the section of speech being analyzed, part (b) shows the resulting prediction error signal, part (c) shows the log magnitude of the DFT of the signal in part (a) (obtained via FFT computation) with the log magnitude of $H(e^{j\omega T})$ superimposed, and part (d) shows the log magnitude spectrum of the error signal (obtained via FFT computation). Figures 8.6 and 8.7 are for 20 msec of an /i/ vowel (as in we) spoken by a male speaker (LRR) using the covariance and autocorrelation methods (with a Hamming window) respectively. The error signal is seen to be sharply peaked at the beginning of each pitch period, and the error spectrum is fairly flat, showing a comb effect due to the effects of the pitch period. Note the rather large predic-

422

tion error at the beginning of the segment in Fig. 8.7 for the autocorrelation method. This is, of course, due to the fact that we are attempting to predict the samples of the signal from the zero valued samples outside the interval $0 \leqslant m \leqslant 199$. The tapering effect of the Hamming window is thus not completely effective in reducing this error.

Figures 8.8 and 8.9 show similar results for 20 msec of an /a/ vowel (as in father) for a female speaker (SAW). For this speaker approximately 5 complete pitch periods are contained within the analysis interval. Thus in Fig. 8.8 the error signal displays a large number of sharp peaks during the analysis interval for the covariance method of analysis. However, the effect of the Hamming window in the autocorrelation method of Fig. 8.9 is to taper the pitch pulses near the ends of the analysis interval; hence the peaks in the error signal due to the pitch pulses are likewise tapered.

The behavior of the error signal shown in the preceding figures would lead one to believe that it would, by itself, be a natural candidate for a signal from which pitch could simply be detected. Unfortunately the situation is not quite so clear for other examples of voiced speech. Makhoul and Wolf [5] have

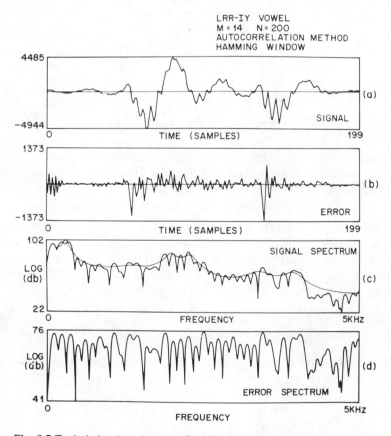

Fig. 8.7 Typical signals and spectra for LPC autocorrelation method for a male speaker. (After Rabiner et al. [16].)

423

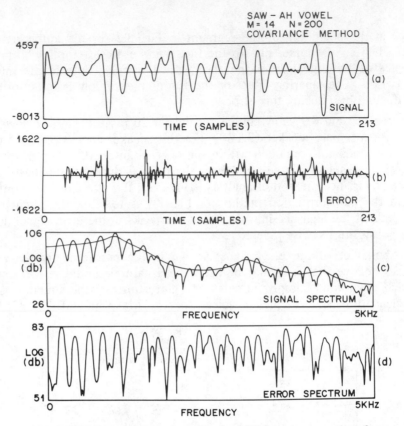

4597

(a)

SIGNAL

-8013

0   TIME (SAMPLES)   213

1622

(b)

ERROR

-1622

0   TIME (SAMPLES)   213

106

LOG
(db)

(c)

SIGNAL   SPECTRUM

26

0   FREQUENCY   5KHz

83

LOG
(db)

(d)

ERROR   SPECTRUM

51

0   FREQUENCY   5KHz

Fig. 8.8 Typical signals and spectra for LPC covariance method for a female speaker. (After Rabiner et al. [16].)

shown that for sounds which are not rich in harmonic structure, e.g., liquids like *r*, *l*, or nasals such as *m*, *n*, the peaks in the error signal are not always very sharp or distinct. Additionally at the junctions between voiced and unvoiced sounds, the pitch markers in the error signal often essentially disappear.

In summary, although the error signal $e(n)$ appears to be an ideal candidate for a pitch detector, it has its own difficulties in locating pitch markers for a wide variety of voiced sounds, and thus cannot be relied on exclusively for this purpose. In Section 8.10.1 we shall discuss one pitch detection scheme based upon the prediction error signal.

8.5.1 Alternative expressions for the normalized mean-squared error

The normalized mean squared prediction error for the autocorrelation method is defined as

$$V_n = \frac{\sum_{m=0}^{N+p-1} e_n^2(m)}{\sum_{m=0}^{N-1} s_n^2(m)} \tag{8.98a}$$

where $e_n(m)$ is the output of the prediction error filter corresponding to the speech segment $s_n(m)$ located at time index n. For the covariance method, the corresponding definition is

$$V_n = \frac{\displaystyle\sum_{m=0}^{N-1} e_n^2(m)}{\displaystyle\sum_{m=0}^{N-1} s_n^2(m)} \qquad (8.98b)$$

Defining $\alpha_0 = -1$, the prediction error sequence can be expressed as

$$e_n(m) = - \sum_{k=0}^{p} \alpha_k s_n(m-k) \qquad (8.99)$$

Substituting Eq. (8.99) into Eq. (8.98) and using Eq. (8.13) it follows that

$$V_n = \sum_{i=0}^{p} \sum_{j=0}^{p} \alpha_i \frac{\phi_n(i,j)}{\phi_n(0,0)} \alpha_j \qquad (8.100a)$$

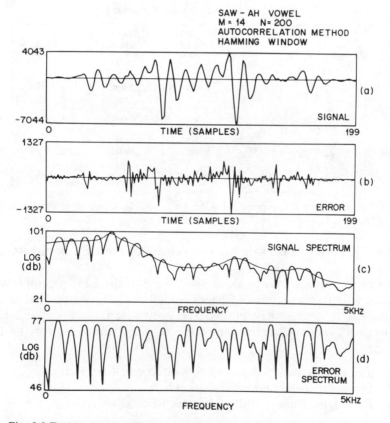

Fig. 8.9 Typical signals and spectra for LPC autocorrelation method for a female speaker. (After Rabiner et al. [16].)

425

and substituting Eq. (8.14) into (8.100) gives

$$V_n = - \sum_{i=0}^{p} \alpha_i \frac{\phi_n(0,i)}{\phi_n(0,0)} \qquad (8.100b)$$

Still another expression for V_n was obtained in the Durbin algorithm; i.e.,

$$V_n = \prod_{i=1}^{p} (1-k_i^2) \qquad (8.101)$$

The above expressions are not all equivalent and are subject to interpretation in terms of the details of a given linear predictive method. For example, Eq. (8.101), being based upon the Durbin algorithm is valid only for the autocorrelation and lattice methods. Also, since the lattice method does not explicitly require the computation of the correlation functions Eqs. (8.100a) and (8.100b) do not apply directly to the lattice method. Table 8.2 summarizes the above expressions for normalized mean-squared error and indicates the scope of validity of each expression. (Note that the subscript n and the superscript p have been eliminated in the table for simplicity.)

Table 8.2 Expressions for the Normalized Error

	Covariance Method	Autocorrelation Method	Lattice Method
$V = \dfrac{\sum_m e^2(m)}{\sum_m s^2(m)}$	Valid	Valid*	Valid
$V = \sum_i \sum_j \alpha_i \dfrac{\phi(i,j)}{\phi(0,0)} \alpha_j$	Valid	Valid**	Not Valid
$V = \sum_i \alpha_i \dfrac{\phi(i,j)}{\phi(0,0)}$	Valid	Valid**	Not Valid
$V = \prod_i (1-k_i^2)$	Not Valid	Valid	Valid

*This expression is computed using the windowed signal and upper limit is $N-1+p$.
**In these cases $\phi(i,j) = R(i-j)$.

8.5.2 Experimental evaluation of values for the LPC parameters

To provide guidelines to aid in the choice of the LPC parameters p and N for practical implementations, Chandra and Lin [15] performed a series of investigations in which they plotted the normalized mean-squared prediction error, for a pth order predictor versus the relevant parameter for the following conditions:

1. The covariance method and the autocorrelation method
2. Synthetic vowel and natural speech
3. Pitch synchronous and pitch asynchronous analysis

where V is defined as in Table 8.2. Figures 8.10-8.15 show the results obtained by Chandra and Lin for the above conditions [15].

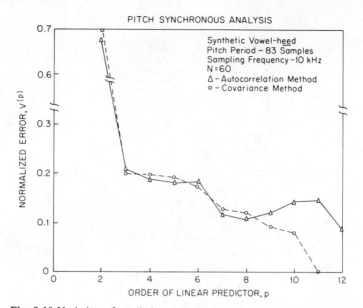

Fig. 8.10 Variation of prediction error with predictor order, p, for voiced section of a synthetic vowel—pitch synchronous analysis. (After Chandra and Lin [15].)

Figure 8.10 shows the variation of V with the order of the linear predictor, p, for a section of a synthetic vowel (/i/ in heed) whose pitch period was 83 samples. The analysis section length N was 60 samples beginning at the beginning of a pitch period — i.e., these results are for a pitch synchronous analysis. For the covariance method the prediction error decreases monotonically to 0 at $p = 11$ which was the order of the system used to create the synthetic speech. For the autocorrelation method the prediction remains at a value of about 0.1 for values of p greater than about 7. This behavior is due to the fact that for the autocorrelation method with short windows ($N = 60$) the prediction error at the beginning of the segment is an appreciable part of the total mean-squared error. This is, of course, not the case with the covariance method, where speech samples from outside the averaging interval are available for prediction.

Figure 8.11 shows the variation of V with the order of the linear predictor for a pitch asynchronous analysis for the same section of speech as used in Fig. 8.10. This time, however, the section length was $N = 120$ samples. For this case the covariance and autocorrelation methods yielded nearly identical values of V for different values of p. Further the values of V decreased monotonically to a value of about 0.1 near $p = 11$. Thus in the case of an asynchronous LPC analysis, at least for the example of a synthetic vowel, both analysis methods appear to yield similar results.

Figure 8.12 shows the variation of V with N (section length) for a linear predictor of order 12 for the synthetic speech section. As anticipated, for values of N below the pitch period (83 samples) the covariance method gives significantly smaller values of V than the autocorrelation method. For values of

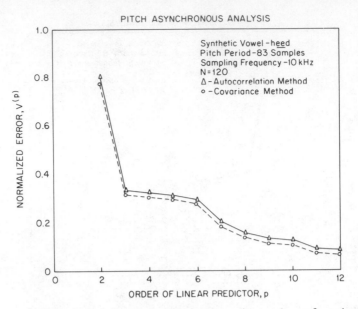

Fig. 8.11 Variation of prediction error with predictor order, p, for voiced section of a synthetic vowel—pitch asynchronous analysis. (After Chandra and Lin [15].)

V at or near multiples of the pitch period, the values of V show fairly large jumps due to the large prediction error when a pitch pulse is used to excite the system. However, for most values of N on the order of 2 or more pitch periods, both analysis methods yield comparable values of V.

Figures 8.13-8.15 show a similar set of figures for the case of a section of natural voiced speech. Figure 8.13 shows that the normalized error for the

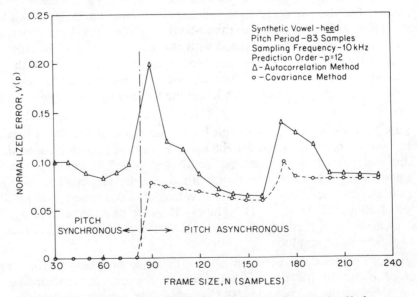

Fig. 8.12 Variation of prediction error with section length, N, for a voiced section of synthetic speech. (After Chandra and Lin [15].)

428

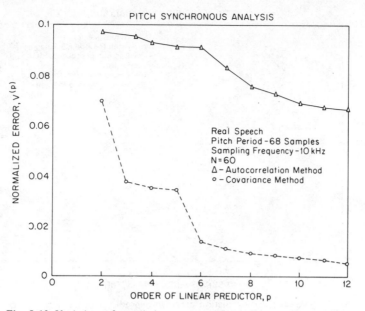

Fig. 8.13 Variation of prediction error with predictor order, p, for a voiced section of a natural vowel—pitch synchronous analysis. (After Chandra and Lin [15].)

covariance method is significantly lower than the normalized error for the auto-correlation method for a pitch synchronous analysis, whereas Figure 8.14 shows that for a pitch asynchronous analysis, the values of V are comparable. Finally Figure 8.15 shows how the values of V vary as N varies for an analysis with $p = 12$. It can be seen that in the region of pitch pulse occurrences, the value of V for the autocorrelation analysis jumps significantly whereas the value of V for the covariance analysis changes only a small amount at these points. Also for large values of N it is seen that the curves of V for the two methods approach each other.

8.5.3 Variations of the normalized error with frame position

We have already shown some properties of the LPC normalized error in Section 8.5.2 — namely its variation with section length N, and with the number of poles in the analysis, p. There remains one other major source of variability of V—namely its variation with respect to the position of the analysis frame. To demonstrate this variability Figure 8.16 shows plots of the results of a sample-by-sample (i.e. the window is moved one sample at a time) LPC analysis of 40 msec of the vowel sound /i/, spoken by a male speaker (LRR). Figure 8.16a shows the signal energy (computed at a 10 kHz rate); Fig. 8.16b shows the normalized mean-squared error (V) (again computed at a 10 kHz rate) for a 14 pole ($p=14$) analysis with a 20 msec ($N=200$) frame size for the covariance method; Fig. 8.16c shows the normalized mean-squared error for the autocorrelation method using a Hamming window; and Fig. 8.16d shows the normalized mean-squared error for the autocorrelation method using a rec-

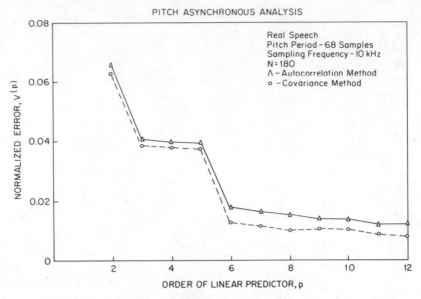

Fig. 8.14 Variation of prediction error with predictor order for a voiced section of a natural vowel—pitch asynchronous analysis. (After Chandra and Lin [15].)

tangular window. The average pitch period for this speaker was 84 samples (8.4 msec); thus about 2.5 pitch periods were contained within the 20 msec frame. For the covariance method the normalized error shows a substantial variation with the position of the analysis frame (i.e., the error is not a smooth function of time). This effect is essentially due to the large peaks in the error signal, $e(n)$, at the beginning of each pitch period as discussed previously. Thus, in this example, when the analysis frame is positioned to encompass 3 sets of error peaks, the normalized error is much larger than when only 2 sets of error peaks are included in the analysis interval. This accounts for the normalized error showing a fairly large discrete jump in level as each new error peak is included in the analysis frame. Each discrete jump of the normalized error is followed by a gradual tapering off and flattening of the normalized error. The exact detailed behavior of the normalized error between discrete jumps depends on details of the signal and the analysis method.

Figures 8.16c and 8.16d show somewhat different behavior of the LPC normalized error for the autocorrelation analysis method using a Hamming window, and a rectangular window respectively. As seen in this figure the normalized mean-squared error shows a substantial amount of high frequency variation, as well as a small amount of low frequency and pitch synchronous variation. The high frequency variation is due primarily to the error signal for the first p samples in which the signal is not linearly predictable. The magnitude of this variation is considerably smaller for the analysis using the Hamming window than for the analysis with the rectangular window due to the tapering of the Hamming window at the ends of the analysis window. Another component of the high frequency variation of the normalized error is related to the position of the analysis frame with respect to pitch pulses as discussed previously for the covariance method. However, this component of the error is much less a factor for the autocorrelation analysis than for the covariance method — especially in the case when a Hamming window is used since new pitch pulses which enter the analysis frame are tapered by the window.

Variations of the type shown in Fig. 8.16 have been found typical for most vowel sounds [16]. The variability with the analysis frame position can be reduced using allpass filtering and spectral pre-emphasis of the signal prior to linear predictive analysis [16].

8.6 Frequency Domain Interpretations of Linear Predictive Analysis

Up to this point we have discussed linear predictive methods mainly in terms of difference equations and correlation functions; i.e., in terms of time domain representations. However, we pointed out at the beginning that the coefficients of the linear predictor are *assumed* to be the coefficients of the denominator of the system function that models the combined effects of vocal tract response, glottal wave shape, and radiation. Thus, given the set of predictor coefficients we can find the frequency response of the model for speech production simply

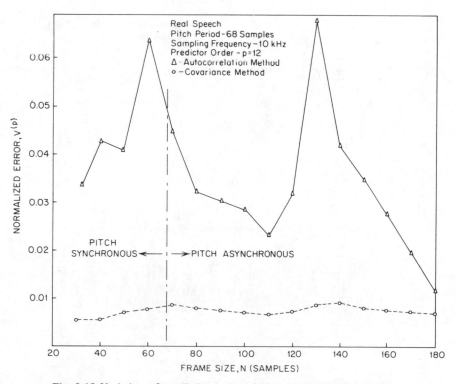

Fig. 8.15 Variation of prediction error with section length for a voiced section of natural speech. (After Chandra and Lin [15].)

431

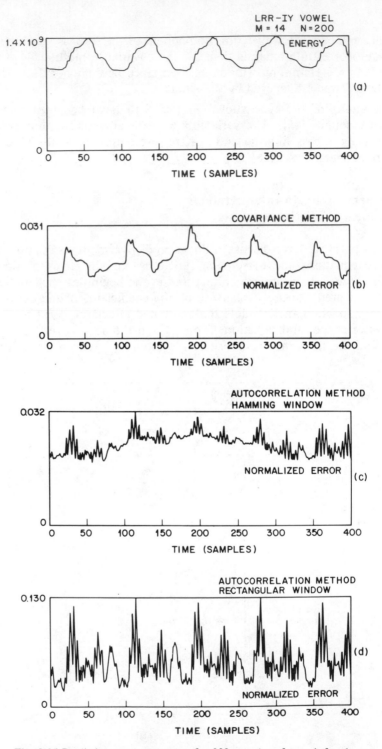

Fig. 8.16 Prediction error sequences for 200 samples of speech for three LPC systems. (After Rabiner et al. [16].)

by evaluating $H(z)$ for $z = e^{j\omega}$ i.e.,

$$H(e^{j\omega}) = \frac{G}{1 - \sum\limits_{k=1}^{p} \alpha_k e^{-j\omega k}} = \frac{G}{A(e^{j\omega})} \qquad (8.102)$$

If we plot $H(e^{j\omega})$ as a function of frequency[7] we should expect to see peaks at the formant frequencies just as we have in spectral representations discussed in previous chapters. Thus linear predictive analysis can be viewed as a method of short-time spectrum estimation. Indeed such techniques are widely applied outside the speech processing field for just this purpose [12]. In this section we shall present a frequency domain interpretation of the mean-squared prediction error and compare linear predictive techniques to other methods of estimating frequency domain representations of speech.

8.6.1 Frequency domain interpretation of mean-squared prediction error

Consider a set of predictor coefficients obtained using the autocorrelation method. In this case, the mean-squared prediction error can be expressed in the time-domain as

$$E_n = \sum_{m=0}^{N+p-1} e_n^2(m) \qquad (8.103a)$$

or in the frequency domain (using Parseval's Theorem) as

$$E_n = \frac{1}{2\pi} \int_{-\pi}^{\pi} |S_n(e^{j\omega})|^2 |A(e^{j\omega})|^2 d\omega \qquad (8.103b)$$

where $S_n(e^{j\omega})$ is the Fourier transform of the segment of speech $s_n(m)$, and $A(e^{j\omega})$ is

$$A(e^{j\omega}) = 1 - \sum_{k=1}^{p} \alpha_k e^{-j\omega k} \qquad (8.104)$$

If we recall that

$$H(e^{j\omega}) = \frac{G}{A(e^{j\omega})} \qquad (8.105)$$

then Eq. (8.103b) can be expressed as

$$E_n = \frac{G^2}{2\pi} \int_{-\pi}^{\pi} \frac{|S_n(e^{j\omega})|^2}{|H(e^{j\omega})|^2} d\omega \qquad (8.106)$$

Since the integrand in Eq. (8.106) is positive it follows that minimizing E_n is equivalent to minimizing the integral of the ratio of the energy spectrum of the speech segment to the magnitude squared of the frequency response of the linear system in the model for speech production.

[7]See Problem 8.2 for a consideration of how to evaluate $H(e^{j\omega})$ using the FFT.

In Section 8.2 it was shown that the autocorrelation function, $R_n(m)$, of the segment of speech, $s_n(m)$, and the autocorrelation function, $\tilde{R}(m)$, of the impulse response, $h(m)$, corresponding to the system function, $H(z)$, are equal for the first $(p+1)$ values. Thus, as $p \rightarrow \infty$ the respective autocorrelation functions are equal for all values and therefore

$$\lim_{p \rightarrow \infty} |H(e^{j\omega})|^2 = |S_n(e^{j\omega})|^2 \tag{8.107}$$

This implies that if p is large enough we can approximate the signal spectrum with arbitrarily small error with the all-pole model, $H(z)$.

It is interesting to note that even though Eq. (8.107) says that as $p \rightarrow \infty$, $|H(e^{j\omega})|^2 = |S_n(e^{j\omega})|^2$, it is not necessarily (or generally) true that $H(e^{j\omega}) = S_n(e^{j\omega})$ — i.e., the frequency response of the model need not equal the Fourier transform of the signal. This is so because $S_n(e^{j\omega})$ need not be minimum phase, whereas $H(e^{j\omega})$ is required to be minimum phase since it is the transfer function of an all-pole filter with poles inside the unit circle.

To illustrate the nature of the spectral modelling capability of linear predictive spectra, Fig. 8.17 (due to Makhoul [7]) shows a comparison between $20 \log_{10} |H(e^{j\omega})|$ and $20 \log_{10} |S_n(e^{j\omega})|$. The signal spectrum was obtained by an FFT analysis of a 20 msec section of speech (sampled at 20 kHz), weighted by a Hamming window as discussed in Chapter 6. The speech sound was the vowel /ae/. The LPC spectrum was that of a 28-pole predictor ($p=28$) obtained by the autocorrelation method [2]. The harmonic structure of the signal spectrum is clearly seen in this figure. A significant feature of the LPC spectral modelling can also be seen in this figure. This is the fact that the LPC spectrum matches the signal spectrum much more closely in the regions of large signal energy (i.e., near the spectrum peaks) than near the regions of low signal energy (i.e., near the spectral valleys). This is to be expected in view of Eq. (8.106) since regions where $|S_n(e^{j\omega})| > |H(e^{j\omega})|$ contribute more to the total error than regions where $|S_n(e^{j\omega})| < |H(e^{j\omega})|$. Thus the LPC spectral error criterion favors a good fit near the spectral peaks, whereas the fit near the spectral valleys is nowhere near as good.

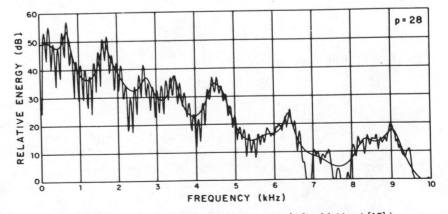

Fig. 8.17 28 pole fit to an FFT signal spectrum. (After Makhoul [17].)

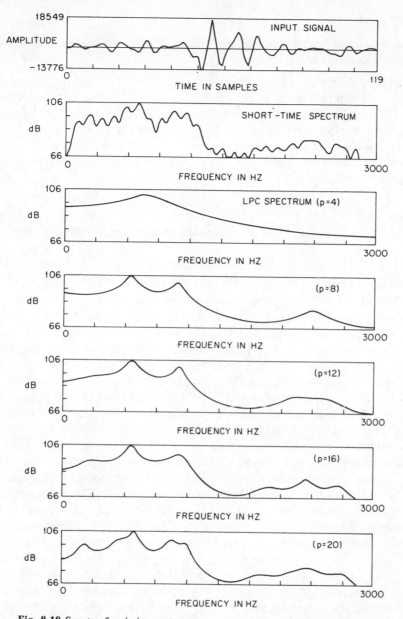

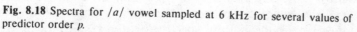

Fig. 8.18 Spectra for /a/ vowel sampled at 6 kHz for several values of predictor order *p*.

The above discussion suggests that the order *p* of the linear predictive analysis can effectively control the degree of smoothness of the resulting spectrum. This is illustrated in Fig. 8.18 which shows the input speech segment, the Fourier transform of that segment and linear predictive spectra for various orders. It is clear that as *p* increases, more of the details of the spectrum are preserved. Since our objective is to obtain a representation of only the spectral effects of the glottal pulse, vocal tract, and radiation, it is clear that we should choose p as discussed before so that the formant resonances and the general spectrum shape are preserved.

It should be pointed out that we have assumed in this discussion that the predictor parameters were computed using the autocorrelation method. This was necessary because only in this case is the Fourier transform of the short-time autocorrelation function equal to the magnitude squared of the short-time Fourier transform of the signal. However this does not preclude the use of H $(e^{j\omega})$ as a spectrum estimate even if the predictor coefficients are estimated by the covariance method.

8.6.2 Comparison to other spectrum analysis methods

We have already discussed methods of obtaining the short-time spectrum of speech in Chapters 6 and 7. It is instructive to compare these methods with the spectrum obtained by linear predictive analysis.

As an example, Fig. 8.19 (due to Zue [10]) shows four log spectra of a section of the synthetic vowel /a/. The first two spectra were obtained using the short-time spectrum method discussed in Chapter 6. For the first spectrum, a section of 512 samples (51.2 msec) was windowed, and then transformed (using a 512 point FFT) to give the relatively narrow band spectral analysis shown at the top of Fig. 8.19. In this spectrum the individual harmonics of the excitation are clearly in evidence due to the relatively long duration of the window. For the second spectrum the analysis duration was decreased to 128 samples (12.8 msec) leading to a wideband spectral analysis. Now the excitation harmonics are not resolved; instead the overall spectral envelope can be seen. Although the formant frequencies are in evidence in this spectrum, it is not a simple matter to reliably locate or identify them. The third spectrum was obtained by homomorphic smoothing as discussed in Chapter 7. The unsmoothed spectrum was obtained from a 300 sample (30 msec) section using the FFT method described above. The smoothed spectrum shown in this figure was obtained by linear smoothing of the log spectrum. For this example the individual formants are well resolved and are easily measured from the smoothed spectrum using a simple peak picker. However, the bandwidths of the formants are not easily obtained from the homomorphically smoothed spectrum due to all the smoothing processes which have been used in obtaining the final spectrum. Finally the bottom spectrum is the result of a linear predictive analysis using $p = 12$ and a section of $N = 128$ samples (12.8 msec). A comparison of the linear prediction spectrum to the other spectra shows that the parametric representation appears to represent the formant structure very well with no extraneous peaks or ripples. This is due to the fact that the linear predictive model is very good for vowel sounds if the correct order, p, is used. Since the correct order can be determined knowing the speech bandwidth, the linear prediction method leads to very good estimates of the spectral properties due to the glottal pulse, vocal tract and radiation.

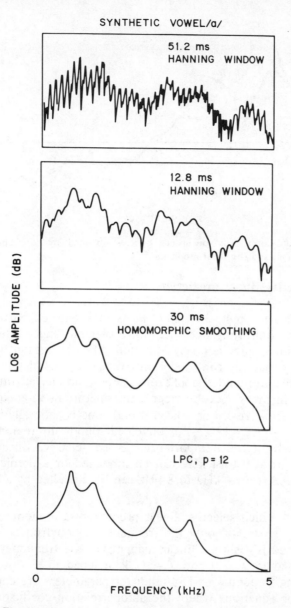

Fig. 8.19 Spectra of synthetic vowel /a/. (Afer Zue [10].)

Figure 8.20 shows a direct comparison of the spectra of a voiced section from natural speech obtained by both homomorphic smoothing and linear prediction. Although the formant frequencies are clearly in evidence in both plots, it can be seen that the LPC spectrum has fewer extraneous peaks than the homomorphic spectrum. This is because the LPC analysis assumed a value of $p = 12$ so that at most 6 resonance peaks could occur. For the homomorphic spectrum no such restriction existed. As noted above, the spectrum peaks from the LPC analysis are much narrower than the spectrum peaks from the homomorphic analysis due to the smoothing of the short-time log spectrum.

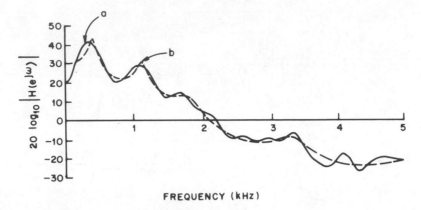

Fig. 8.20 Comparison of speech spectra obtained by (a) cepstrum smoothing; and (b) linear prediction.

8.6.3 Selective linear prediction

It is possible to apply the above ideas to a selected portion of the spectrum, rather than uniformly over the entire spectral range. This idea has been called selective linear prediction by Makhoul [8]. The reason this method is of potential value is that one can model only those regions of the spectrum which are important to the intended application. For example, a sampling rate of 20 kHz is required in many speech recognition applications to adequately represent the spectrum of fricatives. For voiced sounds one is generally interested in the region from 0 to about 4 kHz. For unvoiced sounds the region from 4 kHz to 8 kHz is generally of most importance. Using selective linear prediction the signal spectrum from 0 to 4 kHz can be modelled by a predictor of order p_1; whereas the region from 4 kHz to 8 kHz can be modelled by a different predictor of order p_2.

The way in which selective linear prediction is implemented is relatively straightforward. To model only the frequency region from $f = f_A$ to $f = f_B$, all that is required is a simple linear mapping of the frequency scale such that $f = f_A$ is mapped to $f' = 0$ and $f = f_B$ is mapped to $f' = \omega'/2\pi = 0.5$ (i.e., half the sampling frequency). The predictor parameters are computed by solving the predictor equations where the autocorrelation coefficients are obtained from

$$R'(i) = \frac{1}{2\pi} \int_{-\pi}^{\pi} |S_n(e^{j\omega'})|^2 e^{j\omega'i} d\omega' \tag{8.108}$$

Figure 8.21 (due to Makhoul [8]) illustrates the method of selective linear prediction. The signal spectrum is identical to the one of Fig. 8.17. The region from 0 to 5 kHz is modelled by a 14-pole predictor ($p_1=14$), whereas the region from 5-10 kHz is modelled independently by a 5-pole predictor ($p_2=5$). It can be seen that at 5 kHz, the model spectra show a discontinuity since there is no constraint that they agree at any frequency.

438

8.6.4 Comparison to analysis-by-synthesis methods

As discussed in Chapter 6, the error measure which is normally used in analysis-by-synthesis methods is the log of the ratio of the signal power spectrum to the power spectrum of the model, i.e.,

$$E' = \int_{-\pi}^{\pi} \left\{ \log \left[\frac{|S_n(e^{j\omega})|^2}{|H(e^{j\omega})|^2} \right] \right\}^2 d\omega \qquad (8.109)$$

Thus for analysis-by-synthesis minimization of E' is equivalent to minimizing the mean square error between the two log spectra.

A comparison between the error measures used for LPC modelling and analysis-by-synthesis modelling leads to the following observations:

1. Both error measures are related to the ratio of the signal to model spectra.
2. Both error measures tend to perform uniformly over the whole frequency range.
3. Both error measures are suitable to selective error minimization over specified frequency regions.
4. The error criterion for linear predictive modelling places higher weight on frequency regions where $|S_n(e^{j\omega})|^2 > |H(e^{j\omega})|^2$ than when $|S_n(e^{j\omega})|^2 < |H(e^{j\omega})|^2$, whereas the error criterion for analysis-by-synthesis places equal weight on both these regions.

The conclusion which is drawn from these observations is that when dealing with signal spectra which are unsmoothed (as in Figure 8.17) the linear predictive error criterion yields better spectral matches than the analysis-by-synthesis method [7]. Furthermore the required computation for the linear predictive modelling is significantly less than for the analysis-by-synthesis method. If one is modelling smooth signal spectra (as might be obtained at the

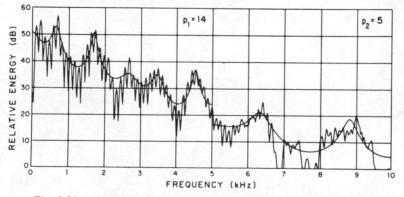

Fig. 8.21 Application of selective linear prediction to the signal spectrum of Fig. 8.17 with a 14-pole fit to the 0-5 kHz region and a 5-pole fit to the 5-10 kHz region. (After Makhoul [2].)

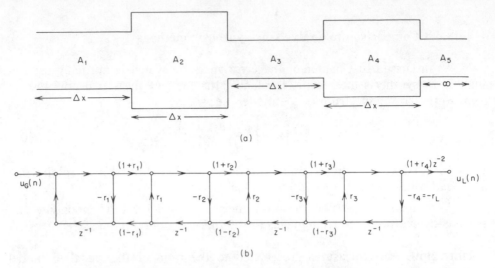

Fig. 8.22 (a) Lossless tube model terminated in infinitely long tube; (b) corresponding signal flow graph for infinite glottal impedance.

output of a filter bank) then both the LPC and analysis-by-synthesis methods give reasonably good fits to the spectra. In practice the analysis-by-synthesis method is applied almost always to this type of signal spectrum.

8.7 Relation of Linear Predictive Analysis to Lossless Tube Models

In Chapter 3 we discussed a model for speech production that consisted of a concatenation of N lossless acoustic tubes as shown in Fig. 8.22. The reflection coefficients r_k in Fig. 8.22b are related to the areas of the lossless tubes by

$$r_k = \frac{A_{k+1} - A_k}{A_{k+1} + A_k} \tag{8.110}$$

In Section 3.3.4, the transfer function of such a system was derived subject to the condition that the reflection coefficient at the glottis was $r_G = 1$, i.e., the glottal impedance was assumed to be infinite. In Section 3.3.4, the system function of a system such as shown in Fig. 8.22 was shown to be

$$V(z) = \frac{\displaystyle\prod_{k=1}^{N} (1+r_k)z^{-N/2}}{D(z)} \tag{8.111}$$

where $D(z)$ satisfies the polynomial recursion

$$D_0(z) = 1 \tag{8.112a}$$

$$D_k(z) = D_{k-1}(z) + r_k z^{-k} D_{k-1}(z^{-1}) \tag{8.112b}$$

$$D(z) = D_N(z) \tag{8.112c}$$

All of this is very reminiscent of the discussion of the lattice formulation in Section 8.3.3. Indeed, there it was shown that the polynomial

440

$$A(z) = 1 - \sum_{k=1}^{p} \alpha_k z^{-k} \qquad (8.113)$$

obtained by linear prediction analysis could be obtained by the recursion

$$A^{(0)}(z) = 1 \qquad (8.114a)$$

$$A^{(i)}(z) = A^{(i-1)}(z) - k_i z^{-i} A^{(i-1)}(z^{-1}) \qquad (8.114b)$$

$$A(z) = A^{(p)}(z) \qquad (8.114c)$$

where the parameters $\{k_i\}$ were called the PARCOR coefficients. By comparing Eqs. (8.112) and (8.114) it is clear that the system function

$$H(z) = \frac{G}{A(z)} \qquad (8.115)$$

obtained by linear prediction analysis has the same form as the system function of a lossless tube model consisting of p sections. If

$$r_i = - k_i \qquad (8.116)$$

then it is clear that

$$D(z) = A(z) \qquad (8.117)$$

Using Eq. (8.110) and Eq. (8.116) it is easy to show that the areas of the equivalent tube model are related to the PARCOR coefficients by

$$A_{i+1} = \left[\frac{1-k_i}{1+k_i} \right] A_i \qquad (8.118)$$

Note that the PARCOR coefficient gives us a ratio between areas of adjacent sections. Thus the areas of the equivalent tube model are not absolutely determined and any convenient normalization will produce a tube model with the same transfer function.

It should be pointed out that the "area function" obtained using Eq. (8.118) cannot be said to be the area function of the human vocal tract. However, Wakita [17] has shown that if pre-emphasis is used prior to linear predictive analysis to remove the effects due to the glottal pulse and radiation, then the resulting area functions are often very similar to vocal tract configurations that would be used in human speech.

8.8 Relations Between the Various Speech Parameters

Although the set of predictor coefficients, α_k, $1 \leqslant k \leqslant p$, is often thought of as the basic parameter set of the linear predictive analysis, it is straightforward to transform this set of coefficients to a number of other parameter sets, to obtain alternative representations of speech. Such alternative representations often are more convenient for applications of linear predictive analysis. In this section we discuss how other useful parameter sets can be obtained directly from LPC coefficients [1,2].

8.8.1 Roots of the predictor polynomial

Perhaps the simplest alternative to the predictor parameters is the set of roots of the polynomial

$$A(z) = 1 - \sum_{k=1}^{p} \alpha_k z^{-k} = \prod_{k=1}^{p} (1 - z_k z^{-1}) \qquad (8.119)$$

That is, the roots $\{z_i, i=1,2,...,p\}$ are an equivalent representation of $A(z)$. If conversion of the z-plane roots to the s-plane is desired, this can be achieved by setting

$$z_i = e^{s_i T} \qquad (8.120)$$

where $s_i = \sigma_i + j\Omega_i$ is the s-plane root corresponding to z_i in the z-plane. If $z_i = z_{ir} + jz_{ii}$ then

$$\Omega_i = \frac{1}{T} \tan^{-1}\left|\frac{z_{ii}}{z_{ir}}\right| \qquad (8.121)$$

and

$$\sigma_i = \frac{1}{2T} \log(z_{ir}^2 + z_{ii}^2) \qquad (8.122)$$

Equations (8.121) and (8.122) are useful for formant analysis applications of LPC analysis systems.

8.8.2 Cepstrum

Another alternative to the LPC coefficients is the cepstrum of the impulse response of the overall LPC system. If the overall LPC system has transfer function $H(z)$ with impulse response $h(n)$ and complex cepstrum $\hat{h}(n)$ then it can be shown that $\hat{h}(n)$ can be obtained from the recursion

$$\hat{h}(n) = \alpha_n + \sum_{k=1}^{n-1} \left(\frac{k}{n}\right) \hat{h}(k) \alpha_{n-k} \quad 1 \leq n \qquad (8.123)$$

where

$$H(z) = \sum_{n=0}^{\infty} h(n) z^{-n} = \frac{G}{1 - \sum_{k=1}^{p} \alpha_k z^{-k}} \qquad (8.124)$$

8.8.3 Impulse response of the all-pole system

The impulse response, $h(n)$, of the all-pole system with the transfer function of Eq. (8.124) can be solved for recursively from the LPC coefficients as

$$h(n) = \sum_{k=1}^{p} \alpha_k h(n-k) + G\delta(n) \qquad 0 \leq n \qquad (8.125)$$

442

where $h(n)$ is assumed (by definition) to be 0 for $n < 0$, and G is the amplitude of the excitation.

8.8.4 Autocorrelation of the impulse response

As discussed in Section 8.2, it is easily shown (see Problem 8.1) that the autocorrelation function of impulse response of the filter defined as

$$\tilde{R}(i) = \sum_{n=0}^{\infty} h(n)h(n-i) = \tilde{R}(-i) \tag{8.126}$$

satisfies the relations

$$\tilde{R}(i) = \sum_{k=1}^{p} \alpha_k \tilde{R}(|i-k|) \quad 1 \leqslant i \tag{8.127}$$

and

$$\tilde{R}(0) = \sum_{k=1}^{p} \alpha_k \tilde{R}(k) + G^2 \tag{8.128}$$

Equations (8.127) and (8.128) can be used to determine $\tilde{R}(i)$ from the predictor coefficients and vice versa.

8.8.5 Autocorrelation coefficients of the predictor polynomial

Corresponding to the predictor polynomial, or inverse filter,

$$A(z) = 1 - \sum_{k=1}^{p} \alpha_k z^{-k} \tag{8.129}$$

is the impulse response of the inverse filter

$$a(n) = \delta(n) - \sum_{k=1}^{p} \alpha_k \delta(n-k)$$

The autocorrelation function of the inverse filter impulse response is

$$R_a(i) = \sum_{k=0}^{p-i} a(k)a(k+i) \quad 0 \leqslant i \leqslant p \tag{8.130}$$

8.8.6 PARCOR coefficients

For the autocorrelation method the predictor coefficients may be obtained from the PARCOR coefficients using the recursion

$$a_i^{(i)} = k_i \tag{8.131a}$$

$$a_j^{(i)} = a_j^{(i-1)} - k_i a_{i-j}^{(i-1)} \quad 1 \leqslant j \leqslant i-1 \tag{8.131b}$$

with Eqs. (8.131a) and (8.131b) being solved for $i = 1, 2, \ldots, p$ and with the final set being defined as

$$\alpha_j = a_j^{(p)} \qquad 1 \leqslant j \leqslant p \tag{8.131c}$$

Similarly the set of PARCORS may be obtained from the set of LPC coefficients using a backward recursion of the form

$$k_i = a_i^{(i)} \tag{8.132a}$$

$$a_j^{(i-1)} = \frac{a_j^{(i)} + a_i^{(i)} a_{i-j}^{(i)}}{1 - k_i^2} \qquad 1 \leqslant j \leqslant i - 1 \tag{8.132b}$$

where i goes from p, to $p-1$, down to 1 and initially we set

$$a_j^{(p)} = \alpha_j \qquad 1 \leqslant j \leqslant p \tag{8.132c}$$

8.8.7 Log area ratio coefficients

An important set of equivalent parameters which can be derived from the PARCOR parameters is the log area ratio parameters defined as

$$g_i = \log\left[\frac{A_{i+1}}{A_i}\right] = \log\left(\frac{1 - k_i}{1 + k_i}\right) \qquad 1 \leqslant i \leqslant p \tag{8.133}$$

The g_i parameters are equal to the log of the ratio of the areas of adjacent sections of a lossless tube equivalent of the vocal tract having the same transfer function as the linear predictive model as discussed in Section 8.7. The g_i parameters have also been found to be especially appropriate for quantization by Makhoul [2] and others [1] because of the relatively flat spectral sensitivity of the $g_i's$.

The k_i parameters may be directly obtained from the g_i by the inverse transformation

$$k_i = \frac{1 - e^{g_i}}{1 + e^{g_i}} \ , \qquad 1 \leqslant i \leqslant p \tag{8.134}$$

8.9 Synthesis of Speech from Linear Predictive Parameters

Speech can be synthesized from the linear predictive analysis parameters in several different ways. The simplest way is to use a system which is the same parametric representation as was used in the analysis. Figure 8.23 shows a block diagram of such a speech synthesizer. The time varying control parameters needed by the synthesizer are the pitch period, a voiced/unvoiced switch, the gain or rms speech value, and the p predictor coefficients. The impulse generator acts as the excitation source for voiced sounds producing a pulse of unit amplitude at the beginning of each pitch period. The white noise generator acts as the excitation source for unvoiced sounds producing uncorrelated, uniformly distributed random samples with unity standard deviation, and zero mean. The selection between the two sources is made by the voiced/unvoiced

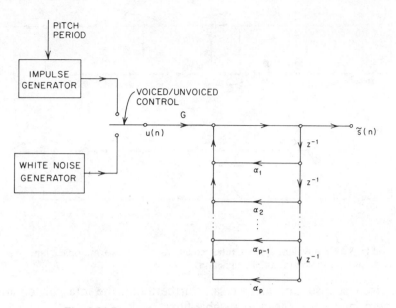

IMPULSE
GENERATOR

VOICED/UNVOICED
CONTROL

G

$u(n)$

$\tilde{s}(n)$

z^{-1}

WHITE NOISE
GENERATOR

α_1

z^{-1}

α_2

α_{p-1}

z^{-1}

α_p

Fig. 8.23 Block diagram of linear predictive synthesizer.

control. The gain control G determines the overall amplitude of the excitation. The synthetic speech samples are determined by

$$\tilde{s}(n) = \sum_{k=1}^{p} \alpha_k \tilde{s}(n-k) + Gu(n) \qquad (8.135)$$

A network which realizes Eq. (8.135) is shown in Fig. 8.23. This direct form network is the most simple and straightforward method for synthesizing speech from the predictor parameters. A total of p multiplies and p adds are required to generate each output sample.

In the synthesis model of Fig. 8.23 the synthesis parameters must be changed with time. Although the parameters are usually estimated at regular intervals during regions of voiced speech, the control parameters are changed at the beginning of each period. For unvoiced speech they are simply changed once per frame (i.e., every 10 msec for a 100 frame/sec rate). The updating of control parameters at the beginning of each pitch period (called pitch synchronous synthesis) has been found to be a much more effective synthesis strategy than the process of updating the parameters once each frame (called asynchronous synthesis). This requires that the control parameters be interpolated to obtain the values at the beginning of each pitch period. Atal has found that the pitch and gain parameters should be interpolated geometrically [3] (i.e., linearly on a log scale); however, due to stability constraints, the predictor parameters themselves cannot be interpolated. This is due to the fact that interpolation between two sets of stable predictor coefficients can lead to an unstable interpolated result. One way around this difficulty, according to Atal, is to interpolate the first p samples of the autocorrelation function of the impulse response of the filter of Fig. 8.21. Using the relations of Section 8.4, the predictor coefficients can be obtained from the first p samples of the autocorrelation of

445

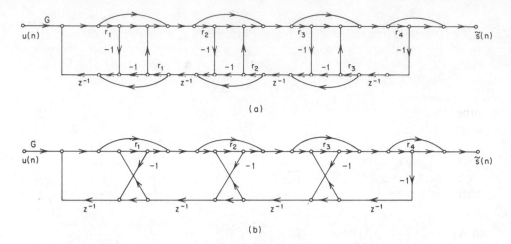

(a)

(b)

Fig. 8.24 Equivalent lossless tube models using (a) two multiplier junctions; and (b) one multiplier junction.

the impulse response, and vice versa. Furthermore, the interpolated autocorrelation coefficients always lead to a stable filter.[8]

The synthesizer of Fig. 8.23 has been used in a wide variety of simulations of LPC systems. Its main advantage is its simplicity and ease of implementation. Its main drawback is that it requires considerable computational accuracy to synthesize the speech because the structure is basically a direct form recursive structure which tends to be quite sensitive to changes in the coefficients. Perhaps the most attractive alternative to synthesis based on the predictor parameters is the use of the reflection coefficients or the PARCOR coefficients in a lossless tube equivalent. In other words, this direct form network in Fig. 8.23 can be replaced by a structure such as Fig. 8.22. The advantage of such a structure is that the multipliers are the reflection coefficients, $r_i = -k_i$, which have the property that they are bounded ($|k_i| < 1$), and also that they can be interpolated directly while maintaining a stable filter. Such structures are also less sensitive to quantization effects in finite word length implementations of the synthesizer than the direct form implementation of Fig. 8.23.

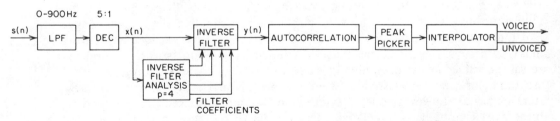

Fig. 8.25 Block diagram of the SIFT algorithm for pitch detection.

[8]Similarly the PARCOR coefficients or the log area ratio coefficients can be interpolated and the resulting system is guaranteed to be stable if the parameter sets which are being interpolated are stable.

446

It is clear from Fig. 8.22b that to implement a pth order synthesis filter as an acoustic tube model requires $4p + 2$ multiplications and $2(p-1)$ additions per sample as compared to p multiplications and p additions for the direct form. In Section 3.3.3 it was shown that the four multiplier junctions in Fig. 8.22 can be replaced by one and two multiplier junctions at the expense of increased number of additions. By making the substitutions indicated in Fig. 3.41, the flow graph of Fig. 8.22b can be transformed into those shown in Fig. 8.24. Figure 8.24a requires $2p - 1$ multiplications and $4p - 1$ additions while Fig. 8.24b requires p multiplications and $3p - 2$ additions. In using lossless tube models for synthesis, the choice of the particular form depends on a variety of factors so that it is not possible to say that any one form is the most efficient.

8.10 Applications of LPC Parameters

As evidenced in the preceding sections of this chapter, the theory of linear prediction is highly developed. Based on this theory, and its implications, a large variety and range of applications of linear predictive analysis to speech processing has evolved. Schemes have been devised for estimating all the basic speech parameters from linear predictive analyses. Based on such analyses, vocoders have been studied extensively, leading to an understanding of the quantization properties of the various LPC representations. Finally these techniques have been used in many speech processing systems for speaker verification and identification, speech recognition, speech classification, speech dereverberation, etc. In the following sections and in Chapter 9 we present outlines of several representative methods for estimating speech parameters using linear predictive analysis.

8.10.1 Pitch detection using LPC parameters

We have already discussed how the error signal $e(n)$ from the LPC analysis can, in theory, be used to estimate the pitch period directly. Although this method will generally be capable of finding the correct period, a somewhat more sophisticated method of pitch detection was proposed by Markel [19]. This algorithm is called the SIFT (simple inverse filtering tracking) method. A similar method was proposed by Maksym [20].

Figure 8.25 shows a block diagram of the SIFT algorithm. The input signal $s(n)$ is lowpass filtered with a cutoff frequency of about 900Hz, and then the sampling rate (nominally 10 kHz) is reduced to 2 kHz by a decimation process (i.e., 4 out of every 5 samples are dropped at the output of the lowpass filter). The decimated output, $x(n)$, is then analyzed using the autocorrelation method with a value of $p = 4$ for the filter order. A fourth order filter is sufficient to model the signal spectrum in the frequency range 0-1 kHz because there will generally be only 1-2 formants in this range. The signal $x(n)$ is then inverse filtered to give $y(n)$, a signal with an approximately flat spectrum.[9]

[9]The output $y(n)$ is simply the prediction error for the fourth order predictor.

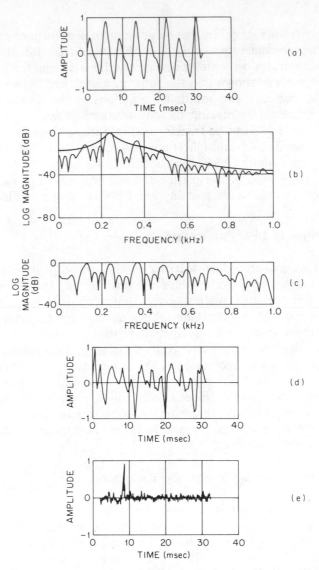

Fig. 8.26 Typical signals from the SIFT algorithm. (After Markel [19].)

Thus the purpose of the linear predictive analysis is to spectrally flatten the input signal, similar to the clipping methods discussed in Chapter 4. The short-time autocorrelation of the inverse filtered signal is computed and the largest peak in the appropriate range is chosen as the pitch period. To obtain additional resolution in the value of the pitch period, the autocorrelation function is interpolated in the region of the maximum value. An unvoiced classification is chosen when the level of the autocorrelation peak (suitably normalized) falls below a given threshold.

Figure 8.26 (due to Markel [19]) illustrates some typical waveforms obtained at several points in the analysis. Figure 8.26a shows a section of the input waveform being analyzed; Fig. 8.26b shows the input spectrum, and the reciprocal of the spectrum of the inverse filter. For this example there appears to be a single formant in the range of 250 Hz. Figure 8.26c shows the spectrum of the signal at the output of the inverse filter, whereas Fig. 8.26d shows the time waveform at the output of the inverse filter. Finally Fig. 8.26e shows the normalized autocorrelation of the signal at the output of the inverse filter. A pitch period of about 8 msec is clearly in evidence.

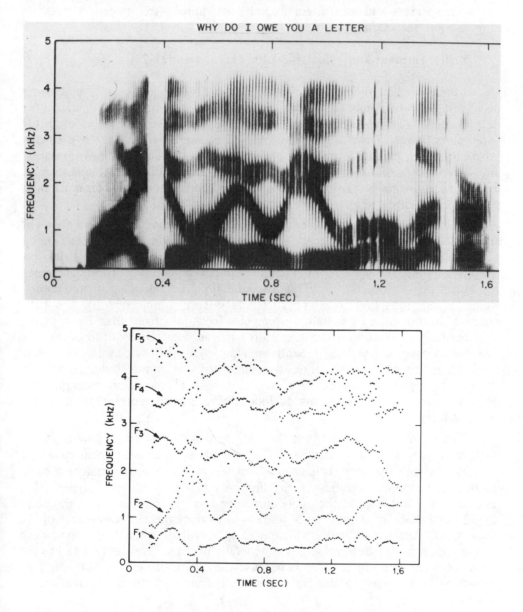

Fig. 8.27 (a) Spectrogram of original speech; (b) center frequencies of complex pole locations for 12th order linear predictive analysis. (After Atal and Hanauer [3].)

The SIFT algorithm uses the linear predictive analysis to provide a spectrally flattened signal to facilitate pitch detection. To the extent that this spectral flattening is successful, the method appears to be a reasonably good one for pitch analysis. However, for high pitched speakers (such as children) the spectral flattening is generally unsuccessful due to the lack of more than one pitch harmonic in the band from 0 to 900 Hz (especially for telephone line inputs). For such speakers and transmission conditions, other pitch detection methods may be more successful.

8.10.2 Formant analysis using LPC parameters [21-23]

Linear predictive analysis of speech has several advantages, and some disadvantages when applied to the problem of estimating the formants for voiced sections of speech. Formants can be estimated from the predictor parameters in one of two ways. The most direct way is to factor the predictor polynomial and, based on the roots obtained, try to decide which are formants, and which correspond to spectral shaping poles [21,22]. The alternative way of estimating formants is to obtain the spectrum, and choose the formants by a peak picking method similar to the one discussed in Chapter 7 [23].

A distinct advantage inherent in the linear predictive method of formant analysis is that the formant center frequency and bandwidth can be determined accurately by factoring the predictor polynomial. Since the predictor order p is chosen a priori, the maximum possible number of complex conjugate poles which can be obtained is $p/2$. Thus the labelling problem inherent in deciding which poles correspond to which formants is less complicated for the LPC method since there are generally fewer poles to choose from than for comparable methods of obtaining the spectrum such as cepstral smoothing. Finally extraneous poles are generally easily isolated in the LPC analysis since their bandwidths are often very large, compared to what one would expect for bandwidths typical of speech formants. Figure 8.27 shows an example that illustrates that the pole locations do indeed give a good representation of the formant frequencies [3].

The disadvantage inherent in the LPC method is that an all-pole model is used to model the speech spectrum. For sounds such as nasals and nasalized vowels, although the analysis is adequate in terms of its spectral matching capabilities, the physical significance of the roots of the predictor polynomial is unclear. It is not clear if the roots correspond to the nasal zeros or the additional nasal poles, or if they are at all related to the expected resonances of the vocal tract. Another difficulty with the analysis is that although the bandwidth of the root is readily determined, it is generally not clear how it is related to the actual formant bandwidth. This is because the bandwidth of the root has been shown to be sensitive to the frame duration, frame position, and method of analysis.

With these advantages and disadvantages in mind, several methods have been proposed for estimating formants from LPC derived spectra using peak picking methods, and from the predictor polynomial by factoring methods.

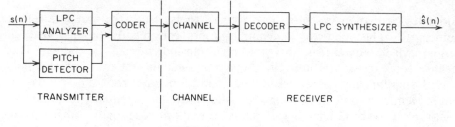

RECTANGULAR WINDOW

Fig. 8.28 Block diagram of LPC vocoder.

Once the candidates for the formants have been chosen, the techniques used to label these candidates — i.e., the assigning of a candidate to a particular formant, are similar to those used for any other analysis method. These include reliance on formant continuity, a need for spectral pre-emphasis to minimize the possibility of close formants merging, and the use of an off the unit circle contour for evaluating the LPC spectrum thereby sharpening the spectral peaks. Discussion of the various methods is given by Markel [21,22], Atal [3], Makhoul and Wolf [5], and McCandless [23].

8.10.3 An LPC vocoder — quantization considerations [24-25]

One of the most important applications of linear predictive analysis has been the area of low bit rate encoding of speech for transmission (the LPC vocoder) and storage (for computer voiced response systems). Figure 8.28 shows a block diagram of an LPC vocoder. The vocoder consists of a transmitter which performs the LPC analysis and pitch detection, and then codes the parameters for transmission, a channel over which the parameters are sent, and a receiver which decodes the parameters and synthesizes the output speech from them. We have already discussed both the analyzer and the synthesizer. We assume, for simplicity, that the channel is an error free transmission medium. Thus in this section we look at the coder and decoder to see which set of parameters is most appropriate for encoding at a given bit rate.

The basic LPC analysis parameters are the set of p predictor coefficients, the pitch period, a voiced/unvoiced parameter, and the gain parameter. Techniques for properly coding pitch, voiced/unvoiced switch, and the gain are fairly well understood. For the pitch period 6 bit quantization is adequate; for the voiced/unvoiced switch, 1 bit is required; and for the gain a total of about 5 bits distributed on a logarithmic scale are sufficient [3].

Although one could consider direct quantization of the predictor coefficients, this approach is not recommended because, to ensure stability of the predictor polynomial, a relatively high accuracy (8-10 bits per coefficient) is required. The reason for this is that small changes in the predictor coefficients can lead to relatively large changes in the pole positions. Thus direct quantization of the predictor coefficients is generally avoided.

This leaves open the question as to an appropriate parameter set for coding and transmission. Among the proposed parameter sets the most reasonable candidates are the predictor polynomial roots, and the set of reflection coefficients. The predictor polynomial roots can readily be quantized in a manner which guarantees that the resulting polynomial is stable. This is because roots inside the unit circle guarantee stability of the predictor polynomial. Using this approach Atal [3] has found 5 bits per root (i.e., 5 bits for the center frequency and 5 bits for the bandwidth) are adequate to preserve the quality of the synthesized speech so as to make it essentially indistinguishable from speech synthesized from the unquantized parameters.

Using such a coding scheme, the overall bit rate for transmission or storage is $72 \cdot F_s$ bits per second where F_s is the number of frames per second which are stored or transmitted. Typical values for F_s are 100, 67, and 33 giving bit rates of 7200, 4800, and 2400 bits per second respectively.

Another interesting parameter set which can be easily quantized and for which stability can be guaranteed is the set of PARCOR coefficients, k_i. The stability condition on the $k_i's$ is $|k_i| < 1$ which is simple to preserve under quantization. Makhoul and Viswanathan [25] have found that the distribution of the reflection coefficients is highly skewed; thus a transformation of these parameters is required to optimally allocate the fixed number of bits in a reasonable manner. Using a spectral sensitivity measure, Makhoul and Viswanathan [25] found the optimal transformation to be of the form

$$g_i = f(k_i) = \log\left(\frac{1-k_i}{1+k_i}\right) = \log\left(\frac{A_{i+1}}{A_i}\right) \qquad 1 \leqslant i \leqslant p \qquad (8.136)$$

where A_i is the area function of a lossless tube representation of the vocal tract. Thus the optimal parameter for linear encoding is the logarithm of the ratio of areas of a lossless tube representation of the vocal tract. It is easily seen that Eq. (8.136) maps the region $-1 \leqslant k_i \leqslant 1$ to $-\infty \leqslant g_i \leqslant \infty$. Using this transformation Atal [27] found that the coefficients g_i had a fairly uniform amplitude distribution, and low inter-parameter correlations; therefore these parameters were quite good for digital transmission. With this parameter set a total of about 5-6 bits per log area ratio is necessary to achieve the same quality synthetic speech as obtained from the uncoded parameters.

In all the above coding schemes it was assumed the parameters were being encoded using some type of PCM representation. It has recently been demonstrated by Sambur [26] that the coding techniques discussed in Chapter 5 can be applied directly to the various LPC parameter sets leading to further decreases in the required bit rates for transmission and storage. Using ADPCM coding of the predictor parameters, Sambur claims good quality speech with bit rates on the order of 1000-2000 bits per second.

8.10.4 Voice-excited LPC vocoders [27,28]

We have already shown that the weakest link in most vocoders is accurate estimation and representation of the excitation function. In Chapter 6 we discussed some vocoder systems which did not require direct estimation of pitch

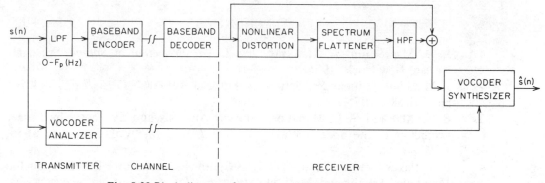

Fig. 8.29 Block diagram of a voice-excited LPC vocoder.

and voiced/unvoiced classification, but instead represented the excitation in terms of the phase (or phase derivative) of the signal. Another approach to avoiding direct estimation of excitation parameters for a vocoder is the voice-excited vocoder. Systems of this type have been studied by Atal et al. [27] and Weinstein [28]. Figure 8.29 shows a block diagram of a voice-excited LPC vocoder. There are two distinct transmission paths in this system; one producing a low frequency band of the direct signal, one producing the normal vocoder parameters (e.g., LPC coefficients, spectral magnitudes, etc.). The low frequency band, which can be coded using any of the methods described in Chapter 5, is used to generate the excitation signal for the synthesizer by an appropriate combination of nonlinear distortion and spectral flattening. The reason this procedure is effective is that the low frequency band contains all the necessary information about the excitation — i.e., it is periodic with the correct period for voiced speech, and it is noise-like for unvoiced speech. Thus, using such a scheme to generate the excitation eliminates the need for methods for estimating pitch, and voiced/unvoiced classification. However, this method has the disadvantage that additional information must be transmitted over the channel to accurately describe the low frequency band of the signal; thus voice-excited vocoders generally require somewhat higher bit rates than conventional vocoders. For example a voice-excited LPC vocoder requires on the order of 3000-4000 bps or about 1000-2000 bps more than the conventional LPC vocoder described in the previous section. The benefit obtained from the higher bit rates is an increased uniformity in the speech quality for different speakers and transmission conditions, due to the elimination of the pitch and voiced/unvoiced detector. The details of implementation of voice excited LPC vocoders are given by Atal et al. [27] and Weinstein [28].

8.11 Summary

In this chapter we have studied the technique of linear prediction of speech. We have primarily focused on the formulations which provide the most insight into the modeling of the process of speech production. We have discussed the issues involved with implementing these systems and have tried to compare the similarities and differences between the basic methods whenever possible.

REFERENCES

1. J. D. Markel and A. H. Gray, Jr., *Linear Prediction of Speech,* Springer-Verlag, New York, 1976.

2. J. Makhoul, "Linear Prediction: A Tutorial Review," *Proc. IEEE,* Vol. 63, pp. 561-580, 1975.

3. B. S. Atal and S. L. Hanauer, "Speech Analysis and Synthesis by Linear Prediction of the Speech Wave," *J. Acoust. Soc. Am.,* Vol. 50, pp. 637-655, 1971.

4. F. I. Itakura and S. Saito, "Analysis-Synthesis Telephony Based Upon the Maximum Likelihood Method," *Proc. 6 th Int. Congress on Acoustics,* pp. C17-20, Tokyo, 1968.

5. J. Makhoul, and J. Wolf, "Linear Prediction and the Spectral Analysis of Speech," *BBN Report No. 2304,* August 1972.

6. F. I. Itakura and S. Saito, "A Statistical Method for Estimation of Speech Spectral Density and Formant Frequencies," *Elec. and Comm. in Japan,* Vol. 53-A, No. 1, pp. 36-43, 1970.

7. J. Makhoul, "Spectral Linear Prediction: Properties and Applications," *IEEE Trans. on Acoustics, Speech, and Signal Proc.,* Vol. ASSP-23, No. 3, pp. 283-296, June 1975.

8. J. Makhoul, "Spectral Analysis of Speech by Linear Prediction," *IEEE Trans. on Audio and Electroacoustics,* Vol. AU-21, No. 3, pp. 140-148, June 1973.

9. J. D. Markel and A. H. Gray Jr., "On Autocorrelation Equations as Applied to Speech Analysis," *IEEE Trans. on Audio and Electroacoustics,* Vol. AU-21, pp. 69-79, April 1973.

10. V. Zue, "Speech Analysis by Linear Prediction," *MIT QPR No. 105, Research Lab of Electronics,* April 1972.

11. J. Makhoul, "Stable and Efficient Lattice Methods for Linear Prediction," *IEEE Trans. Acoustics, Speech, and Signal Proc.,* Vol. ASSP-25, No. 5, pp. 423-428, October 1977.

12. J. Burg, "A New Analysis Technique for Time Series Data," *Proc. NATO Advanced Study Institute on Signal Proc.,* Enschede Netherlands, 1968.

13. M. R. Portnoff, V. W. Zue, and A. V. Oppenheim, "Some Considerations in the Use of Linear Prediction for Speech Analysis," *MIT QPR No. 106., Research Lab of Electronics,* July 1972.

14. H. Strube, "Determination of the Instant of Glottal Closure from the Speech Wave," *J. Acoust. Soc. Am.,* Vol. 56, No. 5, pp. 1625-1629, November 1974.

15. S. Chandra and W. C. Lin, "Experimental Comparison Between Stationary and Non-stationary Formulations of Linear Prediction Applied to Speech," *IEEE Trans. Acoustics, Speech, and Signal Proc.,* Vol. ASSP-22, pp. 403-415, 1974.

16. L. R. Rabiner, B. S. Atal, and M. R. Sambur, "LPC Prediction Error-Analysis of Its Variation with the Position of the Analysis Frame," *IEEE Trans. Acoustics, Speech, and Signal Proc.,* Vol. ASSP-25, No. 5, pp. 434-442, October 1977.

17. H. Wakita, "Direct Estimation of the Vocal Tract Shape by Inverse Filtering of Acoustic Speech Waveforms," *IEEE Trans. on Audio and Electroacoustics,* Vol. AU-21, No. 5, pp. 417-427, October 1973.

18. E. M. Hofstetter, "An Introduction to the Mathematics of Linear Predictive Filtering as Applied to Speech Analysis and Synthesis," *Tech. Note 1973-36, MIT Lincoln Labs,* July 1973.

19. J. D. Markel, "The SIFT Algorithm for Fundamental Frequency Estimation," *IEEE Trans. on Audio and Electroacoustics,* Vol. AU-20, No. 5, pp. 367-377, December 1972.

20. J. N. Maksym, "Real-Time Pitch Extraction by Adaptive Prediction of the Speech Waveform," *IEEE Trans. on Audio and Electroacoustics,* Vol. AU-21, No. 3, pp. 149-153, June 1973.

21. J. D. Markel, "Application of a Digital Inverse Filter for Automatic Formant and F_o Analysis," *IEEE Trans. on Audio and Electroacoustics,* Vol. AU-21, No. 3, pp. 149-153, June 1973.

22. J. D. Markel, "Digital Inverse Filtering — A New Tool for Formant Trajectory Estimation," *IEEE Trans. on Audio and Electroacoustics,* Vol. AU-20, No. 2, pp. 129-137, June 1972.

23. S. S. McCandless, "An Algorithm for Automatic Formant Extraction Using Linear Prediction Spectra," *IEEE Trans. on Acoustics, Speech, and Signal Proc.,* Vol. ASSP-22, No. 2, pp. 135-141, April 1974.

24. J. D. Markel and A. H. Gray Jr., "A Linear Prediction Vocoder Simulation Based Upon the Autocorrelation Method," *IEEE Trans. on Acoustics, Speech, and Signal Proc.,* Vol. ASSP-22, No. 2, pp. 124-134, April 1974.

25. R. Viswanathan and J. Makhoul, "Quantization Properties of Transmission Parameters in Linear Predictive Systems," *IEEE Trans. on Acoustics, Speech, and Signal Proc.,* Vol. ASSP-23, No. 3, pp. 309-321, June 1975.

26. M. R. Sambur, "An Efficient Linear Prediction Vocoder," *Bell Syst. Tech. J.,* Vol. 54, No. 10, pp. 1693-1723, December 1975.

27. B. S. Atal, M. R. Schroeder, and V. Stover, "Voice-Excited Predictive Coding System for Low Bit-Rate Transmission of Speech," *Proc. ICC,* pp. 30-37 to 30-40, 1975.

28. C. J. Weinstein, "A Linear Predictive Vocoder with Voice Excitation," *Proc. Eascon,* September 1975.

PROBLEMS

8.1 Consider the difference equation

$$h(n) = \sum_{k=1}^{p} \alpha_k h(n-k) + G\delta(n) \ .$$

The autocorrelation function of $h(n)$ is defined as

$$\tilde{R}(m) = \sum_{n=0}^{\infty} h(n)h(n+m)$$

(a) Show that $\tilde{R}(m) = \tilde{R}(-m)$

(b) By substituting the difference equation into the expression for $\tilde{R}(-m)$, show that

$$\tilde{R}(m) = \sum_{k=1}^{p} \alpha_k \tilde{R}(|m-k|) \qquad m = 1, 2, \ldots, p$$

8.2 The system function $H(z)$ evaluated at N equally spaced points on the unit circle is

$$H(e^{j\frac{2\pi}{N}k}) = \frac{G}{1 - \sum_{n=1}^{p} \alpha_n e^{-j\frac{2\pi}{N}kn}} \qquad 0 \leqslant k \leqslant N - 1$$

Describe a procedure for using an FFT algorithm to evaluate $H(e^{j\frac{2\pi}{N}k})$.

8.3 Equation (8.30) can be used to reduce the amount of computation required to obtain the covariance matrix in the covariance method.

(a) Using the definition of $\phi_n(i,k)$ in the covariance method, show that

$$\phi_n(i+1,k+1) = \phi_n(i,k) + s_n(-i-1)s_n(-k-1) - s_n(N-1-i)s_n(N-1-k)$$

Suppose that $\phi_n(i,0)$ is computed for $i = 0, 1, 2, \ldots, p$.

(b) Show that the elements on the main diagonal can be computed recursively starting with $\phi_n(0,0)$; i.e., obtain a recurrence formula for $\phi_n(i,i)$.

(c) Show that the elements on the lower diagonals can also be computed recursively beginning with $\phi_n(i,0)$.

(d) How can the elements on the upper diagonal be obtained?

8.4 Linear prediction can be viewed as an optimal method of estimating a linear system, based on a certain set of assumptions. Figure P8.4 shows

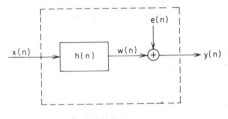

Fig. P8.4

another way in which an estimate of a linear system can be made. Assume that we can observe both $x(n)$ and $y(n)$, and that $e(n)$ is a white Gaussian noise of zero mean, and variance σ_e^2, and $e(n)$ is statistically independent of $x(n)$. An estimate of the impulse response of the linear system is desired such that the squared error

$$\epsilon = E[(y(n) - \hat{h}(n) * x(n))^2]$$

is minimized, where $\hat{h}(n)$, $0 \leqslant n \leqslant M - 1$ is the estimate of $h(n)$.

(a) Determine a set of linear equations for $\hat{h}(n)$ in terms of the auto-correlation function of $x(n)$ and the cross-correlation function between $y(n)$ and $x(n)$.

(b) How would you implement a solution to the set of Eqs. derived in part (a)? How is this related to the LPC method discussed in this chapter?

(c) Derive an expression for ϵ, the minimum mean squared error.

8.5 In deriving the lattice formulation, the ith order prediction error filter was defined as

$$A^{(i)}z = 1 - \sum_{k=1}^{i} \alpha_k^{(i)} z^{-k}$$

The predictor coefficients satisfy

$$\alpha_j^{(i)} = \alpha_j^{(i-1)} - k_i \alpha_{i-j}^{(i-1)} \quad 1 \leqslant j \leqslant i-1$$
$$\alpha_i^{(i)} = k_i$$

Substitute these expressions for $\alpha_j^{(i)}$, $1 \leqslant j \leqslant i$ into the expression for $A^{(i)}(z)$ to obtain

$$A^{(i)}(z) = A^{(i-1)}(z) - k_i z^{-i} A^{(i-1)}(z^{-1}) \ .$$

8.6 Given a section of speech, $s(n)$, which is perfectly periodic with N_P samples, then $s(n)$ can be represented as the discrete Fourier series

$$s(n) = \sum_{k=1}^{M} \left[\beta_k e^{j\frac{2\pi}{N_P}kn} + \beta_k^* e^{-j\frac{2\pi}{N_P}kn} \right]$$

where M is the number of harmonics of the fundamental $(2\pi/N_P)$ which are present. To spectrally flatten the signal (to aid in pitch detection) we desire a signal $y(n)$ of the form

$$y(n) = \sum_{k=1}^{M} \left[e^{j\frac{2\pi}{N_P}kn} + e^{-j\frac{2\pi}{N_P}kn} \right]$$

This problem is concerned with a procedure for spectrally flattening a signal using a combination of LPC and homomorphic processing techniques.

(a) Show that the spectrally flattened signal, $y(n)$, can be expressed as

$$y(n) = \frac{\sin[\frac{\pi}{N_P}(2M+1)n]}{\sin[\frac{\pi}{N_P}n]} - 1$$

Note that this sequence is sketched in Fig. 6.20 for $N_P = 15$ and $M = 2$.

Now suppose that an LPC analysis is done on $s(n)$ using a window that is several pitch periods long and a value p in the LPC analysis

457

such that $p = 2M$. From this analysis the system function

$$H(z) = \frac{1}{1 - \sum\limits_{k=1}^{p} \alpha_k z^{-k}} = \frac{1}{A(z)}$$

is obtained. The denominator can be represented as

$$A(z) = \prod_{k=1}^{p} (1 - z_k z^{-1})$$

(b) How are the $p = 2M$ zeros of $A(z)$ related to the frequencies present in $s(n)$?
The cepstrum, $\hat{h}(n)$, of the impulse response, $h(n)$, is defined as the sequence whose z-transform is

$$\hat{H}(z) = \log H(z) = - \log A(z) .$$

(Note that $\hat{h}(n)$ can be computed from the α_k's using Eq. (8.123).)
Show that $\hat{h}(n)$ is related to the zeros of $A(z)$ by

$$\hat{h}(n) = \sum_{k=1}^{p} \frac{z_k^n}{n} \quad n > 0$$

(c) Using the results of (a) and (b) argue that

$$y(n) = n\,\hat{h}(n)$$

is a spectrally flattened signal as desired for pitch detection.

8.7 The "standard" method for obtaining the short-time spectrum of a section of speech is shown in Fig. P8.7a. A much more sophisti-

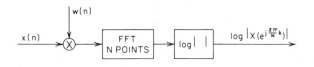

(a)

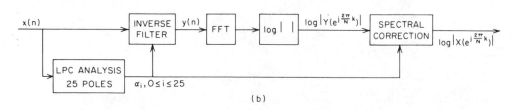

(b)

Fig. P8.7

cated, and computationally more expensive, method of obtaining $\log|X(e^{j\omega})|$ is shown in Fig. P8.7b.

(a) Discuss the new method of obtaining $\log|X(e^{j\frac{2\pi}{N}k})|$ and explain what the spectral correction network should be.

(b) What are the possible advantages of this new method? Consider the use of windows, the presence of zeros in the spectrum of $x(n)$, etc.

8.8 One proposed method for detecting pitch based on LPC processing is to use the autocorrelation function of the LPC error signal $e(n)$. Recall that $e(n)$ can be written as

$$e(n) = \hat{s}(n) - \sum_{i=1}^{p} \alpha_i \hat{s}(n-i)$$

and if we define $\alpha_0 = -1$, then

$$e(n) = -\sum_{i=0}^{p} \alpha_i \hat{s}(n-i)$$

where the windowed signal $\hat{s}(n) = s(n)w(n)$ is nonzero for $0 \leqslant n \leqslant N - 1$, and zero everywhere else.

(a) Show that the autocorrelation function of $e(n)$, $R_e(m)$ can be written in the form

$$R_e(m) = \sum_{l=-\infty}^{\infty} R_a(l) R_{\hat{s}}(m-l)$$

where $R_a(l)$ is the autocorrelation function of the LPC coefficients, and $R_{\hat{s}}(l)$ is the autocorrelation function of $\hat{s}(n)$.

(b) For a speech sampling rate of 10 kHz, how much computation (i.e. multiplies and adds) is required to evaluate $R_e(m)$ for values of m in the interval 3 to 15 msec?

8.9 We have discussed a number of vocoder systems in this book, namely

1. Channel vocoder
2. Serial formant vocoder
3. Parallel formant vocoder
4. Homomorphic vocoder
5. Phase vocoder
6. LPC vocoder

Theoretically speaking, how would you order the quality of the output of these vocoders? Explain your ordering completely. The issues which should be discussed include dependence on a model, information lost in analysis, necessity for pitch tracking, etc.

459

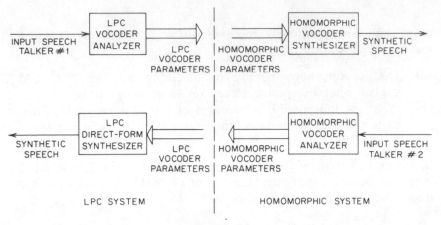

Fig. P8.10

8.10 Suppose that two talkers are trying to communicate using different vocoder systems as depicted in Fig. P8.10. Talker #1 has an LPC vocoder analyzer of the form discussed in Section 8.10.3 and a direct-form LPC vocoder synthesizer as discussed in Section 8.9. Talker #2 has a homomorphic vocoder analyzer and synthesizer as discussed in Section 7.5.

(a) In order for Talker #1 to communicate with Talker #2, the LPC representation must be converted to the homomorphic vocoder representation so that speech can be synthesized using the homomorphic vocoder synthesizer. Devise a method for this conversion.

(b) Devise a method of converting from the homomorphic representation to the LPC representation so that Talker #2 can communicate with Talker #1.

8.11 Consider two (windowed) speech sequences $x(n)$ and $\hat{x}(n)$ both defined for $0 \leqslant n \leqslant N - 1$. (Outside this region the sequences $x(n)$ and $\hat{x}(n)$ are defined to be 0). We perform an LPC analysis (using the autocorrelation method) of each frame. Thus we obtain autocorrelation sequences $R(k)$ and $\hat{R}(k)$ defined as

$$R(k) = \sum_{n=0}^{N-1-k} x(n)x(n+k) \quad 0 \leqslant k \leqslant p$$

$$\hat{R}(k) = \sum_{n=0}^{N-1-k} \hat{x}(n)\hat{x}(n+k) \quad 0 \leqslant k \leqslant p$$

From the autocorrelation sequences we solve for the predictor parameters $\boldsymbol{\alpha} = (\alpha_0, \alpha_1, \ldots, \alpha_p)$ and $\hat{\boldsymbol{\alpha}} = (\hat{\alpha}_0, \ldots, \hat{\alpha}_1, \hat{\alpha}_p)$ $(\alpha_0 = \hat{\alpha}_0 = -1)$.

(a) Show that the prediction (residual) error, defined as

$$E^{(p)} = \sum_{n=0}^{N-1+p} e^2(n) = \sum_{n=0}^{N-1+p} \left[-\sum_{i=0}^{p} \alpha_i x(n-i) \right]^2$$

460

can be written in the form

$$E^{(p)} = \alpha \mathbf{R}_\alpha \alpha'$$

where $\mathbf{R}_\alpha$ is a $(p+1)$ by $(p+1)$ matrix. Determine $\mathbf{R}_\alpha$.

(b) Consider passing the input sequence $\hat{x}(n)$ through the inverse LPC system with LPC coefficients α, to give the error signal $\tilde{e}(n)$, defined as

$$\tilde{e}(n) = -\sum_{i=0}^{p} \alpha_i \hat{x}(n-i)$$

Show that the mean squared error $\tilde{E}^{(p)}$, defined as

$$\tilde{E}^{(p)} = \sum_{n=0}^{N-1+p} [\tilde{e}(n)]^2$$

can be written in the form

$$\tilde{E}^{(p)} = \alpha \mathbf{R}_{\hat{\alpha}} \alpha'$$

where $\mathbf{R}_{\hat{\alpha}}$ is a $(p+1)$ by $(p+1)$ matrix. Determine $\mathbf{R}_{\hat{\alpha}}$.

(c) If we form the ratio

$$D = \frac{\tilde{E}^{(p)}}{E^{(p)}}$$

what can be said about the range of values of D?

8.12 A proposed measure of similarity between two frames of speech with LPC sets α and $\hat{\alpha}$, and augmented correlation matrices $\mathbf{R}_\alpha$ and $\mathbf{R}_{\hat{\alpha}}$ (see Problem 8.11) is

$$D(\alpha, \hat{\alpha}) = \frac{\alpha \mathbf{R}_{\hat{\alpha}} \alpha'}{\hat{\alpha} \mathbf{R}_{\hat{\alpha}} \hat{\alpha}'}$$

(a) Show that the distance function $D(\alpha, \hat{\alpha})$ can be written in the computationally efficient form

$$D(\alpha, \hat{\alpha}) = \left| \frac{(b(0)\hat{R}(0) + 2\sum_{i=1}^{p} b(i)\hat{R}(i))}{\hat{\alpha} \mathbf{R}_{\hat{\alpha}} \hat{\alpha}'} \right|$$

where $b(i)$ is the autocorrelation of the α array, i.e.

$$b(i) = \sum_{j=0}^{p-i} \alpha_j \alpha_{j+i} \quad 0 \leq i \leq p$$

(b) Assume the quantities (i.e. vectors, matrices, scalars) α, $\hat{\alpha}$, $\mathbf{R}$, $\hat{R}$, $(\hat{\alpha} \mathbf{R}_{\hat{\alpha}} \hat{\alpha}')$, $\mathbf{R}_{\hat{\alpha}}$, and $\mathbf{b}$ are precomputed - i.e. they are available at the time the distance calculation is required. Contrast the computation required to evaluate $D(\alpha, \hat{\alpha})$ using both expressions for D given in this problem.

461

9

Digital Speech Processing for Man-Machine Communication by Voice

9.0 Introduction

In the preceding chapters we have focused almost entirely on the essential theoretical framework which is required to understand most modern techniques for digital processing of speech signals. We have not yet discussed the broad area of applications; i.e., the various ways in which the basic models and the associated parameters derived from them are used in an integrated system whose purpose is to transmit or to automatically extract information from the speech signal. This is the purpose of this chapter, i.e., to give representative examples of digital speech processing systems and to show how digital processing techniques are used in such systems.

We wish to emphasize from the outset that we have made no attempt to survey the entire field of speech communication for applications. Specifically, we have selected examples related to man-machine communication by voice. Thus, systems for digital transmission of voice are not discussed, even though this is one of the biggest areas of application. There are several reasons for restricting our attention to man-machine communication. First of all, this area is extremely rich in the use of digital speech processing techniques, and therefore is illustrative of almost all the processing methods described in the preceding chapters. Additionally this area is an extremely important and exciting new applications area which is just developing, and which shows tremendous potential for widespread use in the future.[1] A final consideration in the choice of this area of applications is the knowledge and experience of the authors in the details of designing and implementing the various systems to be discussed in this chapter.

There are generally recognized to be three major areas (modes of communication) within the general area of man-machine communication by voice. These areas include:

1. Voice response systems
2. Speaker recognition systems
3. Speech recognition systems.

Voice response systems are designed to respond to a request for information using spoken messages. Thus voice response systems communicate by voice in one direction only, i.e. from the machine to man.

On the other hand, areas 2 and 3 in the above list deal with systems in which communication is by voice from man to machine. For speaker recognition systems the task of the system is to either verify a speakers identity (i.e., a yes-no decision as to whether the speaker is who he claims to be), or to identify the speaker from some known ensemble. Thus the speaker recognition area is itself broken down into the sub-areas of speaker verification, and speaker identification. We will discuss further the similarities and differences between speaker verification and speaker identification later in this chapter.

The last area, speech recognition, can be subdivided into a large number of sub-areas depending on such factors as the vocabulary size, speaker population, speaking conditions, etc. The basic task of a speech recognition system is either to recognize the entire spoken utterance exactly (e.g., a phonetic or orthographic speech-to-text typewriter system), or else to "understand" the spoken utterance (i.e., to respond in a correct manner to what was spoken). The concept of understanding rather than recognizing the utterance is of most importance for systems which deal with fairly large vocabulary continuous speech input, whereas the concept of exact recognition is of most importance for limited vocabulary, small speaker population, isolated word systems. We will spend some time discussing the various alternatives in speech recognition systems later in this chapter.

In the remainder of this chapter we will present, and discuss in some detail, representative systems from each of the areas of man-machine communication by voice. We will emphasize the digital speech processing parts of the system so as to reinforce the discussion in the previous chapters. However, both for completeness and for understanding, we will also discuss the general information processing aspects of the system since these are often just as important to the successful operation of the entire system.

---

[1]For an excellent introduction to this area the reader is referred to the survey paper by J. L. Flanagan [1].

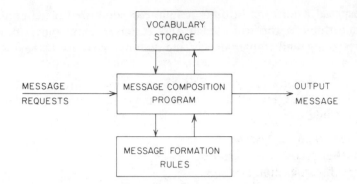

9.1 Voice Response Systems

Figure 9.1 shows a block diagram of a general computer voice response system. The elements of a voice response system include:

1. Provision for storage of a vocabulary for the voice response system.
2. Rules for forming messages from elements of the vocabulary.
3. A program for composing voice response messages.

The input to the voice response system is in the form of a message request, which may be initiated by another information processing system or directly by a human seeking information from the voice response system. The output messages are in the form of speech utterances in response to the message requests. A simple example would be an automated telephone directory assistance system in which improperly dialed telephone numbers would be detected, the type of problem determined (e.g., the telephone has been disconnected, or a new number has been assigned, etc.), and a request for an appropriate message would be sent to a voice response system. For such a system the vocabulary entries are generally entire phrases, as well as a limited number of isolated words (e.g., the digits with various spoken inflections).

As a second example, consider an information retrieval system such as a stock price quotation system where a user could key in a code via a *TOUCH-TONE*® telephone, for the price of a desired stock. The system would decode the touch-tone signals, determine the current price of the stock, and then issue a request to the voice response system to create the appropriate spoken message. For this case the vocabulary would consist of a wide variety of words and phrases.

There are two main approaches to the implementation of a voice response system. One approach is to attempt to build a machine with powers of speech comparable to a human. Such systems (often referred to as speech synthesis-by-rule systems) are based upon the model for speech production as discussed in Chapter 3. In this case the vocabulary storage is essentially a pronouncing

464

dictionary and the message formation rules must generate the required control signals, (e.g., pitch, intensity, and vocal tract response parameters) to control a speech synthesizer based on the speech production model. Such systems are of interest when an exceedingly large vocabulary is required. The implementation of such systems is an extremely challenging research problem, and there is ample opportunity for the application of the signal processing techniques that we have discussed in the final synthesis stage. However, a major concern in such systems is the discovery of rules for controlling the synthesizer. Since this would take us far afield into the domain of linguistics, we shall not attempt to discuss examples of such systems here. Many papers in this area are available to the interested reader [2-6].

The second type of computer voice response system is the limited vocabulary system in which the output message is created by concatenating isolated natural speech elements in the vocabulary storage. Figure 9.2 shows a block diagram of a digital voice response system in which the vocabulary consists of isolated words and phrases which are represented in digital form and stored in digital memory. Messages are created by retrieving the required words and phrases from storage and reproducing them in the proper sequence. There are three major considerations in the design of voice response systems of this type. First, a means of representing and storing the basic vocabulary elements must be selected and a system designed to permit easy access to each element of the vocabulary. Second, a means of editing speech recordings to select the desired vocabulary elements must be provided, along with a means of recording the vocabulary elements onto the storage medium. The third requirement is a system for selecting and reproducing vocabulary elements in prescribed sequence (i.e., the message composition system).

Since the objective of a voice response system is to produce speech utterances that are useful for communication with humans, intelligibility is of paramount importance. However, subjective factors such as quality and naturalness of the speech utterances have a great effect on the usefulness and acceptability of a voice response system. Thus, it is important that the three components of the voice response system be designed so that there is maximum potential for the production of highly intelligible, natural sounding speech.

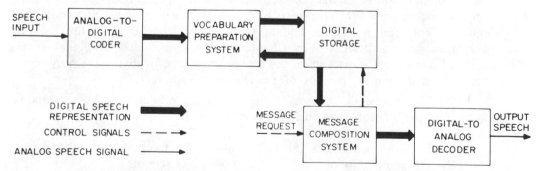

Fig. 9.2 Block diagram of an all digital voice response system. (After Rabiner and Schafer [8].)

465

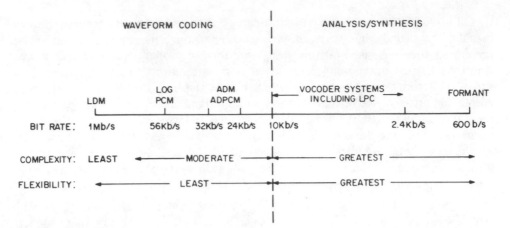

Fig. 9.3 Summary of types of speech encoding methods.

9.1.1 General considerations in the design of voice response systems

Developments in both digital techniques for the representation and processing of speech signals and in the area of digital hardware have made it possible to implement a voice response system using digital techniques throughout. As depicted in Figure 9.2, a digital system first requires an analog-to-digital coder; i.e., a system for obtaining a digital representation of a speech signal. Likewise an appropriate digital-to-analog decoder is required to convert from the digital representation to an analog signal. Once the vocabulary is represented in digital form, it can be stored in digital memory. The message composition system is then required to access the vocabulary entries in the correct sequence and concatenate them into a digital representation of the desired message. This digital representation is in turn fed to the digital-to-analog decoder.

Within this general structure, there is tremendous flexibility in system design. The key factor in the design of a digital voice response system is the choice of the form of digital representation for the speech utterances that make up the vocabulary. We have seen so far that there are abundant possibilities, ranging from the waveform coding methods of Chapter 5 to the analysis-synthesis systems discussed in Chapters 6-8. The choice of digital coding method has a great impact upon the amount and type of digital memory that is required, and upon the method of message composition.

In considering the choice of digital coding method for voice response applications it is helpful to consider three factors:

1. The information rate (bit rate) required for acceptable speech quality.
2. The complexity of the coding and decoding schemes.
3. The flexibility of the representation; i.e., the potential for modification of the vocabulary elements.

Figure 9.3 shows a comparison of a number of the digital representations that were discussed in Chapters 5-8 according to the above three factors. The waveform coding representations clearly require the greatest bit rate and thus would require the most digital storage for a voice response vocabulary. The waveform representations are the simplest in terms of the coding/decoding algorithms. On the other hand, the analysis/synthesis systems, which literally "take the speech signal apart," have the greatest potential for modifying the vocabulary elements in useful ways. The first two considerations, i.e., bit rate and complexity of implementation of the digital representation, impact mainly on the economics of the design of an all-digital voice response. To illustrate these concerns let us consider a typical vocabulary, suitable for many applications, which might involve about 100 words whose average duration would be less than 1 second. Thus, as a relatively conservative estimate of storage requirements, Table 9.1 shows the digital storage required for 100 seconds of speech material for a wide range of bit rates. Even in the case of log PCM coding, the memory requirement is not unreasonable. The main question is the cost of digital storage relative to the cost of coding and decoding hardware for the more complicated coding systems. In the last column of Table 9.1, the value of 0.1¢ per bit[2] is applied to the memory requirements for a 100 second vocabulary to obtain an estimate of memory cost. Realizing that the memory can be shared among a multiplicity of output channels, it can be seen that the cost of the digital coder and decoders must be kept low in order that they not become the dominant cost consideration. It is clear for example that formant synthesizers of the type required to produce high quality speech output would cost significantly more than the cost of the memory for a small vocabulary.

Table 9.1 Memory Requirements For Storage of
Digital Speech Representations

Coding Method	Bit Rate	Digital Storage Required for 100 sec (bits)	Approximate Cost at 0.1¢/bit
PCM	40 kb/s	4,000,000	$4000.00
ADPCM	24 kb/s	2,400,000	2400.00
LPC	2.4 kb/s	240,000	240.00
Formants	0.6 kb/s	60,000	60.00

Vocabulary preparation and editing is another basic concern in the design of automatic voice response systems. In this area, digital techniques have the potential of high efficiency and great flexibility in preparing vocabulary elements and providing high quality voice output. Typically, the words and phrases making up a voice response vocabulary are spoken by a trained speaker and a high quality audio recording is made. A word or phrase is recorded and then converted to a digital representation by the analog-to-digital coder. The digital representation (which may be either of the waveform or analysis-synthesis type) is then stored away temporarily in digital form in the computer. An

---

[2]This is a conservatively high estimate of the true memory costs.

automatic scheme is required to find the beginning and end of the utterance so that the surrounding silence regions can be eliminated. As shown in Chapter 4, the beginning and end of an utterance can be located quite accurately for high quality digital recordings. At this point the computer can determine precisely whether the utterance is of proper duration. Also, the utterance can be played back to the speaker to check the inflection of the word or phrase. The recording process can be easily repeated until satisfactory duration and inflection is obtained.

As the final stage in preparing a voice response vocabulary, an automatic scheme can compare the intensities of all words in the vocabulary and suitably adjust all levels to some uniform level or according to prescribed levels determined by the intended usage of each vocabulary entry. This might involve simply a calculation of peak signal amplitude or a more sophisticated measure of intensity such as short-time energy could be employed.

Once a word or phrase has been suitably recorded, it is stored in its permanent area of the vocabulary memory. This involves simply setting up a system of speech files and a directory of memory addresses that are used by the message composition system to locate the beginning and end of each vocabulary entry.

Given means for preparation and storage of a vocabulary of words and phrases, the voice response system is completed by providing a means for composing speech utterances from the vocabulary elements. Here again, the form of digital representation has a major impact. If a waveform representation is used, all that needs to be done is to simply concatenate the waveforms of the vocabulary elements. This may lead to somewhat unnatural sounding speech utterances if the vocabulary consists mostly of isolated words, but this approach has the virtue that the message composition system can be extremely simple. Indeed such a system can easily be implemented using a microprocessor. An example of such a system is discussed in Section 9.1.2.

On the other hand, representations based on some form of analysis-synthesis offer increased flexibility for altering the properties of the vocabulary elements so that the composite utterance retains some of the properties of natural utterances — e.g., timing, inflection, etc. This advantage is potentially more important than the lower bit-rates that are also possible with analysis-synthesis representations. Because the vocabulary entries are represented in terms of fundamental parameters of speech production, it is possible, for example, to alter the pitch and duration of a word so as to make it fit a particular message context. Even more interesting is the possibility of altering the speech parameters at word boundaries so as to produce synthetic speech utterances that sound more like connected natural speech. To do this in even the simplest situations requires rules for determining appropriate pitch and timing and algorithms for altering the speech parameters to accomplish word duration changes and merging of word boundaries. Because of the low information rate of the parametric representation, a microprocessor would be adequate to implement a fairly sophisticated message composition system based on an analysis-synthesis

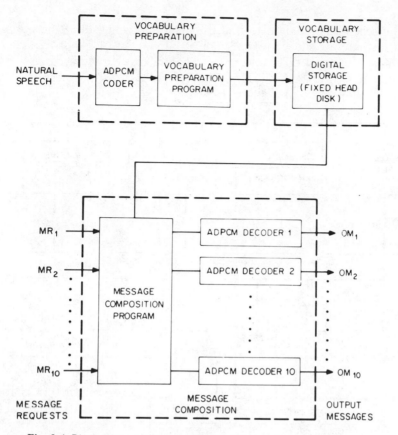

Fig. 9.4 Block diagram of a multiline digital voice response system. (After Rosenthal et al. [7].)

representation. An example of a simple system of this type is discussed in Section 9.1.3.

9.1.2 A multiple-output digital voice response system

Figure 9.4 shows a block diagram of a multiline digital voice response system that has been implemented at Bell Laboratories using a small general purpose computer [7,8]. In this system, the vocabulary elements were represented using ADPCM coding at 24 kb/s. The ADPCM coder and decoders were implemented in hardware. The beginning and end points of each word were automatically located by an algorithm which was based upon the computation of the short-time "energy" of the ADPCM code words [7]. The vocabulary was stored on a fixed head disk for rapid retrieval by the message composition program. This part of the system involves mostly logical operations and data transfers. An important point, however, is that with most computers and memory systems, there exists the possibility that a single vocabulary memory can serve the needs of many message channels. The computer receives message requests through telephone lines or direct digital interfaces to other computers. The message composition program locates the required vocabulary ele-

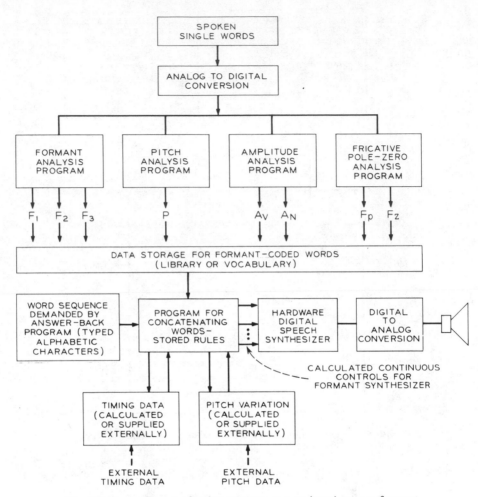

Fig. 9.5 Block diagram of voice response system based upon a formant representation. (After Rabiner et al. [9].)

ments and retrieves them from memory. The digital representations of the desired messages are stored in buffers in the random access memory of the computer. These buffers are accessed by the ADPCM decoders through direct memory access (DMA) channels. In this way the system is able to provide voice output simultaneously on many channels. The system implemented at Bell Laboratories was capable of simultaneous voice response on 10 channels [7,8]. It has been used in a variety of applications as discussed in Section 9.1.4.

9.1.3 Speech synthesis by concatenation of formant-coded words [9]

As an example of the use of an analysis-synthesis representation, consider the block diagram of Figure 9.5. In this case, the vocabulary elements were processed as described in Section 7.4 to obtain a digital representation in terms of pitch period, voiced/unvoiced classification, intensity, and formant frequencies. It was thus possible to store the vocabulary words and phrases using only 600 bits/sec.

In addition to the vocabulary elements, the message composition system requires means for obtaining appropriate word durations and pitch for the desired utterance. Given this information, the formant frequencies of adjacent words are smoothly merged together as occurs in natural connected speech. Figure 9.6 shows an example of how the formant representation provides the capability to modify the vocabulary elements to fit a particular situation. Note in particular that by interpolation, the durations can be changed. Also, note that at the junction between the second and third words, the formant frequencies are readjusted so as to be continuous across the word boundary as they would be in natural speech. Finally, note that the original pitch contours of the vocabulary elements can be discarded in favor of a single pitch contour appropriate for the complete utterance.

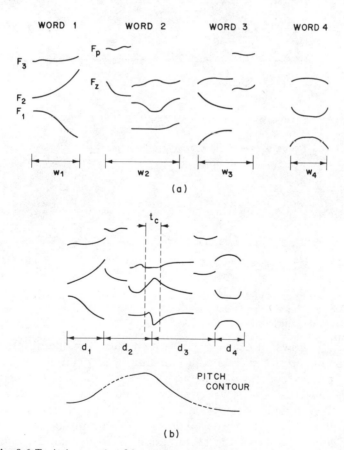

Fig. 9.6 Typical example of how control parameters are generated from the word vocabulary store. A message composed of four words is illustrated. All parameters are functions of time. (After Rabiner et al. [9].)

471

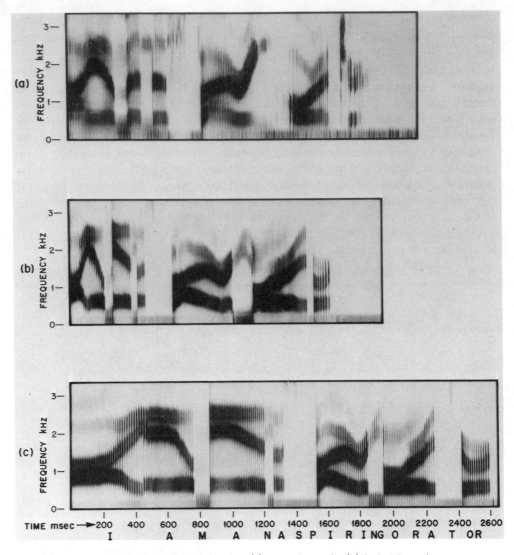

Fig. 9.7 Spectrograms comparing (a) natural speech; (b) isolated words, modified for prosodic timing and pitch; and (c) isolated words, abutted. (After Levinstone [10].)

Figure 9.7 shows an example [10] which illustrates the difference between concatenation of waveforms and concatenation of formant-coded words. Figure 9.7a shows the spectrogram of a natural utterance of "I am an aspiring orator." Figure 9.7c shows the spectrogram of the same utterance created by abutting the individually spoken, formant-coded words with no alteration of the pitch or duration of the individual words. The lack of continuity of the formants is obvious. Figure 9.7b shows the spectrogram of an utterance created from the formant-coded words by merging the formant parameters smoothly at the word boundaries. The pitch contour and word durations used in Figure 9.7b were obtained from data supplied by a speech synthesis-by-rule system developed by

472

Coker and Umeda [3,5,6]. Note that the word durations in Figure 9.7b are quite comparable to those of Figure 9.7a, and the variations of the formants in Figure 9.7b are very similar to the formant variations in Figure 9.7a.

The system depicted in Figure 9.5 was used to synthesize telephone messages of the form, "The number is 135-3201" [9]. In this case pitch and timing rules for the 7-digit number strings were empirically derived from measurements on natural speech. It was found that synthetic speech produced by the system of Figure 9.5 was superior in communicative effectiveness to speech produced simply by abutting the individual words. This was true despite the "machine-like" accent of the synthetic speech. Although such results are encouraging, much more research is required to define synthesis strategies that will produce natural sounding high quality synthetic speech from a vocabulary of digitally coded words and phrases [11].

Finally, it should be noted that there are many possibilities for digital representations of the vocabulary elements that would offer similar flexibility for manipulating the parameters of the speech utterance.

9.1.4 Typical applications of computer voice response systems

The flexibility of the voice response system described in Section 9.1.2 has facilitated its use in a wide variety of experimental applications within Bell Laboratories. The following systems have been implemented and studied to date:

1. A system for producing vocal instructions for wiring communications equipment
2. A directory assistance system
3. A stock price quotation system
4. A data set testing information system
5. A flight information system
6. A speaker verification system

In this section two of these systems are described.

9.1.4a Application of Voice Response to Wiring Communication Equipment

One of the first applications of the voice response system of Section 9.1.3 was as an aid to wiring communications equipment [12]. Conventionally, a wireman works from a printed list that contains the information for each wiring operation. However, in many wiring situations it is awkward for the wireman to divert his eyes from his work in order to consult such a list. In these cases, it is more convenient to record the wire list in spoken form on a cassette tape and allow the wireman to work from a spoken list. Typically, a foot switch is used to start the playback unit, and recorded tones on the tape automatically stop the unit after each wiring instruction.

473

Wiring lists can, of course, be recorded by a human. However, one person must read the list (which may be several hours long), and another must monitor the result for errors. These errors must then be edited and corrected. Even after a wiring list has been recorded successfully, any future updating of the list requires that this entire process be repeated. This could occur several times in the course of a few days or weeks. The resulting tedium tends to increase human errors, making the entire procedure impractical.

Therefore, there are several advantages of using a voice response system for recording wiring lists. These are the following:

1. Wiring instructions are generally simple commands conveying highly noncontextual information such as wire color, wire length, beginning terminal (to which the wire is connected), and ending terminal. Thus, smooth, connected speech is not required.
2. The instruction can be formed from a relatively small vocabulary — on the order of 50 words are adequate for specific pieces of equipment; on the order of 100 words are adequate for all the communications equipment wired by Western Electric using conventional wiring techniques.
3. Wire lists are normally designed by computer — hence they are generally available in a form amenable for use by a digital voice response system.
4. Wiring instructions are frequently modified. Thus, the use of a voice response system simplifies the tedious task of updating wiring lists.

Figure 9.8 shows a block diagram of the overall wire list voice response system. A deck of computer cards is created on a Western Electric Company computer for use as input to the wire-list system. On these cards are punched entries that describe the words to be used in the specific wire list. For example, a typical wiring instruction might be:

RED-PAUSE-20-7-PAUSE-4-R-PAUSE-7-Z-END

which tells the wireman to wire a 27 inch red wire from terminal 4R to terminal 7Z. Using an appropriate vocabulary, the voice response system composes the appropriate message, sends it to the ADPCM decoder, and the spoken message (the wiring instructions) is recorded on a cassette tape recorder.

For this application the telephone line input-output capability of the system is not exploited; however, the multiline capability is exploited by either generating a number of different wiring lists simultaneously, or a long single list is generated in pieces at a faster rate [7,8].

9.1.4b Information Retrieval Systems

In the application of the voice response system to wiring of communication equipment, there is essentially no interaction between the eventual user and the system which composes the output messages. This is because the con-

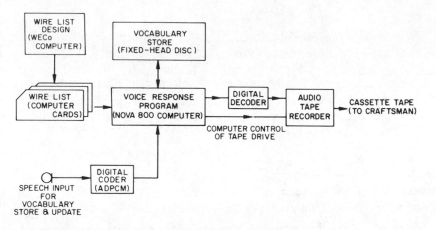

Fig. 9.8 Voice response system for automatic generation of spoken wiring instructions. (After Rabiner and Schafer [8].)

ventional *TOUCH-TONE*® input to the voice response system was replaced by a prepunched set of computer cards which specified the output messages which were required. In the application of the voice response system to problems like directory assistance, credit inquiries, bank balance inquiries, and inventory control, the basic premise is that the voice response system is capable of accessing a data bank of information. Therefore, it can find the desired information and compose the appropriate message to send this information to the user. Figure 9.9 shows a block diagram of such an information retrieval voice response system. In this system it is assumed that the data bank can be updated from both an external source, and from the voice response system.

By way of example, suppose that the data bank was an inventory of the quantity of goods produced by a company and available sale and distribution. If the voice response system was accessed by salesmen in the field, then each time a sale was made the voice response system could acknowledge the sale, and simultaneously reduce the inventory data bank. As more goods are manufactured by the company, the inventory could be externally updated as these goods become available for sale. In this example, the voice response system not only helps keep track of inventory, it also prevents the possibility of several salesmen essentially selling the same item when the inventory of goods is low. It

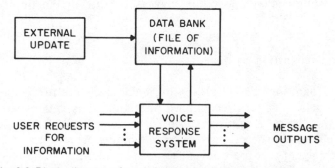

Fig. 9.9 Block diagram of an information retrieval voice response system. (After Rabiner and Schafer [8].)

475

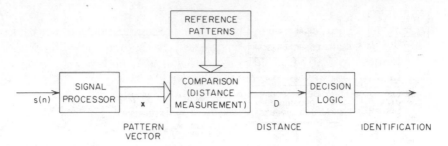

Fig. 9.10 General representation of the speaker recognition problem.

also helps the company keep up-to-the-minute statistics on sales, and therefore be capable of dynamic variations in the manufacturing of goods.

Another interesting computer voice response information retrieval system is a stock price quotation system. For this system the data bank is the current market price of any stock, as well as the market price at the close of the preceding business day. The mechanism for externally updating the stock prices could be a ticker tape, or a high speed paper tape fed directly from the stock exchange board.

A typical scenario for the use of the stock price quotation system is as follows. The user dials the system, which then responds:

"This is the Bell Laboratories stock price quotation system. Prices are quoted as of the close of the last business day. Please enter market abbreviation of the stock desired".

The user keys in

A-T-T-*

and the system responds:

"American Telephone and Telegraph, 62-and 3/8, up 1/4"

Providing such information as stock market prices, without the need for cumbersome teletypes or ticker tapes is one good application of this type of voice response system. Indeed it is clear that the future holds much promise for the widespread use of computer voice response systems, and digital speech processing techniques will no doubt play a key role in the implementation of such systems.

9.2 Speaker Recognition Systems

In speaker recognition, digital processing techniques are often the first step in what is essentially a pattern recognition problem. This is depicted in Fig. 9.10. As seen in this figure, a representation (pattern vector) of the speech signal is obtained using digital speech processing techniques which preserve the features of the speech signal that are relevant to speaker identity. The resulting pattern is compared to previously prepared reference patterns and subsequent decision logic is used to make a choice among available alternatives. There are two dis-

476

tinct subareas of speaker recognition—speaker verification and speaker identification. For speaker verification an identity is claimed by the user, and the decision required of the verification system is strictly binary; i.e., to accept or reject the claimed identity. In order to make this decision, a set of features designed to retain essential information about speaker identity is measured from one or more utterances of the speaker, and the resulting measurements are compared (often using some highly nonlinear measure of comparison) to a set of stored reference patterns for the claimed speaker. Thus for speaker verification only a single comparison between the set (or sets) of measurements, and the reference pattern is required to make the final decision to accept or reject the claimed identity. Generally a distance measure between the given measurements and the stored reference distribution is computed. Based on the relative costs of making the two possible types of error (i.e., verifying an impostor, or rejecting the correct speaker) an appropriate threshold is set on the distance function. It is readily shown that the probability of making the two types of errors described above is essentially independent of N, the number of reference patterns stored in the system, since the reference patterns for all other speakers go into forming the stable distribution which characterizes all speakers. Stated in more mathematical terms, if we denote the probability distribution for the measurement vector $\mathbf{x}$ for the ith speaker as $p_i(\mathbf{x})$, then a simple decision rule for speaker verification might be of the form:

$$Verify\ speaker\ i\ if\ \ p_i(\mathbf{x}) > c_i p_{av}(\mathbf{x})$$

$$Reject\ speaker\ i\ if\ \ p_i(\mathbf{x}) < c_i p_{av}(\mathbf{x}) \tag{9.1}$$

where c_i is a constant for the ith speaker which determines the probabilities of error for the ith speaker, and $p_{av}(\mathbf{x})$ is the average (over all speakers in the ensemble) probability distribution for measurement $\mathbf{x}$. By varying the constant c_i, a simple control of the error mix between the two types of errors is readily obtained.

The problem of speaker identification differs significantly from the speaker verification problem. In this case the system is required to make an absolute identification among the N speakers in the user population. Thus instead of a single comparison between a set of measurements and a stored reference pattern, N complete comparisons are required. The decision rule for such systems is essentially of the form:

$$choose\ speaker\ i\ such\ that\ p_i(\mathbf{x}) > p_j(\mathbf{x}),\ \ \ j = 1, 2, \ldots ,N,\ \ \ j \neq i \tag{9.2}$$

i.e., choose the speaker with the minimum absolute probability of error. In this case it seems plausible that as the user population gets very large, the probability of error must tend to one since an infinite number of distributions cannot remain distinct in a finite parameter space — i.e., it becomes increasingly likely that two or more speakers in the ensemble will have measurement distributions that are extremely close to each other. Under these circumstances, reliable speaker identification essentially becomes impossible.

It can be seen from the above discussion that there exists a great number of similarities, as well as differences, between speaker verification and speaker

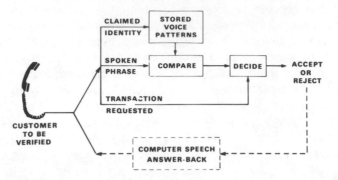

Fig. 9.11 Block diagram of a speaker verification system. (After Rosenberg [13].)

identification systems. Each process requires a speaker to utter one or more test phrases, makes some measurements on these test phrases, and then computes one (or more) distance functions between the measurement vector and the stored reference vector. Thus, in terms of the signal processing aspects of these two problems, the methods are quite similar. The major differences occur in the decision logic. We shall now discuss an example of each type of system and show some typical results obtained with these systems.

9.2.1 Speaker verification system

Figure 9.11 shows a block diagram of an on-line digital speaker verification system developed by Rosenberg and others [13-16]. The person wishing to be verified first enters his claimed identity, and then on request from the verification system (via a computer voice response system of the type discussed in Section 9.1.2) utters his verification phrase, and requests some transaction to be made in the event he is verified. For example, the transaction requested may be access to privileged information from a bank, etc. The spoken utterance is processed to obtain a pattern which is compared to the stored reference patterns for the claimed identity and then on the basis of the transaction requested which determines the error mix constant (c_i in Eq. (9.1)) a decision to accept or reject the customer is made.

Figure 9.12 shows a block diagram of the signal processing parts of the speaker verification system. The sample utterance which occurs somewhere within a preselected time interval must first be accurately pinpointed. This is the job of the endpoint detection system. In this particular implementation the time domain endpoint detector described in Chapter 4 was used to locate the sample utterance. Once the beginning and end of the utterance have been found, a series of measurements and parameter estimates are made to provide patterns which represent the utterance. In particular, a pitch detector is used to measure the pitch contour of the utterance; a short-time energy measurement is made to give energy contours, an LPC analysis is used to give predictor parameter contours, and finally an estimate of the formant locations is made.

(All the parameters shown in Fig. 9.12 have been used at some time in the verification system; however, due to the large amount of computation required for LPC analysis or formant estimation the on-line implementation restricted the measurement set to pitch and intensity measurements.) To illustrate the wide range of digital processing that was used in just the analysis phase of the verification system it is worthwhile mentioning the specific algorithms used by Rosenberg [13] (and others who have worked on this system) to perform these measurements. The pitch detector was the parallel processing time-domain method discussed in Chapter 4. The intensity measurement was the short-time average magnitude as discussed in Chapter 4. The LPC analysis was the auto-correlation method discussed in Chapter 8. Finally, the formant analysis used the homomorphic method described in Chapter 7.

Figures 9.13-9.15 show typical measurement contours for the test utterance "We were away a year ago" spoken by a male speaker. Figure 9.13 shows the pitch period and intensity contours for the utterance [13]. These data are estimated 100 times per second and were smoothed by a 16 Hz lowpass, linear phase, FIR digital filter. There are significantly more variations in the intensity plot than in the pitch period contour for this particular speaker. Figure 9.14 shows plots of the first three formants along with pitch and intensity for a different version of the utterance [15]. The formant data were also smoothed by the same 16 Hz lowpass filter as was used for the pitch period and intensity contours. Finally, Fig. 9.15 shows plots of the first 8 predictor coefficient contours for a 12-pole LPC analysis [16]. It can be seen from this figure that there is a significant amount of redundancy in the LPC coefficient contours for this utterance. Thus in using these data for verification, all the LPC coefficient contours do not give equal contributions to reducing the errors in verification. In

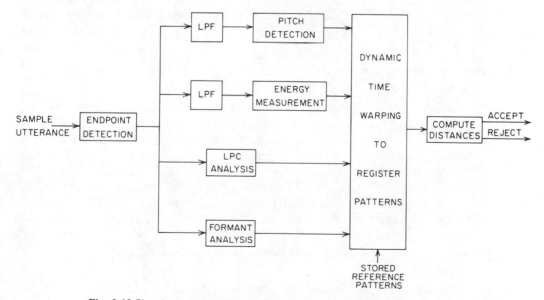

Fig. 9.12 Signal processing aspects of the speaker verification system.

479

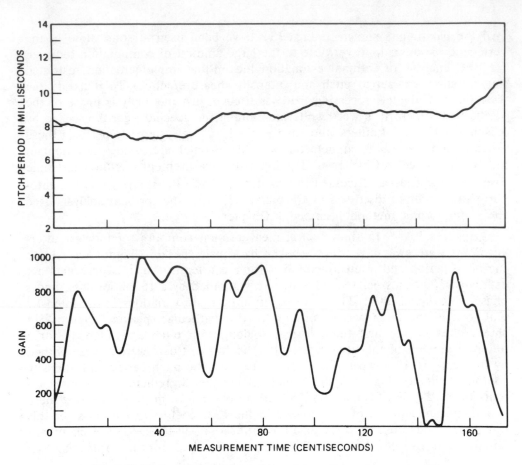

Fig. 9.13 Pitch period and intensity contours of an utterance used in speaker verification. (After Rosenberg [13].)

fact it has been shown that careful selection of the verification measurements yields almost as low an error score as using all the measurements with no regard to their verification efficiency.

After the desired parametric representation has been computed it is necessary to compare it to the corresponding reference patterns for the speaker whose identity is claimed. It is not straightforward to compare temporal patterns such as pitch, intensity and formant variations since a speaker is generally not able to speak at precisely the same rate for different repetitions of the verification phrase. The solution to this difficulty has been to nonlinearly warp the time scale of the input patterns to obtain the best possible registration between the stored reference patterns for the claimed speaker, and the measured patterns for his sample utterance. The process of time warping is an extremely important one and has been used in a variety of speech processing applications.

Conceptually the process of time warping is as illustrated in Fig. 9.16. The idea is to warp the time scale *t* of a reference utterance so that significant

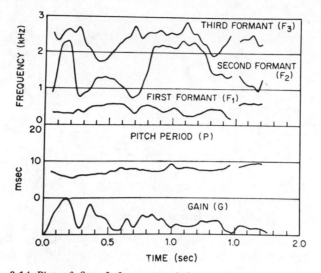

Fig. 9.14 Plot of first 3 formants, pitch and intensity for a speaker verification utterance. (After Lummis [15].)

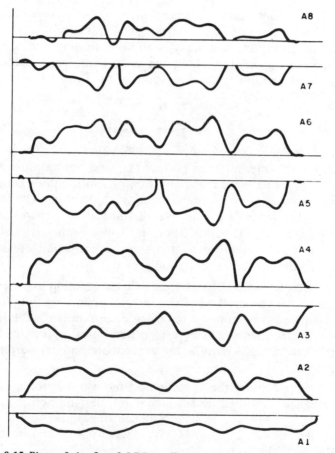

Fig. 9.15 Plots of the first 8 LPC coefficients for a speaker verification utterance. (After Rosenberg and Sambur [16].)

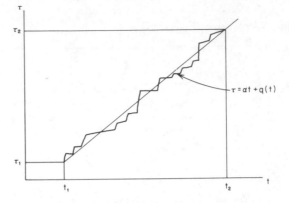

Fig. 9.16 Illustration of time warping.

events in some measurement contour $a(t)$ line up with the same significant events in the reference contour $r(t)$. The warping function is assumed to be of the form

$$\tau = \alpha t + q(t) \tag{9.3}$$

where $q(t)$ is the nonlinear time warp function, and α represents the average slope of the time warping function. Note that omission of $q(t)$ corresponds to simple linear modification. Boundary conditions are imposed to insure that the beginning and ending points of both the sample and reference utterances line up properly. These boundary conditions are of the form

$$\tau_1 = \alpha t_1 + q(t_1) \tag{9.4a}$$

$$\tau_2 = \alpha t_2 + q(t_2) \tag{9.4b}$$

The remaining problem is the choice of the function $q(t)$ and the constant α so as to best align the measurement contours. One very simple approach is to define $q(t)$ to be a piecewise linear function of t with a finite number of breakpoints along the t axis at which the slope of $q(t)$ changes. The breakpoints and the slopes of $q(t)$ (as well as the overall slope α) are then determined by a steepest descent type of optimization in which either a distance measure between the test and reference contours, or a correlation measure is used to control the direction of the optimization search.

A significantly simpler, and much faster solution to the time warping problem is to utilize the method of dynamic programming to optimally choose a constrained warping function. By imposing a continuity condition on the warping function, rather than imposing a piecewise linear approximation, it is relatively straightforward to determine the appropriate optimal warping function for a set of contours [17].

Consider performing the appropriate time warping for a pair of contours which are sampled at a discrete set of points. Let the points in the measured contour be labelled $n = 1, 2, \ldots, N$ and let the points in the reference contour be labelled $m = 1, 2, \ldots, M$. We then wish to choose a time warping function w, such that

$$m = w(n) \tag{9.5}$$

482

The boundary conditions on $w(n)$ are:

$$w(1) = 1 \qquad \textit{beginning points} \tag{9.6}$$

$$w(N) = M \qquad \textit{ending points}$$

If a linear warping were used, then the resulting time warping function w would be of the form

$$w(n) = \left[\left[\left(\frac{M-1}{N-1}\right)(n-1) + 1\right]\right] \tag{9.7}$$

For a more general nonlinear warping function we must consider a strategy for beginning at the point $n = 1$, $m = 1$ and progressing through the discrete grid to end up at the point $n = N$, $m = M$ as required by the boundary conditions. To limit the degree of nonlinearity of the warping function it is reasonable to impose the mild continuity condition that the warping function w cannot change by more than 2 grid points at any index n. More formally we require

$$w(n+1) - w(n) = 0, 1, 2 \qquad \text{if} \quad w(n) \neq w(n-1) \tag{9.8}$$

$$= 1, 2 \qquad \text{if} \quad w(n) = w(n-1)$$

Thus the slope of the warping function is either 0, 1, or 2 if at the previous grid index the warped index changed, or 1 or 2 if at the previous grid index the warped index stayed constant.

To determine which of the conditions of Eq. (9.8) to use at grid index n requires the use of a similarity measure between the reference data measured at

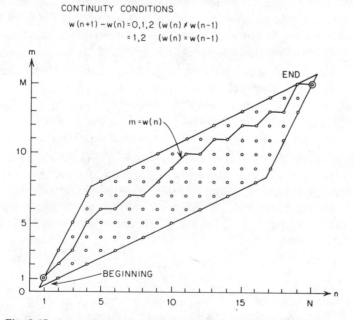

Fig. 9.17 An example of a typical warping function. (After Itakura [17].)

483

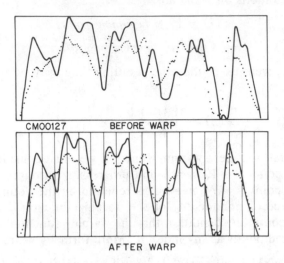

CMO0127   BEFORE WARP

AFTER WARP

Fig. 9.18 An example of the effects of time warping on a speech intensity contour. (After Rosenberg [13].)

grid index n, and the test data measured at grid index m. The similarity measure (or distance measure) is used to determine the path of the warping function which locally minimizes the maximum total distance, subject to the continuity constraints of Eq. (9.8).

The example of Figure 9.17 [17] shows the domain of possible grid coordinates (n,m) and a typical warping function $w(n)$ (the solid line within the grid) for warping a 20 point ($N=20$) reference to a 15 point ($M=15$) test utterance. Because of the continuity constraints the warping function must lie within the heavy parallelogram shown in this figure.

The technique of time warping has been applied both for speaker verification [13] and for speech recognition [17]. As an example, Fig. 9.18 (due to Rosenberg [13]) shows a set of intensity contours of a reference and a test utterance both prior to and following nonlinear time warping. The improvement in registration of these contours is quite striking for this example.

The final steps in the verification process of Fig. 9.12 are to compute some overall distance measure (based on the individual distance measures for the separate measurements), and then to compare the overall distance to an appropriately chosen threshold. The simplest contour distance measure is a normalized sum of squares: e.g., for the jth measurement contour, the distance d_j would be of the form

$$d_j = \sum_i \left[[a_{js}(i) - a_{jr}(i)]/\sigma_{aj}(i) \right]^2 \qquad (9.9)$$

where $a_{js}(i)$ is the value of the jth measurement contour at time i, $a_{jr}(i)$ is the value of the jth measurement of the reference contour at time i, and $\sigma_{aj}(i)$ is the standard deviation of the jth measurement at time i. The overall distance function, D, is generally a weighted sum of squares, i.e.,

$$D = \sum_j w_j d_j \qquad (9.10)$$

where w_j is the jth weight, chosen on the basis of the effectiveness of the jth measurement in verifying the speaker.

The speaker verification system just discussed has been extensively studied and the results of numerous performance evaluations give an idea of the potential for practical application of such systems. A wide variety of tests have been run on the system ranging from high quality utterances by a small set of speakers to telephone bandwidth utterances using a very large set of speakers, to experiments in which trained professional mimics were hired to "beat the system." Tests results have indicated that for high quality speech, the equal error rate score (i.e., when probability of false rejection and false acceptance are made equal) of the system can be made essentially 0 if enough measurements are used, and if the weighting coefficients on the overall distance computation are carefully selected for each speaker. For such a system, the equal error rate using professional mimics was about 4.1%. For the operation of the verification system using telephone quality speech an equal error rate of about 7% was attained using just the pitch period and intensity measurements. The addition of more sophisticated measurements, such as formants, or the LPC parameters, would significantly reduce this error rate.

9.2.2 Speaker identification systems [18-20]

There are many similarities between the problems of speaker verification and speaker identification. In terms of the signal processing aspects, the two problems are treated almost identically in that most of the processing shown in Fig. 9.12 is applied to speaker identification as well as speaker verification. The main differences in processing involve the choice of parameters used to make the distance measurements, and the necessity of making N distance measurements (for identification) rather than 1. Of course the final decision for speaker identification is to choose the speaker whose reference patterns are closest in distance to the sample patterns, whereas for speaker verification the final decision is the binary choice of whether to accept or reject the identity claim based on the magnitude of the distance as compared to some suitably chosen distance threshold.

Although a simple classical distance measurement of the type given in Eq. (9.10) is generally appropriate for most speaker verification systems, a somewhat more sophisticated and robust distance measure has been used by Atal [18,20] and others for speaker identification systems. Recall that the purpose of the distance calculation is to provide a measure of similarity between the reference pattern and the input pattern. The distance measure used by Atal can be derived as follows. Let $\mathbf{x}$ be an L-dimensional column vector representing the input pattern, in which the kth component of $\mathbf{x}$ is the kth measurement. It is assumed that the joint probability density function of the measurements for the ith speaker is a multi-dimensional Gaussian distribution with mean $\mathbf{m}_i$ and covariance matrix $\mathbf{W}_i$. Thus, the L-dimensional Gaussian density function for $\mathbf{x}$ is given by

$$g_i(\mathbf{x}) = (2\pi)^{-1/2}|\mathbf{W}_i|^{-1/2}\exp[-\frac{1}{2}(\mathbf{x}-\mathbf{m}_i)'\mathbf{W}_i^{-1}(\mathbf{x}-\mathbf{m}_i)] \qquad (9.11)$$

where $\mathbf{W}_i^{-1}$ is the inverse of the matrix $\mathbf{W}_i$ (assuming $\mathbf{W}_i$ is nonsingular), $|\mathbf{W}_i|$ is the determinant of $\mathbf{W}_i$, and the t denotes the transpose of a vector. The decision rule which *minimizes the probability of error* states that the measurement vector $\mathbf{x}$ should be assigned to class i if

$$p_i g_i(\mathbf{x}) \geqslant p_j g_j(\mathbf{x}) \quad \text{for } all \quad i \neq j \qquad (9.12)$$

where p_i is the a priori probability that $\mathbf{x}$ belongs to the ith class. Since $\ln y$ is a monotonically increasing function of its argument y, the decision rule of Eq. (9.12) can be considerably simplified and rewritten as:

Decide class i if

$$d_i(\mathbf{x}) = \frac{1}{2}(\mathbf{x}-\mathbf{m}_i)'\mathbf{W}_i^{-1}(\mathbf{x}-\mathbf{m}_i) + \frac{1}{2}\ln|\mathbf{W}_i| - \ln p_i$$

$$\leqslant d_j(\mathbf{x}) \quad \text{for } all \quad i \neq j \qquad (9.13)$$

The last two terms on the right side of Eq. (9.13) do not depend on the measurement vector $\mathbf{x}$ and can be thought of as a constant term representing a bias toward the ith class. As a practical matter it has been found that for most cases of interest, the decision rule which includes the bias term does not provide any significant advantage over a decision rule based exclusively on the first term on the right side of Eq. (9.13). Thus, instead of the decision rule of Eq. (9.13), the distance measure $\hat{d}_i$, defined as

$$\hat{d}_i = (\mathbf{x}-\mathbf{m}_i)'\mathbf{W}_i^{-1}(\mathbf{x}-\mathbf{m}_i) \qquad (9.14)$$

is generally used and the index i is chosen such that $\hat{d}_i$ is minimized over all i.

The decision rule requires the computation of the mean vector $\mathbf{m}_i$ and the covariance matrix $\mathbf{W}_i$ for each class i in the decision set. For a training set of N_i measurement vectors with components $\mathbf{x}_i(n)$ belonging to class i, the mean and covariance matrix are defined by the relations

$$\mathbf{m}_i = \frac{1}{N_i}\sum_{n=1}^{N_i}\mathbf{x}_i(n) \qquad (9.15)$$

and

$$\mathbf{W}_i = \frac{1}{N_i}\sum_{n=1}^{N_i}\mathbf{x}_i(n)\mathbf{x}_i'(n) - \mathbf{m}_i\mathbf{m}_i' \qquad (9.16)$$

Figure 9.19 shows some typical examples of measured distributions of speech parameters along with 1-dimensional Gaussian fits to the data. The quality of the fits is better in some cases than in others.

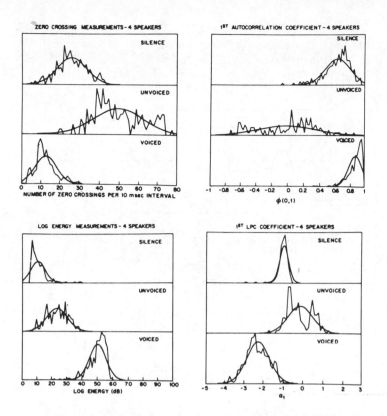

Fig. 9.19 Measured distributions of several speech parameters and Gaussian fits to the data. (After Rabiner and Sambur [27].)

Several comments should be made about the assumptions and the required computation leading to the decision rule of Eq. (9.14). The assumption of a normal distribution for the measurements can be justified from several considerations. First, for the decision rule to be correct, it is not necessary that the distribution be exactly normal. In the case of unimodal distributions, it is sufficient that the distribution be normal in the center of its range — a property often found to be true for physical measurements. Moreover, as mentioned earlier, the decision rule is optimum for a class of probability densities which are related to the Gaussian density through arbitrary monotonic functional relationships. Finally, the decision rule requires information only about the first two moments of the distribution. Accurate estimation of higher-order moments is usually difficult for practical situations.

A major advantage of the distance metric of Eq. (9.14) is that the distance $\hat{d}_i$ is invariant with respect to any arbitrary nonsingular linear transformation of the data [20]. This property of the distance metric of being invariant with respect to arbitrary linear transformations of the measurement space has great significance in that the results for a single parameter set are identical theoretically to the results from any linear transformation of the parameter set, e.g., Fourier transforms of the data set yield identical results to the data set itself. A second important property of the distance metric of Eq. (9.14) is that it is par-

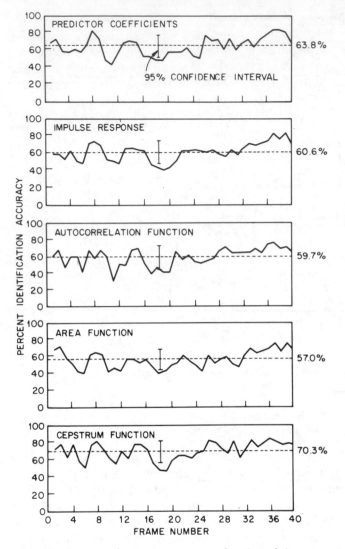

Fig. 9.20 Speaker identification accuracy as a function of the parameter set. (After Atal [20].)

ticularly effective in weighting the different pattern vector components according to their importance [20].

Using the decision rule of Eq. (9.14) Atal has investigated the effectiveness of different parametric representations of the speech signal in terms of the speaker identification problem [20]. He used a population of 10 speakers, each speaking the same test phrase six times. Each utterance was divided into 40 equal length segments thus providing a crude time warping. The average length of each segment was about 50 msec. Then an LPC analysis was performed on each of the 40 frames for each of the 60 sentences. Thus a pattern vector of LPC coefficients was obtained for each frame. From these LPC coefficients an impulse response, an autocorrelation function, a lossless tube area function and a cepstrum were obtained as discussed in Chapter 8. Then speaker identification accuracy was tested by selecting one utterance for each speaker as test utterances and using the remaining five utterances to form reference patterns for each speaker. Then the decision rule of Eq. (9.14) was applied on a frame-by-frame basis for each speaker to obtain an average identification rate for each of the 40 frames. These results are shown in Fig. 9.20 for each of the equivalent parameter sets. Although all these measurement sets yielded fairly good scores for such a short speech segment, the cepstrum measurement yielded somewhat higher accuracies than the other measurement sets. By combining several frames to obtain a higher dimensional pattern vector, higher accuracies can be attained. Figure 9.21 shows curves of the identification accuracy achieved using the cepstrum measurements as a function of the total duration of the speech interval used in making the distance calculation. It can be seen in this figure that identification accuracies on the order of 95% were achieved for this limited ensemble of speakers for speech samples of duration on the order of half a second.

9.3 Speech Recognition Systems [17,21-26]

For speech recognition, as in speaker recognition, digital speech processing techniques are applied to obtain a pattern which is then compared to stored reference patterns. In speech recognition, the goal is to determine what word, phrase, or sentence was spoken.

Unlike the areas of computer voice response, and speaker recognition, in which the problems are generally quite well defined, the area of speech recognition is one in which a large number of options must be specified before the

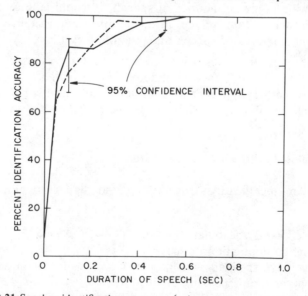

Fig. 9.21 Speaker identification accuracy (using cepstrum parameters) as a function of the speech duration. (After Atal [20].)

489

problem can even be approached [21]. Examples of these options are the following:

1. Type of speech — e.g., isolated words, continuous speech, etc.
2. Number of speakers — e.g., single speaker system, multiple designated speakers, unlimited population.
3. Type of speakers — e.g., cooperative, casual, male, female, child.
4. Speaking environment — e.g., soundproof booth, computer room, public place.
5. Transmission system — e.g., high quality microphone, close talking microphone, telephone.
6. Type and amount of system training — e.g., no training, fixed training set, continuous training.
7. Vocabulary size — e.g., small vocabulary (1-20 words), medium vocabulary (20-100 words), large vocabulary (greater than 100 words).
8. Spoken input format — e.g., constrained text, free spoken format.

It can be seen from the above list that a wide variety of options and alternatives are available in the specification of a speech recognition system. In this section we will restrict our attention to three representative speech recognition systems in which a substantial amount of digital speech processing is used. These systems are all limited vocabulary, context free, recognition systems. Although the more sophisticated type of continuous speech recognition systems [21,22] also use substantial amounts of digital processing in the analysis stages, a large amount of the effort in implementing these systems has been concerned with syntactic and semantic analyses of the utterance. Since these areas are closely allied with linguistic theories of speech, it would take us too far afield to devote time to such systems. Therefore, the interested reader is referred to the references for discussions of speech understanding systems.

The three specific recognition systems to be discussed in this section include a small vocabulary, speaker independent, isolated digit recognition system; a small vocabulary, speaker independent continuous digit recognition system; and a large vocabulary, designated speaker, isolated word recognition system.

9.3.1 Isolated digit recognition system [25]

The system specifications for the isolated digit recognition system were the following:

1. Isolated word vocabulary.
2. Unlimited population of speakers.
3. Cooperative speakers with no restrictions as to sex or age.
4. Computer room speaking environment.
5. Transmission over either a close talking or high quality microphone.
6. No system training.

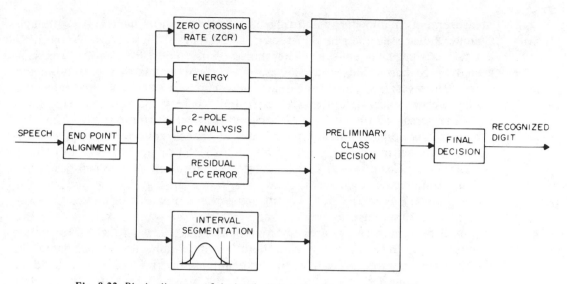

Fig. 9.22 Block diagram of isolated digit recognition system. (After Sambur and Rabiner [25].)

7. Small vocabulary size consisting of the 10 digits (0-9) with 0 pronounced as zero.
8. Single word format with pauses between each spoken input.

Figure 9.22 shows a block diagram of the isolated digit recognition system. As seen in this figure, the basic analysis consists of endpoint detection (as in speaker recognition), processing the utterance to give a pattern or a set of measurements, segmentation of the utterance into intervals, and a preliminary and then a final class decision as to which digit was spoken.

Although a wide variety of representations are candidates for a general speech recognition system, in order to be used in a speaker independent system the measurements must be reasonably robust [25]. Important characteristics include:

1. The measurements can be made simply and unambiguously.
2. The measurements can be used to grossly characterize a large proportion of speech sounds.
3. The measurements can be conveniently interpreted in a speaker-independent manner.

Among the candidates for such robust measurements are the ones shown in Fig. 9.22 which include:

1. Zero-crossing rate.
2. Energy.
3. LPC analysis using $p = 2$ poles.
4. LPC residual error.

Measurements 1 and 2 have already been discussed in Chapter 4. Although Chapter 8 discussed the general issues in LPC analysis, the use of a 2-pole LPC analysis is somewhat unusual. The choice of a 2-pole LPC analysis was suggested by Makhoul and Wolf [26] as an excellent means of representing the gross features of the short-time spectrum. The pole frequency indicates the major energy concentration in the spectrum, whereas the residual error indicates the spread or tilt in the spectrum. By way of illustration, Fig. 9.23 [26] shows comparisons of the spectra of several speech sounds obtained directly using FFT spectrum measurements, with the spectra of the two-pole LPC fit to the spectrum. For a two-pole LPC analysis, the resulting polynomial has either a single complex-conjugate root, or two real roots. In Fig. 9.23a the spectra for the sound /sh/ as in the word "short" are plotted. For this example, the two-pole LPC analysis gave a complex conjugate pole at about 3 kHz — i.e., the region of maximum energy concentration in the spectrum. In Fig. 9.23b similar results are shown for the vowel /a/ where the major concentration of energy in the spectrum is around 800 Hz; In the examples of Figs. 9.23c and 9.23d (a voice bar or voiced stop and the vowel /i/), the major concentration of spectral energy is around 0 Hz; thus the two-pole LPC analysis gives two real poles in the right half z-plane. From Fig. 9.23 it can be seen that the computed pole frequency of the 2-pole LPC analysis gives a fairly good indication of the location of the dominant portion of the spectral energy of the sound, and can thus

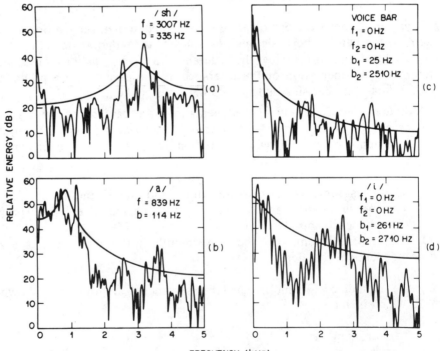

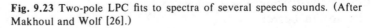

Fig. 9.23 Two-pole LPC fits to spectra of several speech sounds. (After Makhoul and Wolf [26].)

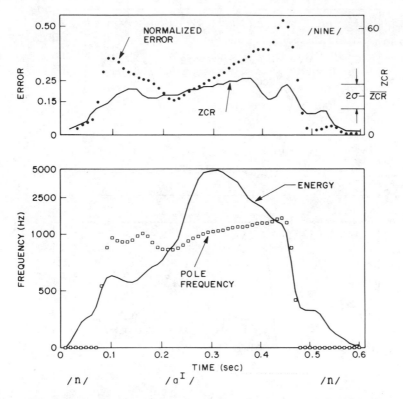

Fig. 9.24 Speech recognition measurements for the isolated digit /nine/. (After Sambur and Rabıner [25].)

be effectively used to characterize sounds with relatively high-frequency or low-frequency concentrations of energy. For example, noise-like sounds are characterized by a relatively high frequency concentration of energy, while nasals and vowels generally have a much lower frequency for the energy concentration.

To illustrate typical analysis measurements, Figs. 9.24 and 9.25 show the complete set of measurements for spoken versions of the digits nine and six. Ultimate recognition of these digits is based on making gross determination of the sound classes (at various points throughout the utterance) for each separate measurement, and then pooling the separate analyses to make the final decision. Thus, for example, the initial nasal section of the nine of Fig. 9.24 is characterized by the low normalized (LPC) error, and the 2-pole location of 0 Hz; whereas the fricative beginning and end of the digit six of Fig. 9.25 are indicated by the relatively high values of the normalized (LPC) error, the pole frequency, and the zero-crossing rate (ZCR) measurement.

Using a parallel processing, tree-like decision structure based on separate indications from each of the measurements, in each of the segmented intervals of the digits, a reliable recognition strategy was derived by Sambur and Rabiner [25]. Recognition accuracies of from 94.4 to 97.3% were obtained across 65 speakers.

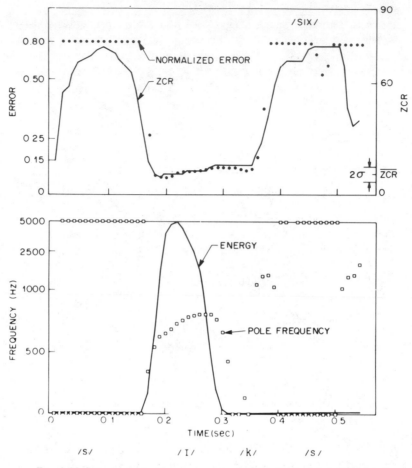

Fig. 9.25 Speech recognition measurements for the isolated digit /six/. (After Sambur and Rabiner [25].)

9.3.2 Continuous digit recognition system [27,28]

In this section the more complex problem of recognizing connected digits, in a speaker independent manner, is illustrated. The specifications for this system are almost identical to those of the isolated digit recognition system of Section 9.3.1, but with one major exception. Specification number 8 is a connected word (digit) format in a fixed digit sequence (i.e., 3 digits per utterance) with no pauses between words.

Although there are a great many similarities between the problems of isolated and continuous digit recognition, the implementations of the recognition systems are significantly different, particularly in the analysis or signal processing phase. This result is attributable to the necessity of segmenting the continuous digit string into individual digits prior to recognition. The segmentation problem is an extremely difficult one for which no simple solutions are generally known.

Figure 9.26 shows a block diagram of the signal processing steps in the continuous digit recognition system. The recorded digit string is first subjected to an endpoint analysis to determine where in the recording interval the speech utterance occurs. The endpoint analysis used is the time domain method described in Chapter 4. Following endpoint detection, the speech signal is processed to given the following measurements (at a rate of 100 times per second).

1. Zero-crossing rate.
2. Log energy.
3. LPC coefficients.
4. LPC log error.
5. First autocorrelation coefficient.

The measured parameters were then used as input to a statistical pattern recognition scheme which classified each 10 msec interval as silence, unvoiced speech, or voiced speech based on a non-Euclidean distance measure (of the type shown in Eq. (9.14)). Following some simple nonlinear smoothing as as described in Chapter 4, the voiced/unvoiced/silence contour is used along with some statistical information about the reliability of the classification of each 10 msec interval, and the speech energy measurements, to segment the connected digit string into the individual digits. The system requires knowledge of the number of digits in the input string for proper segmentation. For all examples to be shown in this section it was assumed that there were exactly three digits in the input string.

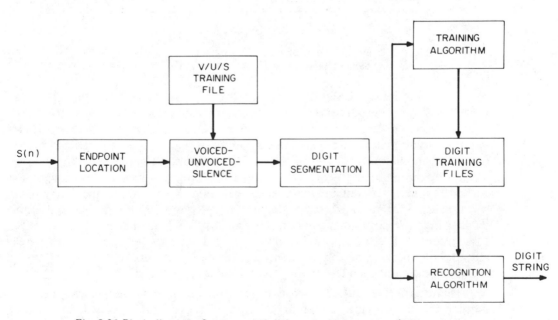

Fig. 9.26 Block diagram of a connected digit recognition system. (After Rabiner and Sambur [27].)

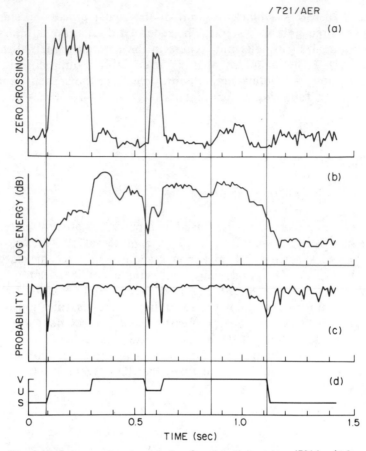

Fig. 9.27 Segmentation boundaries for the digit string /721/. (After Rabiner and Sambur [27].)

The way in which segmentation of the digit string was accomplished was by using known results about the various measurements for each of the 10 digits to find the digit boundaries. For example, it was known that each interval of nonvoiced speech constituted a boundary region between digits since no digit has an internal nonvoiced interval of speech. Also, it was known that strong dips in the energy contour during voiced regions almost always signalled a digit boundary. Based on these observations a set of simple and straightforward rules were used to segment the digit string. Although several cases existed in which accurate segmentation was quite difficult, it was shown possible to segment digit strings with less than 1% gross errors — i.e., cases in which auditory verification of the boundaries indicated that part of the preceding or following digit was included within the boundaries of the current digit.

Figures 9.27 and 9.28 show two examples of digit strings which were segmented by the recognition system. In each of these figures parts a, b, c, and d show plots of the zero-crossing contour, the log energy contour, a statistical measure of the certainty of the voiced/unvoiced/silence analysis, and the voiced/unvoiced/silence contour of the utterance, respectively. The statistical parameter shown in part c is a measure of the probability that the decision made by the voiced-unvoiced-silence analysis is correct, and it varies from 0 to 1.0. The voiced/unvoiced/silence contour of part d is a 3-level contour where level 1 is silence, level 2 is unvoiced speech, and level 3 is voiced speech.

Figure 9.27 shows the segmentation boundaries for the digit string /721/. The initial boundary was placed at the beginning of the first unvoiced region, i.e., the /s/ in seven. The second boundary was placed at the initial interval of the second unvoiced region, corresponding to the /t/ in two. The third boundary was placed in the region of a local minimum of the log energy contour within the second voiced region. The exact boundary location is not at the absolute minimum of the log energy, but instead occurs somewhere within the region of the minimum. The exact location is determined by a series of complex decisions in the segmentation algorithm. Although the correct location of the third boundary is not readily determined, it has been found that precise location of the boundaries within voiced regions is not required for reliable digit recognition. The final digit boundary is located at the beginning of the last silence region.

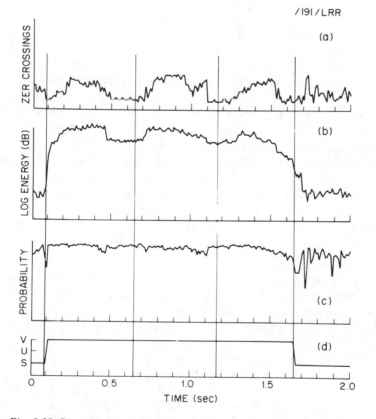

Fig. 9.28 Segmentation boundaries for the digit string /191/. (After Rabiner and Sambur [27].)

It should be pointed out that another possible candidate for a boundary location in Fig. 9.27 is at the strong local minimum in the log energy contour at the /v/ in seven. However, the segmentation rules were able to eliminate this case quite readily and instead choose the minimum in the second voiced region.

Figure 9.28 shows a somewhat more complicated digit string — the string /191/. The input is all voiced; thus there are no convenient boundaries in unvoiced regions. In addition the local minima in the log energy plot are not very strong ones (e.g., the energy dips are not large ones), and the widths of these minima are quite large. Thus the choice of boundary locations was made using logic rules in the segmentation algorithm. Listening tests showed that the location of these boundaries within the all-voiced regions was not critical due to the high degree of coarticulation in the speaking of such all-voiced digit strings.

The next stage in the method is the digit recognition algorithm. For each segmented digit the region of voiced speech (as obtained from the voiced/unvoiced analysis) is analyzed using a 10-pole LPC analysis. The recognition strategy is a statistical approach in which frames of the test utterance (LPC parameter sets) are compared to stored frames of the reference digits and the digit having the greatest similarity (i.e., smallest distance) to the test digit is chosen.

The digit reference files contain a statistical description of the behavior of the LPC coefficients for each frame of each digit. Information about the mean and variation of the LPC coefficients across multiple replications and speakers is contained in these files. For recognition the test digit must be time warped to the duration of each of the reference digits. Any of the time warping methods discussed in Section 9.1 are applicable to this problem. Since all the digits (with the exception of 7) are monosyllables, linear time warping has been used in many practical digit recognition systems. For each reference digit an average distance between its LPC coefficients and those of the test digit is computed, and the reference digit whose average distance is minimum is chosen as the spoken digit. The measure of distance between LPC parameter sets is an important factor in the success of such a recognition system. A variety of LPC distance metrics have been proposed. In the next section we discuss some LPC distance measures and show how they are related to the inherent statistical properties of the LPC coefficients.

Evaluations of systems of the type shown in Figure 9.26 indicate that digit recognition on the order of 98-100% are attainable for connected digits spoken by a designated speaker — i.e., one for whom the reference digit patterns were obtained, and about 95% for a speaker independent system [27-28].

9.3.3 LPC distance measures

We have seen that for both speaker and speech recognition it is necessary to compare in a quantitative and computationally efficient manner, two frames of speech for which LPC analyses have yielded different LPC coefficient sets. Thus we seek a measure $D(\mathbf{a}, \hat{\mathbf{a}})$ where D is the distance between frames of speech with LPC parameter sets $\mathbf{a} = (1, a(1),\ a(2),...,a(p))$, and $\hat{\mathbf{a}} = (1, \hat{a}(1),\ \hat{a}(2),...,\hat{a}(p))$. Since D is a distance measure we require that

$$D(\hat{\mathbf{a}}, \mathbf{a}) \geqslant 0 \tag{9.17}$$

and

$$D(\hat{\mathbf{a}}, \mathbf{a}) = 0 \quad \text{if} \quad \mathbf{a} = \hat{\mathbf{a}} \tag{9.18}$$

A fairly sophisticated distance measure $D(\hat{\mathbf{a}}, \mathbf{a})$ was proposed by Itakura [17]. This distance measure can be obtained by the following reasoning. It can be argued that because of noise, as well as the inexactness of the linear prediction model of speech, it is not possible to measure the true LPC coefficients associated with a segment of speech. It is only possible to estimate (measure) the underlying LPC coefficients for the speech segment. Assume we are given a segment of speech with estimated LPC coefficients $\hat{\mathbf{a}}$. The problem is to determine the probability that $\hat{\mathbf{a}}$ is from a speech segment with true LPC coefficients $\mathbf{a}$. Once this probability is determined, an effective measure for assessing dissimilarity can be obtained.

It has been shown by Mann and Wald [29] that the probability distribution governing the estimates of $\hat{\mathbf{a}}$ is a multi-dimensional Gaussian distribution with mean $\mathbf{a}$ and covariance matrix Λ defined as

$$\Lambda = \frac{\mathbf{R}^{-1}}{N}(\hat{\mathbf{a}}\mathbf{R}\hat{\mathbf{a}}^t) \tag{9.19}$$

where $\mathbf{R}$ is the ($p+1$ by $p+1$) correlation matrix of the speech segment, N is the length of the speech frame in samples, and t denotes the vector transpose. Thus the probability of obtaining the estimate $\hat{\mathbf{a}}$ when the underlying LPC coefficients are $\mathbf{a}$ is

$$P(\hat{\mathbf{a}}/\mathbf{a}) = [(2\pi)^{p/2}|\Lambda|^{\frac{1}{2}}]^{-1}\exp[-0.5(\hat{\mathbf{a}}-\mathbf{a})\Lambda^{-1}(\hat{\mathbf{a}}-\mathbf{a})^t] \tag{9.20}$$

where $|\Lambda|$ is the determinant of the matrix Λ. An appropriate distance measure is obtained by taking the logarithm of Eq. (9.20), and neglecting the bias term due to $|\Lambda|$. The resulting distance measure is

$$D(\hat{\mathbf{a}}, \mathbf{a}) = (\hat{\mathbf{a}}-\mathbf{a})\left[N\frac{\mathbf{R}}{\hat{\mathbf{a}}\mathbf{R}\hat{\mathbf{a}}^t}\right](\hat{\mathbf{a}}-\mathbf{a})^t \tag{9.21}$$

It is readily seen that the greater the probability that $\hat{\mathbf{a}}$ came from the distribution with underlying LPC coefficients $\mathbf{a}$, the smaller the distance computed using the metric of Eq. (9.21). Because of computational considerations Itakura proposed the closely related distance measure

$$D'(\hat{\mathbf{a}}, \mathbf{a}) = \log\left[\frac{\mathbf{a}\mathbf{R}\mathbf{a}^t}{\hat{\mathbf{a}}\mathbf{R}\hat{\mathbf{a}}^t}\right] \tag{9.22}$$

The key assumption in the above analysis is that the ensemble of all possible speech segments derived from the same speech sound are similar in that

499

the underlying LPC coefficients **a** are identical. The differences in the measured LPC coefficients for these speech segments are attributed primarily to the effects of statistical sampling. For a wide variety of systems this assumption is quite reasonable. However, for cases in which the LPC coefficients vary because of known effects such as varying speakers, coarticulation, etc., the underlying or true LPC coefficients are not constant, but instead are better described in terms of a statistical distribution around some mean value.

Thus for a complete characterization of an LPC frame from a given sound it is necessary to determine the distribution of **a** itself. A reasonable assumption is that **a** is Gaussian with mean **m** and covariance matrix **S**. Based on this total characterization of **a** the distance relating $\hat{\mathbf{a}}$ and **a** becomes

$$\hat{D}(\hat{\mathbf{a}}, \mathbf{a}) = (\hat{\mathbf{a}} - \mathbf{m})\mathbf{C}^{-1}(\hat{\mathbf{a}} - \mathbf{m})' \tag{9.23}$$

where **C** is the total covariance matrix and is of the form

$$\mathbf{C} = \mathbf{S} + \frac{\mathbf{R}^{-1}}{N}(\hat{\mathbf{a}}\mathbf{R}\hat{\mathbf{a}}') \tag{9.24}$$

To use the distance measure of Eq. (9.23) requires measurement of the quantities **m** and **S** for each frame of reference data. The quantity $\mathbf{m} = (1, m(1), m(2), \ldots, m(p))$ is the average value of the **a**'s and is defined as

$$m(n) = \frac{1}{J}\sum_{j=1}^{J} \hat{a}_j(n) \quad n = 1, 2, \ldots, p \tag{9.25}$$

where $\hat{a}_j(n)$, $j = 1, 2, \ldots, J$ is a statistical sampling of frames with the same underlying distribution **a**. Similarly the covariance matrix **S**, with component $s(n,p)$ is obtained as

$$s(n,p) = \frac{1}{J}\sum_{j=1}^{J} \hat{a}_j(n)\hat{a}_j(p) - m(n)m(p) \tag{9.26}$$

9.3.4 Large vocabulary word recognition system

The third word recognition system to be described is one in which the vocabulary size is significantly larger than in the previous two systems. However, the tradeoff for the large vocabulary size is that the system is no longer speaker independent, but instead must be trained a priori by each intended user of the system.

In terms of our introductory discussion, the specifications for the large vocabulary word recognition system developed by Itakura [17] were the following:

1. Isolated word vocabulary.
2. Single speaker system, adaptable to any number of speakers with appropriate system training.
3. Cooperative speakers with no limitations as to sex or age.

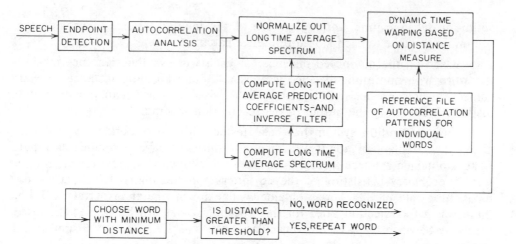

Fig. 9.29 Block diagram of a large vocabulary, speaker dependent, word recognition system.

4. No major restrictions on speaking environment.
5. Telephone transmission system.
6. System training consisting of 1 or more repetitions of each word in the vocabulary.
7. Vocabulary size in the range 100-500 isolated words.
8. Isolated word format with distinct pauses between consecutive words.

Figure 9.29 shows a block diagram of the processing in the word recognition system. For maximum efficiency and minimum processing, Itakura used a sampling rate of 6.67 kHz for the input speech. Since the bandwidth of the input speech was limited to about 3 kHz by the telephone system, such a sampling rate was entirely adequate for this application.

Following endpoint detection (again using the time domain system of Chapter 4) an autocorrelation analysis was performed in which the first 8 autocorrelation coefficients were measured 67 times per second. To compensate for the spectral shaping due to the telephone line, Itakura computed a long-time average spectrum by averaging the autocorrelation coefficients on a frame-by-frame basis and then computing a 2-pole LPC fit to the long-time average spectrum. From the 2-pole fit to the long-time average spectrum, the appropriate inverse filter was computed. The long-time average spectrum was normalized out of the input by convolving the original autocorrelation coefficients and the autocorrelation coefficients of the impulse response of the second order inverse filter. The first six normalized autocorrelation coefficients were then used both to make the reference patterns, and as the basis for recognizing unknown inputs.

After normalizing out the long-time average spectrum the recognition phase begins. The unknown utterance is compared to each utterance in the stored reference file. The basis for comparison is the distance measure of Section 9.3.3 (Eq. (9.22)). This distance measure is also used for computing the optimum dynamic time warping pattern of the input utterance to provide the minimum distance to each stored reference pattern.

Based on the computed distance for each word in the reference catalog, the word with minimum distance is chosen. If the absolute distance exceeds some threshold no decision is made. Otherwise the word with the minimum distance is chosen as the output of the recognition scheme.

This recognition system has been tested with two separate vocabularies. Using a vocabulary of about 120 distinctly different Japanese cities as input, Itakura obtained a correct recognition rate of 97.3% with a 1.65% rejection rate. For a vocabulary consisting of the 26 letters and the digits 0 to 9, a correct recognition rate of 88.6% was obtained. The much higher error rate (11.4%, there was a 0% rejection rate) for this vocabulary was attributable to the great similarity between many of the vocabulary entries — e.g., *b* and *d*, *m* and *n*, or *i* and *y*.

9.4 A 3-Mode Speech Communication System

In order to give a brief look at how man-machine communication might be used in the future, we conclude this chapter with a brief discussion of a 3-mode speech communication system which combines voice response, speaker verification, and speech recognition in a single application. In particular the system we are referring to is an experimental vehicle that is an airlines information and ticketing service.[3] Although this system has been implemented strictly in the laboratory for experimental purposes, it is felt that such a system represents a "real-world" application of the concepts discussed in this chapter, as well as throughout this book.

Rather than describe the system, a simple scenario provides the reader with a feeling for how such a system can work and provide a useful and valuable service. An experimental scenario is as follows:

A frequent credit customer of the airlines dials the computer. The voice response answers with:

Computer: This is ABC Airlines information and ticketing service. Please key-in your account number. (Customer does, computer looks it up, and comes back with):

Computer: Good morning, Mr. XYZ. Please speak your verification phrase for credit charging. (Customer does, computer churns it and, if result favorable, comes back with):

Computer: Thank you, Mr. XYZ, I confirm your charge account. Where would you like to fly to?

Customer: Washington.

[3]This example is taken from the paper by J. L. Flanagan [1].

Computer: From what airport would you like to fly to Washington?[4]
Customer: Newark.
Computer: On what date would you like to fly from Newark?
Customer: (Gives date; or may say no date).
Computer: At what time [on date, or of day] do you wish to depart?
Customer: (Gives time).
Computer: Just a moment, please, I will give you the flights available within two hours of [time]. (Computer goes to the current airlines timetable stored in disc file, looks up flights available, reads them out, and asks):
Computer: Which flight do you wish to book?
Customer: Flight 123.
Computer: Do you wish First Class or Economy on Flight 123?
Customer: First Class.
Computer: How many First Class seats do you wish?
Customer: One
Computer: A moment, please, while I check availability of one seat.
Computer: I am pleased to confirm [reads it all back]. Your ticket will appear on your teletypewriter within the hour. Your bill will be sent at the end of the month.
Computer: Do you wish to book further?
Customer: No.
Computer: Thank you for calling ABC Airlines. (Hangs up.)

A variant of this system is already working on an experimental basis at Bell Laboratories over dialed-up connections and from conventional telephone sets in the local exchange.

9.5 Summary

We have attempted to show how the techniques discussed throughout this book can be used in the implementation of a wide variety of useful and interesting speech processing systems. The more that is learned about digital processing of speech signals, the more sophisticated the speech processing systems become, and the greater the potential applications, both in normal human communication and in communication between humans and machines.

REFERENCES

1. J. L. Flanagan, "Computers that Talk and Listen: Man-Machine Communication by Voice," *Proc. IEEE,* Vol. 64, No. 4, pp. 405-415, April 1976.

[4]Note two things. (1) At each stage the answerback message has a woven-in confirmation of the last recognition. If the last recognition is incorrect, the customer can push a dial button and have the previous question repeated. (2) At each stage the recognition vocabulary is contracted in size to the answer ensemble appropriate to the question. This vocabulary contraction is an important contribution towards robustness of the system.

2.	N. R. Dixon and H. D. Maxey, "Terminal Analog Synthesis of Continuous Speech Using the Diphone Method of Segment Assembly," *IEEE Trans. on Audio and Electroacoustics,* Vol. AU-16, No. 1, pp. 40-50, January 1968.

3.	J. L. Flanagan, C. H. Coker, L. R. Rabiner, R. W. Schafer and N. Umeda, "Synthetic Voices for Computers," *IEEE Spectrum,* Vol. 7, pp. 22-45, January 1970.

4.	J. Allen, "Synthesis of Speech from Unrestricted Text," *Proc. IEEE,* Vol. 64, No. 4, pp. 433-442, April 1976.

5.	N. Umeda, "Linguistic Rules for Text-to-Speech Synthesis," *Proc. IEEE,* Vol. 64, No. 4, pp. 443-451, April 1976.

6.	C. H. Coker, "A Model of Articulatory Dynamics and Control," *Proc. IEEE,* Vol. 54, No. 4, pp. 452-459 April 1976.

7.	L. H. Rosenthal, L. R. Rabiner, R. W. Schafer, P. Cummiskey, and J. L. Flanagan, "A Multiline Computer Voice Response System Utilizing ADPCM Coded Speech," *IEEE Trans. Acoustics, Speech, and Signal Proc.,* Vol. ASSP-22, No. 5, pp. 339-352, October 1974.

8.	L. R. Rabiner and R. W. Schafer, "Digital Techniques for Computer Voice Response: Implementations and Applications," *Proc. IEEE,* Vol. 64, No. 4, pp. 416-433, April 1976.

9.	L. R. Rabiner, R. W. Schafer, and J. L. Flanagan, "Computer Synthesis of Speech by Concatenation of Formant-Coded Words," *Bell System Tech. J.,* Vol. 50, No. 5, pp. 1541-1548, May-June 1971.

10.	D. S. Levinstone, "Speech Synthesis System Integrating Formant-Coded Words and Computer-Generated Stress Parameters," M.Sc. Thesis, Dept. of Electrical Engr., MIT, Cambridge, 1972.

11.	J. P. Olive and L. H. Nakatani, "Rule-Synthesis of Speech by Word Concatenation; A First Step," *J. Acoust. Soc. Am.,* Vol. 55, No. 3, pp. 660-666, March 1974.

12.	J. L. Flanagan, L. R. Rabiner, R. W. Schafer, and J. D. Denman, "Wiring Telephone Apparatus from Computer Generated Speech," *Bell System Tech. J.,* Vol. 51, pp. 391-397, February 1972.

13.	A. E. Rosenberg, "Automatic Speaker Verification: A Review," *Proc. IEEE,* Vol. 64, No. 4, pp. 475-487, April 1976.

14.	G. R. Doddington, "A Method of Speaker Verification," Ph.D. dissertation, Univ. Wisconson, Madison, 1970.

15.	R. C. Lummis, "Speaker Verification by Computer using Speech Intensity for Temporal Registration," *IEEE Trans. on Audio and Electroacoustics,* Vol. AU-21, pp. 80-89, 1973.

16. A. E. Rosenberg and M. R. Sambur, "New Techniques for Automatic Speaker Verification," *IEEE Trans. Acoustics, Speech, and Signal Proc.,* Vol. ASSP-23, pp. 169-176, 1975.

17. F. Itakura, "Minimum Prediction Residual Principle Applied to Speech Recognition," *IEEE Trans. Acoustics, Speech, and Signal Proc.,* Vol. ASSP-23, No. 1, pp. 67-72, February 1975.

18. B. S . Atal, "Automatic Recognition of Speakers from Their Voices," *Proc. IEEE,* Vol. 64, No. 4, pp. 460-475, April 1976.

19. P. D. Bricker et al., "Statistical Techniques for Talker Identification," *Bell System Tech. J.,* Vol. 50, pp. 1427-1454, April 1971.

20. B. S. Atal, "Effectiveness of Linear Prediction Characteristics of the Speech Wave for Automatic Speaker Identification and Verification," *J. Acoust. Soc. Am.,* Vol. 55, pp. 1304-1312, June 1974.

21. A. Newell et al., *Speech Understanding Systems,* Academic Press, New York, 1975.

22. D. R. Reddy, Editor, *Speech Recognition: Invited Papers of the IEEE Symposium,* Academic Press, New York, 1975.

23. T. B. Martin, "Practical Applications of Voice Input to Machines," *Proc. IEEE,* Vol. 64, No. 4, pp. 487-501, April 1976.

24. D. R. Reddy, "Speech Recognition by Machine: A Review," *Proc. IEEE,* Vol. 64, No. 4, pp. 501-531, April 1976.

25. M. R. Sambur and L. R. Rabiner, "A Speaker Independent Digit Recognition System," *Bell System Tech. J.,* Vol. 54, No. 1, pp. 81-102, January 1975.

26. J. Makhoul and J. Wolf, "The Use of a Two-Pole Linear Prediction Model in Speech Recognition," Report 2537, Bolt, Beranek, and Newman, September 1973.

27. L. R. Rabiner and M. R. Sambur, "Some Preliminary Results on the Recognition of Connected Digits," *IEEE Trans. Acoustics, Speech, and Signal Proc.,* Vol. ASSP-24, No. 2, pp. 170-182, April 1976.

28. M. R. Sambur and L. R. Rabiner, "A Statistical Approach to the Recognition of Connected Digits," *IEEE Trans. Acoustics, Speech, and Signal Proc.,* Vol. ASSP-24, No. 6, December 1976.

29. H. B. Mann and A. Wald, "On the Statistical Treatment of Linear Stochastic Difference Equations," *Econometrica,* Vol. 11, Nos. 3 and 4, pp. 173-220, July-October 1943.

Projects

The following three broad categories are suggested for term projects in courses on digital speech processing:

(i)   A literature survey and report
(ii)   A hardware design project
(iii) A computer project

Suggested guidelines for the three types of projects follow.

(i) Literature Survey and Report

In this case the student should choose a topic and consider the following questions:

1. What is the problem?
2. What is the importance of the problem — e.g., application areas, etc.
3. What have been the basic approaches?
4. What has already been accomplished in this area?
5. Are new approaches called for?
6. What are the unsolved problems? What needs more work?
7. What are the impediments to further progress — e.g., technological, lack of basic knowledge, etc.

Some suggested topics for a literature survey are:

1. Pitch detection methods
2. Voiced/unvoiced analysis methods
3. The effects of the telephone line on speech analysis
4. Phonetic feature descriptions of English
5. Physical characteristics and modelling of the sound source for speech production
6. Formant analysis methods
7. Speech synthesis by rule
8. Adaptive quantization methods
9. Vocal tract area function analysis methods
10. Speaker identification methods
11. Computer voice response systems
12. Digit recognition by machine
13. Helium speech translation
14. Speech aids for the hearing impaired
15. Speakerphone dereverberation problems
16. Echo suppression methods
17. LPC synthesis structures
18. Linear prediction and system identification methods
19. Applications of homomorphic speech processing
20. Speeded up and slowed down speech
21. Pole-zero analysis of speech
22. Analysis-by-synthesis processing of speech
23. Articulatory modelling of speech
24. Hardware for waveform coding of speech
25. Speech bandwidth reduction systems.

(ii) Hardware Design Project

Such a project could be carried to the hardware stage if possible, but should at least be advanced to the level of a detailed hardware design — e.g., at the logic level. Some guidelines for this type of project are:

1. What is the problem that you propose to solve? Note that a project of this type allows you to use your creativity to think of a new and better way to do something that may have already been done.
2. What is available to solve the problem — e.g., theory and technology.
3. What are the details of the solution? These should be worked out at as low a level as is feasible with available time.
4. Is the solution presently feasible? If not, then why not?
5. What are the hardware requirements for implementations of the system?

6. If possible a chip count (gates, adders, microprocessors, memory, other storage) should be given, as should a cost estimate.

Some suggested topics for a hardware project are:

1. Design a code converter for PCM to ADPCM, PCM to ADM etc.
2. Design a circuit to detect a tone imbedded in speech
3. Propose a speech processing system which can be implemented on a commercially available microprocessor
4. Design a circuit to detect the presence of speech over a noisy telephone line
5. Design a 4 band speech spectrum analyzer
6. Design a system to display a speech spectrogram
7. Design a parallel formant speech synthesizer
8. Design a speech scrambler or encryption device
9. Design a digital pitch detector
10. Design a voiced/unvoiced detector
11. Design a system for distinguishing speech from noise

(iii) Computer Project

The student should consider this only if he is already facile with an available computer, and can obtain sufficient computer time for this project. The requirements for this project are a short description of the problem containing relevant mathematical theory and objectives of the project, and a listing (with thorough documentation and comments) of the program, and a demonstration that the program works properly. Some suggested topics for a computer project are:

1. Pitch detector—time domain, autocorrelation, cepstrum, LPC, etc.
2. Voiced/unvoiced detector
3. Beginning and endpoint detector
4. Formant analyzer
5. LPC analysis system—signal to LPC to spectrum
6. N channel spectral analyzer—phase vocoder, channel vocoder
7. Waveform coder — e.g. ADM, ADPCM etc.
8. Area function to vocal tract transfer function
9. Examine effect of window shape and duration on energy, autocorrelation or speech spectrogram
10. Speech synthesizer—serial, parallel, direct, lattice
11. Area function to formants program
12. Code converter between any 2 formats
13. Cepstrally smoothed spectrum from speech signal
14. Transform LPC parameters to alternate parameter sets and show statistical properties
15. Compare LPC, FFT, and cepstrally smoothed spectra

Index

Complex cepstrum:
  computation of, 363-65
  properties of, 360-62
  for rational transforms, 360
  of speech, 365-72
    unvoiced, 370-72
    voiced, 367-70
Composite frequency response, 268
Composite impulse response using
    real filters, 285
Compressor u-law, 188
Computational requirements in
    LPC, 417-18
Computer voice response (see Voice
    response)
Concatenation of formant-coded
    words, 470-73
Continuant sounds, 43
Continuous digit recognition,
    494-98
Convolution, descrete, 13
Correlation function:
    long term estimate, 177-78
    (see also Short-time autocorrela-
    tion)
Covariance method of LPC, 403-4
Cross-correlation, 148
CVSD, 223-24
    maximum and minimum step
    sizes, 247

**Decimation and interpolation, 27,
    273-74**
Deconvolution, 355
Delta modulation, 216-25
    adaptive, 221-24
    double integration, 225
    linear, 216-21
Design of digital filter banks (see
    Digital filter bank design)
Differential PCM (DPCM), 225-32
Differential quantization, 208-16
Digital code conversion, 235-38
Digital coding:
    of the cepstrum, 388-89
    of formants, 382-84
    of LPC parameters, 450-53
    of the time dependent Fourier
    transform, 324-34
        using adaptive delta modula-
        tion, 331-32
        using PCM, 332
Digital filter bank design, 282-302
    practical considerations in, 282-90
    using FIR filters, 292-302
    using IIR filters, 290-92

Digital filters, 18-23
    causality of, 19, 20
    frequency response of, 18
    implementation of, 23
    stability of, 19, 20
    system function of, 18
Digital transmission of speech, 7
Diphthongs, 48
Direct form implementation, 23
Discrete Fourier transform, 16-18
Discrete-time model for speech,
    103-5
Distance measures, 484-85, 498-500
Dithering, 242-43
Double integration delta modula-
    tion, 225
DPCM, 225-32
Durbin's method of solution of the
    LPC equations, 411-13
Dynamic range, 187
    of u-law quantizer, 191

**Encoding of quantized samples,
    179-81**
Energy, 119
Enhancement of speech quality, 8
Error, quantization, 182
Expander, 188
Exponential sequence, 11

Fast Fourier Transform (FFT).
    definition of, 18
    use in computing the cepstrum,
    363-65
    use in short-time Fourier analysis,
    303-6
    use in short-time Fourier synthesis,
    306-10
Feedback adaptation, 203-7
Feedback quantization:
    in differential PCM, 227-28
    performance of, 207
Feed-forward adaptation, 199-203
    in differential PCM, 226-27
    performance of, 203
Filter bank summation method,
    266-74
    implementation of, 303-10
Finite impulse response (see FIR)
FIR systems, 20-21
    design of, 20
    linear phase, 20
Formant frequencies, 41, 44
    quantization of, 382-84
    of uniform lossless tube, 65-66

Formant frequency estimation:
    using the cepstrum, 378-85
    using LPC, 442-50
Formant vocoder, 382-85
Forward prediction error, 414
Fourier transform, 15-16
Frequency domain interpretation
    of LPC, 431-40
Frequency resolution dependence
    on window length, 260
Frequency response, 18
Frequency response of telephone
    line, 174
Fricative, 40
    excitation model, 81-82
    unvoiced, 51-52
    voiced, 52

Gain computation in LPC, 404-7
Glottis, 39
    boundary condition at, 63, 81, 87
Granular noise, 219

Hamming window:
    definition of, 121
    Fourier transform of, 122
Homomorphic systems for con-
    volution, 356-65
    canonic form for, 357
    characteristic system, 357
    implementation of, 363-65
Homomorphic vocoder, 385-90

Idle channel noise, 195
IIR systems, 21-23
    design of, 22
    implementation:
        cascade form 23
        direct form, 23
        parallel form, 23
Infinite duration impulse response
    (see IIR)
Information rate, 180
Information rate of speech, 2
Instantaneous quantization, 179-95
Integrator, 217
Interpolation, 28-29
    of LPC synthesis parameters, 445
    of short-time Fourier transform,
    306-10, 324-34
Isolated digit recognition, 490-93

**Laplacian probability density, 176,
    241**
Lattice formulation of LPC, 413-17
LDM-to-PCM conversion, 236-37

510